Birkhäuser

Studies in Universal Logic

Jean-Yves Beziau • Ioannis Vandoulakis
Editors

The Exoteric Square of Opposition

The Sixth World Congress on the Square of Opposition

Birkhäuser

Editors
Jean-Yves Beziau
Federal University of Rio de Janeiro
Rio de Janeiro, Brazil

Ioannis Vandoulakis
The Hellenic Open University
Patras, Greece

ISSN 2297-0282 ISSN 2297-0290 (electronic)
Studies in Universal Logic
ISBN 978-3-030-90822-5 ISBN 978-3-030-90823-2 (eBook)
https://doi.org/10.1007/978-3-030-90823-2

This book is published under the imprint Birkhäuser, www.birkhauser-science.com by the registered company Springer Nature Switzerland AG
The registered company address is: Gewerbestrasse 11, 6330 Cham, Switzerland

Preface

This book gathers 16 papers on the square of opposition, related to the *6th World Congress on the Square of Opposition* which took place at the Orthodox Academy of Crete, November 1–5, 2018.

The square of opposition is a logical structure coming from Aristotelian logic. The topic has been continuously studied for 2000 years. Even Frege, one of the founders of modern mathematical logic, has contributed to the topic.

During the second half of the twentieth century, research on the topic was revived. New extensions and generalization of the square of opposition were suggested, far beyond Aristotle's traditional logic. These advances were accompanied by different configurations (triangle, rhombus, hexagon, octagon, polyhedra, and multi-dimensional objects) that were conceived to illustrate the new theories.

These theories found exciting applications in many fields, ranging from metalogic to highway code, through economics, music, physics, color theory, and theology.

The research in this field is interdisciplinary but anchored on a clear and precise logical theory, symbolized by the traditional diagram of the square of opposition. The theory of the square of opposition is at the same time a traditional and original topic that included new interpretations of the past theories, discoveries on the logical theory, and modern applications in a variety of spheres.

All these aspects are present in this book, with chapters displaying novel aspects of the theory of opposition and its variations that are embraced in a mainstream research project that revitalizes the theory of the square of opposition.

We thank all the contributors to this volume and the work of the referees who have critically analyzed the papers, leading to improved versions.

Rio de Janeiro, Brazil — Jean-Yves Beziau

Patras, Greece — Ioannis Vandoulakis

Contents

Contributors

Régis Angot-Pellissier Lycée Montesquieu, Le Mans, France

Jean-Yves Beziau Federal University of Rio de Janeiro, Rio de Janeiro, Brazil

Roland Bolz Humboldt University Berlin, Berlin, Germany

Saloua Chatti FSHST, University of Tunis, Tunis, Tunisia

Lorenz Demey Center for Logic and Philosophy of Science, KU Leuven, Leuven, Belgium

Anna-Sophie Heinemann Universität Paderborn, Paderborn, Germany

Yiannis Kiouvrekis Department of Public and One Health, University of Thessaly, Volos, Greece and University of Nicosia, Nicosia, Cyprus

Jens Lemanski FernUniversität in Hagen, Institute for Philosophy LG, Hagen, Germany

Alessio Moretti Université degli Studi eCampus, Novedrate, Italy and SSML di Varese, Varese, Italy

Krystyna Mruczek-Nasieniewska Department of Logic, Institute of Philosophy, Faculty of Philosophy and Social Sciences, Nicolaus Copernicus University, Toruń, Poland

Marek Nasieniewski Department of Logic, Institute of Philosophy, Faculty of Philosophy and Social Sciences, Nicolaus Copernicus University, Toruń, Poland

Valentin Pluder University of Siegen, Siegen, Germany

Fabien Schang Universidade Federal de Goiás, Goiânia, Brazil

Hubert Martin Schüler University of Hagen, Hagen, Germany

Gianluigi Segalerba aka – Arbeitskreis Kulturanalyse Universität Wien (Working Group Cultural Analysis University of Vienna), Vienna, Austria and International Mariinskaya Academy named after M.D. Shapovalenko, Moscow, Russia

Sumanta Sarathi Sharma School of Philosophy and Culture, Faculty of Humanities and Social Sciences, Shri Mata Vaishno Devi University, Jammu and Kashmir, India

Petros Stefaneas Department of Mathematics, National Technical University of Athens, Athens, Greece

Julio Michael Stern IME-USP Institute of Mathematicsátics and Statisticsística, University of São Paulo, São Paulo, Brazil

Marcin Trepczyński Faculty of Philosophy, Uniwersytet Warszawski, University of Warsaw, Warsaw, Poland

Ioannis Vandoulakis The Hellenic Open University, Patras, Greece

The Square of Opposition: Past, Present, and Future

Jean-Yves Beziau and Ioannis Vandoulakis

Abstract We first explain the origin and development of the theory of opposition, its generalization to many concepts, and figures of opposition, particularly the hexagon of opposition. We also survey the organization of a series of events on the topic since 2007 in Montreux. We then talk in details about the sixth edition of the world congress on the square emphasizing the fact that it was organized at the Orthodox Academy of Crete. In the third part, we discuss the bright future of the theory.

Keywords Square of opposition · Hexagon of opposition · Theory of opposition · Diagram · Contradiction · Semiotics · Aristotle · Labyrinth · Plato's cave

Mathematics Subject Classification (2000) Primary: 03-06 Secondary 03A05; 03B22;03B45;03B80

J.-Y. Beziau (✉)
Federal University of Rio de Janeiro, Rio de Janeiro, Brazil
e-mail: jyb@ufrj.br

I. Vandoulakis
The Hellenic Open University, Patras, Greece

J.-Y. Beziau, I. Vandoulakis (eds.), *The Exoteric Square of Opposition*, Studies in Universal Logic, https://doi.org/10.1007/978-3-030-90823-2_1

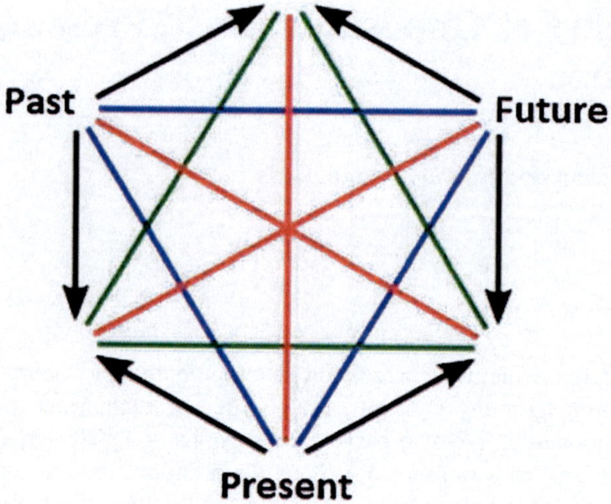

1 The Square of Opposition: A Diagram and a Theory

The square of opposition is a diagram going back to Apuleius (*Peri Hermeneias*) and Boethius (*De Silogismo Categorico*) based on some ideas by Aristotle (*De Interpretatione* 6–7, 17 b 17–26 and *Prior Analytics* I.2, 25 a 1–25) (for details see [32]). The expression "square of opposition" is used today not only to talk about this diagram but also about the theory surrounding it, whose original diagram is just a primitive print.

There are two crucial aspects of the SQUARE theory: on the one hand, the visual aspect, and on the other hand, the logical aspect. The visual aspect is related to geometry, originally a square, and the logical aspect to the notion of opposition, hence the terminology "square of opposition."

Opposition is a single notion (about the concept of notion, see [5]), the reason why the singular is used, although the theory includes three kinds of opposition: contradiction, contrariety, and subcontrariety. These three oppositions are presented in a unified logical framework in the theory of opposition. These are logical concepts involving truth and falsity, going beyond the truth/falsity dichotomy, not by introducing a third value but by combination and interaction of these two values.

These three kinds of opposition were originally defined on the basis of propositions. Two propositions are contradictory iff (if and only if) they cannot be true together and cannot be false together, *contrary* iff they can be false together but not true together, and *subcontrary* iff they can be true together but not false together. The square of opposition was indeed an essential part of the theory of categorical proposition, giving an architecture, a classification of Aristotelian propositions (Fig. 1).

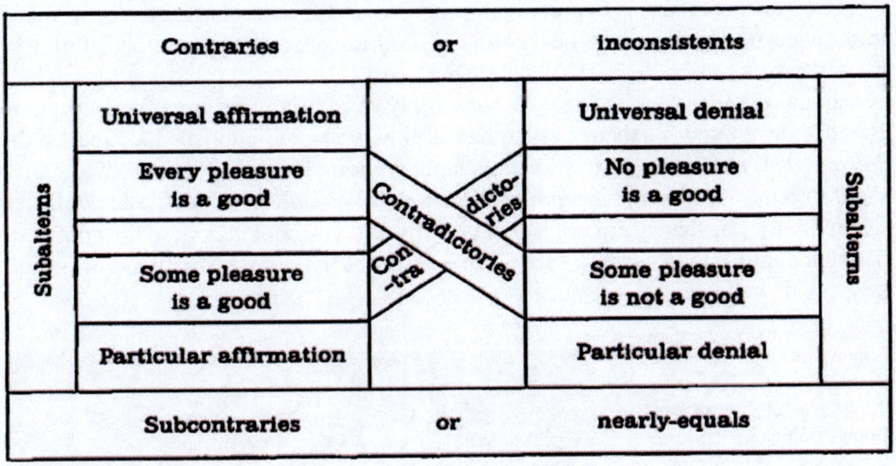

Fig. 1 Apuleius's reconstructed square

But this structure can easily be extended to concepts conceived extensionally or intensionally. For example, extensionally contradiction corresponds to the set-theoretical notion of complementation. This variation to concepts is important and gives rise to numerous applications of the square to deontic notions (prohibition, obligation, and the like) (Fig. 2), color classification, music concepts, and many other topics. The *square of opposition theory* has evolved in particular through this conceptual appraisal, not by the development of new forms of opposition.

Fig. 2 Deontic traffic sign square of opposition

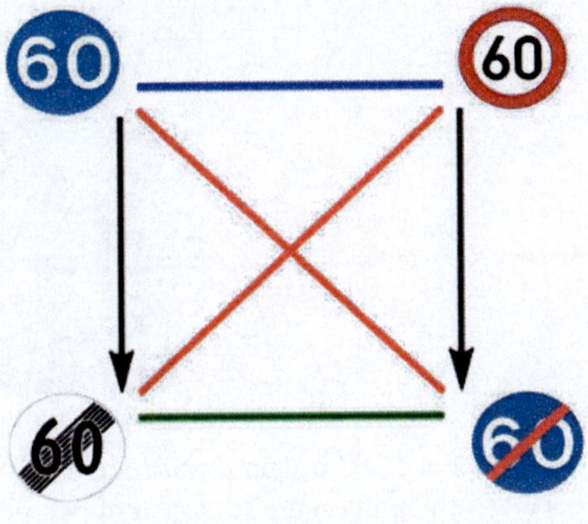

Another significative step in the development of the theory is the versatility of the relations between the three kinds of opposition and the notion of subalternation which escorts these relations. These variations are naturally produced by the geometrical aspect of the theory. From a square we can go to a hexagon, octagon, decagon, a cube, a dodecahedron, and so forth.

A central geometrical figure which emerged and turned out to be a central key to the theory is however the hexagon [3]. The hexagonal theory of opposition was mainly promoted by Robert Blanché (but see [21]). And the basis of Blanché's hexagonal theory is the triangle of contrariety, which is the heart of the hexagon. Blanché developed his theory at the end of the 1950s. He published a paper in the *Journal of Symbolic Logic* [15], but his main work on the topic, his book *Structures intellectuelles - Essai sur l'organisation systématique des concepts*, was published in 1966 [16]. His theory did not have much impact. His book has up to now not been translated into English or into another foreign language except Brazilian Portuguese (Fig. 3).

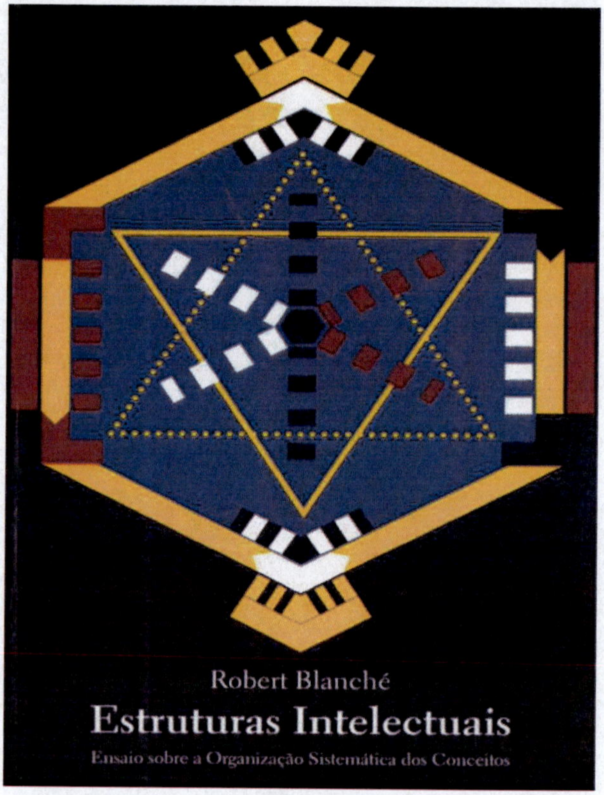

Fig. 3 Brazilian version of Blanché's book. (Cover by Sergio Kon of the Book by Robert Blanché *Estruturas intelectuais: ensaio sobre a organização sistemática dos conceitos.* Reproduction courtesy of EDITORA PERSPECTIVA, https://editoraperspectiva.com.br/produtos/estruturas-intelectuais/)

The theory of the square was revived by the first author when he wrote in 2003 the paper "New light on the square of oppositions and its nameless corner," published in the logic journal of the *Russian Academy of Sciences* in Moscow [2], introducing:

- Coloring.
- A three-dimensional object which is a combination of three hexagons.
- An octagon which like the hexagon is the combination of two dual figures of opposition, a square of contrariety and a square of subcontrariety (Fig. 4).

Fig. 4 Octagon of opposition generalizing Blanché's hexagon. (Photos courtesy of Jean-Yves Beziau)

He was then in touch with both Alessio Moretti, a French-Italian student who did a PhD with him on the topic in Neuchâtel, developing in particular the theory of *n*-opposition [23] corresponding to the third point and Hans Smessaert, a Belgian linguist much interested in polyhedra. Since then these two gentlemen have constantly been working on the topic (e.g., see [18, 24, 25, 29, 30]). They were present at the *first World Congress on the Square of Opposition* (SQUARE) organized by the first author in Montreux in 2007 [7, 8] and also at the sixth edition organized by both authors in Crete in 2018.

Between these two events, there were the second edition of the SQUARE in Corsica in 2010 [6], the third in Beirut in 2012 [10], the fourth at the Vatican in 2014 [11, 12], and the fifth in Easter Island in 2016 [13]. Many distinguished scholars from logic, mathematics, philosophy, computer science, semiotics, theology, and psychology have participated in these events, among them: Pierre Cartier, Pascal Engel, Jan Woleński, Larry Horn, Terence Parsons, Dale Jacquette (Fig. 5), Stephen Read, Peter Schroeder-Heister, Wolfgang Lenzen, and John Woods.

Fig. 5 Dale Jacquette at the third SQUARE in Beirut. (Photos courtesy of Jean-Yves Beziau)

2 The Sixth World Congress on the Square of Opposition: Crete 2018

Most of the papers included in this book are related to talks presented at the sixth edition of the *World Congress on the Square of Opposition* which took place at the *Orthodox Academy of Crete* (OAC) in the picturesque Kolymbari village, near Chania, in Crete, November 1–5, 2018. Other papers that were presented at this event were published in a special issue of *Logica Universalis* edited by the first author and Jens Lemanski [14].

The idea of organizing the event at the OAC in Crete is due to the second author. He was born in Crete and had already organized an event at this academy. The two authors of the present paper and editors of this book were the main organizers of the event (Fig. 6), supported, besides the OAC, on the one hand by Jens Lemanski, from the Institute of Philosophy of the *University of Hagen* in Germany, and on the other hand by Petros Stefaneas and Ioannis Kriouvrekis, both from the Department of Applied Mathematics, of the *National Technical University of Athens*.

Fig. 6 I.Vandoulakis and J.-Y.Beziau organizers of the sixth SQUARE. (Photos courtesy of the Orthodox Academy of Crete)

The Orthodox Academy of Crete was created in 1968 and functions under the auspices of the Ecumenical Patriarchate. It is a research, education, and conference center aiming at promoting the dialog between faith, science, and culture and inspired by the Platonic tradition of συμφιλοσοφεῖν (symphilosophein – philosophizing together). The academy participates in EU (European Union) and national research projects and is a member of the *Ecumenical Association of Academies* and the *Laity Centres in Europe* (Oikosnet Europe).

The academy hosts a unique *Museum of Cretan Herbs* that includes about 6000 herbs of the collection of Cretan herbs, gathered by the French professor of botany Jacques Zaffran, who dedicated his life to the scientific study of the rich flora and especially the endemic species of Crete. Some of them, for instance, the dittany of Crete (*Origanum dictamnus*), is mentioned by Aristotle in his work *History of Animals* (612a4) and his pupil Theophrastus in his work *Enquiry into Plants* (9.16.1). Many important events have been organized by the OAC including international conferences related to philosophy, theology, environmental studies, physics, biology, medicine, computer science, bioethics, and social issues.

The event on the square was organized shortly after the commemoration of the 50th anniversary of the academy and its award of the silver medal by the Academy of Athens. The event lasted for 5 days, starting Wednesday, November 1, 2018, and ending Sunday, November 5, 2018, with invited talks, contributing talks, and tutorials.

On Friday, November 3rd afternoon, an excursion was organized in the old Venetian town and harbor of Chania (*La Canea*). The Venetian architecture is quite visible from the first moment since, historically, after the Arabs and Byzantines,

Crete was conquered by Venetians in 1252. Chania is a place where different civilizations of the East and West have flourished throughout the centuries and left their impact visible today in every step. Besides the Venetian part of the city, there is the Jewish Quarter with the *Etz Hayyim Synagogue*, and the Turkish part called *Splantzia* with a maze of narrow streets leading to the Venetian port with the *Mosque of Kioutsouk Hassan*, the oldest Ottoman building in Crete, erected in 1645.

The excursion also included the seventeenth-century Monastery of *Agia Triada* (Holy Trinity) in the Akrotiri peninsula built by two brothers Jeremiah-Ioannis and Laurentius-Lucas of the Venetian noble Zancaroli family. The church is built in the Byzantine architectural cruciform style with three domes with two large Doric-style columns and one smaller, Corinthian-style column on either side of the main entrance. The facade bears an inscription in Greek, which is dated to 1631.

The Holy Trinity typically forms a *triangle of contrariety*, called "Shield of the Trinity" or *Scutum Fidei* in the Western Christian tradition (see [19]). This trinity was one of the reasons why the fourth edition of the SQUARE was organized at the *Pontifical Lateran University* in the Vatican in 2014 where the participants were welcomed by Bishop Enrico dal Covol, the rector of this university, nicknamed the "Pope University" (cf. [11, 12]).

Amazingly, this triangle can be recognized in the façade of the Monastery of Agia Triada (Fig. 7). This incidence can be possibly explained by the fact that the Zancaroli brothers were converted from the Catholic to the Greek Orthodox faith and were familiar with both religious cultures.

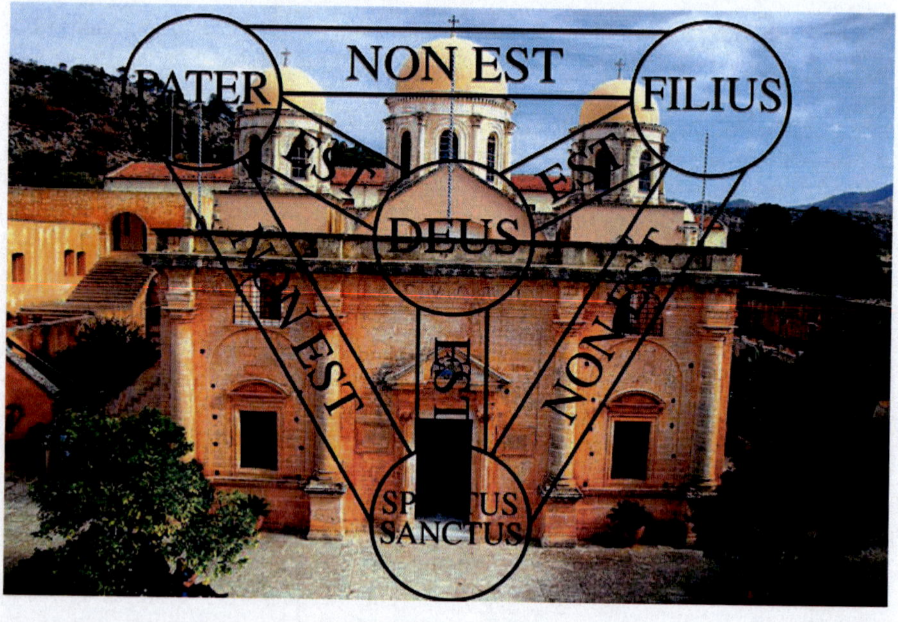

Fig. 7 Monastery of Agia Triada Tsangarolon (of the Zancaroli family) in the Akrotiri peninsula represents fantastic blend of features of different architectural traditions. On its façade, the triangle of contrariety can be recognized. (Photos courtesy of the Orthodox Academy of Crete)

3 The Square of Opposition: An Ongoing Open Project

The adventures of the square of opposition will go on. The seventh edition of SQUARE is scheduled to be held in Leuven, in Belgium in 2022 (organized by KU Leuven, it was delayed due the COVID-19 pandemic) (Fig. 8) and the eighth edition on the Island of Madeira, Portugal.

Fig. 8 The seventh SQUARE will take place in KU Leuven, Belgium, Sept 2022. (Photo courtesy of Jacques Riche)

The future of the square of opposition looks bright, capturing more and more the interest of a great variety of scholars. The theory of opposition grows in depth and complexity and spreads over all fields of knowledge with extensive applications. Much work can be developed concerning the history of the square, its philosophical, logical, mathematical, and semiotic aspects.

Regarding the history of the square, the full story, within various cultures and traditions, still needs to be explored and told [9]. This can be done in different ways. We plan in particular to publish a "diagrammatic book" on the square, collecting the most significant diagrams of the history of the square with lengthy commentaries.

The square has flourished and will be flourishing indefinitely by its internal structure of opposition that grows and expands in many directions like any mathematical theory. There is a continuous interaction between geometrical objects, logical structures and all kinds of concepts. The theory has been applied in particular to color theory [20], music theory (see [26] and Fig. 9), quantum physics [28], painting theory [17], analogy [4, 27], and Kant's theory of antinomies [22].

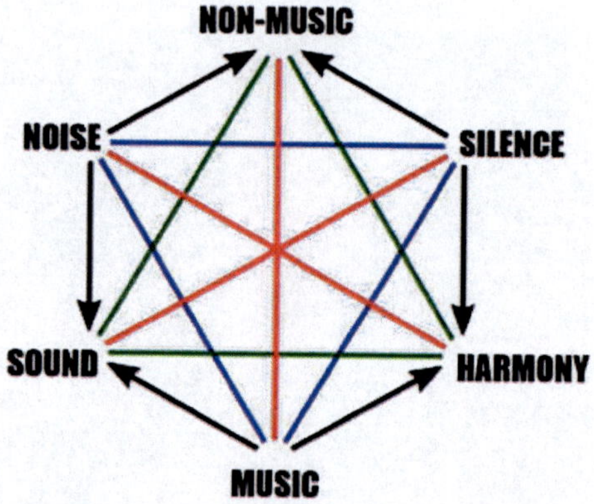

Fig. 9 The hexagon of music

The reason to call the present book *The Exoteric Square of Opposition* is that in the future we expect to have more mysterious things going on, both at the theoretical level and at the level of applications. Then we will publish another book that will be entitled *The Esoteric Square of Opposition*.

The SQUARE project can be considered as a symbol of successful interdisciplinarity. The reason for its success is that it is based on a simple, yet rich, structure understandable by everybody. As with the chess game, there is a good balance/contrast between the simplicity of the basic rules and the complexity of what it is possible to do with these rules. The SQUARE can be seen as a kind of chess game of thought. However, it is a game with no losers, opening infinite possibilities.

This game, this theory, is based on logic, logic which is the foundation of rationality. The SQUARE theory is a good expression/reflection of the four aspects of the *Logos*: reasoning, relation, science, and language, that can themselves be represented by the four-theory of oppositions with a tetrahedron (of subcontrariety), the 3-simplex (Fig. 10).

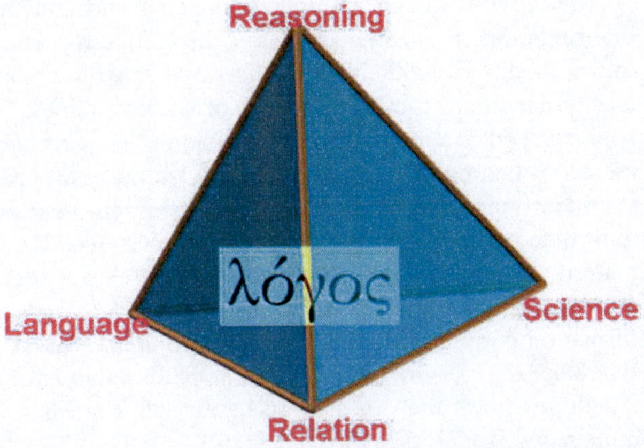

Fig. 10 The tetrahedron of the *Logos*

The organization of the sixth SQUARE in Greece was important for symbolic reasons. First, because the idea of the square was born in Greece in the work of Aristotle. Moreover to have that event in Crete was even more impressive, due to the fact that Crete is the origin of the Greek civilization. There were born the alphabet and Zeus. And in Crete there is Plato's cave and the Labyrinth, two fundamental philosophical symbols.

Crete is the cradle of the Minoan civilization, a Bronze Age Aegean civilization that represents the first advanced civilization in Europe. Several writing systems go back to the Minoan period, most of which remain still undeciphered, such as the Linear A script and the script on the Phaistos disc.

Crete is also associated with the semi-mythical philosopher and poet Epimenides of Knossos or Phaistos (seventh or sixth century BC), famous for his *Epimenides paradox* ("Cretans, always liars"), origin of the liar paradox (for a recent appraisal of this paradox, see [31]). According to Diogenes Laërtius, Epimenides met Pythagoras in Crete, and they went to the cave called *Dictaeon Antron* (the *Psychro* cave) of Mount Ida, where Zeus was raised by a goat named Amalthea. This cave is known to be the origin of one of the most famous texts of philosophy: the *Allegory of the Cave* by Plato (see [1]).

Crete is also famous for the labyrinth, as Arthur Evans called the Knossos palace, because of its architectural complexity. According to the myth, King Minos commanded the skillful craftsman Daedalus (the name Δαίδαλος connotes "labyrinth") to construct a monumental building of interconnected rooms – a *labyrinth* (λαβύρινθος) – to imprison *Minotaur* (a monster with the body of a man and the head of a bull). Minotaur's original name has obvious cosmological connotations: *Asterion*, which is the ruler of the stars; Minotaur also has a connotation to the constellation of *Taurus*.

The labyrinth is an ambiguous concept and construction that allows a double interpretation. From inside, it is a disorientating, chaotic construction; a man is imprisoned in it and unable to understand its structure and to find a way out of it; he is entrapped in a repetitious pattern of wrong choices. From outside or above, it is a sophisticated ordered construction of admirable complexity. Thus, the labyrinth is a metaphor that combines two opposite visions: overt chaotic complexity (internally) vs. underlying order (externally), imprisonment vs. freedom, confusion vs. clarity, multitude vs. unity, and limited perception vs. overall comprehension. The labyrinth pattern also appears in non-European cultures, for instance, in Indian manuscripts and esoteric Buddhist texts, such as the *Chakravyuha* that refers to a military formation narrated in the Hindu epic *Mahabharata*. A prehistoric petroglyph on a riverbank in Goa shows a labyrinthine pattern that has been dated to circa 2500 BC. All these shows that the pattern of the labyrinth is one of the oldest symbols of human civilization (see [33]).

In the European philosophical tradition, the labyrinth was also conceptualized in dynamic terms and used as a metaphor for mental processes. According to Gottfried Wilhelm Leibniz (1646–1716), there are "two famous labyrinths where our reason very often goes astray":

(i) The problem of human freedom.
(ii) The structure of the continuum.

In the twentieth century, the computer scientist and Pulitzer Prize-winner Douglas R. Hofstadter represented the mind by the metaphor of an ant colony, i.e., a labyrinth of rooms, with endless rows of doors flinging open and slamming shut; a network of intricate domino chains, branching apart and rejoining, with little timed springs to stand the dominoes back up.

The SQUARE helps us circulate along the labyrinth of thought and to escape the cave's darkness and illusion to reach understanding (Fig. 11).

Fig. 11 Escaping the cave with the labyrinth of thought

Acknowledgments We thank all the participants of SQUARE 2018 and all people who helped to organize it. Special thanks to Dr. Antonis Kalogerakis, responsible for the local organization of the event at the OAC, and Dr. Konstantinos Zorbas, general director of the *Orthodox Academy of Crete*.

References

1. J.-Y.Beziau, *D'une caverne à l'autre*, MD in Philosophy, supervised by Sarah Kofman, University Paris 1, Panthéon-Sorbonne, 1988.
2. J.-Y.Beziau, "New light on the square of oppositions and its nameless corner", *Logical Investigations*, 10, (2003), pp.218-232.
3. J.-Y.Beziau, "The power of the hexagon", *Logica Universalis*, vol.6 (2012), pp.1-43.
4. J.-Y.Beziau, "An Analogical Hexagon", *International Journal of Approximate Reasoning* , Volume 94, March 2018, Pages 1–17.
5. J.-Y.Beziau, "The Pyramid of Meaning", in J.Ceuppens, H.Smessaert, J. van Craenenbroeck and G.Vanden Wyngaerd (eds), *A Coat of Many Colours - D60*, Brussels, 2018.
6. J.-Y. Beziau and D.Jacquette (eds), *Around and beyond the square of opposition*. Birkhäuser, Basel, 2012.
7. J.-Y.Beziau and G.Payette (eds), Special Issue on the Square of Opposition, *Logica Universalis*, Issue 1, Volume 2 (2008).

8. J.-Y. Beziau and G.Payette (eds), *The square of opposition - A general framework for cognition*, Peter Lang, Bern, 2012.
9. J.-Y.Beziau and S.Read (eds), Special issue of *History and Philosophy of Logic* on the square of opposition, 4 (2014).
10. J.-Y.Beziau and S.Gerogiorgakis (eds), *New dimension of the square of Opposition*, Philosophia, Munich, 2016
11. J.-Y. Beziau and G.Basti (eds), *The square of opposition, a cornerstone of thought*, Birkhäuser, Basel, 2016.
12. J.-Y.Beziau and R.Giovagnoli (eds), Special Issue on the Square of Opposition. Logica Universalis, Issue 2-3, Volume 10 (2016).
13. J.-Y.Beziau (ed), special issue on the Square of Opposition, *South American Journal of Logic*, Volume 3, 2017.
14. J.-Y.Beziau and J.Lemanski, "The Cretan Square", *Logica Universalis*, Issue 1, Volume 14 (2020), pp.1-5.
15. R.Blanché, "Sur la structuration du tableau des connectifs interpropositionnels binaires", *Journal of Symbolic Logic*, 22 (1957), pp.17–18.
16. R.Blanché, *Structures intellectuelles. Essai sur l'organisation systématique des concepts*, Vrin, Paris, 1966. Translation in Protuguese: *Estruturas intelectuais: ensaio sobre a organização sistemática dos conceitos*, Perspectiva, São Paulo, 2012.
17. C.Chantilly and J.-Y.Beziau, "The Hexagon of Paintings", *South American Journal of Logic*, Volume 3, 2017, pp.369-388.
18. L.Demey and H.Smessaert, "Metalogical decorations of logical diagrams", *Logica Universalis*, 10 (2016), pp.233-292.
19. J. Dupuis (ed.), *The Christian Faith in the Doctrinal Documents of the Catholic Church.* Alba House, New York, 1962. Including the English translation of *Enchiridion symbolorum, definitionum et declarationum de rebus fidei et morum.*
20. D.Jaspers, "Logic and Colour", *Logica Universalis*, volume 6 (2012), pp.227–248.
21. D.Jaspers and P.A.M. Seuren, "The Square of opposition in catholic hands: A chapter in the history of 20th-century logic", *Logique et Analyse*, 59 (2016), pp.1-35.
22. P.McLaughlin & O.Schlaudt, "Kant's Antinomies of Pure Reason and the 'Hexagon of Predicate Negation'", *Logica Universalis*, volume 14 (2020), pp.51–67.
23. A.Moretti, *The geometry of logical opposition*, PhD Thesis, University fo Neuchâtel, 2009.
24. A.Moretti, "Why the Logical Hexagon?", *Logica Universalis*, volume 6 (2012), pp.69–107.
25. A.Moretti, "Was Lewis Carroll an Amazing Oppositional Geometer?", *History and Philosophy of Logic*, 35, (2014), pp.383-409.
26. F.Nicolas, "The hexagon of opposition in music", in [8] p.299-320
27. H.Prade & G.Richard, "From Analogical Proportion to Logical Proportions", *Logica Universalis*, volume 7 (2013), pp.441–505.
28. C.de Ronde, H.Freytes and G.Domenech, "Quantum Mechanics and the Interpretation of the Orthomodular Square of Opposition", in [8] pp.221-240.
29. H.Smessaert, "On the 3d visualisation of logical relations", *Logica Universalis*, 3 (2009), pp.303-332.
30. H.Smessaert, "The classical Aristotelian hexagon versus the modern duality hexagon", *Logica Universalis*, 6 (2012), pp.171-199.
31. José-Luis Usó-Doménech, Josué-Antonio Nescolarde-Selva, Lorena Segura-Abad, Kristian Alonso-Stenberg and Hugh Gash, "Mathematical Perspectives on Liar Paradoxes", *Logica Universalis*, vol.15, issue 3 (2021).
32. I.M.Vandoulakis and T.Yu Denisova,. "On the Historical Transformations of the Square of Opposition as Semiotic Object", *Logica Universalis 14* (2020), 7-26.
33. I.M.Vandoulakis and D.Nagy (eds), Symmetry and Labyrinth - Proceedings of the 9th Interdisciplinary Symmetry Congress-Festival of the International Society for the Interdisciplinary Study of Symmetry – Crete, Greece, September 9-15, 2013 - With a Forward by Dan Shechtman, 2011 Nobel Prize in Chemistry. Special issue of the *Journal of the International Society for the Interdisciplinary Study of Symmetry*, 1-4 (2013).

Division of Entities and Foundations of Reality: Aristotle's Ontological Square

Gianluigi Segalerba

Abstract In my paper, I analyse some aspects regarding Aristotle's interpretation of the organisation of ontology. In my opinion, Aristotle is looking for a new ontology in many of his works. Hence, in his investigation, Aristotle aims to discover the correct components of the ontology and to put these components in their due ontological place. Being qua being, categories, substance among the categories, universals, form, matter and so on are analysed and defined by Aristotle throughout his works.

In this analysis, I concentrate my attention on two schemes of reality which, in Aristotle, precede, at least as regards some aspects, the other structures of reality. These schemes, which constitute the first frame of reality, are the two-district scheme and the four-domain scheme. The two-district scheme is the structure of reality composed by individual entities and by universal entities; the four-domain scheme consists in the structure of reality composed by individual substantial entities, individual non-substantial entities, universal substantial properties and universal non-substantial properties. The four-domain scheme, which is representable in the form of the ontological square, is an extension of the two-district scheme.

The mentioned districts and domains correspond to Aristotle's realms of reality. These realms are mutually incompatible in the sense that any entity can belong only to one of these realms, but they are not mutually isolated, since, for example, individual entities are instances of the corresponding universal entities. The discovery, explanation and analysis of these schemes are keys to determining the position of the different entities in the reality. Without a correct understanding of the position of the entities in the reality, no ontology can function correctly. For example, entities which have, in the reality, the position of instances must always

G. Segalerba (✉)
aka – Arbeitskreis Kulturanalyse Universität Wien (Working Group Cultural Analysis University of Vienna), Vienna, Austria

International Mariinskaya Academy Named After M.D. Shapovalenko, Moscow, Russia
e-mail: gianluigisegalerba@gmail.com

J.-Y. Beziau, I. Vandoulakis (eds.), *The Exoteric Square of Opposition*, Studies in Universal Logic, https://doi.org/10.1007/978-3-030-90823-2_2

be distinguished from entities which, within the reality, do not have the position of instances; individual entities must always be distinguished by common entities and by universals. Through the distinction regarding entities which are instances and entities which are not instances, Aristotle is opening the fields of existence to different realms of entities. Not only does numerically one entity exist, universal entities exist too.

Ontological square (four-domain scheme) and two-district scheme represent the basic structure, the very framework of Aristotle's ontology, since they are the general rule of Aristotle's ontology. Any interpretation of ontology which does not respect the distinction existing between entities and which is expressed through these schemes leads to the collapse of the ontology itself or paves the way for this collapse, as witnessed, for example, by the Third Man regress. The inquiry on Aristotle's four-domain scheme is completed by a comparison between Aristotle's position and the positions of E. J. Lowe's ontological square.

I contend that Aristotle interprets individual entities as instances of properties (or as instantiated properties). The basic status of the individual entity consists in its being an instance of a property. Furthermore, Aristotle considers universal properties as being programmes/dispositions concretised in the individual entities (at least as regards biological properties). Within Aristotle's ontology, the particular existence field of the instances is always constituted by individuals (by individual entities), while the whole field of existence is constituted both by individuals (by individual entities) and by universal properties. Hence, reality does not consist exclusively of individuals. Properties, at least biological properties like "being man" or "being animal", are programmes/dispositions being concretised through their instances. Their existence does not depend on the existence of one particular instance; it does not depend, likewise, on the existence of a determined plurality, but it does depend on the existence of at least one instance: properties do not transcend the dimension of the individual entities.

In order to explain the consequences deriving from a misunderstanding of the realms of reality, I analyse the arguments of "the One Over Many" and of the "Third Man" from Aristotle's lost work *De Ideis*. In the Third Man Argument, it becomes clear that, if within an ontology an entity, which is not an instance, is interpreted as an instance, the consequence of this mistake is the collapse of the whole ontology.

Keywords Realms of reality · Aristotle · *Categories* · *Metaphysics* · *De Ideis* · Typological ontology · Two-district ontology · Four-domain ontology · Ontological square · Substance · Universal properties · Individuals · Particulars · Universals · Lowe · Kung · Liske · One Over Many · Third Man

Mathematics Subject Classification (2000) Primary 03A05·Secondary 03A10, 03B38, 03B65, 03B80

1 Introduction

In my essay, I analyse some aspects of Aristotle's ontology. The description of the four domains of entities in *Categories* 2 and of the two districts of entities in *Categories* 5 will serve to introduce Aristotle's strategy of differentiation of entities. The difference between entities which are instances and entities which are not instances will prove to be one of the foundations of Aristotle's ontology. The basic structure of individuals as instances of properties will emerge, in this study, as the fundamental status of individuals themselves. Individuals are not simply individuals: they are, constitutively, instances of something. Aristotle does not accept a theory of bare substratum: every individual is an individual something.[1]

In many passages of his works (e.g. in *Categories* 5, 3b10–21; in *Metaphysics Zeta* 8, 1033b19–1034a8; and in *Metaphysics Zeta* 13, 1038b34–1039a3), Aristotle distinguishes entities being (having the ontological status of) a "this something (τόδε τι)" or being (having the ontological status of) a "this such (τόδε τοιόνδε)" from entities being (having the ontological status of) a "quality (ποιόν)" or being (having the ontological status of) a "such (τοιόνδε)". Entities having the ontological status of "this something/this such", on the one hand, and entities having the ontological status of "quality" and of "such", on the other hand, constitute realms of reality which should not be confused with each other. These kinds of entities belong to realms of reality that are reciprocally incompatible.

Aristotle aims to put order in the ontology. The general distinction existing between entities which are instances and entities which are not instances serves to determine the ontological realm to which the different entities belong. Moreover, this distinction serves to avoid any confusion between types of entities. In particular, through his distinction strategy, Aristotle is opening spaces of reality for determined

[1] At the beginning of my contribution, I would like to mention three studies which I regard as central for my way of analysing Aristotle's texts: Joan Kung's essay *Aristotle on Thises, Suches and the Third Man Argument*; Michael-Thomas Liske's book *Aristoteles und der aristotelische Essentialismus: Individuum, Art, Gattung*; and Edward Jonathan Lowe's book *The Four-Category Ontology. A Metaphysical Foundation for Natural Science*. I would like to briefly introduce the themes dealt with in the mentioned pieces of research that have proved to be the most important ones for me. Kung's article opened a new view of Aristotle for me with her inquiry on the presence of a typological ontology in Aristotle. Kung interprets the difference between individuals and universals as a difference of types of entities. Liske's book gave me new perspectives regarding the aspects that can be assigned to Aristotelian essences as vital forces. Lowe's investigation gave me, through his interpretation of a four-category ontology, a new way of interpreting Aristotle's strategy of differentiation between entities in *Categories* 2 and a new way of connecting Aristotle's ontology to the discovery of models for natural sciences. As regards my interpretation of Aristotle's relation between particular and universal and between object and concept, I gained important ideas while analysing G. Frege's works *Über Begriff und Gegenstand, Funktion und Begriff, Über Sinn und Bedeutung* and *Die Grundlagen der Arithmetik. Eine logisch-mathematische Untersuchung über den Begriff der Zahl*. The responsibility for the interpretation which I present in my paper is, of course, mine alone.

entities. He is opening realms of reality for entities which, although they are not instances, are nonetheless real entities. One of the foundations of reality and one of the general rules of ontology are represented in Aristotle by the two-district scheme, which is constituted by individual entities and by universal entities and by the four-domain scheme – a complexification of the two-district scheme – which is constituted by individual substantial entities, individual non-substantial entities, universal substantial properties and universal non-substantial properties. Both schemes imply an opposition between the entities composing them. A precise opposition exists between the two districts, on the one hand, and between the four domains, on the other hand, since their features are mutually incompatible. Any entity of the reality can, respectively, only belong to one of the districts and, correspondingly, only to one of the domains. Any entity can be either individual or universal, on the one hand; it can be only individual substantial, or individual non-substantial, or universal substantial, or universal non-substantial, on the other hand. Incompatibility does not mean isolation between districts or between domains, though, since individual entities are instances of the corresponding universal entities.

A new conception of opposition should therefore be introduced in the interpretation of Aristotle. This new concept of opposition, which precedes every other possible opposition, is the concept of the ontological square of opposition (i.e. the concept of the ontological four-domain opposition) and of the ontological two-district opposition.

The opposition between different entities, both substantial and non-substantial individuals, on the one hand, and substantial and non-substantial universals, on the other hand, steadily shows up in Aristotle's works. Without the correct interpretation of this basic opposition, which requires a correct ontological distinction, there is no space for any entity, and there is no space for other forms of opposition.

The ontological square of oppositions should become, in my opinion, the first authentic square of opposition in Aristotle, inasmuch as it represents the most general rule of ontology: no ontology is possible without this distinction, and everything stops if this distinction is not respected. Without the ontological square, there is, for Aristotle, no beginning of ontology.

Correspondingly, the regress of the Third Man shows that the mistake represented by regarding entities which cannot be instances as if they actually were instances provokes the collapse of the whole ontology. Hence, the correct interpretation of the status of the entities will turn out to be indispensable for the cohesion and coherence of the ontology. Aristotle's ontological enterprise possesses, therefore, different aspects which are always mutually interconnected. The investigation into the components of ontology, the determination of the features of these components and the analysis of the consequences which occur if the features of these components are not correctly interpreted are part of the same ontological strategy: without a general distinction between individual and universal and without respecting the incompatibility between individual and universal, there is no way of preventing the collapse of the ontology.

A healthy ontology, first of all, ought to be able to distinguish between entities which are individual/particular and entities which are universal/common. Thus, it ought to be able to distinguish between entities which are, as regards their own position in the reality, instances and entities which are not instances (but have, that notwithstanding, a position in the reality). On the basis of Aristotle's insistence on this difference,[2] I believe that it is legitimate to say that this differentiation capacity is one of the most important requirements of a healthy ontology. Of course, the mentioned ontological schemes do not represent the whole of Aristotle's ontology because they are only one of the foundations of it. The other ontological fields represented, for instance, by being qua being, by the categories, by substance as category and by the functions of potentiality and actuality are still to be explored.

2 Terminology, Definitions, Translations

Before directly dealing with Aristotle's ontological project, I shall specify the translations of the main concepts of Aristotle which will be addressed in this study. Moreover, I shall share a few remarks on my interpretation of Aristotle's substance, since substance will be mentioned in the essay. Due to the centrality of this concept for Aristotle's ontology, I feel it is appropriate to comment briefly on my own interpretation of substance.

(a) In this text, the ancient Greek word "οὐσία" will be translated with "substance".
(b) In this text, the ancient Greek expression "τὸ τί ἦν εἶναι" will be translated with "essence".
(c) In this text, the ancient Greek expression "τόδε τι" will be translated with "this something".[3]
(d) I am of the opinion that substance (οὐσία), in the main, signifies[4]:

- Entity belonging to the biological field and being able to independently exist, like an individual man, an individual horse and an individual tree[5]; substance corresponds, therefore, to the complex organisms, to the complex entities of the biological dimension like animals and plants.

[2] See footnote 49 for a partial list of Aristotle's passages in which the incompatibility between particulars (and related concept) and universals (and related concepts) is asserted.

[3] For an analysis of the expression "τόδε τι", I refer to J. A. Smith's essay *Tóde ti in Aristotle*.

[4] The values for substance that are being mentioned in my essay are not the only ones that Aristotle's concept assumes. I refer to my study *Semantik und Ontologie: Drei Studien zu Aristoteles* for my positions regarding the different values substance can, in my opinion, have.

[5] As regards the value of substance as organism, see, for example, *Categories* 4, 1b27–28 (man and horse); *Categories* 5, 2b13–14 (tree); and *Metaphysics Zeta* 7, 1032a18–19 (man and plant).

This is, in my interpretation, the sense that should be attributed to the first substance of the *Categories*.[6] In other works of Aristotle such as in *De Anima*,[7] substance can either maintain this value[8] or assume the following value:

- Form, essence and soul of an entity belonging to the biological field; substance is, in this case, the factor that directs the individual entity during its own life and that leads the whole individual entity to its own development[9] (the soul of the individual man e.g. directs the whole life of the individual man and leads the whole individual man to his own development).[10] Within the biological dimension, essence, soul, nature, form and substance of the living entity are to be seen, in my opinion, as different expressions for the same entity.

(e) The terms "individual" and "particular", whenever they refer to the realm of reality of the instances, are, in this text, reciprocally interchangeable.
(f) Property is, in this study, universal property. Universal property is a programme for realisation in the instances, without being itself an instance.

---

[6] As regards the editions, the translations and the commentaries regarding Aristotle's works, I only mention editions, translations and commentaries corresponding to the works of Aristotle which are actually quoted in my essay. The editions of Aristotle's works which I used for my analysis are the following ones: for the *Categories* I used the edition of L. Minio-Paluello. For the lost work *De Ideis*, I used the edition of W. D. Ross contained in the volume *Aristotelis Fragmenta Selecta. Recognovit Brevique Adnotatione Instruxit W. D. Ross* and the edition of D. Harlfinger contained in W. Leszl, *Il "De Ideis" di Aristotele e la teoria platonica delle idee. Edizione critica del testo a cura di Dieter Harlfinger*; G. Fine in her book *On Ideas: Aristotle's Criticism of Plato's Theory of Forms* follows the edition of D. Harlfinger as regards the passages of *De Ideis* which will be quoted and analysed in my essay. For the *Metaphysics* I used the edition of W. Jaeger and the edition of W. D. Ross.

[7] I consulted the following translations of Aristotle's works: for the *Categories* I consulted the translation of J. L. Ackrill; for *De Ideis* I consulted the translation of G. Fine; for the whole *Metaphysics* I consulted the translation of W. D. Ross (contained in Barnes J. (ed.), *The Complete Works of Aristotle. The Revised Oxford Translation, Volume Two*) and the translation of H. Tredennick; for the books *Gamma* and *Delta* of the *Metaphysics*, I consulted the translation of Ch. Kirwan; for the books *Mu* and *Nu* of the *Metaphysics*, I consulted the translation of J. Annas. I would like to add that I consulted these translations without, however, entirely following any of them. I always tried to find my own translation of the texts of Aristotle which are quoted in my analysis. I of course assume the responsibility for my translations.

[8] As regards the particular question of the continuity or discontinuity, in Aristotle, of the interpretation of substance, I limit myself to the following remarks: I do not agree with all the positions maintaining the presence of a caesura between Aristotle's interpretation of substance in the *Categories*, on the one hand, and Aristotle's interpretation of substance in the central books of the *Metaphysics*, on the other hand. The value of substance as an individual entity belonging to the biological field is, in my opinion, never abandoned as a primary value for substance by Aristotle. This value of substance remains, at least in my opinion, a primary value for substance (i.e. it is not relegated to a secondary role).

[9] See, for these values of substance, the chapter *De Anima* II 1.

[10] See, for example, the chapter *De Anima* II 1 as regards the value of substance as form, soul and essence of a living entity.

(g) The concepts "universal" and "universal property", in this text, are either equivalent, or "universal" represents a name, a linguistic deputy for "universal property". In either case, the problem of the existence or non-existence of universals in Aristotle depends, in my opinion, on the question whether in Aristotle universal properties exist or not, as I shall thereafter reveal.

(h) In my analysis, when I use the concept "property", I refer mostly to properties belonging to the biological field, such as "being man" and "being horse" (for properties corresponding to species – in these cases, to the species "man" and to the species "horse"), on the one hand, and such as "being animal" (for properties corresponding to genera), on the other hand. Hence, property mainly refers to substantial biological properties. Aristotle considers, in my opinion, all biological properties as properties belonging to reality (i.e. they do not correspond to mere instruments of classification invented by the speaking subjects). The property "being man" exists, even though it does not exist at the same ontological level as the ontological level at which the instances of this same property (e.g. individual men) exist.

(i) By using the concept "property" in my analysis, I am not referring to fictitious properties. The properties that I refer to are properties belonging to the objective reality. These properties exist independently of their being acknowledged, or of their being thought of, or of their being known by a (thinking, speaking, knowing) subject; these kinds of properties exists independently of whichever subject.

(j) I consider terms like "universals", "common entities", "that which is said universally", "that which belongs universally" and "that which is predicated in common", which can be found, for example, in *Metaphysics Zeta* 13, as reciprocally equivalent; they refer to universal entities and are opposed to individual entities.

(k) Through the distinction between the realm of individual entities and the realm of universal entities, Aristotle establishes a typological ontology consisting of individuals as instances of universal properties, on the one hand, and of universal properties, on the other hand.

3 Division of Entities: Aristotle's Ontological Square

Coming now to the ontological organisation of the entities, I would like, first of all, to say that this organisation precedes any particular property and any particular entity we could have in the reality. Independent of which properties, in particular, exist (we could have in the reality, e.g. other biological properties and, as a consequence, other biological species than those we actually have), we shall have, all the same, the organisation of reality expressed in the two districts and in the four domains together with the reciprocal relations which hold, respectively, between the entities belonging to the two districts and between the entities belonging to the four domains of the whole field of existence.

Aristotle steadily works on the discovery of the basic ontological structures: the curtains of this ontological enterprise are opened in different works, so that the interpretative reconstruction of Aristotle's strategy must be built on the basis of different passages from different works. This aspect, of course, makes the analysis of the structure of reality in Aristotle much more difficult than if Aristotle's observations were concentrated in the same work. On the other hand, the fact that Aristotle's observations concerning the very foundations of reality are present in different works is a testament to the importance, for Aristotle, of the discovery of the foundations of ontology and to the relevance of the correct interpretation of these foundations.[11]

The passage corresponding to *Categories* 2, 1a20–1b9, can be illuminating as regards the division and organisation of reality[12]:

> Of the entities some are said of a subject, but are in no subject (Τῶν ὄντων τὰ μὲν καθ' ὑποκειμένῳ τινὸς λέγεται, ἐν ὑποκειμένῳ δὲ οὐδενί ἐστιν), as, for example, man is said of a subject, the individual man, but is in no subject; some entities, then, are in a subject, but are said of no subject (τὰ δὲ ἐν ὑποκειμένῳ μέν ἐστι, καθ' ὑποκειμένου δὲ οὐδενὸς λέγεται) – I call in a subject that which, being in something not as a part, cannot exist separately from that which it is in – (ἐν ὑποκειμένῳ δὲ λέγω ὃ ἔν τινι μὴ ὡς μέρος ὑπάρχον ἀδύνατον χωρὶς εἶναι τοῦ ἐν ᾧ ἐστίν), as, for example, the individual knowledge of grammar is in a subject, the soul, but is said of no subject; and the individual white is in a subject, the body – for every colour is in a body –, but is said of no subject, some entities, then, are both said of a subject and in a subject (τὰ δὲ καθ' ὑποκειμένου τε λέγεται καὶ ἐν ὑποκειμένῳ ἐστίν), as, for example, knowledge is in a subject, the soul, and is also said of a subject, knowledge of grammar; some entities, then, neither are in a subject nor are said of a subject (τὰ δὲ οὔτε ἐν ὑποκειμένῳ ἐστὶν οὔτε καθ' ὑποκειμένου λέγεται), like, for example, the individual man or the individual horse (οἷον ὁ τὶς ἄνθρωπος ἢ ὁ τὶς ἵππος) – for none of such entities either is in a subject or is said of a subject. Entities that are individual and numerically one (ἁπλῶς δὲ τὰ ἄτομα καὶ ἓν ἀριθμῷ) are generally said of no subject (κατ' οὐδενὸς ὑποκειμένου λέγεται), but nothing prevents that some of them are in a subject (ἐν ὑποκειμένῳ δὲ ἔνια οὐδὲν κωλύει εἶναι): for the individual knowledge of grammar is one of the entities in a subject.

The structure of reality is organised in four different subdivisions:

(i) Entities which are said of a subject, but which are not in a subject (an example is man – man is predicated of the individual men).

[11] I used the following commentaries of Aristotle's works: for the *De Ideis*, I used W. Leszl's *Il "De Ideis" di Aristotele e la teoria platonica delle idee. Edizione critica del testo a cura di Dieter Harlfinger* and G. Fine's *On Ideas: Aristotle's Criticism of Plato's Theory of Forms*; for the *Categories* I used J. L. Ackrill's *Aristotle's Categories and De Interpretatione, Translated with Notes*; for the whole *Metaphysics*, I used W. D. Ross' *Aristotle's Metaphysics. A Revised Text with introduction and commentary by W. D. Ross*; for *Metaphysics Gamma* and *Delta*, I used Ch. Kirwan's Aristotle *Metaphysics. Books Γ, Δ, and E.*

[12] Within my essay, I shall not entirely quote the original text of Aristotle; I only quote those expressions and those concepts of Aristotle's text which, in my opinion, are the most relevant ones.

(ii) Entities which are in a subject but which are not said of a subject (examples are individual knowledge of grammar or individual white – the individual knowledge of grammar is in the soul; the individual white is in the body).
(iii) Entities which are said of a subject and which are in a subject (an example is knowledge – knowledge is in the soul and is said of the knowledge of grammar).
(iv) Entities which are not in a subject and which are not said of a subject (examples are the individual man or the individual horse).

The entity being said of a subject constitutes an essential property of the subject to which it is referred, whereas the entity being in a subject does not constitute an essential property of the subject. The feature "being numerically one" belongs both to what neither is in a subject nor is said of a subject and to what is in a subject but is not said of a subject. We can observe the presence of a differentiation between entities which are numerically one and, therefore, are individual and entities which are not numerically one and, therefore, are not individual.

4 Lowe's Ontological Square

The comparison of Aristotle's ontological construction with the ontological construction proposed by Lowe is, in my opinion, relevant for the understanding of Aristotle's ontological scheme. Lowe derives his own ontological square directly from Aristotle's division of entities in *Categories* 2[13]; his own ontological scheme (compared with Aristotle's) is as follows[14]:

[13] See, for example, *The Possibility of Metaphysics: Substance, Identity, and Time*, pp. 203–204, *The Four-Category Ontology: A Metaphysical Foundation for Natural Science*, p. 21, and *More Kinds of Being: A Further Study of Individuation, Identity, and the Logic of Sortal Terms*, pp. 8–11.

[14] See, for Lowe's description of the four-category ontology, Chaps. 1 and 2 of *The Four-Category Ontology: A Metaphysical Foundation for Natural Science*. See also Lowe's *More Kinds of Being: A Further Study of Individuation, Identity, and the Logic of Sortal Terms* (Lowe describes, in the paragraph *New Developments* of the Introduction of this last book, his conversion to the four-category ontology as the most significant change in his metaphysical thinking since the writing of his book *Kinds of Being*). Lowe's meditation represents an example of a four-domain scheme. Different schemes of ontology are represented, in the contemporary ontology, by D. M. Armstrong, who presents a two-domain scheme constituted by individual entities and by universal entities (see *Universals & Scientific Realism, Volume I: Nominalism and Realism; Volume II: A Theory of Universals*) and by K. Campbell, who presents an ontology of tropes (see *Abstract Particulars*). A comparison of Aristotle's ontology with these ontological interpretations must, unfortunately, be reserved for a further analysis.

- Lowe's individual substances (objects)[15] = Aristotle's entities which are in a subject and are not said of a subject.
- Lowe's property/relation instances (modes) = Aristotle's entities which are in a subject but which are not said of a subject.
- Lowe's substantial universals (kinds) = Aristotle's entities which are said of a subject but which are not in a subject.
- Lowe's non-substantial universals (property/relations)[16] = Aristotle's entities which are said of a subject and which are in a subject[17].

Substances represent the basis of the whole reality, since every other domain of entities depends, either directly or indirectly, on substances for its own existence.

I think that Lowe's interpretation of reality as consisting of substantial and non-substantial universals, on the one hand, and of individual substances and property/relation-instances, on the other hand, represents a highly valuable instrument in understanding the relation existing between individual entities and universal entities in Aristotle, in spite of the fact that Lowe's square is, actually, not a commentary of Aristotle's ontological positions.

Likewise, Lowe's interpretation of substantial and non-substantial universals as dispositions, on the one hand, and of individual substances and of property/relation instances as occurrences,[18] on the other hand, corresponds, in my opinion, to Aristotle's aims when Aristotle speaks of individual entities like the individual man and of properties like being man. In particular, Lowe's concept of dispositions as a complex of properties which determine the individual entities having them can

[15] There are probably some differences between Aristotle and Lowe as regards the interpretation of substance and of natural kinds; for example, Lowe interprets gold as a natural kind (see *The Four-Category Ontology. A Metaphysical Foundation for Natural Science*, p. 21), whereas I am not sure that Aristotle would consider gold as being a natural kind. The important aspect of similarity between Aristotle and Lowe nonetheless consists in the suggestion that a basic framework exists already before the concrete appearance of entities in the reality. This framework precedes anything else. Moreover, Lowe's distinction between disposition and occurrence corresponds, at least in certain aspects, to Aristotle's distinction between potentiality and actuality, between potential faculties of entities and actualisation of these faculties. Both thinkers individuate in the structure of reality a field of actualisations/occurrences, i.e. of concrete cases of laws (e.g. of natural laws), and a field of potentialities/dispositions, which constitutes the general range of possibilities for the concretisations. This second field of reality is the range of laws constituting the frame of reality of which individuals/particulars are the concretisation. Reality is not only made up of individual/particular cases; there is a realm of reality constituted by general laws which direct the existence of the individual entities.

[16] See *The Four-Category Ontology. A Metaphysical Foundation for Natural Science*, p. 22.

[17] I tend to agree with Lowe's square, but I do not think that substantial individuals are instances of kinds; I consider them as instances of properties. Substantial individuals are, in my opinion, members of kinds, since they are instances of properties defining the corresponding kinds. The substantial individual "man" is an instance of the property "being man", which delimits the kind "man".

[18] See, for the definition of the concepts "occurrence/occurrent" and "disposition/dispositional", *The Four-Category Ontology: A Metaphysical Foundation for Natural Science*, Chap. 1.

correspond, in my opinion, to Aristotle's conception of biological properties as programmes for the development of the biological entity.[19]

The scheme of a four-category ontology[20] as such represents a model of metaessentialism applied to the structure of the reality[21]: no matter which entities we can meet (men, dinosaurs, Martians and so on), the deep structure of the four domains is always present; it precedes every particular nature; it precedes any manifestation, any presence and any appearance of concrete entities. This structure is the framework of reality as such.

5 A Proposal of Adaptation of Lowe's Scheme

In order to adapt Lowe's ontological scheme to my previous distinctions of Aristotle's entities, I would propose the following four-domain scheme for Aristotle:

(i) Substances (in the sense of entities belonging to the biological realm – men, horses, trees and so on).
(ii) Non-substantial individual entities (like particular qualities and particular quantities).
(iii) Universal substantial properties (like being man, being horse, being tree)[22].
(iv) Universal non-substantial properties (like being a quality, being a quantity)[23].

While Lowe, if I have correctly understood his positions, considers individual substances as instances of kinds, it seems to me that Aristotle considers substances as instances of universal substantial properties and as members of species and genera

[19] Within Lowe's ontological scheme, the relationships between the entities of the different categories are instantiation, characterisation and exemplification (actually, Lowe presents different schemes showing slight differences in terminology with each other in his book *The Four-Category Ontology: A Metaphysical Foundation for Natural Science*). Individual entities (both individual substances and property/relation instances) instantiate, respectively, substantial universals (kinds and non-substantial universals (properties/relations)). Individual substances are characterised by property/relation instances (modes). Substantial universals (kinds) are characterised by non-substantial universals (properties/relations). Furthermore, individual substances exemplify non-substantial universals (properties/relations). As regards Aristotle's square and as regards the relations within Aristotle's works between particulars/individuals and universals (universal properties), I use the concept "instantiation" as applying to the result of the concretisation of a property (the essence) in its instances.

[20] For my analysis, I prefer to use concepts like "four-domain scheme" in order to avoid any interference between the four mentioned categories of Lowe and the categories of Aristotle.

[21] I use the word "metaessentialism" in order to express that the deep structure of reality consists of the four domains of substantial instances, non-substantial instances, universal kinds and non-substantial universals.

[22] These entities express the essential identification of substances.

[23] These entities express the essential identification of the instances of non-substantial entities.

(kinds).[24] As I shall afterwards show in relation to some aspects of Aristotle's defence of the validity of the principle of contradiction, I believe that Aristotle considers reality as being immediately composed of properties: properties are one of the bearing structures of his ontology. Instances are always to be seen, in my opinion, in their immediate connection to their essential properties. In general, individual entities are instantiated properties for Aristotle. They belong to kinds but are, as such, instances of properties defining kinds.

Universal substantial properties determine kinds (species and genera) in the sense that they represent the condition, for possible members of the kinds, for belonging or not belonging to a determined kind. For example, the property "being man" determines the kind (the species) "man" and distinguishes the kind (the species) "man" from all other species (kinds); the property "being man", furthermore, represents the condition for belonging or not belonging to the species "man". Individual entities which are men belong to the species "man"; individual entities which are not men do not belong to this same species.

I believe that the conception of the ontological square can represent a useful extension of the conception of the typological ontology, which is introduced by Aristotle in order to free the ontology from the dangers of the Third Man regress. I think that Aristotle's introduction of the typological ontology, in general, and Aristotle's insistence on the incompatibility existing between features of the individual entities and features of the universal entities, in particular, correspond to Aristotle's aims of describing a healthy ontology. A healthy ontology has, among its characteristics, the presence of the distinction between the entities which are instances and the entities which are not instances. Both the two-district ontology and the four-domain ontology constitute the framework of the reality, which is, as such, independent of the concrete entities present in the reality. Before any entity whatsoever appears in the world, the deep structure of reality is already constituted, in Aristotle's view, by the two districts:

(a) Individual entities.
(b) Universal properties.

and by the four domains:

(i) Substances.
(ii) Non-substantial instances.
(iii) Universal substantial properties.
(iv) Universal non-substantial properties.

These divisions of entities are useful both to understand the description of reality Aristotle aims to reach and the interpretation of reality Aristotle wishes to avoid. In particular, no confusion between entities which are instances and entities which are

---

[24] Universal substantial properties individuate through their own definition the corresponding kinds. Kinds (both species and genera) function, in my opinion, as classes of the individual entities.

not instances can be accepted; the mistake represented by confusing the realms of reality with each other means the destruction of the whole ontology.[25]

Any entity appearing in the world belongs to one of these districts and to one of these domains. An entity must not necessarily belong to precisely one of these two districts and of these four domains, but any entity whatsoever certainly belongs to one, and only one, of these districts and to one, and only one, of these domains. The conditions for the way of presence in the reality of the entities are given: any entity will belong to one, and only one, of these fields.

6 Foundations: Two-District Ontology

The general aim of Aristotle's ontology, in my opinion, consists in assigning the individual entities, on the one hand, and universal properties, on the other hand, to the right realms of existence. At the same time, Aristotle aims to correctly determine the relation of a substance with the factor that constitutes the essence of the substance itself. This factor does not constitute an entity which exists apart or is separated from the entities to which it is related. The factor due to which any entity whatsoever is essentially that which it is, is not a further entity existing apart or being separated from the first entity. The problems that are to be dealt with by Aristotle are, therefore, the following ones:

(i) The general determination of the ontological realms of the different entities.
(ii) The determination of the realm of existence of the universal properties and, in particular, of the essences of the individual entities.

As we shall see, the realms of reality of the individual entities and of the universal properties are mutually incompatible. The corresponding kinds of entities should not be confounded with each other. The organisation of entities consists, therefore, in the following realms of reality:

(a) Realm of the individual entities.
(b) Realm of the universal properties.

I would like to mention, first of all, Aristotle's differentiation between universals and particulars expressed in *Metaphysics Beta* 4, 999b33–1000a1:

> For there is no difference between saying numerically one or saying particular: for we call the particular in this way, the numerically one, but we call universal what is said of these[26] (τὸ γὰρ ἀριθμῷ ἓν ἢ τὸ καθ' ἕκαστον λέγειν διαφέρει οὐθέν. οὕτω γὰρ λέγομεν τὸ καθ' ἕκαστον, τὸ ἀριθμῷ ἕν, καθόλου δὲ τὸ ἐπὶ τούτων).

The division of entities which we meet in this passage is the following one:

[25] Of course, incompatibility between realms of reality does not mean mutual isolation between these realms.

[26] That is, of these particulars.

- Particular is the same thing as numerically one.
- What is predicable of a plurality of particulars is the universal. The universal is something belonging to a plurality or is predicated of a plurality.

The existence of a universal presupposes the existence of a plurality to which it is referred. Universals are, therefore, distinguished from, or opposed to, the entities that are numerically one.[27] The presence of and the difference between the two realms of reality begins, therewith, to loom.

In order to now give a more articulated example of Aristotle's typological ontology than the previous one, I am going to quote the text *Categories* 5, 3b10–21, where the differentiation between first substance and second substance can be observed. In this text, Aristotle's strategy of differentiating:

- between entities that are instances of properties, and that, therefore, are numerically one, on the one hand,
- and entities that only express the essential identification[28] of the entities which are instances of properties, without being themselves instances of properties and without being, therefore, themselves numerically one or particular, on the other hand,

comes to light:

> Every substance seems to signify a this something (Πᾶσα δὲ οὐσία δοκεῖ τόδε τι σημαίνειν). Certainly, as regards the first substances, it is indisputable and true that it signifies a this something (τόδε τι): for the entity revealed is individual (ἄτομον) and numerically one (ἓν ἀριθμῷ). But, as regards the second substances, it appears, on the one hand, because of the form of the name, whenever one speaks of man or of animal, that a second substance likewise signifies a this something (ἐπὶ δὲ τῶν δευτέρων οὐσιῶν φαίνεται μὲν ὁμοίως τῷ σχήματι τῆς προσηγορίας τόδε τι σημαίνειν); this is not really true, but, rather, it signifies a certain quality (ποιόν τι), – for the subject is not, as the primary substance is, one (οὐ γὰρ ἕν ἐστι τὸ ὑποκείμενον), but the man and the animal are said of many entities (κατὰ πολλῶν ὁ ἄνθρωπος λέγεται καὶ τὸ ζῷον); – however, it does not signify simply a certain quality, as the white does; the white signifies nothing

[27] For a further consideration of the function of a universal, see, for example, this passage contained in *Metaphysics Delta* 26, 1023b29–32: "For the universal, and that which is said in a whole way, as being a whole (τὸ μὲν γὰρ καθόλου, καὶ τὸ ὅλως λεγόμενον ὡς ὅλον τι ὄν), is universal in the sense that it contains many entities because it is predicated of each (οὕτως ἐστὶ καθόλου ὡς πολλὰ περιέχον τῷ κατηγορεῖσθαι καθ' ἑκάστου), and because all, each respectively, are one (καὶ ἓν ἅπαντα εἶναι ὡς ἕκαστον), for example man, horse, God, because they are all animals (οἷον ἄνθρωπον ἵππον θεόν, διότι ἅπαντα ζῷα)". Universals are both containers of entities and units of measurement of the entities themselves.

[28] With the expression "essential identification", I mean the function, exercised by a second substance or by a universal, of expressing the essence of an entity without therewith expressing the whole composition of the essence. A second substance like man expresses the essential identification of the individual man, since it expresses the name of the essence of the individual man. A second substance like animal expresses a less informative essential identification of the individual animal, since the individual animal is identified not specifically, but only generically, by animal. See *Categories* 5, 2b7–14, and 2b29–34 for the comparison between giving the account of the species and giving the account of the genus of an entity.

> but a quality, but the species and the genus determine the quality concerning substance (τὸ δὲ εἶδος καὶ τὸ γένος περὶ οὐσίαν τὸ ποιὸν ἀφορίζει), – for they signify substance of a certain quality (ποιὰν γάρ τινα οὐσίαν σημαίνει).

In the text we can observe Aristotle's intention of assigning the features of τόδε τι only to entities which are numerically one. Entities which are not numerically one are excluded from this range of entities. Furthermore, we can observe the following contrapositions between the features belonging to first substances and the features belonging to second substances:

- First substance is this something, is numerically one, signifies a τόδε τι and is individual.
- Second substance is not this something, is said of many entities, is not numerically one and signifies a ποιόν.

This text conveys different contents which are of great importance. Aristotle is focussing on a determined feature of a substance qua substance in general and not qua substance of a particular kind. The following correlation between the first substances and the features belonging to the first substances qua first substances holds:

(x)(first substance (x) → individual and numerically one (x) → this something (τόδε τι)(x)[29])

This implication holds for every substance qua substance, no matter which particular substance is dealt with. The following correspondence holds:

- Instance ↔ numerically one.[30]

As regards second substances qua second substances, the following correlation between the second substances and their own features holds:

(x)(second substance (x) → not numerically one (x) → not this something (not τόδε τι)(x) → quality (ποιόν)(x))

We reach, therefore, the following results:

- What is said of many is not numerically one.
- Since entities being said of many are not numerically one, they cannot be put in the same realm of entities together with the entities that are numerically one; they belong to another realm of reality.
- Second substances are not individual entities existing besides the entities of which they are predicated. They presuppose the existence of the entities of

[29] It should be noted that entities which are in something and are not said of something are numerically one and individual too, without themselves being a τόδε τι (see *Categories* 2, 1b6–9).

[30] This correspondence holds not only for substances but also for every individual entity, that is, it is not limited to the category of substance.

which they are predicated: their existence presupposes the existence of the first substances.

- Second substances express the way of being of the entities of which they are predicated: they are a sort of synthesis of the complex of properties of the individual substances.
- The quality which is signified by the second substance is not simply a quality; it expresses a substance of a certain quality. Aristotle does not want second substances to be reduced to qualities. The position of the second substances in the field of reality is different from the position of mere qualities.[31]

The proposition:

– Socrates is man.

assigns to Socrates the essential property "being man". Socrates instantiates the property "being man": the proposition states that the property "being man" is instantiated in Socrates. The essential property "being man" attributed to Socrates is not a numerically one entity existing besides Socrates. Being man exists as a universal property having a determined content of faculties. This property is inscribed in the reality as one of the properties individuating biological species. The property is not a further man; it is the complex of dispositions which, if there is a man, will be instantiated by this man.

By drawing a distinction between entities that have the ontological structure of "this something (τόδε τι)"[32] or of "this such (τόδε τοιόνδε)",[33] on the one hand, and entities that have the ontological position of "such (τοιόνδε)"[34] or, alternatively, of "quality (ποιόν)",[35] on the other hand, Aristotle is, in my opinion, aiming to draw a distinction between different fields of existence. Likewise, he is aiming to assign the entities to their own realm of reality. Hence, Aristotle is not therewith aiming to exclude either universal properties or universals from the field of existence.

[31] The relation between first substances and second substances could be interpreted, among other things, as a relation of dependence from the first substances on the second substances. The first substance Socrates would depend on the second substance for its being something. I do not agree, though, with this hypothesis. I think that a first substance is immediately an instantiated property. To be is to be something. To exist, for any first substance, is to be an instantiated property of a determined biological kind. There is no individual entity which first of all exists and only thereafter is a determined property; if an individual entity exists, it is an instantiated property. Existence cannot be detached or distinguished from the act of instantiating a complex of biological properties. Biological entities are the complex of properties contained in their own soul; without these properties they do not exist.

[32] See, for example, *Categories* 5, 3b10–21; *Metaphysics Beta* 6, 1003a8–9; and *Metaphysics Zeta* 13, 1038b34–1039a3.

[33] See, for example, *Metaphysics Zeta* 8, 1033b19–1034a8.

[34] See, for example, *Metaphysics Beta* 6, 1003a8–9; *Metaphysics Zeta* 8, 1033b19–1034a8; and *Metaphysics Zeta* 13, 1038b34–1039a3.

[35] See, for example, *Categories* 5, 3b10–21.

7 Instances and Properties

Generally speaking, every individual entity is an instance of a property. There is no individual bare entity; there is no individual entity that can be neutral in relation to all its properties. For an individual entity, to exist means instantiating a property or a complex of properties. This property or this complex of properties is the essence of the individual entity itself; without essence there is no entity. Individual entities are never bare entities. Individual entities are instances which, as instances, constitute the realisation of a determinate range of properties. Instances cannot assume and lose any property since they cannot lose their essential properties. If they were to lose their essential properties, they would disappear from the realms of existence. Furthermore, they are not an addition of accidental properties. There is, on the contrary, a complex of constitutive properties which makes up the individual entity and is the way of existence of the individual entity.

The property "being man" can exist, in spite of the absence of a particular instance or in spite of the disappearance of a particular instance. This same property, though, like every other biological property, cannot exist if there is no instance of it. There are no non-instantiated properties in Aristotle.[36] First substances constitute instances of properties (the individual man Socrates represents, e.g. an instance of the property "man"). Both entities, that is, substances and properties, exist; they do not belong, nonetheless, to the same field of existence. As such when it comes to their ontological status, first substances are instances of properties. Properties are – at least biological properties – programmes (dispositions) which, once instantiated, direct every aspect of the development of an entity.

The fields of entities which are numerically one and entities which are not numerically one are rigidly distinguished from each other. Reality is divided in two fields:

(i) Individuals.
(ii) Universals.

Universals (at least universals corresponding to a biological class) and second substances[37] possess the following features:

- They correspond to a class (the universal "man" corresponds to the class of men).
- They refer to a property that is instantiated by the individual entities (the universal "man" corresponds to the property "being man", which is instantiated in every member of the class "man").

[36] Aristotle's refusal of the existence of non-instantiated properties can be found in *Categories* 11, 14a6–10.

[37] Since both second substances and universals are predicated of a plurality of entities (see the quoted passage of *Categories* 5, 3b10–21 and, for example, *De Interpretatione* 7, 17a38–b1) and since "man", for example, can be both a second substance and a universal, I think that second substances and universals can be considered as equivalent, at least as regards universals expressing names of biological species and genera.

- If they are considered as classes, they have, as their own extension, the individual entities that belong to the class.
- If they are considered as the reference to an intension, they are the name of the property that states the condition for belonging to a class[38].
- They are not instances (universals belong to a realm of reality which is different from the realm of reality to which the individual entities belong; individual entities and universals – or individual entities and properties as programmes for instances – belong to different realms of reality).
- They are, furthermore, immanent to the plurality and not transcendent in relation to the plurality, that is, universals must have at least one instance in order to exist[39].
- They presuppose the existence of the instances of which they are predicated.

8 Some Remarks on the Relevance of the Essence in Aristotle's Ontological System

The status of the individual entities being instances of properties has been mentioned. Instances have determined properties as their own essences: the general structure of instances/properties is the kernel of reality. It is relevant for the analysis of Aristotle's vision of ontology to analyse some consequences that follow in the event that the principle of contradiction does not hold (*Metaphysics Gamma* 4, 1007a20–33), so that we can observe Aristotle's absolute refusal of the hypothesis that existence of the essence is jeopardised. Properties and essences turn out to be indispensable within ontology:

> And in general, those who use this argument do away with substance and essence (οὐσίαν καὶ τὸ τί ἦν εἶναι). For it is necessary that they say that all attributes are accidents, and that there is no being essentially man or being essentially animal. For, if being essentially man (τι ὅπερ ἀνθρώπῳ εἶναι) is something, this will not be being not man (μὴ ἀνθρώπῳ εἶναι) or not being man (μὴ εἶναι ἀνθρώπῳ) (and yet these are negations of it); for that which it meant was one thing, and this was the substance of something (τινος οὐσία). Signifying substance is that, for it, the essence is not something else (τὸ δ' οὐσίαν σημαίνειν ἐστὶν ὅτι οὐκ ἄλλο τι τὸ εἶναι αὐτῷ). But if, for it, being essentially man is either being essentially not man or essentially not being man (εἰ δ' ἔσται αὐτῷ τὸ ὅπερ ἀνθρώπῳ εἶναι ἢ ὅπερ μὴ ἀνθρώπῳ εἶναι ἢ ὅπερ μὴ εἶναι ἀνθρώπῳ), the essence will be something else (ἄλλο τι ἔσται), so that it is necessary for them to say that there will not be such notion of anything,[40] but that all attributes are accidental; for in this aspect

[38] See, for example, *Categories* 5, 2a14–19: species and genera as second substances are the entities to which first substances belong. See also *Metaphysics Delta* 26, 1023b29–32 for the function of the universal as the entity which contains a plurality of entities.

[39] For the difference which exists between immanent universals and transcendent universals, I refer to the analyses of Armstrong in *Universals & Scientific Realism, Volume I: Nominalism and Realism*, p. 128.

[40] That is, there is no essence.

> substance and accident are distinguished from each other: for the white is accidental to the man, since man is white, but he is not what white is (τὸ γὰρ λευκὸν τῷ ἀνθρώπῳ συμβέβηκεν ὅτι ἔστι μὲν λευκὸς ἀλλ' οὐχ ὅπερ λευκόν).

If there is no principle of contradiction, there is no essence. Since this consequence is unacceptable, the principle of contradiction must retain its validity. In the case of the non-validity of the principle of contradiction, any content of essence is annulled, as it can be seen on the basis of the example represented by the essence of man. Actually, properties as such completely disappear since their contents are annulled by the non-validity of the principle of contradiction. Aristotle sees reality as consisting of properties. Reality is organised in properties, which are essences of determined individual entities. Aristotle is so convinced of this structure that he uses it as the basis of the validity of the principle of contradiction: a reality without properties is simply not acceptable. Reality would remain without properties if the principle of contradiction had no validity.

If the essence of man has a determined content, the essence of man will be this determined content. If the principle of contradiction does not have validity, the essence of man will be the negations, too, of this content, that is, it will be the essence of being not man or it will be the essence of not being man, that is, it will have a content of properties which is other than the original essence of man and which is incompatible with the original essence of man. Every essential content is therewith annulled: essence as essence does not exist. In general, the content of any property whatsoever collapses (the collapse of the principle of contradiction is not only a problem for the essences, it is a problem regarding all properties as such).

If the principle of contradiction has no validity, then there is no possibility of existence for properties and for essences. The individual entity cannot have an essence, since no essence has a determined content. We have no essences, so that we cannot have individual entities that are determined by essences. Any individual entity whatsoever, in the absence of validity for the principle of contradiction, would be, in general, essentially f, but it would be not f, too. Therefore, there would be no sense in speaking of an essence that should determine the entity as such and that should differentiate this individual entity from other entities in the reality.

Actually, within an ontological condition characterised by the absence of validity of the principle of contradiction, there is simply no sense in speaking of differences between entities, since, as everything can have every property and everything can be denied every property, there is no way of distinguishing the entities from each other. There is no difference at all any longer between entities, because there is no longer any determined property content.

Aristotle's use of concepts like essence and substance implies that he has an ontological scheme that contains properties, plurality, distinguishable plurality and entities as instances of essences. In other words, using the concept of essence within one of the defence strategies for the validity of the principle of contradiction, Aristotle does not use only essence. He also uses a whole apparatus of entities and of concepts connected to essence. There are essences, for Aristotle, since Aristotle has considered individual entities as being, as such, instances of properties. Determined properties are essences since individual entities are instances of them. The status of individual entities as instances of properties is indispensable so that properties are regarded as being essences of determined entities.

By resorting to the essence in order to defend the validity of the principle of contradiction, Aristotle is resorting to a constituent of reality that, in his opinion, cannot be refused in a healthy ontology; an ontological system cannot function without essences. Reality would no longer be reality if the essence disappears. Reality is constituted of properties, which are essences of determined entities. No matter which entities we have in the reality, essences exist. No matter which individual entities exist, individual entities have essences. If the consequence of the fall of the principle of contradiction is that essence and substance of entities disappear, this consequence should be refused so that the principle of contradiction could maintain its own validity. Substance and essence are simply not expendable in a healthy ontological system: they belong to the very structure of reality.

9 *Metaphysics Mu* 10: Instances and Universals

Coming back to particular entities and to universals, the passage contained in *Metaphysics Mu* 10, 1087a4–21 provides more insight into Aristotle's strategy on the relations which exist between particulars and universals. Thanks to this passage, we can see that the structure of instances and universals corresponds to the very foundations of reality. Aristotle's strategy consists, in my opinion, in showing that reality is composed both of individuals and of universals. Moreover, the status of individuals is shown to be that of "instances of ...". The interpretation of individuals as instances of properties aims to show the modality of relation between individuals and properties:

> And now, all these difficulties follow with good reason, whenever they make the ideas out of elements (ὅταν ἐκ στοιχείων τε ποιῶσι τὰς ἰδέας) and maintain that a separated unity exists apart from the substances which have the same form (παρὰ τὰς τὸ αὐτὸ εἶδος ἐχούσας οὐσίας [καὶ ἰδέας] ἕν τι ἀξιῶσιν εἶναι κεχωρισμένον); but if, as in the case of the elements of speech (ὥσπερ ἐπὶ τῶν τῆς φωνῆς στοιχείων), nothing prevents that many alphas and betas exist[41] (πολλὰ εἶναι τὰ ἄλφα καὶ τὰ βῆτα), and if nothing prevents that no alpha itself and no beta itself exist apart from the many[42] (μηθὲν εἶναι παρὰ τὰ πολλὰ αὐτὸ ἄλφα καὶ αὐτὸ βῆτα), in consequence of this there will be infinite similar[43] syllables[44] (ἔσονται ἕνεκά γε τούτου ἄπειροι αἱ ὅμοιαι συλλαβαί). The statement that all knowledge is universal (καθόλου), so that it is necessary both that the principles of entities (τὰς τῶν ὄντων ἀρχὰς) are universal (καθόλου) and are not separated substances

[41] Alternative translation: "nothing prevents that the alphas and the betas are many".
[42] That is, "apart from the many alphas and betas".
[43] Alternative translation: "same".
[44] Alternative translation: "the similar (same) syllables will be infinite".

(οὐσίας κεχωρισμένας),[45] presents indeed, of all the points that were mentioned, the greatest difficulty, nonetheless the statement is, in a sense, true, but, in a sense, it is not true. For knowledge, like knowing, has two senses (ἡ γὰρ ἐπιστήμη, ὥσπερ καὶ τὸ ἐπίστασθαι, διττόν), one of which is in potentiality (τὸ μὲν δυνάμει), the other of which is in actuality (τὸ δὲ ἐνεργείᾳ). The potentiality, being, as matter, universal and indefinite, deals, then, with the universal and indefinite (ἡ μὲν οὖν δύναμις ὡς ὕλη [τοῦ] καθόλου οὖσα καὶ ἀόριστος τοῦ καθόλου καὶ ἀορίστου ἐστίν); but the actuality, being definite, deals with a definite entity, being a this something, it deals with a this something (ἡ δ' ἐνέργεια ὡρισμένη καὶ ὡρισμένου, τόδε τι οὖσα τοῦδέ τινος); the sight, accidentally, sees universal colour, though, because this colour which it sees is colour (ἀλλὰ κατὰ συμβεβηκὸς ἡ ὄψις τὸ καθόλου χρῶμα ὁρᾷ ὅτι τόδε τὸ χρῶμα ὃ ὁρᾷ χρῶμά ἐστιν) and this alpha which the grammarian studies is alpha (καὶ ὃ θεωρεῖ ὁ γραμματικός, τόδε τὸ ἄλφα ἄλφα)...

Many aspects of this passage deserve an appropriate analysis. At the moment, though, I am interested in the relation existing between particulars and universals. In this passage the structure of the entities as instances of universals is proposed as the solution to the problem of the relationship between particulars and universals. Individual entities have the structure of "this alpha" or of "this colour". This structure gives, in my opinion, a solution to the question concerning the connection between particulars and universals. The status of the individual entity is the status of "instance of ...". The particular entity is interpreted as an instance of a universal; any individual entity is a τόδε τι, a this something, in the sense that it is an instance (τόδε) of a property (τι).

Aristotle's example of the individual colour and of the individual alpha can be extended, in my opinion, to every instance which belongs to reality; every individual entity is the concretisation of a property. The individual alpha is an alpha, the individual colour is a colour. As an extension of this organisation, the individual man is a man, that is, the individual man is the instance of the universal "man" and of the property "being man". We have, therefore, the following reality elements:

- Universal properties such as "being an alpha" or "being a colour".
- Instances of properties such as "the individual alpha" or "the individual colour".
- Universals as properties or as predicates that represent properties, that are names of properties (the universal "man" represents the property "being man"; it is a deputy for the property "being man").

The predication of the universals is the consequence of the instantiation of a property in a particular entity and in a plurality of entities. Since a plurality of entities instantiates a property, the universal corresponding to this property can be predicated of the members of this plurality.

[45] Aristotle must avoid the ontological constellation in which entities have universal principles. In the concretised reality, everything is individual; the principles of individual entities must be individual (e.g. see *Metaphysics Lambda* 5, 1071a19–29).

The field of instances is represented by particular entities.[46] However, even the field of existence of substantial and non-substantial universals possesses a right of citizenship in the reality. Substantial and non-substantial universals – like the universals "man" and "colour" – exist (they are not constructions of the mind; the mind finds them; the mind does not invent them).[47] The field of instances is certainly always represented by particulars. This notwithstanding, the whole field of existence is constituted by both particulars and universals, even though the universals' way of existence is different from that of the instances.

Particulars and universals belong to mutually different realms of reality. They correspond to reciprocally different ontological realms. Universals are the elements of the laws of reality, whereas individuals are the elements of the corresponding concretisations of the laws of reality. We thus have the following realms of reality:

- Individuals/particulars as instances of properties, numerically one entity[48].
- Universal properties, not numerically one entity[49].

---

[46] It does not matter, in the present context, whether the substance is a material or an immaterial one. Both material substances and immaterial substances are individual. The status of being individual is the common aspect they possess. The difference between realms of reality is constituted, in Aristotle, by the difference between individual entities and universal entities, not by the difference between material substances and immaterial substances.

[47] At least as regards biological properties, biological species and biological genera, Aristotle seems to consider these entities as indestructible. Biological species and genera are eternal (see *De Generatione Animalium* II 1, 731b22–732a1). Biological properties are already given in the reality. Individual biological entities will instantiate this or that property; individual biological entities will not necessarily instantiate a determined property, but the range of the properties they concretise is already given in the reality.

[48] The feature "being numerically one" could be applied to the universal properties, too, even though, in this case, any universal property would be a numerically one entity existing at another level than the level at which each of the instances exists. However, Aristotle is ready to attribute the feature of being numerically one exclusively to the instances; to be numerically one is considered by Aristotle, at least in the contexts mentioned in this study, as equivalent to the status of instance.

[49] The difference between individuals and universals represents, in my opinion, one of the foundations of Aristotle's ontological system. It represents a way for avoiding the Third Man regress. Aristotle's intention of avoiding the Third Man regress is, in my opinion, both one of the origins and one of the mainsprings of the particular way in which Aristotle's ontological and predicative systems are built. Throughout the *Categories*, the *De Interpretatione*, the *Sophistical Refutations*, the *Posterior Analytics* and the *Metaphysics*, there are many assertions that, in my opinion, can witness this aim, since these assertions express the attention paid by Aristotle in order that any confusion whatsoever between individuals (and features belonging to individuals qua individuals) and universals (and features of universals qua universal) could be avoided. The oppositions and incompatibilities are, for example:

- Between τόδε τι and ποιόν (first substance and second substance, *Categories* 5, 3b10–21).
- Between entities which are numerically one and entities which are not numerically one, generally (i.e. concerning not only substances but also, e.g. qualities, as in *Categories* 2).
- Between entities which are numerically one and entities which are not numerically one (as regards the different values for substance – first substance and second substance – *Categories* 5, 3b10–21).
- Between particular and universal (*De Interpretatione* 7, 17a38–b1).

The system of relationships between entities is fundamental. Aristotle shows, through this system, that particular entities, regardless of whether they are substantial or non-substantial entities, are always concretisations of properties. No matter which entities we concretely have, we have, in any case, this organisation of entities in individuals as instances and in universal properties.

The ontological model in Aristotle is therefore constituted not only by individual entities but also by programmes for instantiation in the instances. Every individual entity is, as regards its ontological position, an instance of something (i.e. it cannot be considered only as an individual entity, its position as an instance is ontologically basic), while every universal property is a potentiality for realisation in its instances (i.e. it cannot be detached by its being a potentiality for instantiation; it does not exist without instances; it is immanent and not transcendent in relation to its instances).

- Between τόδε τι and τοιόνδε (*Sophistical Refutations* 22, 178b36-179a10).
- Between entities which are apart from many and entities which hold of many (i.e. contraposition between ideas and universals and mutual incompatibility between ideas and universals) (*Posterior Analytics* I 11, 77a5–9).
- Between particular/numerically one and universal (*Metaphysics Beta* 4, 999b33–1000a1).
- Between universals and substances (*Metaphysics Beta* 6, 1003a5–17).
- Between common entities and τόδε τι (*Metaphysics Beta* 6, 1003a8–9).
- Between substance as τόδε τι and common entities as signifying a τοιόνδε (*Metaphysics Beta* 6, 1003a8–9).
- Between τόδε τοιόνδε and τοιόνδε (*Metaphysics Zeta* 8, 1033b19–26).
- Between substance and universal (*Metaphysics Zeta* 13, 1038b8–16).
- Between entities universally said and substance (*Metaphysics Zeta* 13, 1038b8–16).
- Between entities which belong universally and substance (*Metaphysics Zeta* 13, 1038b35).
- Between entities predicated in common and τόδε τι (*Metaphysics Zeta* 13, 1038b35-1039a1).
- Between τόδε τι and τοιόνδε (*Metaphysics Zeta* 13, 1038b34–1039a3).
- Between universals and the feature of existing separately besides the particulars (*Metaphysics Zeta* 16, 1040 b25–27).
- Between substance and one over many (*Metaphysics Zeta* 16, 1040b16–1041a5).
- Between substance and predicate (*Metaphysics Iota* 2, 1053b16–24).
- Between this something and universal (*Metaphysics Mu* 10).

These incompatibilities aim to avoid, for example, the risk of the Third Man regress, since they pose a rigid border between entities that are instances and, therefore, are individual and numerically one, on the one hand, and entities that are not instances and are not numerically one, on the other hand; they also avoid other incongruencies which could appear if the realms of reality are not distinguished from each other. An incongruency is, for example, the fact that, if the entity predicated in common were interpreted as a this something, then an entity like Socrates would be a plurality (see *Metaphysics Beta* 6, 1003a9–12). Another incongruency is, for example, the fact that if the universal is interpreted as being the substance of an entity, the plurality of all the entities of which the universal is substance will be annulled (see *Metaphysics Zeta* 13, 1038b9–15). Thus, the Third Man is not the only consequence of an incorrect interpretation of the foundations of ontology; further grounds of destruction of ontology are explained by Aristotle. The Third Man regress is undoubtedly the most striking one, but there are also other incongruences. The realms of reality of individual entities and of entities that are not individual must, therefore, be rigidly distinguished from each other. I refer to Kung's article for further elements regarding this distinction (see especially pp. 207–208).

10 On Properties

In order to deal now with the subject of properties, I think that, for Aristotle, within the biological dimension, the complex of properties which makes up, for example, the property "being man", has a determined content of capacities and of faculties which is realised in the instances of this complex of properties. This complex of properties holds universally for all instances of the property (i.e. for all members of the species). The content of the properties is a programme of life development for all members of a species. This content, which directs any development of the individual, is transmitted through the generation. For whichever instance of a biological property we have, this instance will concretise in itself the property corresponding to the essence of the instance. Hence, the content of the biological property will be always identical for every member of a biological species corresponding to the property (there does not appear to be any concept of evolution in Aristotle).

The property will universally hold for all members of the species. This is the sense in which, at least within the biological dimension, universal properties exist; they are programmes which, if instantiated, will always direct the whole development process of the instances, thus bringing about – under normal circumstances – individuals with the same life development and with the same faculties. They are universal since they are, as programmes, identical for every individual entity instantiating them. Hence, they hold in the same way for all their instances. Therefore, they hold universally for all the instances. The universal property is not, itself, an instance; it is the complex of faculties that will be realised in an individual.

In a healthy ontology, universal properties should not be regarded as being instances. An entity's being universal implies a completely different position in the reality in comparison with the position of individuals. One of the problems we are compelled to face when we discuss the concept of universal in Aristotle is the question regarding the existence or non-existence of universals in Aristotle. I personally think that the question we have to face should rather be whether properties which hold universally exist or do not exist in Aristotle. This means that the question should rather be whether a property such as "being man", which is identical for every man, exists or does not exist in Aristotle's view. If, for Aristotle, a property that is identical for a given plurality exists (e.g. if an identical property "being man" exists), then, as a consequence of the existence of this property, the universal that corresponds to the property exists (in this case, if an identical property "being man" exists, then the universal "man" exists).

In other words, any question regarding the existence of universals should be methodologically preceded, in my opinion, by the question of whether universal properties (at least universal properties belonging to the biological field) exist. In this case, individuals instantiating the same universal property (e.g. individuals instantiating the property "being man" or the property "being animal") are connected with each other by a sameness relationship (i.e. individuals are specifically or generically the same; they are, of course, not numerically the same). In the event that these properties do not exist, individuals are not connected with each other by any sameness relationship.

I believe, therefore, that the central question should not look at the existence or non-existence of universals. The primary question ought rather to determine the way properties are interpreted by Aristotle. The question regarding the existence or non-existence of universal properties is decisive for the destiny of universals. The existence of universals proves to be, in my opinion, a consequence of the existence of universal properties. Conversely, the non-existence of universals would turn out to be a consequence of the non-existence of universal properties.

Aristotle, in my opinion, pleads for the existence of universal properties like being man or being animal.[50] I think that, for Aristotle, properties, at least biological properties, constitute the natural world; they are rooted into the natural world. Every instance of a biological property is the same (specifically or generically) as every other instance of the same biological property. Every instance of the biological property "being man" is specifically the same as every other instance of the property "being man". It is the same in the sense that, under normal conditions, it possesses the same functions determining the species "man" as every other instance of the property "being man". Every instance of the biological property "being animal" is generically the same as every other instance of the property "being animal". It is the same in the sense that, under normal conditions, it possesses the same functions determining the genus "animal" as every other instance of the property "being animal".

I think that the existence of universal properties in Aristotle could find support in Aristotle's texts like *De Generatione* et *Corruptione* II 6, where Aristotle pleads for the existence of a nature which dictates an identical development for the members of the same biological class. All members of a biological class have and will have, under normal conditions, the same development because they instantiate the same biological property (e.g. all men instantiate the biological property "being man", and all men have, under normal circumstances, the same life development and the same faculties). Biological properties are not and should not be deemed as classifications invented by speaking subjects. Properties corresponding to natural species (and to natural genera) and natural species as the natural species "horse" or the natural species "man" exist, in Aristotle's view, in a mind-independent way.

---

[50] As regards the problem of the existence or non-existence of universals, I do not think that Aristotle's assertions expressing what the universal is not, prove that Aristotle therewith aims to deny the very existence of universals. Aristotle rather aims to explain the ontological features that a universal, or an entity which is universally said, or an entity which belongs universally cannot have, such as being substance, being the substance of an entity or existing separately (e.g. see *Metaphysics Zeta* 13 and *Metaphysics Zeta* 16). In my opinion, Aristotle does not plead for the non-existence of universals. Saying what the universal is not, is not the same, in my opinion, as saying that the universal does not exist. I think that Aristotle's strategy consists in eliminating, from the features belonging to universals qua universals, all the features which wrong ontological positions have attributed to the universals themselves. Aristotle's aim consists (at least in my opinion) in denying the existence of the wrong features of universals qua universals, not in denying the existence of universals as such.

Every biological property corresponding to a species or to a genus is a potential programme which is concretised, realised and actualised in its instances. As a consequence, instances belonging to a species or to a genus have an identical development. The uniformity of development of the members belonging to a biological class attests the existence of a universal nature that is the same for all members of the biological class. This nature is universal since it is identical for every member of the class, so that it holds in the same way for all the member of the class. The universal nature is a complex of faculties which will be realised, under normal circumstances, whenever an individual entity instantiating that nature exists.[51]

11 Aristotle's Polemical Targets: The One over Many Argument and the Third Man Argument

The divisions of entities in the ontological square (i.e. in the four domains) and in the two districts ought to be respected. If these divisions are not respected, the collapse of ontology follows. I would now like to analyse some arguments of Aristotle's lost work *De Ideis*, to show which consequences arise if the distinction between realms of reality is not respected. For this purpose, I am going to consider the One Over Many Argument and the Third Man Argument.[52] The argument of the One Over Many gives an answer to the problem of the uniform predication of predicates such as "man" and "animal". The contents of the argument are the following ones (see *De Ideis* 80.9–16)[53]:

> They also use such an argument to establish that there are ideas. If each of the many men is man, and if each of the many animals is animal, and the same applies in the other cases; and if, in the case of each of them, there is not something which is predicated, itself, of itself (καὶ οὐκ ἔστιν ἐφ' ἑκάστου αὐτῶν αὐτὸ αὑτοῦ τι κατηγορούμενον),[54] but there

[51] In my opinion, Aristotle's analyses on the nature, on the life development and on the faculties of biological entities, exposed in texts like *Metaphysics Zeta* 7, 8; *De Generatione et Corruptione* II 6; *Physics* II 1, 7, 8; and *De Anima* II 1, 2, 3, 4, suggest that, in Aristotle's view, these entities follow a determined development which holds true for every entity of the same species. The identity of the development of the members of a species is the consequence of the existence of a determined line of development for all the members of a species. A species is individuated by a property like being man, which entails a determined identical programme of development for every member of the species.

[52] I decided to analyse not only the Third Man Argument but also the One Over Many Argument in order to show the elements of continuity in the interpretation of predication maintained by the positions criticised by Aristotle.

[53] For studies concerning the *De Ideis*, see, for example, the book of W. Leszl *Il "De Ideis" di Aristotele e la teoria platonica delle idee. Edizione critica del testo a cura di Dieter Harlfinger* and the book of G. Fine *On Ideas: Aristotle's Criticism of Plato's Theory of Forms.*

[54] This sentence implies, in my opinion, that no entity is what it is in virtue of itself. Therefore, no member of the given plurality (in this particular case, no member of the plurality of men and no

is something which is predicated of all of them, without being the same as any of them (ἀλλ' ἔστι τι ὃ κατὰ πάντων αὐτῶν κατηγορεῖται οὐδενὶ αὐτῶν ταὐτὸν ὄν),[55] then it exists this which is besides the particular beings, separated from them and everlasting (εἴη ἄν τοῦτο[56] παρὰ τὰ καθ' ἕκαστα ὄντα ὂν κεχωρισμένον αὐτῶν ἀίδιον).

For it is in every case predicated in the same way (ὁμοίως) of all the numerically successive particulars. And what is a one in addition to many, separated[56] from them, and everlasting (ὃ δὲ ἓν ἐστιν ἐπὶ πολλοῖς κεχωρισμένον τε αὐτῶν καὶ ἀίδιον), this is an idea. Therefore, there are ideas.

As regards the logic of the argument, the argument, in my opinion, functions in the following way:

(i) A plurality of entities has a property.
(ii) An entity corresponding to the property of the plurality is predicated in the same way of all members of the plurality and is different from all members of the plurality (since this entity is predicated in the same way of the members of the given plurality, this entity is different from all the members of the given plurality).

Then:

(iii) An entity exists which is besides the members of the plurality, which is separated from these members and which is everlasting.
(iv) The entity that is separated from the members of the plurality, that is one entity in addition to the members of the plurality and that is everlasting coincides, as regards its own features, with the idea and its own features.
(v) Therefore, ideas exist.

Since the entity is predicated in the same way of the entities of which it is predicated, then this same entity cannot be other than separated, everlasting and one in addition to many. The predicated entity must be neutral in relation to the given plurality of which it is predicated; therefore, it must be outside this plurality.

The entity which is predicated exists, no matter whether the particular members of the plurality exist or not. The ascription of everlastingness is a testament to this

member of the plurality of animals) is the property which is predicated of it, in virtue of itself (in this particular case, no man and no animal of the given pluralities are men or animal in virtue of themselves). The predicated entity is always something else in relation to the entity (to the entities) of which it is predicated (a related question regards the aspect because of which the predicated entity is different from the entities of which it is predicated).

55 The entity which is predicated is different from the members of the plurality. It does not coincide with the members of the plurality. The presence of the difference between the entity which is predicated and the members of the plurality of which it is predicated is common to Aristotle too, since, for example, second substances are different from first substances and universals are different from the particulars of which they are predicated. The question is, though, that the sense of this difference is explained in different ways in this argument and in Aristotle's. In the argument, the difference consists, for example, in the separation of the entity predicated, whereas, for Aristotle, the difference consists in both universals and second substances not being instances.

56 Aristotle's statements in *Metaphysics Zeta* 1, 1028a18–34 oppose this concept of separation. The properties attributed to the substance are not separable from the substance. Only substance possesses the feature of existing separately.

entity's complete independence of the particular members of the plurality, and it represents the guarantee that a predication will always be uniform as regards the future predication of the future individual entities.[57] The entity which is predicated in the present argument must transcend, and actually transcends, the particular entities of which it is predicated.[58]

Throughout the argument, the attempt at building conditions for a uniform predication is clearly expressed. The reason why the predicated entity is not a member of the plurality of which it is predicated and the reason why the plurality is not self-predicated in each individual case of the plurality lie in the necessity of establishing a uniformity of predication. This uniformity would not be the case:

- Both if an entity of the plurality were predicated of the entities of the plurality, since the relationship of the entity which is predicated with itself would not be the same as the relationship between the entity which is predicated and the other components of the plurality, there would be a privileged predication in the case of the self-predication of an entity and a secondary predication in all the other cases of the plurality itself. Thus, there would not be any uniformity of predication in all the cases of the plurality.[59]
- And if a self-predication in every case of the plurality takes place – since the entity which is predicated would change in every case – the entity "a" would be predicated of the entity "a", the entity "b" would be predicated of the entity "b", and so forth.[60]

---

[57] At the basis there is a different interpretation of the conditions for uniformity of predication, a different interpretation of the concept of group, a different interpretation of properties and a different interpretation of the conditions for belonging to a group.

[58] In the quoted passage from *Categories* 5, second substances like man (which is predicated of a plurality of entities):

- Are not a this something.
- Are not individual.
- Are not numerically one.
- Are a quality (connected to substance).

In the argument of the One Over Many, the entity predicated (like man):

- Is besides the particular beings.
- Is not identical with any member of the plurality of which it is predicated.
- Is separated from the particular beings.
- Is everlasting.

The two interpretations of predication are completely different from each other, since the presuppositions of the two ontologies are different from each other.

[59] In other words, the entity could not be predicated in the same way of the members of the plurality, if it were itself a member of the plurality.

[60] I think that those who maintain the validity of the One Over Many Argument for the existence of ideas, against whom Aristotle's criticism is directed in the *De Ideis*, aim to exclude both the hypothesis of a generalised self-predication (i.e. every single member of the plurality would be predicated of itself) and the hypothesis that one of the entities of the plurality is predicated of all the members of the plurality. They exclude both hypotheses since, in both cases, the predication would

Therefore, a new entity should be found which can show a neutral relationship towards the plurality. This neutrality can exist only if the entity which is predicated exists besides the plurality itself and only if the entity which is predicated is different from any member of the given plurality. The uniformity of predication is guaranteed by the existence of the entity besides the plurality. If the entity is predicated in the same way of a plurality of entities, this entity must be besides the plurality; therefore it must be separated from the plurality. Hence, the predicated entity is an entity that is one in addition to the plurality.[61] In order to have a uniformity of predication, a further entity, separated from the plurality, is needed.[62] The existence of the uniformity of predication is presupposed throughout the argument.

Within this argument, the interpretation of the conditions for uniformity of predication is compelled to assume the existence of an entity which is separated from the entities of the plurality. This interpretation is compelled to assume the existence of an entity which is besides the members of the plurality. The new entity is, therefore, added to the given plurality. The difference between the entities of the given plurality and the predicated entity consists in the everlastingness of the entity which is predicated. The entity which is predicated does not belong to another realm of reality which could be comparable to the realm to which, for example, Aristotle's second substances belong.

Further elements regarding the interpretation of uniformity of predication can be gained from the Third Man Argument. Moreover, thanks to the Third Man

---

not be uniform. If the predication has to be uniform, the entity which is predicated of the plurality must be outside the plurality; it cannot belong to the plurality. It must be, therefore, separated from the plurality itself. It is in general correct that the entity predicated is different from the entities of which it is predicated; in my opinion, Aristotle would nonetheless object to this position that this difference should be individuated in the entity's predicated belonging to another realm of reality than that to which the members of the plurality belong. To say that the predicated entity is different from the members of the given plurality is not enough; the way of difference ought to be clarified, since problems can begin right here.

61 Within this particular interpretation of the predication of an entity, the condition for the uniformity of predication as such requires that the entity predicated is outside the plurality. If the entity which is predicated belonged to the plurality, the predication could not be the same for all entities (we would have a self-predication in one case). Hence, the entity which is predicated of the plurality cannot be the same as any entity whatsoever belonging to the given plurality.

62 Ontological elements and conditions for the uniformity of predication are different from each other within this system and within Aristotle's system. The universal is not an entity in addition to and on the same level as the given plurality of which the universal is predicated. The universal is an entity belonging to a different realm of reality in comparison with the given plurality. On the contrary, as we can see from the Third Man Argument, the predicated entity belongs to the same realm of reality of the given plurality of which the predicated entity is predicated. One can count the entity which is predicated of the plurality and the members of the plurality as though they belonged to the same class; one can put them together. This cannot happen in the case of the universal; universals, common entities that which are said universally, that which belong universally and that which is predicated in common (these are entities which can be found, for instance, in the chapters *Metaphysics Beta* 6 and *Zeta* 13), belong to a realm of reality which is different from the realm of reality to which the entities having the ontological status of substance and of this something belong; the two kinds of entities cannot belong to the same realm of reality.

Argument, we can see how the false interpretation of the status of the predicated entity directly leads to the infinite multiplication of entities and, therewith, to the collapse of the whole ontology. Hence, we can see the reasons why it is indispensable to correctly determine the field of existence to which any entity belongs. The contents of the Third Man Argument are the following ones (see *De Ideis*, 84.22–85.3)[63]:

> The third man is also proved in this way. If what is predicated truly of some plurality of entities is also another entity besides the entities of which it is predicated (εἰ τὸ κατηγορούμενόν τινων πλειόνων ἀληθῶς καὶ ἔστιν ἄλλο παρὰ τὰ ὧν κατηγορεῖται), being separated from them (κεχωρισμένον αὐτῶν) (for this is what those who posit the ideas think they prove: for this is why, according to them, there is a man-itself (αὐτοάνθρωπος), because the man is predicated truly of the particular men, these being a plurality, and it is another entity than the particular men (ὅτι ὁ ἄνθρωπος κατὰ τῶν καθ' ἕκαστα ἀνθρώπων πλειόνων ὄντων ἀληθῶς κατηγορεῖται καὶ ἄλλος τῶν καθ' ἕκαστα ἀνθρώπων ἐστίν)) – but if this is so, there will be a third man (ἔσται τις τρίτος ἄνθρωπος). For if the man predicated is another entity than the entities of which it is predicated ((εἰ γὰρ ἄλλος ὁ κατηγορούμενος ὧν κατηγορεῖται)), and subsists on its own (κατ' ἰδίαν ὑφεστώς), and if the man is predicated both of the particulars and of the idea (κατηγορεῖται δὲ κατά τε τῶν καθ' ἕκαστα καὶ κατὰ τῆς ἰδέας ὁ ἄνθρωπος), then there will be a third man besides the particulars and the idea (ἔσται τις τρίτος ἄνθρωπος παρά τε τὸν καθ' ἕκαστα καὶ τὴν ἰδέαν). In this way, there will also be a fourth man predicated of this, of the idea, and of the particulars, and in the same way also a fifth, and this on to infinity.

The main features of the ideas are within the argument as follows:

[63] The question of the consistency or of the inconsistency of Plato's argument in the *Parmenides* cannot, unfortunately, be discussed in the present context, since the analysis of this question would require at least a whole study. On the problem of the inconsistency or consistency of the Third Man Argument as it is exposed in the *Parmenides* (the argument should perhaps be called the "Third Large Argument" due to the example which Plato actually uses in his text), I only wish to say, in the present context, that I side with the line of interpretation began by W. Sellars and then followed (with modifications), for example, by S. M. Cohen, by H. Teloh – D. J. Louzecky and by G. Fine. I consider the argument as consistent, since I too think that the non-identity premise should be interpreted as "If *x* is *F*, then x is not identical with *the F-ness* by virtue of which it is *F*" (see Sellars' *Vlastos and the "Third Man"*, p. 418). Hence, if F-ness is F, it can be identical with itself. The non-identity premise only implies that F-ness is not identical with the F-ness due to which the first F-ness is F. Therefore, I do not agree with Vlastos' thesis of the inconsistency of the argument (see *The Third Man Argument in the Parmenides*). In the present context, however, I focus exclusively on Aristotle's version of the argument as it is reported by Alexander of Aphrodisias. I shall specifically deal with Plato's argument and with the problems connected to it in another study. The majority of the studies on the regress deals either with Plato's version or with Aristotle's version. I would like, therefore, in the present context to express my deep gratitude both to G. Fine' analysis of the argument contained in her book *On Ideas: Aristotle's Criticism of Plato's Theory of Forms*, since she examines both Plato's and Aristotle's version of the argument, and to I. M. Vandoulakis, who in his article *Plato's "Third Man" Paradox: its Logic and History* investigated Plato's version of the argument, Proclus' observations regarding Plato's argument and Aristotle's version of the argument.

(a) The entity which is predicated is something different from the plurality of which it is predicated ("if what is predicated truly of some plurality of entities is also another entity besides the entities of which it is predicated").[64]
(b) The entity which is predicated is separated from the plurality of which it is predicated ("being separated from them").[65]
(c) The entity which is predicated has an independent existence ("it subsists on its own").[66]
(d) The idea is itself a subject of predication, that is, it is not only predicated of other entities but is a subject of predication, too ("if the man is predicated both of the particulars and of the idea").[67]

Within the argument, the property of separation for the entity which is predicated is mentioned. This is an important point, since Aristotle contends that only substances are separated, whereas other entities do not possess this particular feature.[68] In this argument an entity which is predicated of a plurality of entities is separated from the plurality, while, in Aristotle, a universal never exists separately from the particulars.[69]

Coming now to the general reconstruction of the Third Man Argument, the premises which are necessary to produce the Third Man regress seem to be the following ones (in spite of the fact that they are not clearly expressed or are not expressed at all in the argument itself):

- One over many: whenever a plurality of entities is f,[70] they are f in virtue of having some one entity, the f, truly predicated of them[71].
- Non-identity: the entity which is predicated of a plurality of entities is an entity which is besides the entities of which it is predicated (this premise implies, together with the one-over-many premise, that nothing which is f is f in virtue of itself. For the entity predicated, due to which the members of the plurality

[64] Universals are different from the entities of which they are predicated. To be different, though, can have different meanings and different implications.

[65] Universals are not separated from the entities of which they are predicated.

[66] Universals do not have an independent existence.

[67] This implies that the idea is itself a bearer of the property. Universals are not bearers of the properties which they assign to the entities of which they are predicated.

[68] See, for example, *Metaphysics Zeta* 1, 1028a33–34, and *Physics* I 2, 185a31–32.

[69] See, for example, *Metaphysics Zeta* 16, 1040b26–27.

[70] Actually, as the text of *Metaphysics Zeta* 6, 1031b28–30 testifies, we can already have an infinite regress if we divide entity and essence and if we then attribute an essence to the essence itself. An infinite regress can be provoked just beginning with one entity alone; there is no need of a plurality of entities. A regress to the infinite can be produced also beginning with only one entity, provided that the essence of an entity is given an essence which is considered as divided from the first essence. Hence, we can have a regress to the infinite, if we do not correctly interpret the position of the essence in the reality, regarding essence, for example, as an entity which has an essence too.

[71] This premise is, in my opinion, expressed only partially, since it is not clearly said that the plurality of entities possessing a property depends on the fact that one and the same entity is predicated of these entities.

have a determined property, is always different from the members of the given plurality. Hence, the cause of the members of a plurality being f is due to an entity which is different from all the members of the plurality).[72]

- Property exemplification[73]: the entity which is predicated of a given plurality of entities which are f is itself f.

Through the application of the premises represented by the non-identity and by the property exemplification, a regress to the infinite can be reached. The logic of the argument functions, in my opinion, in the following way:

- There is a plurality possessing a property "f" ("man").
- In correspondence with the first given plurality having the property "f" (e.g. the property "man"), there is an entity, "f-itself" (e.g. "man-itself"), which is predicated of this given plurality (this is due to the one-over-many premise).
- The "f-itself" (e.g. "man-itself"), which is predicated of the first plurality, is different from all the entities of which it is predicated (this difference is due to the non-identity premise).
- The f-itself (e.g. "man-itself") independently exists of that of which it is predicated (the entity which is predicated is separated from all the members of the given plurality due to the fact that the entity is different from all the members of the plurality).
- Therefore, there is an entity besides the first given plurality.
- This entity is itself "f" (e.g. the entity which is predicated of the first plurality is itself man due to the property-exemplification premise).
- The f-itself is not f in virtue of itself (it is f in virtue of something else; this is due to the non-identity premise).
- Therefore, there will be another entity ("the third man" besides the first given plurality and the first predicated entity). This entity is predicated of the first plurality and of the first predicated entity (due to a further application of the one-over-many premise).
- Thus, there will be another entity besides the first plurality and besides the entity that corresponds to the first predicated entity.
- This new entity is itself f (this is due to the property-exemplification premise).
- Because of the non-identity premise, the new predicated entity is not f in virtue of itself (it is f in virtue of something else).
- There will then be another entity (a fourth man), which is predicated of the first given plurality, of the first predicated entity and of the second predicated entity.

[72] We are compelled to complete this premise.

[73] Usually, this premise is named with the expression "self-predication" or "self-exemplification". I prefer to use the expression "property exemplification"; the entity which is predicated of the given plurality is itself a concretisation of the property which is attributed to the given plurality through the predication of this entity.

Through further applications of the one-over-many premise, of the property-exemplification premise and of the non-identity premise, an infinite regress and an infinite multiplication of entities come about.

A different reconstruction of the argument could be the following one:

1. For the premise of the one over many, if there is a given plurality with a property f, there is an entity, f itself, which is predicated of the plurality.
2. This entity is f, for the premise of the property exemplification.
3. This entity is not f in virtue of itself (it is f in virtue of something else), for the premise of the non-identity.
4. There is, therefore, another entity, which is predicated of the first plurality and of the first predicated entity, in virtue of which the first plurality and the first predicated entity (the first f-itself) are f, for the one-over-many premise.
5. For the property-exemplification premise, the new entity, too, which is predicated of the first plurality and of the first predicated entity, is f.
6. For the non-identity premise, the second predicated entity is not f in virtue of itself (it is f in virtue of something else).
7. There will be, then, another entity, which is predicated of the first plurality, of the first predicated entity and of the second predicated entity, for the one-over-many premise.
8. Through further applications of the one-over-many premise, of the property-exemplification premise and of the non-identity premise, an infinite regress is brought about.

It follows that, in order to explain a plurality (or also a single entity) having a property, one must introduce an infinite series of entities. This contrasts with one of the reasons for introducing ideas, which consisted in giving a unique factor (the idea) explaining a plurality having a property. Two kinds of critiques can, therefore, be expressed against those who maintain the existence of ideas:

- Epistemologically, it can be said that in order to know a property possessed by a plurality of entities, one must know an infinite series of entities. Instead of being instruments to explain reality, ideas make reality not understandable, since one has to know an infinite number of entities in order to know the concrete entities.
- Ontologically, one can say that, if there is an entity in virtue of which a plurality has a property, then there will be an infinite series of things in virtue of which a plurality has the same property.

In general, through the argument, it is shown that if there is one idea, then there are infinite ideas. Ideas were introduced, though, in order to find a unitary explanation of entities possessing a property.[74] Therefore, if ideas cannot represent a unitary explanation, ideas are to be abandoned.[75]

---

[74] See, for example, *Phaedo* 100b1-e4.

[75] In different passages of his works, Aristotle explains what the universal is not. For example, *Metaphysics Zeta* 13 and *Metaphysics Zeta* 16 determine what universals, entities which belong

Aristotle refuses, in *Categories* 5, 3b10–21, as a feature of the entity which is predicated, the property exemplification of the predicated entity itself. Second substances are not individual entities themselves, that is, they are not instances of the properties that the second substances themselves express. This strategy clearly shows his intention to distance himself from the ontological incongruities which lead to the Third Man regress. Since the entity which is predicated is not a τόδε τι but it is exclusively a ποιόν, it follows that the predicated entity cannot be the property it expresses; it only corresponds to a property without being this property, without instantiating this property. Both the ontological square (i.e. the four-domain division) and the two-district division accomplish the ontological duty consisting in the distinction of instances from non-instances.

Aristotle's refusal of the hypothesis that entities predicated in common have the status of this something, as we can see in *Metaphysics Zeta* 13, 1038b35–1039a1, is part of the same strategy. Aristotle does not accept that the entity which is predicated is itself an instance of the property it represents.

The difference existing between individual entities and entities predicated should, therefore, be correctly interpreted. The difference between members of the plurality and the entities which are predicated of the plurality in the argument of the One Over Many and in the argument of the Third Man is not the kind of difference which should exist between members of the plurality and entities which are predicated of the plurality. The correct difference between subjects of predication and predicated entities is exclusively the difference between individuals and universals, i.e. between the realms of reality represented, respectively, by instances and by non-instances.

Right ontology is not only a matter of stating the presence of a difference between entities but also a matter of correctly determining the sense of this difference. The difference of universals or of second substances in comparison with the entities of which they are predicated does not consist in the everlastingness, or in being separated, or in being besides the members of the plurality. Moreover, entities which are predicated do not subsist on their own. Aristotle puts the root of the difference of universals and of second substances in their not being instances of the property they assign to the members of the plurality of which they are predicated.

12 Conclusions

The distinction of entities in instances and non-instances, which can be expressed through the two-district scheme or through the four-domain scheme, is the very basis of Aristotle's ontology: without the determination of the right position in the

universally, entities predicated in common and entities said universally, are not. Aristotle's strategy concerning what the universal is not is as important as his strategy concerning what the universal is: an incorrect interpretation of universals could provoke a multiplication of entities.

reality of the entities, no interpretation of the entities can function. The concept of the typological ontology, which can be extended and specified into the four-domain ontology, represents, in my opinion, the basic organisation of Aristotle's ontology. The distinction between individual entities and universal properties pre-exists to any particular concrete entities and to any concrete state of affairs. Entities will always have a determined place in the framework of the typological ontology. The relations between the fields of the framework will always be the relations between the different entities.

The main items of Aristotle's new ontological proposal are as follows:

- Individual entities are instances of universal properties.
- Reality is organised in the scheme of a typological ontology which divides the entities into individual entities as instances of essences and into properties constituting the essences of the individual entities.
- Universals are properties or are predicates representing names of properties (the universal "man" represents the property – is the deputy for the property – "being man").
- The fact that universals are predicated of a plurality of individuals is the consequence of these individuals instantiating the same universal property.

The central differentiation of Aristotle's ontology regards, in my opinion, individuals as instances of properties, on the one hand, and properties as programmes which are realised in their instances, on the other hand. The elements and features belonging to the opposition between individuals as instances and properties as programmes are the following ones:

- Individuals, substance (first substance), numerically one, this something, this such.
- Universals (universal properties), such, quality, not numerically one, universal, common, universally said, predicated in common, second substances.

At least as regards the biological field, every individual (every individual entity) is the instance of a property that represents the essence (i.e. the way of existence) of the instance itself. The individual entity "Socrates" is the instance of the property "being man" (i.e. it is the result of the process of instantiation of this same property).

The field of instances, in Aristotle, is always represented by individuals, while the whole field of existence is constituted by both individuals and universal properties (universals), even though the way of existence for universals is different from the way of existence possessed by instances of properties. As a consequence of this difference, individuals and universal properties exist in different realms of reality; they correspond to different ontological types. We have a two-district ontology, i.e. a kind of typological ontology.

- The realm of instances is constituted, as such, by individual entities (by numerically one entities). Individual entities, at least individual biological entities, are immediately instances of properties. They are concretised, instantiated properties. The very status of individual entities consists in their being instances

of properties. For Aristotle, no bare entity exists. To exist, for an individual biological entity, is to concretise a determined complex of biological properties. These biological properties will dictate the life development of the corresponding individual entity. Without these properties, the individual entity does not exist.
- The realm of existence contains both individuals (individual entities) and universal properties (universal entities). The realm of instances does not exhaust the field of existence; it is only a part of it. Hence, in Aristotle's ontology, not only particular entities exist.

The scheme of the two-district ontology can be further specified in the four domains represented by individual substantial entities, individual non-substantial entities, universal substantial properties and universal non-substantial properties.

Acknowledgements This essay is a revision of the text I presented at the sixth World Congress on the Square of Opposition, which was held at the Orthodox Academy of Crete, Kolympari, from 1 to 5 November 2018.

I take this opportunity to thank very much all the members of the organising committee of the congress for accepting my paper proposal.

I also wish to thank very much the Orthodox Academy of Crete for its wonderful hospitality.

I would like very much to thank Miss Kathrin Bouvot, MA, MA, BA, BA, for her valuable observations, her helpful comments and her inestimable support.

I very much wish to thank Prof Dr Jean-Yves Beziau and Prof Dr Ioannis Vandoulakis for publishing this essay.

I would like to thank very much the two anonymous reviewers who analysed the first version of my study for their observations, remarks and critiques.

References

1. J. L Ackrill, *Aristotle's Categories and De Interpretatione, Translated with Notes*. Oxford University Press, 1963.
2. R. E. Allen (ed.), *Studies in Plato's Metaphysics, edited by R. E. Allen*. Routledge & Kegan Paul, 1965.
3. J. Annas, *Aristotle's Metaphysics. Books M and N*. Oxford University Press, 1976.
4. Aristotle, *Aristotelis Categoriae et Liber De Interpretatione. Recognovit Brevique Adnotatione Critica Instruxit L. Minio-Paluello*. Oxford University Press, 1949.
5. Aristotle, *Aristotelis Fragmenta Selecta. Recognovit Brevique Adnotatione Instruxit W. D. Ross*. Oxford University Press, 1955.
6. Aristotle, *Aristotelis Metaphysica. Recognovit Brevique Adnotatione Critica Instruxit W. Jaeger*. Oxford University Press, 1957.
7. Aristotle, *Aristotle in 23 Volumes, Vols.17, 18, translated by Hugh Tredennick*. Harvard University Press; William Heinemann Ltd., 1933, 1989.
8. D. M. Armstrong, *Universals & Scientific Realism, Volume I: Nominalism and Realism*. Cambridge University Press, 1978; *Volume II: A Theory of Universals*. Cambridge University Press, 1978.
9. J. Barnes (ed.), *The Complete Works of Aristotle. The Revised Oxford Translation. Edited by Jonathan Barnes*. Volume One. Princeton University Press, 1984.
10. J. Barnes (ed.), *The Complete Works of Aristotle. The Revised Oxford Translation. Edited by Jonathan Barnes*. Volume Two. Princeton University Press, 1984.

11. H. Bonitz, *Index Aristotelicus*. Wissenschaftliche Buchgesellschaft, 1960. Photomechanischer Nachdruck der Ausgabe von 1870.
12. D. Bostock, *Aristotle Metaphysics Book* Z *and* H. *Translated with a Commentary by David Bostock*. Oxford University Press, 1994.
13. K. Campbell, *Abstract Particulars*. Blackwell Publishers, 1990.
14. S. M. Cohen, *The Logic of the Third Man*. The Philosophical Review **80**, 4 (October 1971), 448-475.
15. G. Fine, *On Ideas: Aristotle's Criticism of Plato's Theory of Forms*. Oxford University Press, 1993.
16. G. Frege, *Die Grundlagen der Arithmetik. Eine logisch-mathematische Untersuchung über den Begriff der Zahl*. Verlag von Wilhelm Koebner, 1884; or: Georg Olms Verlagsbuchhandlung, 1961.
17. G. Frege, *Funktion, Begriff, Bedeutung. Fünf logische Studien. Herausgegeben und eingeleitet von G. Patzig*. 7., bibliographisch ergänzte Aufgabe, Vandenhoeck & Ruprecht in Göttingen, 1994.
18. G. Frege, *Function und Begriff, Vortrag gehalten in der Sitzung vom 9. Januar 1891 der Jenaischen Gesellschaft für Medicin und Naturwissenschaft*. Verlag von Hermann Pohle, 1891; or in: *Funktion, Begriff, Bedeutung. Fünf logische Studien. Herausgegeben und eingeleitet von G. Patzig*. 7., bibliographisch ergänzte Aufgabe, Vandenhoeck & Ruprecht in Göttingen, 1994, 17-39.
19. G. Frege, *Über Begriff und Gegenstand*. Vierteljahrsschrift für wissenschaftliche Philosophie **16. Jahrgang**, Nr. 2, 1892, 192–205; or in: *Funktion, Begriff, Bedeutung. Fünf logische Studien. Herausgegeben und eingeleitet von G. Patzig*. 7., bibliographisch ergänzte Aufgabe, Vandenhoeck & Ruprecht in Göttingen, 1994, 66-80.
20. G. Frege, *Über Sinn und Bedeutung*. Zeitschrift für Philosophie und philosophische Kritik **100** (1892), 25-50; or in: *Funktion, Begriff, Bedeutung. Fünf logische Studien. Herausgegeben und eingeleitet von G. Patzig*. 7., bibliographisch ergänzte Aufgabe, Vandenhoeck & Ruprecht in Göttingen, 1994, 40-65.
21. Ch. Kirwan, *Aristotle Metaphysics. Books Γ, Δ, and E*. Oxford University Press, 1971.
22. J. Kung, *Aristotle on Thises, Suches and the Third Man Argument*. Phronesis **XXVI**, No. 3 (1981), 207-247.
23. W. Leszl, *Il "De Ideis" di Aristotele e la Teoria Platonica delle Idee. Edizione critica del testo a cura di Dieter Harlfinger*. Leo S. Olschki Editore, 1975.
24. H. G. Liddell – R. Scott, *A Greek-English lexicon: with a revised supplement 1996 / compiled by H. G. Liddell and R. Scott. Rev. and augm. throughout by H. Stuart Jones*. 9 ed., new suppl. Added, Oxford University Press, 1996.
25. M.-Th. Liske, *Aristoteles und der aristotelische Essentialismus: Individuum, Art, Gattung*. Verlag Karl Alber, 1985.
26. E. J. Lowe, *The Possibility of Metaphysics: Substance, Identity, and Time*. Clarendon Press, 1998.
27. E. J. Lowe, *The Four-Category Ontology: A Metaphysical Foundation for Natural Science*. Clarendon Press, 2006.
28. E. J. Lowe, *More Kinds of Being: A Further Study of Individuation, Identity, and the Logic of Sortal Terms* [New, expanded ed.]. Wiley-Blackwell, 2009.
29. W. D. Ross, *Aristotle's Metaphysics. A Revised Text with Introduction and Commentary by W. D. Ross, 2 vols*. Oxford University Press, 1924.
30. G. Segalerba, *Semantik und Ontologie: Drei Studien zu Aristoteles*. Peter Lang Verlag, 2013.
31. W. Sellars, *Vlastos and the "Third Man"*. The Philosophical Review **64**, 3 (1955), 405-437.
32. J. A. Smith, *Tóde ti in Aristotle*. Classical Review **35** (1921), 19.
33. H. Teloh – D. J. Louzecky, *Plato's Third Man Argument*. Phronesis **XVII**, 1 (1972), 80–94.
34. I. M. Vandoulakis, *Plato's "Third Man" Paradox: its Logic and History*. Archives Internationales d'Historie des Sciences **59**, 162 (Juin 2009), 3-51.

35. G. Vlastos, *The Third Man Argument in the Parmenides*. The Philosophical Review **63**, 3 (July 1954), 319-349; or in: *Studies in Plato's Metaphysics, edited by R. E. Allen*, 231-263 (*with Addendum (1963)*).
36. G. Vlastos, *Addenda to the Third Man Argument: A Reply to Professor Sellars*. The Philosophical Review **64**, 3, (July 1955), 438-448.
37. G. Vlastos, *Plato's "Third Man" Argument (Parm. 132A1-B2): Text and Logic*. The Philosophical Quarterly **19**, 77 (October 1969), 289-301; or as: *Plato's "Third Man" Argument (Parm. 132A1-B2): Text and Logic (TMA II)*, in: *Platonic Studies* (with: *Appendix I: Recent Papers on the TMA. Appendix II: The First Regress Argument in Parm. 132A-B2*), 342-365.
38. G. Vlastos, *Platonic Studies*. Princeton University Press, 1973.

Logical Oppositions in Avicenna's Hypothetical Logic

Saloua Chatti

Abstract In his hypothetical logic, Avicenna introduces new kinds of hypothetical propositions by using quantifications ranging over situations and distinguishing between universal and particular, affirmative and negative and connected (*muttaṣila*) or disjunctive propositions. In Sect. 7, Chap. 1, of *al-Qiyās* (pp. 361–372), he goes further by considering hypothetical connected propositions where the clauses are themselves quantified propositions of the form **A**, **E**, **I** and **O**. When combining their **A**, **E**, **I** or **O** clauses in all possible ways, he lists 16 universal hypothetical affirmative propositions, 16 universal negatives, 16 particular affirmatives and 16 particular negatives and says that the logical relations of contradiction, contrariety, subcontrariety and subalternation hold between all these hypothetical propositions.

In this paper, I will analyse the logical relations between all of these quantified hypothetical connected propositions. Now given the import of all affirmative propositions and the lack of import of all negative ones in Avicenna's frame, both in categorical logic and in hypothetical logic, the 16 universal affirmatives should be different from the 16 universal negatives, likewise for the16 particular affirmatives and the 16 particular negatives. So the total number of distinct propositions is 64. These propositions give rise to several octagons of Buridan's kind, then many octagons of Johnson-Hacker's kind containing other kinds of propositions and finally some octagons of a *new and original* kind, which we don't find in the medieval or modern literature. The octagons can also be grouped two by two, which gives rise to several figures containing 16 vertices and allows for more relations between the propositions. Other figures of different sizes – not considered here – can also be drawn, although Avicenna himself did not draw any figure at all.

Keywords Hypothetical logic · Quantified hypothetical propositions · Conditional propositions · Contradiction · Contrariety · Subcontrariety and subalternation · Octagons of oppositions · Complex figures of oppositions

---

S. Chatti (✉)
FSHST, University of Tunis, Tunis, Tunisia

J.-Y. Beziau, I. Vandoulakis (eds.), *The Exoteric Square of Opposition*, Studies in Universal Logic, https://doi.org/10.1007/978-3-030-90823-2_3

Mathematics Subject Classification (2000) Primary 01A30·03A05; Secondary 03B10·03C80

1 Introduction

Avicenna's hypothetical logic contains two parts, which are very different from each other. The first one uses conditional and disjunctive propositions and states the usual Stoic indemonstrables called by Avicenna the *istithnā'i* syllogisms plus some variants. The second one is entirely new because it quantifies over connected (conditional) and disjunctive propositions. This last theory adds new moods containing quantified disjunctive, conditional and predicative propositions to the usual syllogistic moods. Avicenna introduces universal and particular hypothetical connected propositions as well as disjunctive propositions, containing predicative elements. But he also uses all **A, E, I** and **O** predicative propositions as clauses of these quantified conditional and disjunctive propositions. He provides the complete lists of all these propositions and says that all of them obey the relations of the square of oppositions.

The problem is then the following: what are the logical relations between these quantified hypothetical propositions? Which propositions are contradictories? Which ones are contrary or subcontrary to each other? What propositions are the subalterns of what other ones?

In what follows, I will formalize the aforementioned propositions by taking into account Avicenna's conceptions and I will show that one can construct several octagons of different kinds, which can be assembled two by two giving rise to figures with 16 vertices each.

2 The Quantified Hypothetical Propositions

The hypothetical system that we are considering here is the one where the hypothetical propositions, whether connected (*muttasil*) or disjunctive, are quantified. The connected propositions are those which contain 'if ... then'. Among these propositions, we will consider only the implicative (*luzūmī*) propositions, which express a relation of following from. We won't consider what Avicenna calls the *ittifāqī* ones, where there is no such relation of following from and whose truth conditions are different from those of the implicative ones.

The quantified hypothetical implicative propositions are expressed as follows:

- **A**c: Whenever (*kullamā*) A is B, then H is Z ([2], p. 265).
- **E**c: Never if A is B, then H is Z ([2], p. 280).
- **I**c: It happens that (*qad yakūn*) when every A is B, then every H is Z ([2], p. 278).
- **O**c: Not whenever A is B, then H is Z.

In these sentences, the expressions 'whenever' (*kullamā*), 'it happens that' (*qad yakūn*), 'never' (*laysa al-battata*) and 'not whenever' (*laysa kullamā*) express the universal affirmative quantifier, the particular affirmative quantifier, the universal negative quantifier and the particular negative quantifier, respectively.

Now how can one formalize these quantified propositions?

The usual formalizations stated by Rescher, for instance, are the following (where 't' stands for time and 'A_t' is read as 'A is true in t'):

"*A* (U.A.) $\begin{cases} (t)\,(A_t \supset C_t) \\ (t) \sim (A_t \& \sim C_t) \end{cases}$

E (U.N.) $(t) \sim (A_t \& C_t)$

I (P.A.) $(\exists t)\,(A_t \& C_t)$

O (P.N.) $(\exists t)\,(A_t \& \sim C_t)$" ([11], p. 51, Rescher's notation)

The two pairs of contradictories are, as usual, the pairs **A/O** and **E/I**.

Rescher says that the universal quantifier, for instance, is expressed by the word 'always' and means 'at all times' or 'in all cases' ([11], p. 52). He also says that the conditional used in both *A* and *E* is the 'Diodorian implication "If A, then C", [which] amounts to "At each and every time *t:* If A-at-t, then C-at-t"' ([11], p. 50). This Diodorian implication is opposed by Łukasiewicz to the Philonian implication, which is a material implication (see [9], p. 15). In Avicenna's view, there must be a semantic or causal link between the antecedent and the consequent of the implication, especially when it is a real implication – called a *luzūm* – i.e. a relation of *following from*, which makes the consequent really *follow* semantically or causally from the antecedent.

In addition, since the word used by Avicenna in this particular context is not the word 'time', but the word *ḥāl* (plural *aḥwāl*), which could be translated by 'situation' or 'state' or 'case', we can slightly modify these formalizations by making the quantifiers range not on times but rather on situations or state. We get thus the following formalizations [where the letter 'c' in **Ac**, **Ec**, **Ic** and **Oc**, stands for 'connected']:

- **Ac** : $(\forall s)\,(Ps \rightarrow Qs)$
- **Ic** : $(\exists s)\,(Ps \wedge Qs)$
- **Ec** = ~ **Ic** = $\sim (\exists s)\,(Ps \wedge Qs)$
- **Oc** = $(\exists s)\,(Ps \wedge \sim Qs)$ (see [4])

However, the above formalization of **Ac** does not validate **Ac** conversion, nor *Darapti* and *Felapton*, which are both held by Avicenna in his hypothetical logic too. For the hypothetical *Darapti* is expressed as follows:

- Whenever C is D, then H is Z.
- Whenever C is D, then A is B.

Therefore, it happens that when H is Z, then A is B.

If we formalize **Ac** as above, *Darapti* is formalized as follows (where 'P' stands for 'C is D', 'R' for 'H is Z' and 'Q' for 'A is B'): '$(\forall s)(Ps \rightarrow Rs) \wedge (\forall s)(Ps \rightarrow Qs)] \rightarrow (\exists s)(Rs \wedge Qs)$'.

Unfortunately, this formula is *not* valid. So the formalization of the **A**c propositions should be revised in order to validate this mood, and to validate *Darapti Felapton* and **A**c conversion, which are all admitted by Avicenna.

So the problem is the following: how should one account for the validity of these two moods and of **A**c conversion in Avicenna's hypothetical logic? To answer this question, let us first consider what Avicenna himself says in his various texts.

In *al-Qiyās*, Sect. 5, Chap. 4, while analysing the hypothetical propositions, Avicenna says what follows:

'When we say: "If A is B, then H is Z", we assume from this (*nūjibu min hāḍha*) that at any time where "A is B" *is the case* and when A is B then H is Z, as if the fact that H is Z follows the fact that A is B, *in so far as in effect* A is B (*min hayṭhu huwa kā'inun A [huwa] B*), and it does not contain other conditions such as those that "whenever" contains, which we will mention'. ([2], p. 263.8–9, my emphasis).

Thus he stresses the idea that the *antecedent* of the conditional proposition containing 'if... then' (where 'if' translates the Arabic particle *in*) *must be true* in order for the proposition itself to be true. This particle expresses the strongest kind of conditional, which Avicenna calls *luzūm*, i.e. a real implication, where the consequent really follows from the antecedent. In the end of the quotation, he says that the word 'whenever' (*kullamā*) involves further conditions, which, he says, will be mentioned [afterwards]. But he does not say that the above condition does not apply to the connected propositions containing 'whenever'. Now, one of the conditions involved in the sentences containing 'whenever' (*kullamā*) is already mentioned a few lines before the above quotation, where Avicenna claims 'If what is said is "whenever this is so", then the proposition is a connected *universal* (*fa al-qaḍīyya muttaṣila kullīya*)' ([2], p. 263.3, my emphasis). So this condition is *universality*, that is, when someone uses this word, he expresses a universal proposition, which is not always the case with propositions containing only the expression 'if... then'.

One page later in the same chapter, he returns back to the word '*kullamā*' and gives further precisions about what he means by this word, for he claims:

'When we say "Whenever C is B, then H is Z" we don't only mean by "whenever" the generalizing of what is intended (*ta'amīm al-murād*), so that what is expressed is like saying "Every time where C is B, then H is Z"; rather it involves generalizing every situation (or state: *ḥāl*) connected (*yaqtarinu*) to the sentence "Every C is B" so that any situation or condition related to [that sentence], which makes "C is B" *true* (*mawjūdan*) cannot do so without also making "H is Z" true. For it might happen that the antecedent is something that does not occur repeatedly (*laysa lahu takarrurun wa 'awdun*); rather it is something that is *certain* (*thābitun*) and *true* (*mawjūdun*), not intended (*lā murāda lahu*)'. (*al-Qiyās*, p. 265.1–5, emphasis added).

So here too, he seems to stress the idea that a sentence containing 'whenever' requires the truth of the antecedent, for the conditional is true if the consequent is true in all cases where the antecedent is itself true.

Elsewhere, he says, talking about the universal affirmative hypothetical proposition:

'The universal hypothetical proposition is universal if the consequent follows every positing (*kulla waḍ'in*) of the antecedent, not only by what is intended (*lā fi al-murādi faqat*), but in the [real] states (*fī al-aḥwāl*)... that is, in all states that require assuming (*faraḍa*) the antecedent...' (*al-Qiyās*, p. 272.14–15).

This being so, we could consider that the **Ac** propositions should be formalized by adding what Wilfrid Hodges calls the 'augment' (See [8]). The formalization of **Ac** would thus be the following: Ac: '$(\exists s)Ps \wedge (\forall s)(Ps \rightarrow Qs)$'.

Thus formalized, **Ac** is comparable to the categorical **A** *with import*. With this formalization, **Ac** conversion, *Darapti* and *Felapton* are all valid. Naturally, **Oc** must be formalized accordingly. As to **Ec**, it does not need an augment. Consequently the formalizations of the hypothetical conditional propositions should be the following:

– $$\textbf{Ac} : (\exists s)\, Ps \wedge (\forall s)\, (Ps \rightarrow Qs)$$

– $$\textbf{Ec} : (\forall s)\, (Ps \rightarrow \sim Qs)$$

– $$\textbf{Ic} : (\exists s)\, (Ps \wedge Qs)$$

– **Oc** : $\sim[(\exists s)Ps \wedge (\forall s)(Ps \rightarrow Qs)]$ [$= \sim(\exists s)Ps \vee (\exists s)(Ps \wedge \sim Qs)$] (see [5])[1]

With these formalizations in mind, let us now consider the quantified propositions whose elements are themselves quantified.

3 The Quantified Hypothetical Propositions with Quantified Clauses

In *al-Qiyās*, Sec. 7 ([2], pp. 361–373), Avicenna presents four lists of quantified hypothetical conditional propositions whose clauses are themselves quantified. At pages 374–384, he presents the disjunctive propositions with quantified elements. He also says that the four relations of the square of oppositions hold for these quantified propositions in what follows:

> And you know [what] contradiction, contrariety, subcontrariety and subalternation [signify], so we don't need to tell you again about them, for they are defined as they are in the case of the predicative propositions ([2], p. 362.6).

So we can reasonably assume that all the oppositional relations between **Ac**, **Ec**, **Ic** and **Oc** (where the letter 'c' stands for 'connected', which translates the word *muttaṣil* used by Avicenna) are valid in the hypothetical logic too. These relations are all defined in *al-'Ibāra* (*De Interpretatione*) (see [1], pp. 47–48), where Avicenna defines all the relations of the classical square of oppositions (though

---

[1] Despite the fact that some people criticized this formalization (see [12]), it seems to be the only one to be able to account adequately for all of Avicenna's claims and proofs (see [6])

without drawing the square). Since as we just saw they are also evoked in this section of *al-Qiyās*, we can say that Avicenna validates them even in his hypothetical logic. But this validation is not obvious, if we take into account Avicenna's further explanations about the equivalences between some **Ac** and **Ec** propositions with opposed consequents. These explanations raise problems for they presuppose that the **Ec** propositions imply the **Ac** ones when the two propositions have the same antecedents but opposed consequents. For instance, the **Ec** proposition 'Never if every A is B, then every C is D′ (*laysa al-battata idhā kāna kull A B fa-kull C D*) implies, according to him, the **Ac** proposition 'Whenever every A is B, then not every C is D′ (*Kullamā kāna kull A B, fa-laysa kull C D*) ([2], p. 366). But this implication is not correct, when we formalize the propositions as above, since '$(\forall s)(Ps \rightarrow \sim Qs)$' does not imply '$(\exists s)Ps \wedge (\forall s)(Ps \rightarrow \sim Qs)$', even if this **Ac** does imply that **Ec**. But if **Ac** does not contain the augment, the implication does hold as Avicenna says, for then the two propositions would both be expressed as '$(\forall s)(Ps \rightarrow \sim Qs)$'. Unfortunately, in that case, i.e. if Ac is formalized as '$(\forall s)(Ps \rightarrow \sim Qs)$', then neither contrariety, subcontrariety and subalternation, nor **Ac** conversion, nor *Darapti* and *Felapton* hold, unlike what Avicenna says. So if we choose one of these formalizations, but not the other one, Avicenna's claims become incompatible with each other and the whole system would be inconsistent. Consequently, if one wishes to account for all Avicenna's claims, one has to admit both formalizations of **Ac** and to distinguish them clearly, since Ac *with* the augment validates all the relations of the square, **Ac** conversion plus all third figure moods, while **Ac** *without* the augment validates the equivalences between some **Ec** and some **Ac** propositions stated by Avicenna, the equivalences between some Ac propositions and some A disjunctive ones and the principle of contraposition, i.e. the following equivalence: '$(\forall s)(Ps \rightarrow Qs) \equiv (\forall s)(\sim Qs \rightarrow \sim Ps)$', which is also admitted by Avicenna (see [2], p. 385).

Now since the logical relations are validated by the Ac which contains the augment, let us provide the corresponding formalizations. Otherwise, no Aristotelian relation apart from contradiction holds, and we would have no figure at all, just as in modern logic; neither the square nor any other figure holds for the usual formalizations of the categorical quantified propositions.

Let us now provide the four lists. The 16 **Ac** propositions are the following:

1. Whenever Every A is B, then Every C is D.
2. Whenever Every A is B, then Some C is D.
3. Whenever Some A is B, then Every C is D.
4. Whenever Some A is B, then Some C is D.
5. Whenever No A is B, then No C is D.
6. Whenever No A is B, then Not every C is D.
7. Whenever Not every A is B, then No C is D.
8. Whenever Not every A is B, then Not every C is D.
9. Whenever Every A is B, then Not No C is D.
10. Whenever Every A is B, then Not every C is D.
11. Whenever Some A is B, then No C is D.

12. Whenever Some A is B, then Not every C is D.
13. Whenever No A is B, then Every C is D.
14. Whenever No A is B, then Some C is D.
15. Whenever Not every A is B, then Some C is D.
16. Whenever Not every A is B, then Every C is D ([2], pp. 363–364).

The 16 **Ec** propositions are the following:

1. Never when Every A is B, then Every C is D.
2. Never when Every A is B, then Some C is D.
3. Never when Some A is B, then Every C is D.
4. Never when Some A is B, then Some C is D.
5. Never when No A is B, then No C is D.
6. Never when No A is B, then Not every C is D.
7. Never when Not every A is B, then No C is D.
8. Never when Not every A is B, then Not every C is D.
9. Never when Every A is B, then No C is D.
10. Never when Every A is B, then Not every C is D.
11. Never when Some A is B, then No C is D.
12. Never when No A is B, then Every C is D.
13. Never when No A is B, then Some C is D.
14. Never when Not every A is B, then Every C is D.
15. Never when Not every A is B, then Some C is D.
16. Never when Some A is B, then Not every C is D ([2], pp. 364–365).

While the 16 **Ic** propositions are the following:

1. It happens that when Every A is B, then Every C is D.
2. It happens that when Every A is B, then Some C is D.
3. It happens that when Some A is B, then Every C is D.
4. It happens that when Some A is B, then Some C is D.
5. It happens that when No A is B, then No C is D.
6. It happens that when No A is B, then Not every C is D.
7. It happens that when Not every A is B, then No C is D.
8. It happens that when Not every A is B, then Not every C is D.
9. It happens that when Every A is B, then No C is D.
10. It happens that when Some A is B, then No C is D.
11. It happens that when Every A is B, then Not every C is D.
12. It happens that when Some A is B, then Not every C is D.
13. It happens that when No A is B, then Every C is D.
14. It happens that when Not every A is B, then Every C is D.
15. It happens that when No A is B, then Some C is D.
16. It happens that when Not every A is B, then Some C is D ([2], pp. 369–370).

And the 16 **Oc** propositions are the following:

1. Not whenever Every A is B, then Every C is D.
2. Not whenever Every A is B, then Some C is D.

3. Not whenever Some A is B, then Every C is D.
4. Not whenever Some A is B, then Some C is D.
5. Not whenever No A is B, then No C is D.
6. Not whenever Not Every A is B, then No C is D.
7. Not whenever No A is B, then Every C is D.
8. Not whenever Not Every A is B, then Not every C is D.
9. Not whenever Every A is B, then No C is D.
10. Not whenever Every A is B, then Not every C is D.
11. Not whenever Some A is B, then No C is D.
12. Not whenever Some A is B, then Not every C is D.
13. Not whenever No A is B, then Every C is D.
14. Not whenever No A is B, then Some C is D.
15. Not whenever Not Every A is B, then Every C is D.
16. Not whenever Not Every A is B, then Some C is D ([2], pp. 370–371).

4 How to Formalize these Propositions

How can we formalize these propositions? We will name the quantified clauses by their usual vowels **A**, **E**, **I** and **O** and add numerals to distinguish between the antecedents and the consequents. Thus, in each proposition:

- Every A is B = $\mathbf{A}_1$; Every C is D = $\mathbf{A}_2$.
- Some A is B = $\mathbf{I}_1$; Some C is D = $\mathbf{I}_2$.
- No A is B = $\mathbf{E}_1$; No C is D = $\mathbf{E}_2$.
- Not every A is B = $\mathbf{O}_1$; Not every C is D = $\mathbf{O}_2$.

Given this device, we can formalize the different kinds of propositions as appears in the sequel. However, before providing the formalizations, let us first clarify the notation that we will use. First we will call the universal affirmative hypothetical propositions **Ac**, the universal negative hypothetical ones **Ec**, the particular affirmative hypothetical ones **Ic** and the particular negative hypothetical ones **Oc** (where c stands for 'connected'). Their clauses, which are categorical quantified propositions, will be called as usual **A**, **E**, **I** and **O**. Second our notation is the following (where the small letters stand for the clauses):

All **Ac** propositions are noted as follows: Aaa, Aai, Aii and so on.
All **Ec** propositions are noted as follows: Eaa, Eae, Eai and so on.
All **Ic** propositions are noted as follows: Iaa, Iae, Iai and so on.
All **Oc** propositions are noted as follows: Oaa, Oae, Oai and so on.

Let us now consider the formalizations of the different **Ac** propositions. These are the following:

1. $$(\exists s)\ \mathbf{A}_1 s \wedge (\forall s)\ (\mathbf{A}_1 s \rightarrow \mathbf{A}_2 s)\quad \text{[noted Aaa]}$$

2. $(\exists s)\, \mathbf{A}_1 s \wedge (\forall s)\, (\mathbf{A}_1 s \rightarrow \mathbf{I}_2 s)$ [noted Aai]

3. $(\exists s)\, \mathbf{I}_1 s \wedge (\forall s)\, (\mathbf{I}_1 s \rightarrow \mathbf{A}_2 s)$ [noted Aia]

4. $(\exists s)\, \mathbf{I}_1 s \wedge (\forall s)\, (\mathbf{I}_1 s \rightarrow \mathbf{I}_2 s)$ -----

5. $(\exists s)\, \mathbf{E}_1 s \wedge (\forall s)\, (\mathbf{E}_1 s \rightarrow \mathbf{E}_2 s)$ ------

6. $(\exists s)\, \mathbf{E}_1 s \wedge (\forall s)\, (\mathbf{E}_1 s \rightarrow \mathbf{O}_2 s)$ -----

7. $(\exists s)\, \mathbf{O}_1 s \wedge (\forall s)\, (\mathbf{O}_1 s \rightarrow \mathbf{E}_2 s)$

8. $(\exists s)\, \mathbf{O}_1 s \wedge (\forall s)\, (\mathbf{O}_1 s \rightarrow \mathbf{O}_2 s)$

9. $(\exists s)\, \mathbf{A}_1 s \wedge (\forall s)\, (\mathbf{A}_1 s \rightarrow \mathbf{E}_2 s)$

10. $(\exists s)\, \mathbf{A}_1 s \wedge (\forall s)\, (\mathbf{A}_1 s \rightarrow \mathbf{O}_2 s)$

11. $(\exists s)\, \mathbf{I}_1 s \wedge (\forall s)\, (\mathbf{I}_1 s \rightarrow \mathbf{E}_2 s)$

12. $(\exists s)\, \mathbf{I}_1 s \wedge (\forall s)\, (\mathbf{I}_1 s \rightarrow \mathbf{O}_2 s)$

13. $(\exists s)\, \mathbf{E}_1 s \wedge (\forall s)\, (\mathbf{E}_1 s \rightarrow \mathbf{A}_2 s)$

14. $(\exists s)\, \mathbf{E}_1 s \wedge (\forall s)\, (\mathbf{E}_1 s \rightarrow \mathbf{I}_2 s)$

15. $(\exists s)\, \mathbf{O}_1 s \wedge (\forall s)\, (\mathbf{O}_1 s \rightarrow \mathbf{I}_2 s)$

16. $(\exists s)\, \mathbf{O}_1 s \wedge (\forall s)\, (\mathbf{O}_1 s \rightarrow \mathbf{A}_2 s)$

The **Ec** propositions noted Eaa, Eae, Eai, etc.... are formalized as follows:

1. $(\forall s)\, (\mathbf{A}_1 s \rightarrow\sim \mathbf{A}_2 s)$ $[= (\forall s)\, (\mathbf{A}_1 s \rightarrow \mathbf{O}_2 s)]$ [noted Eaa]

2. $(\forall s)\, (\mathbf{A}_1 s \rightarrow\sim \mathbf{I}_2 s)$ $[= (\forall s)\, (\mathbf{A}_1 s \rightarrow \mathbf{E}_2 s)]$ [noted Eai]

3. $(\forall s)\, (\mathbf{I}_1 s \rightarrow\sim \mathbf{A}_2 s)$ $[= (\forall s)\, (\mathbf{I}_1 s \rightarrow \mathbf{O}_2 s)]$ [noted Eia]

4. $(\forall s)\, (\mathbf{I}_1 s \rightarrow\sim \mathbf{I}_2 s)$ $[= (\forall s)\, (\mathbf{I}_1 s \rightarrow \mathbf{E}_2 s)]$ -----

5. $(\forall s)\, (\mathbf{E}_1 s \rightarrow\sim \mathbf{E}_2 s)$ $[= (\forall s)\, (\mathbf{E}_1 s \rightarrow \mathbf{I}_2 s)]$ -----

6. $(\forall s)\, (\mathbf{E}_1 s \rightarrow\sim \mathbf{O}_2 s)$ $[= (\forall s)\, (\mathbf{E}_1 s \rightarrow \mathbf{A}_2 s)]$ -----

7. $(\forall s)(\mathbf{O}_1 s \rightarrow\sim \mathbf{E}_2 s)$ [= $(\forall s)(\mathbf{O}_1 s \rightarrow \mathbf{I}_2 s)$]

8. $(\forall s)(\mathbf{O}_1 s \rightarrow\sim \mathbf{O}_2 s)$ [= $(\forall s)(\mathbf{O}_1 s \rightarrow \mathbf{A}_2 s)$]

9. $(\forall s)(\mathbf{A}_1 s \rightarrow\sim \mathbf{E}_2 s)$ [= $(\forall s)(\mathbf{A}_1 s \rightarrow \mathbf{I}_2 s)$]

10. $(\forall s)(\mathbf{A}_1 s \rightarrow\sim \mathbf{O}_2 s)$ [= $(\forall s)(\mathbf{A}_1 s \rightarrow \mathbf{A}_2 s)$]

11. $(\forall s)(\mathbf{I}_1 s \rightarrow\sim \mathbf{E}_2 s)$ [= $(\forall s)(\mathbf{I}_1 s \rightarrow \mathbf{I}_2 s)$]

12. $(\forall s)(\mathbf{E}_1 s \rightarrow\sim \mathbf{A}_2 s)$ [= $(\forall s)(\mathbf{E}_1 s \rightarrow \mathbf{O}_2 s)$]

13. $(\forall s)(\mathbf{E}_1 s \rightarrow\sim \mathbf{I}_2 s)$ [= $(\forall s)(\mathbf{E}_1 s \rightarrow \mathbf{E}_2 s)$]

14. $(\forall s)(\mathbf{O}_1 s \rightarrow\sim \mathbf{A}_2 s)$ [= $(\forall s)(\mathbf{O}_1 s \rightarrow \mathbf{O}_2 s)$]

15. $(\forall s)(\mathbf{O}_1 s \rightarrow\sim \mathbf{I}_2 s)$ [= $(\forall s)(\mathbf{O}_1 s \rightarrow \mathbf{E}_2 s)$]

16. $(\forall s)(\mathbf{I}_1 s \rightarrow\sim \mathbf{O}_2 s)$ [= $(\forall s)(\mathbf{I}_1 s \rightarrow \mathbf{A}_2 s)$]

As to the **Ic** propositions, they are noted Iaa, Iai, Iae, etc.... and formalized as follows:

1. $(\exists s)(\mathbf{A}_1 s \wedge \mathbf{A}_2 s)$ [noted Iaa]

2. $(\exists s)(\mathbf{A}_1 s \wedge \mathbf{I}_2 s)$ [noted Iai]

3. $(\exists s)(\mathbf{I}_1 s \wedge \mathbf{A}_2 s)$ [noted Iia]

4. $(\exists s)(\mathbf{I}_1 s \wedge \mathbf{I}_2 s)$ -----

5. $(\exists s)(\mathbf{E}_1 s \wedge \mathbf{E}_2 s)$ -----

6. $(\exists s)(\mathbf{E}_1 s \wedge \mathbf{O}_2 s)$ ----

7. $(\exists s)(\mathbf{O}_1 s \wedge \mathbf{E}_2 s)$

8. $(\exists s)(\mathbf{O}_1 s \wedge \mathbf{O}_2 s)$

9. $(\exists s)(\mathbf{A}_1 s \wedge \mathbf{E}_2 s)$

10. $(\exists s)(\mathbf{I}_1 s \wedge \mathbf{E}_2 s)$

11. $(\exists s)(\mathbf{A}_1 s \wedge \mathbf{O}_2 s)$

12. $(\exists s)\ (\mathbf{I}_1 s \wedge \mathbf{O}_2 s)$

13. $(\exists s)\ (\mathbf{E}_1 s \wedge \mathbf{A}_2 s)$

14. $(\exists s)\ (\mathbf{O}_1 s \wedge \mathbf{A}_2 s)$

15. $(\exists s)\ (\mathbf{E}_1 s \wedge \mathbf{I}_2 s)$

16. $(\exists s)\ (\mathbf{O}_1 s \wedge \mathbf{I}_2 s)$

And the **Oc** propositions are noted Oaa, Oai, Oae, and so on, and formalized as follows:

1. $\sim [(\exists s)\ \mathbf{A}_1 s \wedge (\forall s)\ (\mathbf{A}_1 s \rightarrow \mathbf{A}_2 s)\ ]$ [noted Oaa]
2. $\sim [(\exists s)\ \mathbf{A}_1 s \wedge (\forall s)\ (\mathbf{A}_1 s \rightarrow \mathbf{I}_2 s)\ ]$ [noted Oai)]
3. $\sim [(\exists s)\ \mathbf{I}_1 s \wedge (\forall s)\ (\mathbf{I}_1 s \rightarrow \mathbf{A}_2 s)\ ]$ [noted Oia]
4. $\sim [(\exists s)\ \mathbf{I}_1 s \wedge (\forall s)\ (\mathbf{I}_1 s \rightarrow \mathbf{I}_2 s)]$ -----
5. $\sim [(\exists s)\ \mathbf{E}_1 s \wedge (\forall s)\ (\mathbf{E}_1 s \rightarrow \mathbf{E}_2 s)]$ -----
6. $\sim [(\exists s)\ \mathbf{E}_1 s \wedge (\forall s)\ (\mathbf{E}_1 s \rightarrow \mathbf{O}_2 s)]$ -----
7. $\sim [(\exists s)\ \mathbf{O}_1 s \wedge (\forall s)\ (\mathbf{O}_1 s \rightarrow \mathbf{E}_2 s)]$
8. $\sim [(\exists s)\ \mathbf{O}_1 s \wedge (\forall s)\ (\mathbf{O}_1 s \rightarrow \mathbf{O}_2 s)]$
9. $\sim [(\exists s)\ \mathbf{A}_1 s \wedge (\forall s)\ (\mathbf{A}_1 s \rightarrow \mathbf{E}_2 s)]$
10. $\sim [(\exists s)\ \mathbf{A}_1 s \wedge (\forall s)\ (\mathbf{A}_1 s \rightarrow \mathbf{O}_2 s)]$
11. $\sim [(\exists s)\ \mathbf{I}_1 s \wedge (\forall s)\ (\mathbf{I}_1 s \rightarrow \mathbf{E}_2 s)]$
12. $\sim [(\exists s)\ \mathbf{I}_1 s \wedge (\forall s)\ (\mathbf{I}_1 s \rightarrow \mathbf{O}_2 s)]$
13. $\sim [(\exists s)\ \mathbf{E}_1 s \wedge (\forall s)\ (\mathbf{E}_1 s \rightarrow \mathbf{A}_2 s)]$
14. $\sim [(\exists s)\ \mathbf{E}_1 s \wedge (\forall s)\ (\mathbf{E}_1 s \rightarrow \mathbf{I}_2 s)]$
15. $\sim [(\exists s)\ \mathbf{O}_1 s \wedge (\forall s)\ (\mathbf{O}_1 s \rightarrow \mathbf{I}_2 s)]$
16. $\sim [(\exists s)\ \mathbf{O}_1 s \wedge (\forall s)\ (\mathbf{O}_1 s \rightarrow \mathbf{A}_2 s)]$

Now what are the logical relations between all these propositions? How can we calculate the exact number of pairs inside the whole set of propositions? How can we determine the logical relations between the members of each of these pairs? Can we find some additional relations apart from the classical contradictions **Ac/Oc** and **Ec/Ic**, contrarieties **Ac/Ec**, subcontrarieties **Ic/Oc** and subalternations **Ac/Ic** and **Ec/Oc**? This will be examined in the next section.

5 The Logical Relations Between these Propositions

To calculate the number of pairs between the 64 propositions, we will first consider the pairs of propositions of different kinds (for instance, the pairs Ac/Ec or Ac/Ic, etc.), and then we will calculate the pairs of propositions of the same kind, for instance, the pairs **Ac/Ac** or **Ec/Ec** propositions, whose clauses are different.

Since there are four kinds of propositions in total, namely, **Ac**, **Ec**, **Ic** and **Oc** propositions, we have six possible combinations between these four kinds, when each element of the combination is different from the other one. The six combinations are the following: **Ac/Ec, Ac/Ic, Ac/Oc, Ec/Ic, Ec/Oc** and **Ic/Oc**. In all these cases, we have $16 \times 16 \times 2 = 512$ ordered pairs for each kind of combination, i.e. 16 Ac/Ec, 16 Ec/Ac, 16 Ac/Ic, etc. So the total number of *ordered* pairs of the different propositions is the following: $512 \times 6 = 3072$. But we will show below that we need only the number of *unordered* pairs (= 1536) despite the asymmetric character of subalternation.

Let us first show how to do the calculation of the ordered pairs, and then we will get the unordered pairs just by dividing this number by 2. Take the 16 **Ac** propositions, numbered 1, 2, 3, etc., and the 16 **Ec** propositions, numbered 1′, 2′, 3′, etc., then the total number of ordered pairs **Ac/Ec** is the following[2]:

(1,1′); (1,2′); (1,3′); (1,4′); (1,5′); (1,6′); (1,7′); (1,8′); (1,9′); (1,10′); (1,11′); (1,12′); (1,13′); (1,14′); (1,15′); (1,16′)

(1′,1); (1′,2); (1′,3); (1′,4); (1′,5); (1′,6); (1′,7); (1′,8); (1′,9); (1′,10); (1′,11); (1′,12); (1′,13); (1′,14); (1′,15); (1′,16)

(2,1′); (2,2′); (2,3′); (2,4′); (2,5′); (2,6′); (2,7′); (2,8′); (2,9′); (2,10′); (2,11′); (2,12′); (2,13′); (2,14′); (2,15′); (2,16′)

(2′,1); (2′,2); (2′,3); (2′,4); (2′,5); (2′,6); (2′,7); (2′,8); (2′,9); (2′,10); (2′,11); (2′,12); (2′,13); (2′,14); (2′,15); (2′,16)

\- - - - - -

\- - - - - -

\- - - - - -

(16,1′); (16,2′); (16,3′); (16,4′); (16,5′); (16,6′); (16,7′); (16,8′); (16,9′); (16,10′); (16,11′); (16,12′); (16,13′); (16,14′); (16,15′); (16,16′).

[2] We only give here the first lines plus the very last lines just to give an idea about the method of calculus.

(16′,1); (16′,2); (16′,3); (16′,4); (16′,5); (16′,6); (16′,7); (16′,8); (16′,9); (16′,10); (16′,11); (16′,12); (16′,13); (16′,14); (16′,15); (16′,16)

The same calculus applies to all the pairs **Ac/Ic**, **Ac/Oc**, **Ec/Ic**, **Ec/Oc** and **Oc/Ic**.

But we don't need ordered pairs, i.e. the pairs such that $(1,1') \neq (1',1)$. All we need is the number of *unordered* pairs, i.e. the pairs such that $\{1,1'\} = \{1',1\}$. So the real number of pairs needed for this kind of combinations is $3072/2 = 1536$. Let us explain this more clearly. We are talking here about the Aristotelian relations of contradiction, contrariety, subcontrariety and subalternation. To these we will add the independence relation, but we will not consider equivalence relations, which do not occur in the square. Now the Aristotelian relations are all symmetrical, *except* subalternation (about which more will be said below), and the independence relation is also symmetrical. For instance, contradiction is a valid exclusive disjunction, and the exclusive disjunction is symmetrical, since '$(P \veebar Q) \equiv (Q \veebar P)$'. Likewise contrariety is the validity of a negated conjunction, and since the conjunction is also commutative, the following equivalence holds: '$\sim(P \wedge Q) \equiv \sim(Q \wedge P)$'. Subcontrariety is the validity of an inclusive disjunction, and this disjunction too is commutative, since '$(P \vee Q) \equiv (Q \vee P)$'. As to the independence relation, it is symmetrical, for if a proposition is independent from another one, the other one is also independent from it, since independence means that the propositions can be both true or both false or one of them is true while the other one is false, regardless of their order.

What remains is subalternation. This relation is indeed asymmetric, since it is an implication whose antecedent is a universal proposition and whose consequent is a particular one. And as all implications, it is not symmetrical, since '$(P \rightarrow Q) \not\leftrightarrow (Q \rightarrow P)$'. Still when we find a subalternation between two propositions (e.g. '$Ac \rightarrow Ic$'), the relation that we get when we change the order of the two propositions cannot be another subalternation, since the two propositions considered here are not equivalent, given the way we have formalized them; it cannot be a subcontrariety either, since two propositions related by a subalternation can be false together, nor can it be a contrariety, since both such propositions can be true together, and it cannot be an independence relation, since there is no case where the consequent (e.g. Ic, here) is false while the antecedent (e.g. Ac, here) is true. So the only remaining possibility is that the ordered pair (Ic, Ac) expresses a non-implication. This means that once we know that the relation between the members of the ordered pair (Ac, Ic) is a subalternation, we know that the relation between the members of the ordered pair (Ic, Ac) is a non-implication, provided the two propositions are not equivalent, which is the case with the propositions of Avicenna's lists. This is why we only need *unordered* pairs, even if one of the relations is asymmetric.

But we have to add the number of unordered pairs between the propositions of the same kind, namely, all the unordered pairs **A/A**, **E/E**, **I/I** and **O/O**. For these too, the calculus should *not* be based on the following formula:

$$\frac{m!}{(m-n)!}$$

which calculates the number of ordered pairs. Rather we need to use the formula below (where $m = 16$ and $n = 2$), which provides the number of unordered pairs:

$$\frac{m!}{(m-n)!n!} = \frac{16!}{(16-2)!2!} = \frac{16 \text{ x} 15 \text{ x } \cancel{14!}}{\cancel{14!}\ 2!} = \frac{16 \text{ x } 15}{2!} = 8 \text{ x } 15 = 120$$

This is 'the formula for the number of n-element subsets of a set of m elements',[3] i.e. in our case, the combination of all pairs inside the set of 16 propositions, regardless of the order of their elements. Note that the calculus with this formula does not count the pairs of identical propositions, i.e. the pairs {1,1}, {2,2}, etc., since we don't need these pairs, because we are not searching for equivalences.

This calculus can be applied to the unordered pairs **Ac/Ac**, **Ec/Ec**, **Ic/Ic** and **Oc/Oc**.

For instance, for the **Ac/Ac** pairs, when the propositions are numbered 1, 2, 3, 4, . . . , 16, the lines are the following:

{1,2}; {1,3}; {1,4}; {1,5}; {1,6}; {1,7}; {1,8}; {1,9}; {1,10}; {1,11}; {1,12}; {1,13}; {1,14}; {1,15}; {1,16}

{2,3}; {2,4}; {2,5}; {2,6}; {2,7}; {2,8}; {2,9}; {2,10}; {2,11}; {2,12}; {2,13}; {2,14}; {2,15}; {2,16}[4]

{3,4}; {3,5}; {3,6}; {3,7}; {3,8}; {3,9}; {3,10}; {3,11}; {3,12}; {3,13}; {3,14}; {3,15}; {3,16}.

- - - - - -

- - - - -

- - - -

{14,15}; {14,16}

{15,16}

Likewise we have 120 **Ec/Ec** pairs, 120 **Ic/Ic** pairs and 120 **Oc/Oc** pairs, which means that the total number of such unordered pairs between the same kind of propositions is the following: $120 \times 4 = 480$ pairs.

If we consider the logical relations between these 480 unordered pairs and between the 1536 previous unordered pairs of propositions, we get the following number of pairs: $1536 + 480 = 2016$, the members of which are related by Aristotelian relations or independence ones.

We must thus consider all these relations between the 64 propositions in order to find out which ones are Aristotelian, i.e. are either *contradictions* or *contrarieties* or *subcontrarieties* or *subalternations.* These Aristotelian relations are defined as follows:

1. Two propositions α and β are contradictory if and only if they are never true nor false together (i.e. if and only if '$\alpha \underline{\vee} \beta$' is valid = tautological).

[3] This is the explanation given by one of the referees, whom I thank for his valuable remarks and suggestions.

[4] Note here that we have omitted {2,2}, for the reasons evoked above.

2. Two propositions α and β are contrary if and only if they are never true together but can be false together (i.e. if '$\sim(\alpha \wedge \beta)$' is valid).
3. Two propositions α and β are subcontrary if and only if they are never false together but can be true together (i.e. if '$\alpha \vee \beta$' is valid).
4. The proposition β is the subaltern of the proposition α if and only if when α is true, β is true too, and when β is false, α is false too (i.e. if '$\alpha \rightarrow \beta$' is valid).

The remaining relations between pairs of propositions will be considered as cases of *independence*, where the two elements of the same pair can be both true together, both false together or one of them is true while the other one is false. In these cases, no specific link between these propositions holds, and no deduction can be made starting from one or the other of the propositions. As to the non-implications, they are just parallel to the subalternations and will not be mentioned in the figures below, since they are never mentioned in the classical squares, hexagons or octagons.

Let us start by the most obvious relation, namely, *contradiction*. It is clear that contradiction holds between all **A**c and **O**c propositions and between all **E**c and **I**c propositions with the *same* clauses. Since we have 16 propositions for each kind of quantified propositions, we will have 16 pairs of contradictories **A**c/**O**c and 16 pairs **E**c/**I**c. So we get 32 pairs of contradictories.

Likewise, we can say that the other Aristotelian relations hold between the propositions which have the same clauses. This means that:

All **A**c and **E**c propositions with the *same* clauses are contrary. So here too, we should have 16 pairs of contrarieties, at first sight.
All **I**c and **O**c propositions with the *same* clauses are subcontrary, which means that we should have 16 pairs of subcontrarieties, at first sight.
All **A**c propositions imply the **I**c propositions with the *same* clauses, which gives rise to 16 subalternations, at first sight.
All **E**c propositions imply the **O**c propositions with the *same* clauses, which also gives rise to 16 other subalternations, at first sight.

This leads to the following number of relations: 32 pairs of contradictories +16 pairs of contrarieties +16 pairs of subcontrarieties +32 subalternations = 96 Aristotelian relations between the propositions with the *same* clauses. But we could also have contrarieties, subcontrarieties and subalternations between propositions which do not have the same clauses, as we will see below.

The combinations between the 16 **A**c propositions and the 16 **O**c propositions and between the 16 **E**c propositions and the 16 **I**c propositions give rise to one contradiction in each line. The remaining pairs in these combinations of the different **E**c and **I**c and **A**c and **O**c are either contrarieties or subcontrarieties or independence relations. To illustrate this, let us provide the first line of the combinations **I**c/**E**c. This line is the following:

Iaa/Eaa; *Iaa/Eai*; *Iaa/Eia*; *Iaa/Eii*; Iaa/Eee; Iaa/Eeo; Iaa/Eoe; Iaa/Eoo; Iaa/Eae; Iaa/Eao; Iaa/Eie; Iaa/Eea; Iaa/Eeo; Iaa/Eoa; Iaa/Eoi; Iaa/Eio.

In this line, only the first pair (in bold) is a pair of contradictories. The other ones are either independent propositions or pairs of contraries. The pairs of contraries (in italics) are *Iaa/Eai, Iaa/Eia* and *Iaa/Eii*. All the remaining relations are independence relations. We can thus see from the start that the independence relations are much more numerous than the Aristotelian ones.

The same holds with the 15 other lines where there is exactly one contradiction and some contrarieties or subcontrarieties depending on the line and the propositions involved plus many independence relations. The contrarieties, subcontrarieties and subalternations can hold between propositions which do not have the same clauses, unlike the contradictions.

Likewise, the propositions which are of the same kind (both being **Ac** propositions, for instance) but have opposed consequents can be either contrary or subcontrary, as claimed by Avicenna in the following passage, where talking about **Ac** propositions with contradictory consequents, he says that they are not contradictory but rather contrary:

> ... Thus the universal affirmatives whose consequents are contradictory are [themselves] contrary, for they [can be] both false but they are not contradictory. This is so because one of these affirmatives has the power of a universal negative, which is opposed to the [previous] universal by contrariety ([2], p. 368.15–17).

Elsewhere, he says, talking about the two particular propositions:

> ... The opinion according to which the contradiction of the consequents makes the conditional [propositions] contradictory is false, for these particulars can be true together. But the power of the negative one is the power of an affirmative whose consequent is contradictory to that of the affirmative, while the power of the affirmative one is that of a negative whose consequent is contradictory to that of the negative one. So they are two affirmatives with two contradictory consequents and they are true together, [or] they are two negatives with also [two contradictory consequents] and they are both true ([2], p. 371.14–17).

Thus he stresses the subcontrariety of the two particular propositions whose consequents are contradictory, while their antecedents are the same. So the **Ic** propositions whose consequents are contradictory can be true together, according to him, and the same can be said about the **Oc** propositions whose consequents are contradictory, when their antecedents are the same.

Let us first state the usual Aristotelian relations between all these propositions. The 32 usual pairs of contradictories **Ac/Oc** and **Ec/Ic** are the following:

Aaa/Oaa, Aai/Oai, Aao/Oao, Aae/Oae, Aii/Oii, Aia/Oia, Aie/Oie, Aio/Oio, Aee/Oee, Aea/Oea, Aei/Oei, Aeo/Oeo, Aoa/Ooa, Aoe/Ooe, Aoi/Ooi, Aoo/Ooo, Eaa/Iaa, Eae/Iae, Eai/Iai, Eao/Iao, Eea/Iea, Eee/Iee, Eei/Iei, Eeo/Ieo, Eia/Iia, Eie/Iie, Eii/Iii, Eio/Iio, Eoa/Ioa, Eoe/Ioe, Eoi/Ioi, Eoo/Ioo.

The 16 pairs of *usual* contraries **Ac/Ec**, where the contrary propositions have the *same* clauses, are the following:

Aaa/Eaa; Aee/Eee; Aii/Eii; Aoo/Eoo; Aae/Eae; Aao/Eao; Aai/Eai; Aea/Eea; Aei/Eei; Aeo/Eeo; Aia/Eia; Aie/Eie; Aio/Eio; Aoa/Eoa; Aoi/Eoi; Aoe/Eoe.

The 16 pairs of *usual* subcontrarieties **Ic/Oc**, where the propositions have the same clauses, are the following:

Iaa/Oaa, Iae/Oae, Iai/Oai, Iao/Oao, Iea/Oea, Iee/Oee, Iei/Oei, Ieo/Oeo, Iia/Oia, Iie/Oie, Iii/Oii, Iio/Oio, Ioa/Ooa, Ioe/Ooe, Ioi/Ooi, Ioo/Ooo.

The 16 *usual* subalternations between **Ac** and **Ic** propositions with the same clauses are the following:

Aaa → Iaa; Aai → Iai; Aao → Iao; Aae → Iae; Aee → Iee; Aea → Iea; Aei → Iei; Aeo → Ieo; Aii → Iii; Aia → Iia; Aie → Iie; Aio → Iio; Aoo → Ioo; Aoa → Ioa; Aoe → Ioe; Aoi → Ioi.

Those between the **Ec** and **Oc** propositions with the same clauses are the following:

Eaa → Oaa; Eae → Oae; Eai → Oai; Eao → Oao; Eea → Oea; Eee → Oee; Eei → Oei; Eeo → Oeo; Eia → Oia; Eie → Oie; Eii → Oii; Eio → Oio; Eoa → Ooa; Eoe → Ooe; Eoi → Ooi; Eoo → Ooo.

But the number of pairs is far bigger than that, as the calculations above show. At first sight, the number of contradictory pairs should remain the same (= 32), since all contradictory propositions must have the same clauses and since the change of the order of the clauses does not add anything. But we should find other pairs of *contrarieties*, of *subcontrarieties* and of *subalternations* between propositions which do not have the same clauses. And this must be checked by verifying what kind of relation holds between these propositions where the clauses are not the same.

So let us first check what would be these contrarieties and the pairs of propositions involved. It seems obvious that the following **Ac** and **Ec** propositions are contrary:

Aaa/Eai; Aaa/Eia; Aaa/Eii; Aee/Eeo; Aee/Eoe; Aee/Eoo; Aae/Eao; Aae/Eie; Aae/Eio; Aao/Eio; **Aai/Eii**; Aea/Eei; Aea/Eoa; Aea/Eoi; Aei/Eoi; Aeo/Eoo; Aia/Eii; Aie/Eio; Aoa/Eoi; Aoe/Eoo.

For since the consequent of the **Ec** proposition is either the contrary or the contradictory of the consequent of the **Ac** proposition, they cannot be true together. Given that the antecedents of the **Ec** propositions are either the same as those of the **Ac** ones or subaltern to those of the **Ac** ones, the whole **Ac** and **Ec** propositions can never be true together. They are thus contrary. For instance, let us take as an example the pair Aai/Eii (in bold in the list). This pair can be stated as follows:

$$(\exists s)\, \mathbf{A}_1 s \wedge (\forall s)\, (\mathbf{A}_1 s \rightarrow \mathbf{I}_2 s)\ /\ (\forall s)\, (\mathbf{I}_1 s \rightarrow \sim \mathbf{I}_2 s)$$

It is not a contradictory pair because its clauses are different. But can it be a pair of contrary propositions? These **Ac** and **Ec** propositions would be contrary if they can never be true together but could be false together. Consider, for instance, the case where the **Ac** proposition is true; this means that its main conjunction is true; so $\mathbf{A}_1 s$ is true. Consequently its usual subaltern, namely, $\mathbf{I}_1 s$, which is here the antecedent of the **Ec** proposition, must be true too. But if we want the **Ac** proposition to be true, its second conjunct, namely, '$(\forall s)(\mathbf{A}_1 s \rightarrow \mathbf{I}_2 s)$', should be true too, and since the antecedent of this conditional is true, its consequent (viz. $\mathbf{I}_2 s$) must be true. Unfortunately, in this case, its contradictory (= $\sim \mathbf{I}_2 s$), which is the consequent of the **Ec** proposition must be false, which means that the whole **Ec** proposition is false (its antecedent being itself true). So the two propositions cannot be true together.

Likewise, we could show that when the **Ec** proposition is true, the **Ac** proposition cannot be true. Consider the case where it is the **Ec** proposition that is supposed true;

we get the following line of the table (where we consider two situations s_1 and s_2):

$$\{(\mathbf{A}_1 s_1 \vee \mathbf{A}_1 s_2) \wedge [(\mathbf{A}_1 s_1 \rightarrow \mathbf{I}_2 s_1) \wedge (\mathbf{A}_1 s_2 \rightarrow \mathbf{I}_2 s_2)] / [(\mathbf{I}_1 s_1 \rightarrow\sim \mathbf{I}_2 s_1) \wedge (\mathbf{I}_1 s_2 \rightarrow\sim \mathbf{I}_2 s_2)]$$

0 0 0 **0** 0 1 0   1 0 1 0 1 1 1   1 0 1 1

This line shows that if the two conditionals formalizing the **Ec** proposition are true and if their consequents are true, whatever truth value their antecedents may have, then $\mathbf{I}_2 s_1$ and $\mathbf{I}_2 s_2$ must both be false, in which case both $\mathbf{A}_1 s_1$ and $\mathbf{A}_1 s_2$ must be false in order for the two conditionals of the **Ac** proposition to be true. In this case, the disjunction in the left side is also false, which means that when the **Ec** proposition is true, then the **Ac** proposition is false. If we consider that the disjunction is true, this means that one of the two propositions $\mathbf{A}_1 s_1$ and $A_1 s_2$ should be true; but if one of them is true and if one of the **I** propositions ($\mathbf{I}_2 s_1$ and $\mathbf{I}_2 s_2$) is false, then one of the conditionals of the **Ac** proposition would be false; consequently the whole conjunction would be false too. So whatever supposition we make, we always have a case of falsity under one proposition whenever the other proposition is true, which means that they can never be true together.

They could however be false together, as is shown by the following line of the table, where the formula corresponding to **Ac** in the left side is false together with the one corresponding to **Ec** in the right side:

$$\{(\mathbf{A}_1 s_1 \vee A_1 s_2) \wedge [(\mathbf{A}_1 s_1 \rightarrow \mathbf{I}_2 s_1) \wedge (\mathbf{A}_1 s_2 \rightarrow \mathbf{I}_2 s_2)] / [(\mathbf{I}_1 s_1 \rightarrow\sim \mathbf{I}_2 s_1) \wedge (\mathbf{I}_1 s_2 \rightarrow\sim \mathbf{I}_2 s_2)]$$

1 1 1 **0** 1 0 0 0 1 1 1 1 1 1   **0** 1 0 0

The same proof holds for all similar pairs given above. We can thus say that the 20 pairs above are all pairs of contrary propositions.

Consequently, their contradictories are subcontrary. The pairs of subcontrary propositions are the following:

Oaa/Iai; Oaa/Iia; Oaa/Iii; Oee/Ieo; Oee/Ioe; Oee/Ioo; Oao/Iao; Oae/Iie; Oae/Iio; Oao/Iio; Oai/Iii; Oea/Iei; Oea/Ioa; Oea/Ioi; Oia/Iii; Oei/Ioi; Oeo/Ioo; Oie/Iio; Ooa/Ioi; Ooe/Ioo.

However, the following **Ac** and **Ec** propositions are independent:

Aii/Eia; Aii/Eaa; Aii/Eai; Aoo/Eoe; Aoo/Eee; Aoo/Eoo; Aao/Eae; Aao/Eie; Aai/Eaa; Aia/Eaa; Aai/Eia; Aia/Eaa; Aia/Eai; Aei/Eea; Aei/Eoa; Aeo/Eee; **Aeo/Eoe**; Aie/Eae; Aie/Eao; Aio/Eie; Aio/Eao; Eio/Eae; Aoa/Eei; Aoa/Eea; Aoe/Eee; Aoe/Eeo; Aoi/Eoa; Aoi/Eea; Aoi/Eei.

For take, for instance, the following pair: Aeo/Eoe (in bold in the list above). This pair of propositions is formalized as follows: $(\exists s)\, \mathbf{E}_1 s \wedge (\forall s)(\mathbf{E}_1 s \supset \mathbf{O}_2 s)$ / $(\forall s)(\mathbf{O}_1 s \supset {\sim}\mathbf{E}_2 s)$.

If we consider two situations, the formalization is the following:

$$(\mathbf{E}_1 s_1 \vee \mathbf{E}_1 s_2) \wedge [(\mathbf{E}_1 s_1 \rightarrow \mathbf{O}_2 s_1) \wedge (\mathbf{E}_1 s_2 \rightarrow \mathbf{O}_2 s_2)] / [(\mathbf{O}_1 s_1 \rightarrow \sim \mathbf{E}_2 s_1) \wedge (\mathbf{O}_1 s_2 \rightarrow \sim \mathbf{E}_2 s_2)]$$

0 0 0 0 0 1 1 1 0 1 1 **1 0 0 0** 1 1 1 Line 1

1 1 0 **1** 1 1 1 1 0 1 1 1 1 1 1 0 1 0 Line 2

1 1 0 **0** 1 0 0 **0** 0 1 1 1 1 1 **1** 0 1 1 Line 3

1 1 0 **1** 1 1 1 1 0 1 0 **1 0 0 0** 1 1 1 Line 4

Now when we test the possible values of these propositions, we find that the four lines above show that the two propositions can be false together [when both $\mathbf{E}_1 s_1$ and $\mathbf{E}_1 s_2$ are false and ~ $\mathbf{E}_2 s_1$ is false too (Line 1)] and that they can also be true together [when $\mathbf{E}_1 s_1$ is true while $\mathbf{E}_1 s_2$ is false (Line 2)] or one of them can be true while the other one is false, for instance, when $\mathbf{E}_1 s_1$ is true while $\mathbf{O}_2 s_1$ is false and ~ $\mathbf{E}_2 s_1$ is true while $\mathbf{O}_1 s_2$ is false (Line 3) or when $\mathbf{E}_1 s_1$ is true while ~$\mathbf{E}_2 s_1$ is false (Line 4). Likewise, all other pairs of the above group are independent, since the propositions involved have the same kind of sstructure.

Consequently, their contradictories are also independent, for from the fact that 'p' and 'q' are independent, we can deduce that '~p' and '~q' are independent too, given that they too can be either both true or both false or one of them is true while the other one is false.

So the following Oc and Ic propositions are independent too:

Oii/Iia; Oii/Iaa; Oii/Iai; Ooo/Ioe; Ooo/Iee; Ooo/Ioo; Oao/Iae; Oao/Iie; Oai/Iaa; Oia/Iaa; Oai/Iia; Oia/Iaa; Oia/Iai; Oei/Iea; Oei/Ioa; Oeo/Iee; Oeo/Ioe; Oie/Iae; Oie/Iao; Oio/Iie; Oio/Iao; Oio/Iae; Ooa/Iei; Ooa/Iea; Ooe/Iee; Ooe/Ieo; Ooi/Ioa; Ooi/Iea; Ooi/Iei.

Can we have other contrarieties between other kinds of propositions?

As stressed by Avicenna in the quotations above ([2], p. 368), two **A**c propositions can be contrary when their consequents are contradictory. We can add the same claim for the propositions whose consequents are contrary. So the following pairs of propositions should be contrary:

Aaa/Aae; Aaa/Aao; Aaa/Aio; Aaa/Aie; Aai/Aie; Aae/Aia; Aae/Aii; Aae/Aai; Aai/Aie; Aao/Aia; Aee/Aea; Aee/Aei; Aee/Aoa; Aee/Aoi; Aea/Aeo; Aea/Aoo; Aea/Aoe; Aia/Aie; Aia/Aio; Aie/Aii; Aoa/Aeo; Aoa/Aoo; Aoa/Aoe; Aoe/Aei; Aoe/AoiIn these pairs of propositions, the consequents are sometimes contradictory as in the pair 'Aaa/Aao' but they can also be contrary as in the pair 'Aaa/Aae'. In both cases, when the antecedent of both propositions is the same and is true, the second proposition cannot be true, given that its consequent is either the contradictory of the consequent of the other one or its contrary. If the antecedents are not the same, but are both true, and the consequents are either contradictory or contrary to each other, as in the pairs 'Aaa/Aio' or 'Aae/Aia', the same holds for in these cases, if the antecedent of 'Aaa' is true, the antecedent of 'Aio', which is an I proposition, i.e. a subaltern of A, will be true too, but when the consequent of the first Ac proposition is true, the consequent of the second Ac proposition, which is an O proposition, will be false, being the contradictory of the first consequent. This means that the two Ac propositions can never be true together. We can formalize these two **Ac** propositions as follows: $(\exists s)\ \mathbf{A}_1 s \wedge (\forall s)(\mathbf{A}_1 s \rightarrow \mathbf{A}_2 s)\ /\ (\exists s)\ \mathbf{I}_1 s \wedge (\forall s)(\mathbf{I}_1 s \rightarrow \mathbf{O}_2 s)$.

If we consider only one situation, we get the following formulas, where if the first **Ac** (in the left) is true, the second **Ac** (in the right) cannot be true:

$$[\mathbf{A}_1 s_1 \wedge (\mathbf{A}_1 s_1 \rightarrow \mathbf{A}_2 s_1)]/[\mathbf{I}_1 s_1 \wedge (\mathbf{I}_1 s_1 \rightarrow \mathbf{O}_2 s_1)]$$

1 **1** 1 1 1 1 **0** 1 0 0

If on the other hand, the second **Ac** proposition is true, can the first **Ac** be true? The following lines of the table show that this too cannot hold, whether the antecedent of the first Ac (in the left) is false or true, as shown in the two lines below:

$$[\mathbf{A}_1 s_1 \wedge (\mathbf{A}_1 s_1 \rightarrow \mathbf{A}_2 s_1)]/[\mathbf{I}_1 s_1 \wedge (\mathbf{I}_1 s_1 \rightarrow \mathbf{O}_2 s_1)]$$

0 **0** 0 1 0 1 **1** 1 1 1

1 **0** 1 0 0 1 **1** 1 1 1

So these two **Ac** propositions cannot be true together. However, they can be false together, as shown by the following line:

$$[\mathbf{A}_1 s_1 \wedge (\mathbf{A}_1 s_1 \rightarrow \mathbf{A}_2 s_1)]/[\mathbf{I}_1 s_1 \wedge (\mathbf{I}_1 s_1 \rightarrow \mathbf{O}_2 s_1)]$$

0 **0** 0 1 1 1 **0** 1 0 0

They are thus contrary. The same can be said about all the pairs of the same group mentioned above. These contrarieties are added to the usual ones and involve only **Ac** propositions.

As a consequence, the following pairs of **Oc** propositions are pairs of subcontrary propositions, since when two propositions are contrary, their contradictories are subcontrary:

Oaa/Oae; Oaa/Oao; Oaa/Oio; Oaa/Oie; Oai/Oie; Oae/Oia; Oae/Oii; Oae/Oai; Oai/Oie; Oao/Oia; Oee/Oea; Oee/Oei; Oee/Ooa; Oee/Ooi; Oea/Oeo; Oea/Ooo; Oea/Ooe; Oia/Oie; Oia/Oio; Oie/Oii; Ooa/Oeo; Ooa/Ooo; Ooa/Ooe; Ooe/Oei; Ooe/Ooi.

All these propositions are negative particular propositions. Their subcontrariety confirms what is stressed by Avicenna in the quotation above to the effect that some **Oc** propositions are subcontrary. They can thus be true together, whether their antecedents are the same or related by subalternation, while their consequents are either contrary or contradictory.

Some contrarieties hold between some **Ac** and **Ic** propositions. These are the following:

Aaa/Iae; Aaa/Iao; Aae/Iaa; Aae/Iai; Aai/Iae; Aao/Iaa; Eea/Iee; Eea/Ieo; Aee/Iea; Aee/Iei; Aei/Iee; Aeo/Iea; Aia/Iie; Aia/Iio; Aii/Iie; Aio/Iia; Aie/Iia; Aie/Iii; Aoa/Ioe; Aoa/Ioo; Aoe/Ioa; Aoe/Ioi; Aoo/Ioa; Aoi/Ioe.

The **Ac** and **Ic** propositions involved are those where the antecedents are the same, while the consequents are either contradictory or contrary. For if the antecedent of the **Ac** proposition is true, it is also true in the **Ic** proposition, but if the consequent of the **Ac** proposition is true, the consequent of the **Ic** proposition cannot be true, since it is contradictory or contrary to that first consequent. So both propositions can never be true together. However they can be false together if the two antecedents of both propositions are false. So these propositions are contrary.

Consequently, their contradictories are subcontrary. These subcontrarieties are the following:

Oaa/Eae; Oaa/Eao; Oae/Eaa; Oae/Eai; Oai/Eae; Oao/Eaa; Iea/Eee; Iea/Eeo; Oee/Eea; Oee/Eei; Oei/Eee; Oeo/Eea; Oia/Eie; Oia/Eio; Oii/Eie; Oio/Eia; Oie/Eia; Oie/Eii; Ooa/Eoe; Ooa/Eoo; Ooe/Eoa; Ooe/Eoi; Ooo/Eoa; Ooi/Eoe.

However, when the antecedent of the **Ic** proposition is the subaltern of the antecedent of the **Ac** one, even if their consequents are contrary or contradictory, they are not themselves contrary; rather they are independent. So the following pairs of propositions are *not* contrary, since they can be true together:

Aaa/Iie; Aaa/Iio; Aae/Iii; Aae/Iia; Aai/Iie; Aao/Iia; Aea/Ioe; Aea/Ioo; Aee/Ioi; Aee/Ioa; Aei/Ioe; Aeo/Ioa.

Take, for instance, the first pair, namely, Aaa/Iie. This pair can be expressed as follows:

$$(\exists s)\ \mathbf{A}_1 s \wedge (\forall s)\ (\mathbf{A}_1 s \rightarrow \mathbf{A}_2 s)\ /\ (\exists s)\ (\mathbf{I}_1 s \wedge \mathbf{E}_2 s)\,.$$

When we consider two situations, we get the following formulas:

$$(\mathbf{A}_1 s_1 \vee \mathbf{A}_1 s_2) \wedge [(\mathbf{A}_1 s_1 \rightarrow \mathbf{A}_2 s_1) \wedge (\mathbf{A}_1 s_2 \rightarrow \mathbf{A}_2 s_2)\,]/[\,(\mathbf{I}_1 s_1 \wedge \mathbf{E}_2 s_1) \vee (\mathbf{I}_1 s_2 \wedge \mathbf{E}_2 s_2)]$$

0 1 1 **1** 0 1 0 1 1 1 1 1 1 1 **1** 1 0 0

The line of the table shows that the two propositions can be true together. So they are not contrary; likewise for all similar pairs listed with this one. These pairs can also be false together (precisely when both $\mathbf{A}_1 s_1$ and $\mathbf{A}_1 s_2$ are false, while $\mathbf{A}_2 s_1$ is true and $\mathbf{I}_1 s_1$ and $\mathbf{I}_1 s_2$ are both false), and one of them can be true while the other one is false (precisely when $\mathbf{A}_1 s_1$, $\mathbf{A}_1 s_2$, $\mathbf{A}_2 s_1$, $\mathbf{A}_2 s_2$, $\mathbf{I}_1 s_1$ and $\mathbf{I}_2 s_1$ are all true, while $\mathbf{E}_2 s_1$ and $\mathbf{E}_2 s_2$ are both false for the first case, and when $\mathbf{A}_1 s_1$ and $\mathbf{A}_1 s_2$ are false, while $\mathbf{I}_1 s_1$ and $\mathbf{E}_2 s_1$ are both true). So they are independent.

Consequently, their contradictories should also be independent. These pairs of independent propositions are the following:

Oaa/Eie; Oaa/Eio; Oae/Eii; Oae/Eia; Oai/Eie; Oao/Eia; Oea/Eoe; Oea/Eoo; Oee/Eoi; Oee/Eoa; Oei/Eoe; Oeo/Eoa.

Some contrarieties hold also between **Ic** and **Ec** propositions such as the following:

Iaa/Eia; Iaa/Eai; Iaa/Eii; Iae/Eao; Iae/Eie; Iae/Eio; Iai/Eii; Iao/Eio; Iea/Eei; Iea/Eoa; Iea/Eoi; Iei/Eoi; Iia/Eii; Iia/Eaa; Iia/Eai; Iie/Eio; Iee/Eoe; Iee/Eeo; Iee/Eoo; Ieo/Eoo; Ioa/Eoi; Ioe/EooTake, for instance, the pair Iaa/Eia (in bold in the list), its formalization is the following:

$$(\exists s)\ (\mathbf{A}_1 s \wedge A_2 s)\ /\ (\forall s)\ (I_1 s \rightarrow\sim A_2 s)$$

If we consider only one situation, we get the following formula:

$$(\mathbf{A}_1 s_1 \wedge A_2 s_1)\ /\ (\mathbf{I}_1 s_1 \rightarrow\sim \mathbf{A}_2 s_1)$$

1 **1** 1   1 **0** 0 [Line 1] (prop 1 : true, prop 2 : false)
1 **0** 0   1 **1** 1 [Line 2] (prop 1 : false, prop 2 : true)
0 **0** 1   1 **0** 0 [Line 3] (both props : false)

The values under the propositions in Line 1 show that when the first proposition is true, the second one cannot be true, since its antecedent is true but its consequent is false. If on the other hand we suppose that the second proposition is true as in Line 2, then the first one cannot be true, since in that case, Line 2 shows that when $\sim A_2 s_1$ is true, its contradictory $A_2 s_1$ is false; consequently the whole conjunction in the left side will be false; if both $\sim A_2 s_1$ and $I_1 s_1$ are false, then $\mathbf{A}_1 s_1$ must be false too, since it implies $I_1 s_1$, which makes the whole conjunction false.

However, these propositions can be false together, as shown in Line 3, since when $\mathbf{I}_1 s_1$ is true, $\mathbf{A}_1 s_1$, which implies it, can be false, in which case, the whole conjunction in the left is false, together with the conditional in the right.

Consequently their contradictories are subcontrary. These pairs of subcontrarieties are the following:

Eaa/Iia; Eaa/Iai; Eaa/Iii; Eai/Iii; Eee/Ioe; Eee/Ieo; Eee/Ioo; Eeo/Ioo; Eae/Iao; Eae/Iie; Eae/Iio; Eao/Iio; Eea/Iei; Eea/Ioa; Eea/Ioi; Eei/Ioi; Eie/Iio; Eoa/Ioi; Eia/Iii; Eoe/Ioo; Eia/Iai; Eia/Iaa; Eie/Iio; Eia/Iii; Eoa/Iei.

Likewise, the following **Ac** and **Oc** propositions are contrary:

Aaa/Oai; Aae/Oao; Aea/Oei; Aee/Oeo; Aia/Oii; Aie/Oio; Aoa/Ooi; Aoe/Ooo.

Consequently, their contradictories are subcontrary:

Oaa/Aai; Oee/Aeo; Oae/Aao; Oie/Aio; Oea/Aei; Ooa/Aoi; Ooe/Aoo; Oia/Aii.

However, the remaining **Ic** and **Ec** propositions and Ac and Oc propositions are independent, for instance, the following: Iaa/Eee; Iaa/Eeo; Iaa/Eoe; Iaa/Eoo; Iai/Eia; Iee/Eii; Aaa/Oee; Aaa/Oeo; etc.

As we can see, there are much less contrarieties Ac/Oc than contrarieties Ic/Ec. This is so because the structures of the Ac and Oc propositions are different from the structures of the propositions Ic and Ec.

As to the subalternations, apart from the usual ones which we already mentioned, we find the following subalternations between these **A**c and **I**c propositions:

Aaa → Iai; Aaa → Iia; Aaa → Iii; Aai → Iii; Aao → Ioo; Aae → Iio; Aae → Iao; Aae → Iio; Aae → Iie; Aee → Ieo; Aee → Ioe; Aee → Ioo; Aea → Iei; Aea → Ioi; Aea → Ioa; Aeo → Ioo; Aia → Iii; Aie → Iio; Aoa → Ioi; Aoe → Ioo; Aoa → Ioi; Aei → Ioi.

Consequently, by contraposition, we can say that the contradictories of their consequents, which are Ec propositions, imply the contradictories of their antecedents, i.e. their Oc subalterns:

Eai → Oaa; Eia → Oaa; Eii → Oaa; Eii → Oai; Eoo → Iao; Eio → Oae; Eao → Oae; Eio → Oae; Eie → Oae; Eeo → Oee; Eoe → Oee; Eoo → Oee; Eei → Oea; Eoi → Oea; Eoa → Oea; Eoo → Oeo; Eii → Oia; Eio → Oie; Eoi → Ooa; Eoo → Ooe; Eoi → Ooa; Eoi → Oei.

However, the following **Ac** and **Ic** propositions are independent:

Aai/Iaa; Aai/Iia; Aao/Iae; Aao/Iie; Aeo/Ioe; Aeo/Iee; Aia/Iaa; Aia/Iai; Aii/Iia; Aii/Iaa; Aii/Iai; Aie/Iea; Aie/Iao; Aio/Iae; Aio/Iao; Aio/Iie; Aoa/Iea; Aoa/Iei; Aoe/Iee; Aoi/Iei; Aoi/Iea; Aoo/Iee; Aoo/Ieo; Aoo/IoeFor instance, the first pair is formalized as follows: $(\exists s)A_1s \wedge (\forall s)(A_1s \rightarrow I_2s)$ / $(\exists s)(A_1s \wedge A_2s)$.

With two situations, we get the following formulas:

$$\Big\{(A_1s_1 \vee A_1s_2) \wedge \Big[(A_1s_1 \rightarrow I_2s_1) \wedge (A_1s_2 \rightarrow I_2s_2)\Big]\Big\} / \Big[(A_1s_1 \wedge A_2s_1) \vee (A_1s_2 \wedge A_2s_2)\Big]$$

1	1	1	**1**	1	1	1	1	1	1	1	1	1	1	**1**	1	1	1	[Line 1]
0	0	0	**0**	0	1	1	1	0	1	1	0	0	1	**0**	0	0	1	[Line 2]
1	1	1	**1**	1	1	1	1	1	1	1	1	0	0	**0**	1	0	0	[Line 3]
1	1	1	**0**	1	0	0	0	1	1	1	1	0	0	**1**	1	1	1	[Line 4]

Thus the two propositions can be true together when all propositions are true as in Line 1; they can be false together when A_1s_1 and A_1s_2 are both false; one of them can be false while the other one is true, for instance, when A_1s_1, A_1s_2, I_2s_1 and I_2s_2 are true while A_2s_1 and A_2s_2 are false as in Line 3 or when A_1s_1 is true while I_2s_1 and A_2s_1 are both false, as in Line 4. Note that with one situation this last case is not available. But Avicenna's formulas are quantified, so one should take into account the formula with two situations because they show more clearly the relations between all kinds of propositions.

Likewise, all similar pairs, i.e. the pairs of propositions where the consequent of **Ac** is a particular proposition while the second element of **Ic** is a universal one, are independent. The same thing can be said about the pairs where the **Ac** propositions have a particular antecedent or a particular antecedent and a particular consequent, while the **Ic** propositions have one or more universal elements.

Consequently, the pairs Oc/Ec, whose clauses are the contradictories of those of the pairs Ac/Ic above, are also independent for the same reasons. These are the following pairs:

Oai/Eaa; Oai/Eia; Oao/Eae; Oao/Eie; Oeo/Eoe; Oeo/Eee; Oia/Eaa; Oia/Eai; Oii/Eia; Oii/Eaa; Oii/Eai; Oie/Eea; Oie/Eao; Oio/Eae; Oio/Eao; Oio/Eie; Ooa/Eea; Ooa/Eei; Ooe/Eee; Ooi/Eei; Ooi/Eea; Ooo/Eee; Ooo/Eeo; Ooo/EoeNow we can have subalternations between some **Ac** propositions and some **Ec** ones as in the following list:

Aaa → Eao; Aaa → Eae; Aae → Eai; Aae → Eaa; Aea → Eeo; Aai → Eae; Aao → Eaa; Aee → Eei; Aia → Eio; Aia → Eie; Aie → Eii; Aie → Eia; Aii → Eie; Aio → Eia; Aea → Eeo; Aee → Eea; Aee → Eei; Aei → Eee; Aeo → Eea; Aoa → Eoo; Aoa → Eoe; Aoe → Eoi; Aoe → Eoa; Aoi → Eoe; Aoo → Eoa; Aoa → Eoe.

Consequently, by contraposition, we have the following subalternations between **Ic** and **Oc** propositions:

Iae → Oai; Iaa → Oao; Iae → Oai; Iee → Oei; Iea → Oeo; Iei → Oee; Ieo → Oea; Iia → Oio; Iie → Oii; Iao → Oaa; Iae → Oaa; Iai → Oae; Iaa → Oae; Iio → Oia; Iie → Oia; Iii → Oie; Iia → Oie; Ioo → Ooa; Ioe → Ooa; Ioi → Ooe; Ioa → Ooe; Ieo → Oea; Iea → Oee; Ioa → Ooo; Ioe → Ooi; Ioe → Ooa.

Consequently, the following subalternations between **Ac** and **Oc** propositions hold too:

Aaa → Oao; Aaa → Oae; Aaa → Oie; Aaa → Oio; Aai → Oae; Aai → Oie; Aae → Oai; Aae → Oii; Aae → Oaa; Aao → Oaa; Aao → Oia; Aee → Oei; Aee → Oea; Aee → Ooa; Aee → Ooi; Aea → Oeo; Aea → Oee; Aei → Oee; Aei → Ooe; Aeo → Ooa; Aia → Oio; Aia → Oie; Aie → Oii; Aie → Oia; Aie → Oai; Aii → Oie; Aii → Oae; Aio → Oia; Aio → Oaa; Aoa → Ooo; Aoa → Ooe; Aoa → Oee; Aoe → Ooi; Aoe → Ooa; Aoi → Ooe; Aoi → Oee; Aoo → Ooa; Aoo → Oea.

There are subalternations between **Ac** propositions alone as in the following list:

Aaa → Aai; Aae → Aao; Aea → Aei; Aee → Aeo; Aia → Aii; Aie → Aio; Aoe → Aoo; Aoa → Aoi.

By contraposition, we have the following subalternations between Oc propositions alone:

Oai → Oaa; Oao → Oae; Oei → Oea; Oeo → Oee; Oii → Oia; Oio → Oie; Ooo → Ooe; Ooi → Ooa.

The subalternations between **Ec** propositions alone are the following:

Eao → Eae; Eai → Eaa; Eao → Eae; Eeo → Eee; Eei → Eea; Eia → Eaa; Eie → Eae; Eii → Eia; Eii → Eai; Eii → Eaa; Eio → Eae; Eio → Eie; Eoa → Eea; Eoe → Eee; Eoi → Eea Eoi → Eoa; Eoo → Eee; Eoo → Eeo; Eoo → Eoe.

By contraposition, we get the following subalternations between their contradictory Ic propositions alone as in the following:

Iaa → Iai; Iaa → Iia; Iaa → Iii; Iai → Iii; Iae → Iao; Iae → Iie; Iae → Iio; Iee → Ieo; Iee → Ioe; Iee → Ioo; Iea → Ioa; Iea → Iei; Iea → Ioi; Ieo → Ioo; Iea → Ioi; Iia → Iii; Iie → Iio; Ioe → Ioo; Ioa → Ioi.

However, the remaining **Ic** propositions are independent as, for instance, the following: Iaa/Iee, Iae/Iea, Iio/Ioi, etc.

The subalternations between **Oc** propositions alone are the following:

Oai → Oaa; Oao → Oae; Oei → Oea; Oeo → Oee; Oii → Oia; Oio → Oie; Ooo → Ooe; Ooi → Ooa.

However, the following pairs of **Ac** propositions are not related by subalternation; rather they are all independent:

Aaa/Aia; Aaa/Aii; Aaa/Aea; Aaa/Aee; Aaa/Aei; Aaa/Aeo; Aaa/Aoa; Aaa/Aoi; Aaa/Aoe; Aaa/Aoo; Aai/Aea; Aai/Aee; Aai/Aao; Aai/Aea; Aai/Aei; Aai/Aeo; Aai/Aio; Aai/Aoa; Aai/Aoe; Aai/Aoi; Aai/Aoo; Aae/Aio; Aae/Aea; Aae/Aei; Aae/Aoi; Aae/Aie; Aae/Aoa; Aao/Aea; Aao/Aei; Aao/Aoi; Aao/Aoa; Aao/Aie; Aao/Aio; Aia/Aee; Aia/Aeo; Aia/Aoe; Aia/Aoo; Aia/Aea; Aia/Aei; Aia/Aoi; Aia/Aoa; Aii/Aee; Aii/Aoe; Aii/Aoo; Aii/Aeo; Aii/Aao; Aii/Aio; Aii/Aea; Aii/Aei; Aie/Aea; Aie/Aei; Aie/Aoi; Aie/Aoa; Aio/Aea; Aio/Aei; Aio/Aoi; Aio/Aoa;

Aee/Aoe; Aee/Aoo; Aee/Aae; Aee/Aao; Aee/Aie; Aee/Aio; Aeo/Aoe; Aeo/Aoo; Aeo/Aae; Aeo/Aao; Aeo/Aie; Aeo/Aio; Aeo/Aoi; Aoe/Aae; Aoe/Aao; Aoe/Aie; Aoe/Aio; Aoo/Aae; Aoo/Aao; Aoo/Aie; Aoo/Aio; Aoo/Aei; Aoo/Aoi; Aea/Aoi; Aea/Aoa; Aei/Aeo; Aei/Aoa.

The propositions in these pairs can be either both true or both false, or one of them is true while the other one is false. Take, for instance, the first pair of propositions, namely, Aaa/Aia. When formalized, this pair is expressed as follows:

$$(\exists s)\, A_1 s \wedge (\forall s)\, (A_1 s \rightarrow A_2 s) \,/\, (\exists s)\, I_1 s \wedge (\forall s)\, (I_1 s \rightarrow I_2 s)$$

If we consider two situations, we get the following formulas:

$$\{(A_1 s_1 \vee A_1 s_2) \wedge [(A_1 s_1 \rightarrow A_2 s_1) \wedge (A_1 s_2 \rightarrow A_2 s_2)]\} \,/\, \{(I_1 s_1 \vee I_1 s_2) \wedge [(I_1 s_1 \rightarrow A_2 s_1) \wedge (I_1 s_2 \rightarrow A_2 s_2)]\}.$$

The first line of the table shows that there is no subalternation between the first proposition and the second one, since there is a case of falsity:

$$\{(A_1 s_1 \vee A_1 s_2) \wedge [(A_1 s_1 \rightarrow A_2 s_1) \wedge (A_1 s_2 \rightarrow A_2 s_2)]\} \supset \{(I_1 s_1 \vee I_1 s_2) \wedge [(I_1 s_1 \rightarrow A_2 s_1) \wedge (I_1 s_2 \rightarrow A_2 s_2)]\}.$$

1 1 0 **1** 1 1 1 1 0 1 0**0** 1 1 1 **0** 1 1 1 0 1 0 0

But the two propositions can also be true together, when all their elements are true, and they can be false together when '$A_1 s_1 \vee A_1 s_2$' and '$I_1 s_1 \vee I_1 s_2$' are both false, and the first one can be false while the second one is true when '$A_1 s_1 \vee A_1 s_2$' is false while '$I_1 s_1 \vee I_1 s_2$' is true and both '$A_2 s_1$' and '$A_2 s_2$' are true.

The same can be said about all the pairs above, where the first proposition does not imply the second one despite what one might think.

Consequently their contradictories give rise to the same number of independent pairs.

6 The Octagons of Oppositions with these Propositions

Given that all 64 propositions are distinct from each other, we can first construct 8 octagons[5] by combining 2 **A**c propositions with 2 **E**c propositions, 2 **I**c propositions and 2 **O**c propositions, for each octagon.

We will thus have at first sight the simplest kind of octagons, which we will call 'octagons of kind 1', namely, the ones containing Ac, Ec, Ic and Oc propositions. These are the following eight octagons:

[5] In modal logic, the analysis was made in terms of hexagons rather, since the bilateral possible and its negation have been added to the usual vertices of the modal square (see [3]).

1. **Ac** (1+2); **Ec** (1+2); **Ic** (1+2); **Oc** (1+2)
2. **Ac** (3+4); **Ec** (3+4); **Ic** (3+4); **Oc** (3+4)
3. **Ac** (5+6); **Ec** (5+6); **Ic** (5+6); **Oc** (5+6)
4. **Ac** (7+8); **Ec** (7+8); **Ic** (7+8); **Oc** (7+8)
5. **Ac** (9+10); **Ec** (9+10); **Ic** (9+10); **Oc** (9+10)
6. **Ac** (11+12); **Ec** (11+16); **Ic** (11+12); **Oc** (11+12)
7. **Ac** (13 + 14); **Ec** (12 + 13); **Ic** (13+15); **Oc** (12+13)
8. **Ac** (15+16); **Ec** (15 +14); **Ic** (14+16); **Oc** (15 +14)

These will be called the first kind of octagons. We will see in the sequel that they are comparable to the see [10] for Buridan's modal octagon.

Then, we can construct a second kind of octagons which contain some **Ac** propositions, their **Ec** subalterns and the contradictories of both. These are the following (of kind 2):

1. **Ac** (1 + 2); **Ec** (10+9); **Ic** (11+9); **Oc** (1+2)
2. **Ac** (3+4); **Ec** (16+11); **Ic** (12+10); **Oc** (3+4)
3. **Ac** (5+6); **Ec** (13+12); **Ic** (15+13); **Oc** (5+6)
4. **Ac** (7+8); **Ec** (15+14); **Ic** (16+14); **Oc** (7+8)
5. **Ac** (9+10); **Ec** (2+1); **Ic** (2+1); **Oc** (9+10)
6. **Ac** (11+12); **Ec** (4+3); **Ic** (4+3); **Oc** (11+12)
7. **Ac** (13+14); **Ec** (6+5); **Ic** (6+5); **Oc** (13+14)
8. **Ac** (15+16); **Ec** (7+8); **Ic** (7+8); **Oc** (15+16)

Likewise, we can construct octagons with **Ic** propositions, their **Oc** subalterns and the contradictories of both. We can also construct a third kind of octagons with four **Ac** props (2 **Ac** props + their **Ac** subalterns) and their **Oc** contradictories. These are the following (of kind 3):

1. **Ac** (1 + 2 + 3 + 4); **Oc** (1 + 2 + 3 + 4)
2. **Ac** (5 + 6 + 7 + 8); **Oc** (5 + 6 + 7 + 8)
3. **Ac** (9 + 10 + 11 + 12); **Oc** (9 + 10 + 11 + 12)
4. **Ac** (13 + 14 + 15 + 16); **Oc** (13 + 14 + 15 + 16)

We can also have two **Ac** props, their two **Ac** contraries, plus their contradictories. These are the following (of kind 4):

1. **Ac** (1 + 3 + 9 + 11) + their **Oc** contradictories
2. **Ac** (5 + 7 + 13 + 16) + their **Oc** contradictories

The **Ic** propositions (2 **Ic** props + their **Ic** subalterns) and their **Ec** contradictories give rise to the following octagons (of kind 5):

1. **Ic** (1 + 2 + 3 + 4); **Ec** (1 + 2 + 3 + 4)
2. **Ic** (5 + 6 + 7 + 8); **Ec** (5 + 6 + 7 + 8)
3. **Ic** (9 + 10 + 11 + 12); **Ec** (9 + 10 + 11 + 16)
4. **Ic** (13 + 14 + 15 + 16); **Ec** (12 + 13 + 14 + 15).

In kind 1, we can have the following examples of octagons (Fig. 1):

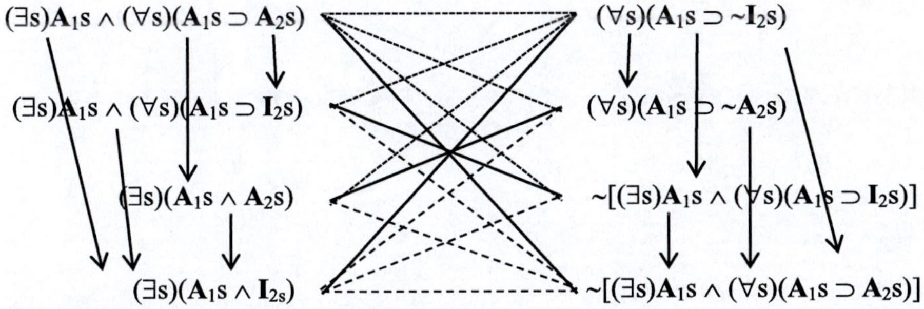

Fig. 1 Octagon1

Octagons **A**c, **E**c, **I**c, **O**c (2 of each).
Octagons **A**c, **E**c, **I**c, **O**c (2 of each) (Fig. 2).
In kind 2, we have the following examples (Figs. 3 and 4):

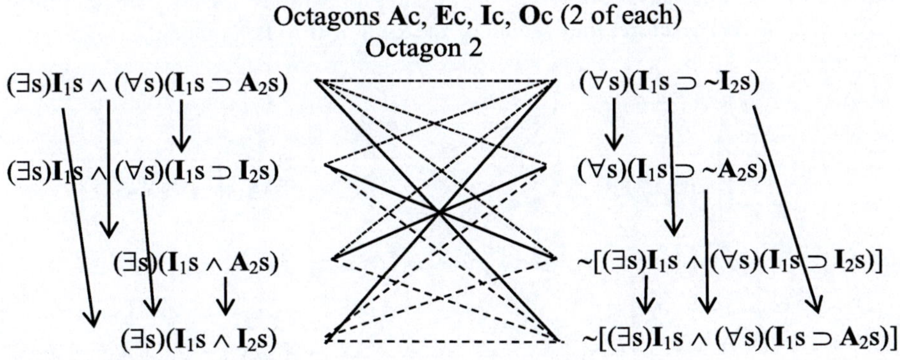

Fig. 2 Octagon 2

Octagons **Ac** + their **Ec** subalterns + contradictories

Octagons **Ac** + their **Ec** subalterns + their contradictories

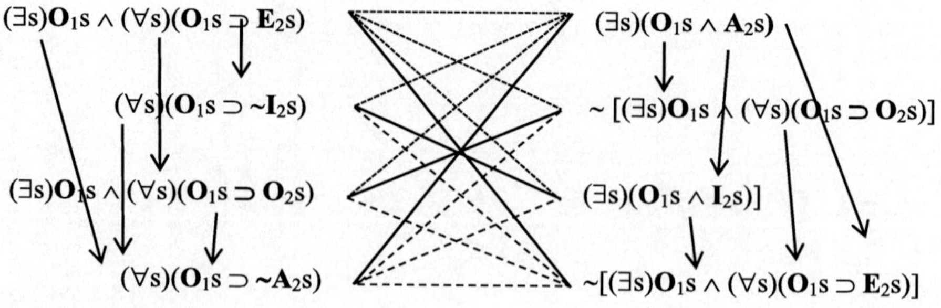

Fig. 3 Octagon 3

In that same kind 2 of octagons, we can have octagons containing **Ic** propositions, their **Oc** subalterns and their contradictories. For instance, the following octagon meets these conditions (Fig. 5):

Octagons **Ic** + **Oc** subalterns + their contradictories

These octagons have the same structure as the ones above in terms of the logical relations they contain and the number of independent propositions, since they all contain five subalternations in each side but the propositions of Line 2 do not imply those of Line 3. Therefore, they are all of the kind of Buridan's modal octagon.

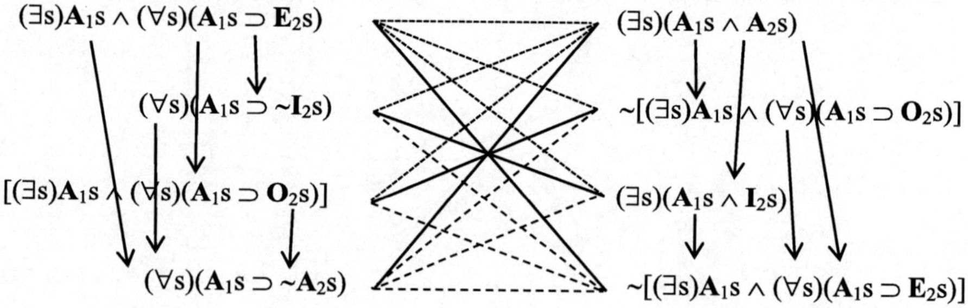

Fig. 4 Octagon 4

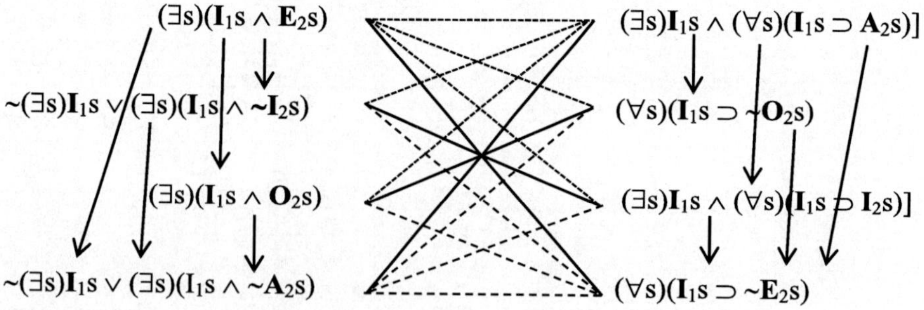

Fig. 5 Octagon 5

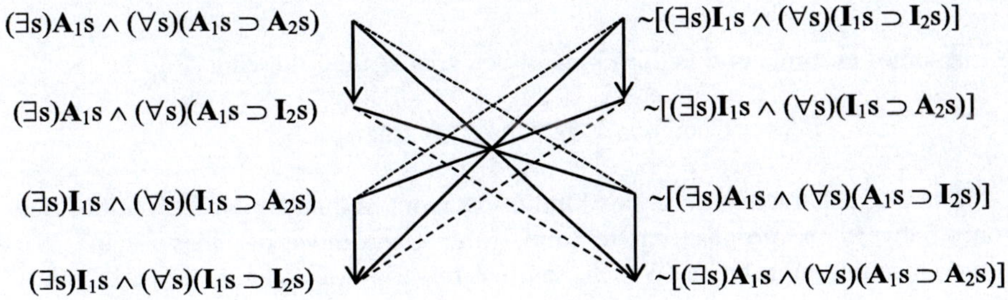

Fig. 6 Octagon 6

The octagons of kind 3 contain four **Ac** and four Oc. They are exemplified by the following (Fig. 6):

Octagons 4 **Ac** + their contradictories

These octagons are very different from the preceding ones, since they contain only two squares relating the top propositions to the bottom ones by the contradictions. It is a really new figure which we don't find in any of the medieval or the modern writings. The independent propositions in it are more numerous than in a usual Buridan's kind of octagons. Apart from the contradictions, whose number is the same as in the above octagons, there are exactly four subalternations, two contrarieties and two subcontrarieties.

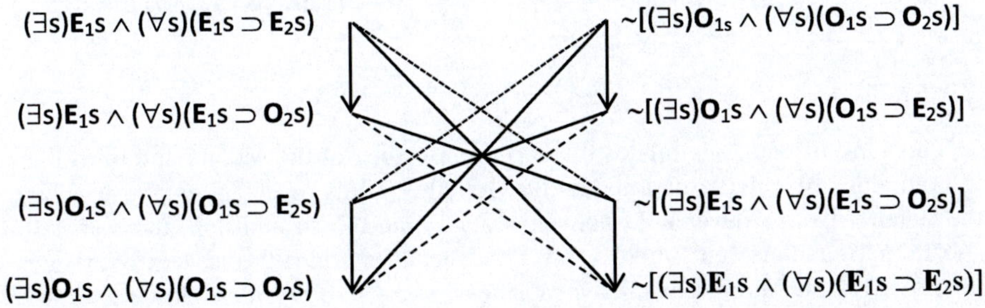

Fig. 7 Octagon 7

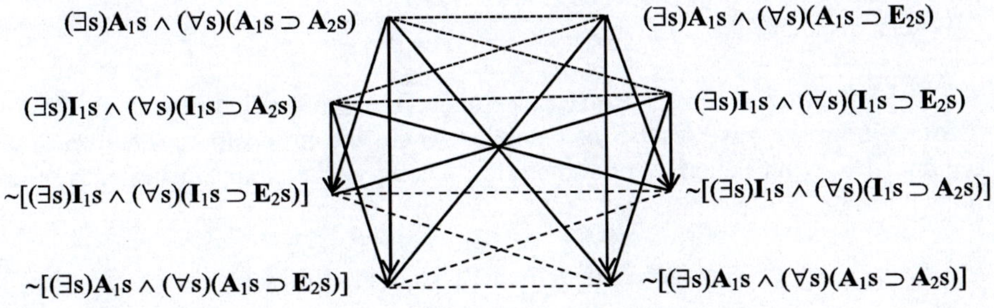

Fig. 8 Octagon 8

Another example of this third kind of octagons is the following (Fig. 7):

Octagons 4 **Ac** + their contradictories

The main characteristic of this kind of octagons is that the subalternations hold only between the propositions of Line 1 and 2 and those of Lines 3 and 4. No other subalternation holds. We can say that this kind of octagons is the 'lightest' kind, since it contains much more independent relations and much less Aristotelian relations than the other kinds of octagons.

On the other hand, we can also have two Ac propositions and their two contrary Ac propositions plus their contradictories. The octagons constructed with these propositions are of kind 4. The following are examples of this kind 4 of octagons (Fig. 8):

Octagons 2 **Ac** + 2**Ac** (contraries) + their contradictories

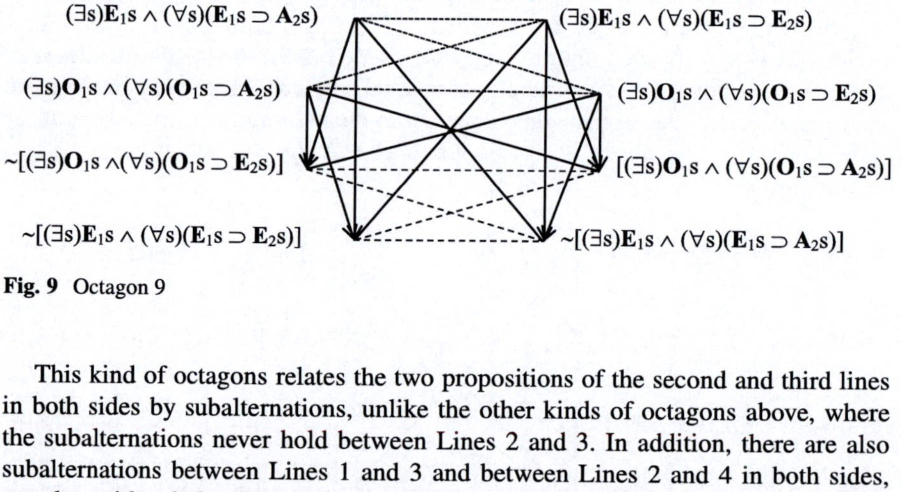

Fig. 9 Octagon 9

This kind of octagons relates the two propositions of the second and third lines in both sides by subalternations, unlike the other kinds of octagons above, where the subalternations never hold between Lines 2 and 3. In addition, there are also subalternations between Lines 1 and 3 and between Lines 2 and 4 in both sides, together with subalternations between Lines 1 and 4 in both sides. However, unlike the octagons of kinds 1 and 2, this kind of octagons does not contain subalternations between Lines 1 and 2 and between Lines 3 and 4.

We can also have the following octagons (Fig. 9):

Octagons 2**Ac** + 2**Ac** (contraries) + their contradictories

In addition, we have kind 5 of octagons where the propositions involved are Ic propositions and their Ec contradictories. These octagons are illustrated by the following (Fig. 10):

Octagons 4**Ic** + their **Ec** contradictories

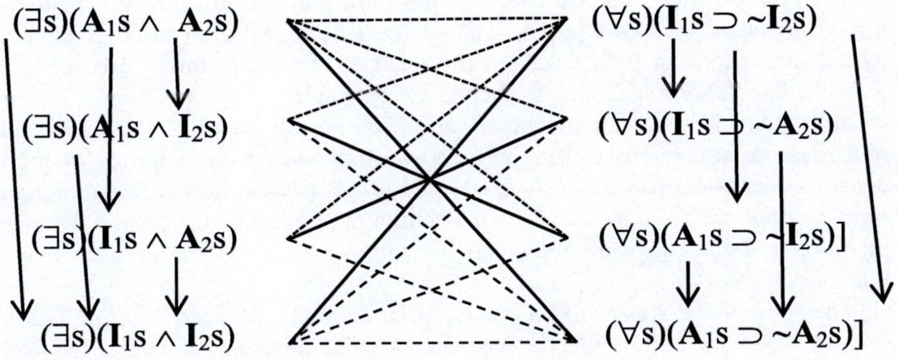

Fig. 10 Octagon 10

This kind too is like Buridan's modal octagon, for the propositions of Lines 2 and 3 are independent while the propositions of Lines 1 and 2, 1 and 3, 1 and 4, 2 and 4 and 3 and 4 are all related by subalternations in both sides.

Another example of such octagons is the following (Fig. 11):

Octagons 4 **Ic** + their contradictories

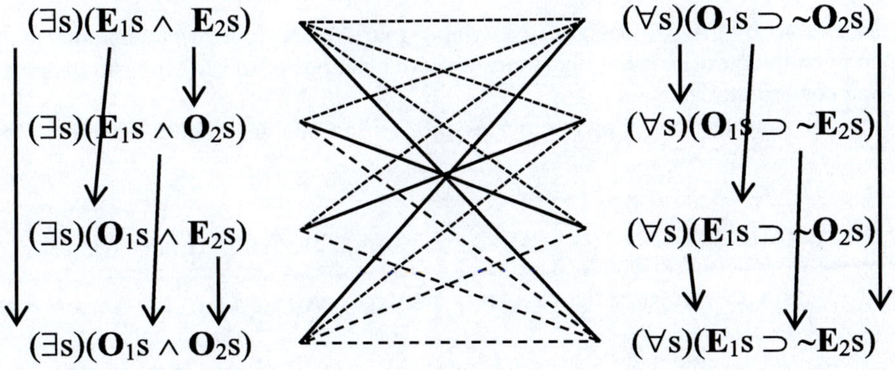

Fig. 11 Octagon 11

So we can say that most octagons are of Buridan's kind (See [10]), but some of them, like octagons 8 and 9 of kind 4 above, are of Johnson-Hacker's kind (see [7]), while some others, like the octagons 6 and 7 of kind 3 above, are different from all kinds of known octagons.

In Buridan's kind of octagons, the propositions in the second and the third lines are independent. But in Johnson-Hacker's kind of octagons, both the first and the second lines' propositions and the third and fourth lines' propositions are independent, while the propositions of the second and third lines are related by subalternations in both sides. In the latter kind of octagons, there are less Aristotelian relations and more independent propositions than in the former.

In the third kind of octagons, there are much more independent relations and much less Aristotelian ones than in the two other kinds. This kind of octagons contains only two squares which relate the top left propositions with the bottom right ones and the top right ones with the bottom left propositions. As far as I know, no octagon of that kind is evoked in the literature. So it seems to be an entirely new kind of octagons.

Note, however, that none of these octagons is drawn by Avicenna himself. I have drawn them on the basis of what Avicenna says about the relations of the square and the propositions themselves, but Avicenna never combined these propositions in his text to construct octagons or any other kind of figures. Nevertheless, these figures complement Avicenna's analysis because they show very clearly all the relations that hold between all kinds of propositions. So drawing them does not in any sense distort the text. Rather it makes it more precise and much clearer.

Now, we can also combine these octagons two by two and construct figures of 16 vertices, to see what relations this figure can contain. This will be done in the next section.

7 Combining the Octagons Two by Two

Let us start by the two octagons containing four **Ac** and four **Ic** propositions and their contradictories. The propositions involved can be Ac and Ic propositions plus their contradictories.

Let us consider the following four **Ac** propositions and their contradictories:

–

– $(\exists s)\,\mathbf{A}_1 s \wedge (\forall s)\,(\mathbf{A}_1 s \supset \mathbf{A}_2 s)$ [= **AAA**] / $\sim [(\exists s)\,\mathbf{A}_1 s \wedge (\forall s)\,(\mathbf{A}_1 s \supset \mathbf{A}_2 s)]$ [=∼ (**AAA**)]

– $(\exists s)\,\mathbf{A}_1 s \wedge (\forall s)\,(\mathbf{A}_1 s \supset \mathbf{I}_2 s)$ [= **AAI**] / $\sim [(\exists s)\,\mathbf{A}_1 s \wedge (\forall s)\,(\mathbf{A}_1 s \supset \mathbf{I}_2 s)]$ [=∼ (**AAI**)]

– $(\exists s)\,\mathbf{I}_1 s \wedge (\forall s)\,(\mathbf{I}_1 s \supset \mathbf{A}_2 s)$ [= **IIA**] / $\sim [(\exists s)\,\mathbf{Il} s \wedge (\forall s)\,(\mathbf{I}_1 s \supset \mathbf{A}_2 s)]$ [=∼ (**IIA**)]

$(\exists s)\,\mathbf{I}_1 s \wedge (\forall s)\,(\mathbf{I}_1 s \supset \mathbf{I}_2 s)$ [= **III**] / $\sim [(\exists s)\,\mathbf{Il} s \wedge (\forall s)\,(\mathbf{I}_1 s \supset \mathbf{I}_2 s)]$ [=∼ (**III**)]

And the following four **Ic** propositions and their contradictories:

– $(\exists s)\,(\mathbf{A}_1 s \wedge \mathbf{A}_2 s)\ [= \mathbf{AA}]\,/\sim (\exists s)\,(\mathbf{A}_1 s \wedge \mathbf{A}_2 s)\ [= \mathbf{A} \sim \mathbf{A}]$

– $(\exists s)\,(\mathbf{A}_1 s \wedge \mathbf{I}_2 s)\ [= \mathbf{AI}]\,/\sim (\exists s)\,(\mathbf{A}_1 s \wedge \mathbf{I}_2 s)\ [= \mathbf{A} \sim \mathbf{I}]$

– $(\exists s)\,(\mathbf{I}_1 s \wedge \mathbf{A}_2 s)\ [= \mathbf{IA}]\,/\sim (\exists s)\,(\mathbf{I}_1 s \wedge \mathbf{A}_2 s)\ [= \mathbf{I} \sim \mathbf{A}]$

– $(\exists s)\,(\mathbf{I}_1 s \wedge \mathbf{I}_2 s)\ [= \mathbf{II}]\,/\sim (\exists s)\,(\mathbf{I}_1 s \wedge \mathbf{I}_2 s)\ [= \mathbf{I} \sim \mathbf{I}]$

With these propositions, which have given rise to 2 octagons of Buridan's kind, we can construct a complex figure grouping these 2 octagons and containing 16 vertices. This figure is the following (Fig. 12):

In this figure, as we can see, the biggest amount of Aristotelian relations is found at the bottom and the top, while the middle of the figure is relatively empty. This is due to the fact that the octagons grouped in this figure are of Buridan's kind, since in these octagons, the relations between the second and the third lines are independence relations.

Now what happens if we choose other kinds of octagons and group them?

Let us, for instance, consider these eight **Ac** propositions and their contradictories.

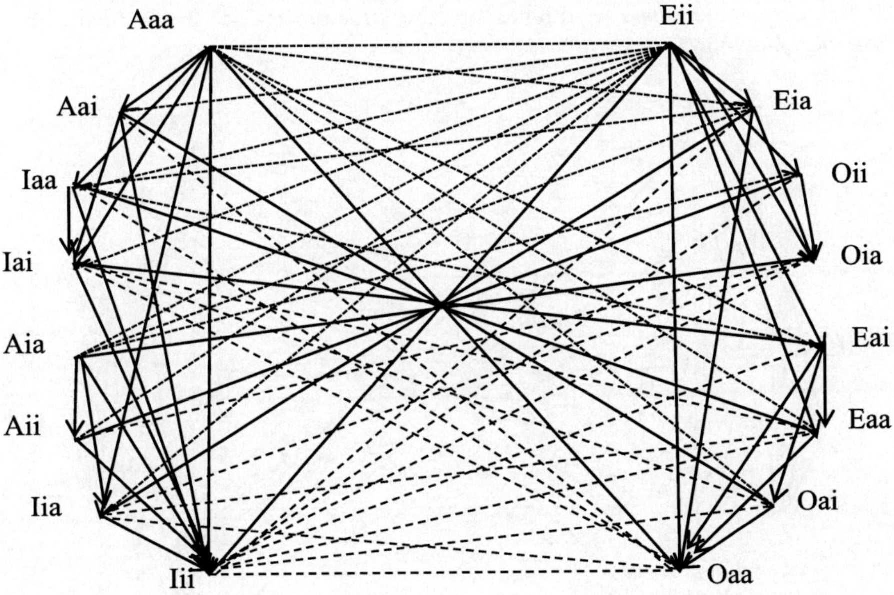

Fig. 12 The first figure with 16 vertices (two octagons)

The propositions and their contradictories are the following:

- $(\exists s)\, \mathbf{A}_1 s \wedge (\forall s)\, (\mathbf{A}_1 s \rightarrow \mathbf{A}_2 s)$ [= **AAA**] / $\sim [(\exists s)\, \mathbf{A}_1 s \wedge (\forall s)\, (\mathbf{A}_1 s \rightarrow \mathbf{A}_2 s)]$ [=~ (**AAA**)]
- $(\exists s)\, \mathbf{A}_1 s \wedge (\forall s)\, (\mathbf{A}_1 s \supset \mathbf{I}_2 s)$ [= **AAI**] / $\sim [(\exists s)\, \mathbf{A}_1 s \wedge (\forall s)\, (\mathbf{A}_1 s \rightarrow \mathbf{I}_2 s)]$ [=~ (**AAI**)]
- $(\exists s)\, \{\mathbf{I}\}_1 s \wedge (\forall s)\, (\mathbf{I}_1 s \supset \mathbf{A}_2 s)$ [= **IIA**] / $\sim [(\exists s)\, \mathbf{I}_1 s \wedge (\forall s)\, (\mathbf{I}_1 s \supset \mathbf{A}_2 s)]$ [=~ (**IIA**)]
- $(\exists s)\, \mathbf{I}_1 s \wedge (\forall s)\, (\mathbf{I}_1 s \supset \mathbf{I}_2 s)$ [= **III**] / $\sim [(\exists s)\, \mathbf{I}_1 s \wedge (\forall s)\, (\mathbf{I}_1 s \supset \mathbf{I}_2 s)]$ [=~ (**III**)]
- $(\exists s)\, \mathbf{A}_1 s \wedge (\forall s)\, (\mathbf{A}_1 s \supset \mathbf{E}_2 s)$ [= **AAE**] / $\sim [(\exists s)\, \mathbf{A}_1 s \wedge (\forall s)\, (\mathbf{A}_1 s \supset \mathbf{E}_2 s)]$ [=~ (**AAE**)]
- $(\exists s)\, \mathbf{A}_1 s \wedge (\forall s)\, (\mathbf{A}_1 s \supset \mathbf{O}_2 s)$ [= **AAO**] / $\sim [(\exists s)\, \mathbf{A}_1 s \wedge (\forall s)\, (\mathbf{A}_1 s \supset \mathbf{O}_2 s)]$ [=~ (**AAO**)]
- $(\exists s)\, \mathbf{I}_1 s \wedge (\forall s)\, (\mathbf{I}_1 s \supset \mathbf{E}_2 s)$ [= **IIE**] / $\sim [(\exists s)\, \mathbf{I}_1 s \wedge (\forall s)\, (\mathbf{I}_1 s \supset \mathbf{E}_2 s)]$ [=~ (**IIE**)]
- $(\exists s)\, \mathbf{I}_1 s \wedge (\forall s)\, (\mathbf{I}_1 s \supset \mathbf{O}_2 s)$ [= **IIO**] / $\sim [(\exists s)\, \mathbf{I}_1 s \wedge (\forall s)\, (\mathbf{I}_1 s \supset \mathbf{O}_2 s)]$ [=~ (**IIO**)]

The first four **Ac** propositions are contrary to the last ones. Consequently their contradictories are subcontrary. But what are the relations between the other propositions in the figure?

Let us construct it and see if it has the same structure as Fig. 12. The figure looks like the following (Fig. 13):

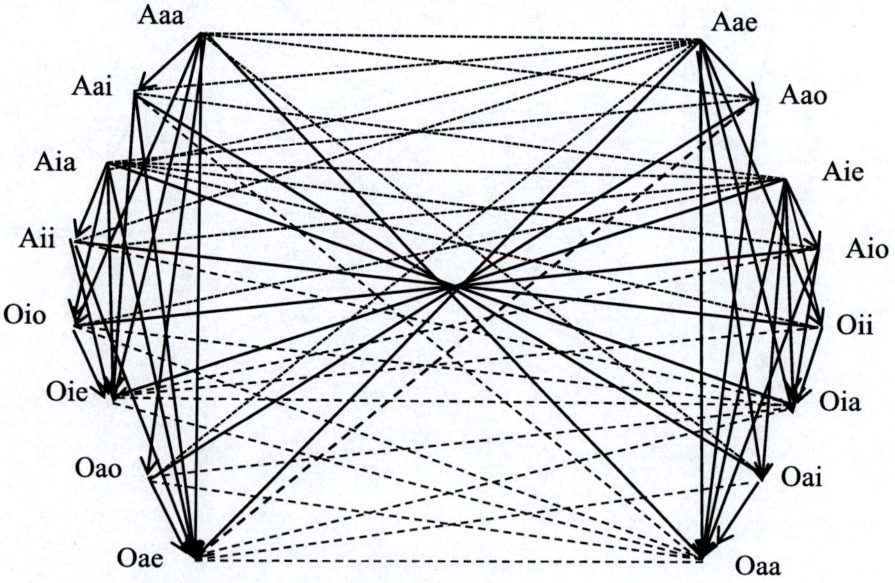

Fig. 13 The second figure with 16 vertices (two octagons)

This figure is different from the above one, since the biggest amount of Aristotelian relations is in the middle, while the top and the bottom of the figure contain less Aristotelian relations. This is so because of the kind of propositions involved, which are those which gave rise to both the third and new kind of octagons and to Johnson-Hacker's kind of octagons, whose structures are different from that of Buridan's modal octagon.

There are certainly much more things to say about the nature of the two figures and their differences, but this would extend the limits of this paper, which focuses on Avicenna's propositions and their relations, without entering into more details with regard to the kind of geometrical figures that they lead to.

As a matter of fact, these octagons and figures containing 16 vertices are just a sample of all the possible combinations that the propositions above can lead to. Absolutely speaking, we could have much more kinds of figures which would have 12 (one octagon plus one square), 14 (one octagon plus one hexagon), 18 (three hexagons), 24 (three octagons) and so on. But the analysis of such figures is not our aim here. It could be the subject of another paper, specifically dedicated to the geometry of oppositions.

As we said above, Avicenna himself never draw any kind of figure although he provided the definitions of all the relations and showed in this part of his text some awareness of the presence of Aristotelian relations that hold between various kinds of propositions and are not limited to the usual relations between **Ac**, **Ec**, **Ic** and **Oc** propositions. As we saw above, he does hold subalternations between some **Ac** and **Ec** propositions, contrarieties between some **Ac** propositions, subcontrarieties between some **Oc** propositions and so on. What I did in this paper is just systematize all these relations and group them inside some geometrical figures. But more can be done if we just focus on the geometrical figures themselves.

8 Conclusion

The oppositions between the quantified hypothetical conditionals give rise to several octagons and several other figures containing 16 vertices or more. The octagons are of various kinds: some are like Buridan's modal octagon, while others are more like Johnson-Hacker's octagon. But there is a third and new kind of octagons, which contains much more independent relations and much less Aristotelian ones and has never been evoked in the literature, as far as I know. This kind of octagons seems thus radically *new* and *original*. It is an additional kind of octagons, which enriches the analysis of this specific figure.

As to the figures containing 16 vertices, they are different with regard to their relations, depending on the propositions they contain. Some of them relate the propositions of the middle, while other ones relate rather the top and bottom ones. All these figures and other possible ones deserve a more detailed analysis which could be made in another paper specifically dedicated to the theory of oppositions in general. One can combine the octagons with squares or with hexagons, which would give rise to figures of different sizes. This is interesting for the theory of oppositions in general, but we cannot say that it has been explored at length by Avicenna himself, who did not draw any figure. This is why I did not enter into more details with regard to these different kinds of figures in the analysis above and just provided a minimal account of these possibilities. But one can see Avicenna's propositions as a historical basis for a great number of geometrical figures of different sizes.

Acknowledgments I would like to thank the anonymous referees for their valuable and helpful comments and suggestions. I am also grateful to the organizers of the workshop 'Logical Geometry'(Unilog 2018) and all the participants for their questions and remarks.

Bibliography

1. Avicenna. *al-Shifā', al-Mantiq 3*: *al-'Ibāra*, ed M. El Khodeiri, rev and intr. by I. Madkour, Cairo,1970.
2. Avicenna. *al-Shifā', al-Mantiq* 4: *al-Qiyās*, ed. S. Zayed, rev. and intr. by I. Madkour, Cairo, 1964.
3. Chatti, Saloua. "Avicenna on possibility and necessity", *History and Philosophy of Logic*, 35:4, 332-353, 2014.
4. Chatti, Saloua. "Avicenna (Ibn Sīnā), Logic", *Encyclopedia of Logic*, in *Internet Encyclopedia of Philosophy*, 2016.
5. Chatti, Saloua. *Arabic Logic from al-Fārābī to Averroes*, Birkhaüser, Springer, 2019.
6. Chatti, Saloua, "On some Ambiguities in Avicenna's Analysis of the Hypothetical Quantified Propositions", forthcoming in *Arabic Sciences and Philosophy*, March 2022.
7. Hacker, E. "The Octagon of Opposition", *Notre Dame Journal of Formal Logic*, volume XVI, n° 3, July 1975.
8. Hodges, Wilfrid, *Mathematical Background to the Logic of Ibn Sīnā*, forthcoming.
9. Łukasiewicz, Jan, 1972, "Contribution à l'histoire de la logique des propositions", translated in French and published in Largeault, Jean (ed) *Logique mathématique, textes*, Editions Armand Colin, Paris, pp. 9-25.
10. Read, Stephen. 2012, "The medieval theory of consequence", *Synthese*, **187**, 899–912.
11. Rescher, Nicholas, 1963, "Avicenna on the Logic of "Conditional" Propositions", *Notre Dame Journal of Formal Logic*, Volume IV, Number 1, pp. 48-58.
12. Strobino, Riccardo, "Ibn Sīnā's Logic", *The Stanford Encyclopedia of Philosophy* in Edward N. Zalta (ed), https://plato.stanford.edu/archives/fall2018/entries/ibn-sina-logic, 2018.

Incommensurability and Inapplicability of the Squares of Opposition

Sumanta Sarathi Sharma

Abstract The relations expressed through the "Traditional Square" stand partially refuted. This refutation may be attributed to whether our universe of discourse does or does not contain referents (truth conditions). In this paper, seven worlds concerning these truth conditions are proposed. These worlds are considered as separate paradigms. It is argued that the squares so formed are incommensurable. In addition to this, the paper also develops the notion of inapplicability (with respect to the relation of "contradiction") in the squares.

Keywords Square of opposition · Existential import · Kuhn · Paradigm · Incommensurability · Thought experiment

Mathematics Subject Classification (2010) Primary 03A10; Secondary 03A05

1 Background

The traditional square of opposition is a brilliant tool to understand relation(s) between propositions. Although the relation(s) expressed in the square are debated and contested, there are two points, which are settled now. First, Aristotle gave us (assertoric) propositions, worked extensively on their interrelations but has not represented it diagrammatically [12], in the form of a square in *De Interpretatione* (or anywhere else) [1, 21, 33]. Second, the oldest depiction of the traditional square of opposition is found in Apuleius of Maduara's commentary (called *Peri Hermeneias*) on *De Interpretatione* [18, 19]. Buridan (among others, in the medieval period) has extensively used this tool [32] to teach Aristotelian propositions and his syllogistic.

S. S. Sharma (✉)
School of Philosophy and Culture, Faculty of Humanities and Social Sciences, Shri Mata Vaishno Devi University, Jammu and Kashmir, India
e-mail: ss.sharma@smvdu.ac.in

J.-Y. Beziau, I. Vandoulakis (eds.), *The Exoteric Square of Opposition*, Studies in Universal Logic, https://doi.org/10.1007/978-3-030-90823-2_4

In the nineteenth century, the developments in logic brought down traditional square from its stellar reputation. Some accepted relations between propositions (like contrary, sub-contrary, and subaltern) were rejected (after Boolean interpretation of propositions [9]) though the "square" (due to the relation of "contradiction") was not entirely overthrown. The revised square, so formed, is formidable and the relation of "contradiction" remains as the common link between both the squares. It has almost become a commonplace to conceive the revised square as a chastened traditional square. In this paper, I challenge this commonplace conception. In other words, I argue that the traditional and revised squares belong to separate paradigms, rather the former being supplanted by the latter. Furthermore, the squares are incommensurable, and the relation of 'contradiction' is found inapplicable in some instances.

This article is divided as follows: In the first part, we re-examine the debates associated with the transformation of traditional square into the revised square. This takes into account the relations expressed in both the squares along with the formulation of propositions and its meaning. In the second section, after providing a summary of Kuhnian conceptions of "normal science," "paradigms," and "incommensurability," we establish the traditional and revised squares as separate paradigms using the notion of possible worlds. In the third section, we show that the squares are incommensurable. The concluding section of the paper devises some thought experiments to show that the relation of "contradiction" does not hold in certain paradigms.

2 The Controversy of Squares

The controversies related to the squares are manifold. In this section, I briefly discuss a few of the problems to understand the difficulty associated with it.

2.1 *Transformation of Traditional to Revised Square*

An agreed depiction of the Traditional Square of Opposition as found in Parsons' [26] is as follows (Fig. 1):

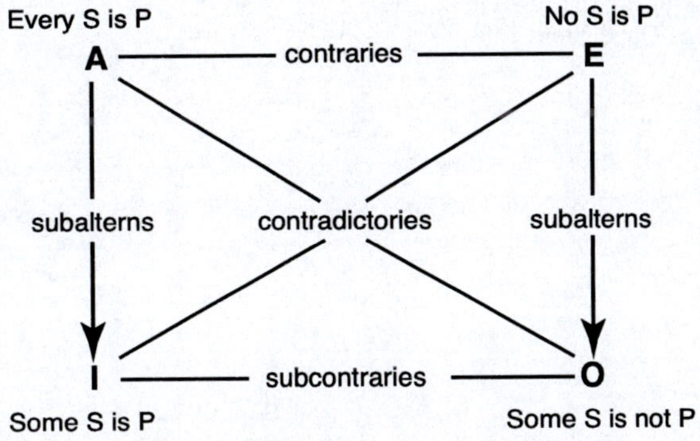

Fig. 1 The Traditional Square of Opposition

In the Traditional Square, **A** and **E** are contraries, i.e., they cannot both be true, though they can be false together. **I** and **O** are sub-contraries, which mean that they cannot be false together, although they can be true together. **A** and **O** along with **E** and **I** are contradictories; thus, they can neither be true nor be false together. Whereas **A** and **I** together with **E** and **O** form subaltern pairs, and hence, if **A/E** is true, then **I/O** is true and if **I/O** is false, then **A/E** is false, respectively, but not vice versa. It is often argued that these relations stood intuitively correct before the "orthodox criticisms" [37] of modern logic came to fore.

In modern logic, statements are often replaced by their symbolic interpretation using insights from Peano and Frege [29]. These symbolic interpretations and representations had helped logic enormously to grow as a system of reasoning. Symbolically, the Aristotelian propositions are represented as follows:

A, i.e., "Every S is P" is represented as $(\forall x)[Sx \rightarrow Px]$
E, i.e., "No S is P" is represented as $(\forall x)[Sx \rightarrow \neg Px]$
I, i.e., "Some S is P" is represented as $(\exists x)[Sx \wedge Px]$
O, i.e., "Some S is not P" is represented as $(\exists x)[Sx \wedge \neg Px]$

Now suppose, if the subject class is empty, then Sx will be false. If Sx is false, then **A** and **E** will be true because a false proposition implies any proposition whatsoever. However, their subaltern counterparts, namely **I** and **O** will be false, since a conjunction is false if at least one of its conjuncts is false. Thus, we obtain **A** and **E** as true, whereas **I** and **O** as false. With this, the relations of subalternation, contrariety, and sub-contrariety fall flat.

It is plain from the above that none of the relations (from the traditional square) survived except contradiction. The revised square first formalized in the functional calculus of Frege's *Begriffsschrift* [10] is as under (Fig. 2):

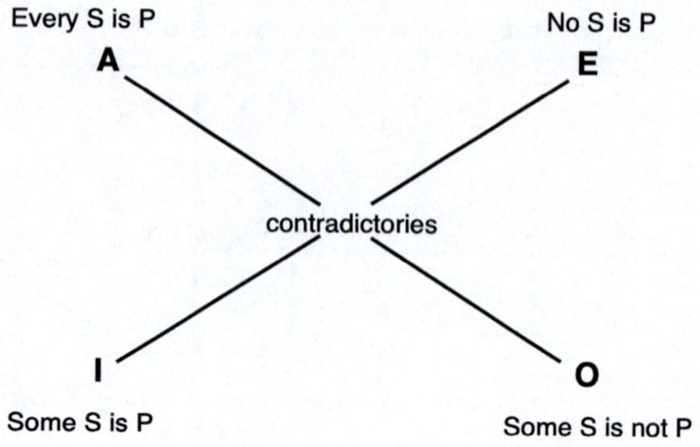

Fig. 2 The Revised Square of Opposition

2.2 *Some Problems*

Parsons [27] mentions the problem as follows. In the Traditional Square, the particular negative proposition **O** is expressed as "Some S is not P." This is a diachronic and lingual error, attributed to Boëthius, who used "Not every S is P" and "Some S is not P" synonymously while translating Greek to Latin. If we accept the above equivalence, then "Every S is P" which can be vacuously true leads to "Some S is not P" true as well, after applying subalternation—contradiction—subalternation, respectively.

In other words, if the subject term is empty, then "Every S is P" is vacuously true, but its subaltern "Some S is P" will be false since S is empty. If "Some S is P" is false, its contradictory "No S is P" is going to be true. Again, if "No S is P" is true, then its subaltern "Some S is not P" has to be true. This defies the relation of contradiction.

Parsons offer at least a couple of defense to the above anomaly, which I found important here. First that a universal proposition is vacuously true if its subject term is empty is a natural language nuance [of English language] which is not endorsed by many logicians. Second, a particular negative proposition needs to be symbolized as the conjunction of "Some S is S" and "Some S is not P," which will be false, if there are no Ss [27].

Read while criticizing Łukasiewicz is correct in pointing out that Aristotle commonly (though not invariably) expresses the **O** proposition as "Not every S is P" (or as he usually puts it: "P does not belong to every S"), and he treats "P does not belong to every S" as equivalent to "P does not belong to some S." Moreover, Aristotle places no requirement that the terms be non-empty.Moreover, "Existential commitment goes with quality, not quantity, thus satisfying all the demands of the [Traditional] Square of Opposition" [30].

Recently, Corkum has suggested that the existential import of universal affirmations and the semantic profile of predications with empty terms follow from mereological truth conditions. For example, "Socrates is pale" is true just in case "Socrates" is a part of the mereological sum of pale things [3]. Strawson defended the traditional Square by suggesting that proposition whose subject term is empty is neither true nor false [37]. In his painstaking survey on this problem from George Boole to P.F. Strawson, Wu has opined that the problem [of existential import] lies in the gap between "logic" as pure abstraction and "logic" as a method applied to existence or human experience [40].

There is no unified approach among the logicians about the question of existential import. A discussion between Uchenko and Northrop, who are at loggerheads, represented two powerful intuitions. Uchenko believed that the question of internal consistency is relative to information at hand. Therefore, the notion of correct and incorrect remains system dependent [38]. On the other hand, for Northrop, consistency is a consequence of certain formal principles that cannot vary from system to system [24, 25]. Thus, two separate interpretations cannot be consistent together.

2.3 Summary

The problem of existential import is perennial, and the discussion is endless. In this section, we witnessed that the problems associated with the traditional square started with the symbolic interpretation of propositions. These symbolic interpretations need further justifications or add-ons to reinstate the traditional square [3, 26, 27, 30] or we end up with a cross of opposition [36]. There is a widespread belief that Aristotle's logic is not equipped to deal with empty terms, despite this belief being erroneous and baseless [30]. Given the above state of affairs, it is unlikely that there can ever be a solution to this stalemate. However, there is a way to address this impasse. In what follows, I endeavor to show that the above two systems are not inconsistent, but rather they belong to two separate paradigms.

3 The Paradigms of Squares

Thomas Kuhn is the most celebrated "philosopher of science" and his signature theory of "paradigm" is an influential one. In a general sense, "paradigm" can be understood as the commonly accepted views of a particular discipline or an area at a given time [17]. Kuhn also called it "disciplinary matrix" [16]. In this section, I first briefly introduce the Kuhnian notions of "normal science," "paradigms," and "incommensurability" (along with some other related notions) and then apply the notion of "paradigms" to the squares.

3.1 Kuhnian Notions

According to Kuhn, "[a] "normal science" means research firmly based upon one or more past scientific achievements, achievements that some particular scientific community acknowledges for a time as supplying the foundation for its further practice" [17]. Thus, normal science is an established body of work to do science (or any enquiry). The purpose of normal science is "puzzle-solving." Kuhn uses the term "puzzle" to distinguish it from "problem" as the latter may not have a solution, but the former has.

Furthermore, he declares, "I shall henceforth refer to as 'paradigms', a term that relates closely to normal science. By choosing it, I mean to suggest that some accepted examples of actual scientific practice—examples which include law, theory, application, and instrumentation together—provide models from which spring particular coherent traditions of scientific research" [17]. He has used the notion of paradigm in several ways [20]. However, there are two senses [5]—it can best be understood. The first is the broad way, where a paradigm is understood as a package of ideas and methods, which, when combined, makes up both a view of the world and a way of doing science. Second is the narrow way, where a paradigm is understood as an exemplar of a field of study.

Normal science is often questioned and challenged with the passage of time and new developments, although the fundamental ideas of the paradigm remain intact. With these, scientists endeavor to expand and extend the paradigm theoretically and experimentally. But, there are times, when a "puzzle" resists a solution. This is called as an "anomaly." Every science faces such challenges. If such resistance is enduring and scientists are unable to find solutions, they start losing faith in the paradigm. This is called as a "crisis" situation.

According to Kuhn, the rejection of a paradigm takes place if the following two conditions are satisfied. First, a critical mass of anomalies has arisen. Second, a rival paradigm has appeared [5]. The rival paradigm offers solutions to the problem—the problem becomes a puzzle again and subsequently solved—and the rival paradigm is established as the new normal science. Interestingly, Kuhn points out that, "since new paradigms are born from old ones, they ordinarily incorporate much of the vocabulary and apparatus, both conceptual and manipulative, that the traditional paradigm had previously employed. But they seldom employ these borrowed elements in quite the traditional way [17]." The old "normal science" and the new "normal science" may have some commonalities. But, even if concepts, ideas, and (some) principles in the ousted paradigm seemingly match with that of the new paradigm—the paradigms are "incommensurable." Kuhn puts it beautifully as, "Two men who perceive the same situation differently but nevertheless employ the same vocabulary in its discussion must be using words differently. They speak, that is, from what I have called 'incommensurable' viewpoints [17]." If we undertake a study of the "squares," we find that the "traditional square" was a "normal science" and a "paradigm," which was challenged and rejected by a rival "paradigm" of the "modern square."

Before analyzing the "squares," it is prudent to review Aristotle's original propositions which gave us the traditional square. We know that Aristotle represented universal affirmative, universal negative, particular affirmative, and particular negative as "P belongs to all S," "P belongs to no S," "P belongs to some S," and "P does not belongs to some S," respectively. Moreover, he did consider singular and indefinite propositions [30], which are absent in the modern interpretation and analysis. The rewording of particular negative proposition by Boëthius is an unintended slip, Strawson's defense is a correction in vain and the interpretation of Łukasiewicz is a misapprehension labeled against Aristotle [26, 27, 30, 31].

It is important to note here that the assertoric propositions considered by Aristotle and the modern interpretation of propositions following Boole and others, is unalike. In other words, they are constituents of different paradigms. The application of the principles of one paradigm on another is unfair and incorrect. For instance, both Euler and Venn used circles to test the validity of syllogisms. They have over-lapping circles. However, they work on different principles. Moktefi and Shin points out that "Euler uses circles to divide the space into subdivisions that are assumed to exist and which are topologically related in the same way as the classes they represent do ... In Venn diagrams, however, none of the subdivisions is assumed to exist. Strictly speaking, Venn does not represent the classes at all, but rather compartments, which when marked, tells whether the corresponding class is empty or occupied" [23]. Thus, Euler and Venn diagrams belong to separate paradigms even though they use "closed curves." Similarly, assertoric propositions of Aristotle and the categorical propositions expounded by Boëthius (and taken over by the Booleans) are different from each other.

Moreover, there is a coherent account of [assertoric] syllogistic propositions which satisfies all the relationships in the traditional square of opposition and at the same time allows the inclusion of empty and universal terms [30, 31]. The existential commitments of the traditional logicians as compared to moderns are dissimilar as well. In this context, Keynes has outlined four possible views one can adopt for an analysis of the subject and predicate terms in these propositions [11]. They are as follows:

1. Every proposition implies the existence of both its subject and predicate terms.
2. A proposition implies the existence of those things denoted by its subject term only.
3. No proposition necessarily implies the existence of things denoted either by its subject term or by its predicate term.
4. Universal propositions do not imply the existence of objects denoted by their subject term, but the particular propositions do.

Of the four positions listed above, modern logicians or those with the revised square adopt the fourth view. In contrast, if we endorse either of the views except the fourth one, the resultant is the traditional square. The solution to the problem of existential import is to some extent a matter of convention and we are guided partly by the ordinary usage of language and partly by considerations of logical convenience and suitability [11]. Moreover, the purpose of logic is not to mirror all of the subtleties of natural language [27]. The point here is to understand that unless we consider the traditional and the modern squares as two separate paradigms, which came out of two separate set of beliefs and principles—there is no way out.

3.2 Possible Worlds

Keynes analysis of subject predicate terms is remarkable. On these lines, and further expanding this idea, let us consider possible worlds delimiting certain truth conditions. Suppose a possible world can have only subject true but predicate true or false. Similarly, there can be world, where both subject and predicate are true. Thus, different lexicons give access to different sets of possible worlds, largely but never entirely over-lapping [14]. Following this trail, I propose the following seven possible worlds and their truth conditions:

STPTF-World. S is always True and P can be either True or False
PTSTF-World. P is always True and S can be either True or False
SFPTF-World. S is always False and P can be either True or False
PFSTF-World. P is always False and S can be either True or False
STPT-World. S is always True and P is always True
SFPF-World. S is always False and P is always False
STFPTF-World. S and P can be either True or False

In the above worlds, we have different underlying assumptions. Thus, the way of doing science (or logic) in **STPTF-World** will also be different from that of **SFPTF-World**. However, since they are guided by ideas where we find some similarities and dissimilarities, it is an attractive proposition, to compare and contrast them. Let us see what we will get then.

In **STPTF-World**, there are two possibilities—STPT and STPF. In STPT, **A** and **I** will be true, whereas **E** and **O** will be false. In STPF, **A** and **I** will be false, while **E** and **O** will be true. The relationship of contrariety and sub-contrariety will be replaced by contradiction since neither they can be false together nor can they be true together, respectively. The following is the state of affairs it portrays (Fig. 3):

Fig. 3 STPTF-World

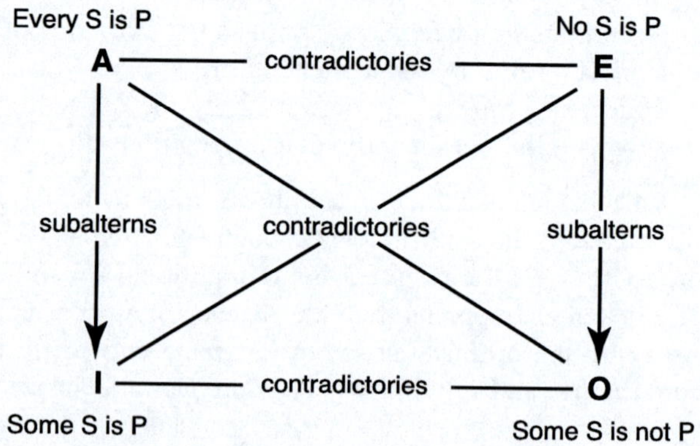

In **PTSTF-World**, there are two possibilities—PTST and PTSF. In PTST, **A** and **I** will be true, whereas **E** and **O** will be false. But in PTSF, **A** and **E** will be true, while **I** and **O** will be false. The relationship of contrariety sub-contrariety and subalternation are gone and what is left is the revised square.

The **SFPTF-World** has two possibilities, namely SFPT and SFPF. These are the truth conditions of the revised square. Thus, we will obtain nothing but the relation of contradictory omitting contrariety, sub-contrariety, and subaltern.

In **PFSTF-World**, there are two possibilities—PFST and PFSF. In the first possibility, **A** and **I** are false, whereas **E** and **O** are true. In the second, **A** and **E** are true, while **I** and **O** are false. Thus, this **PFSTF-World** too goes with the revised square and rejects contrariety sub-contrariety and subalternation.

These three subsequent worlds depict the same state of affairs, i.e., the revised square (Fig. 4):

Fig. 4 PTSTF, SFPTF, PFSTF World

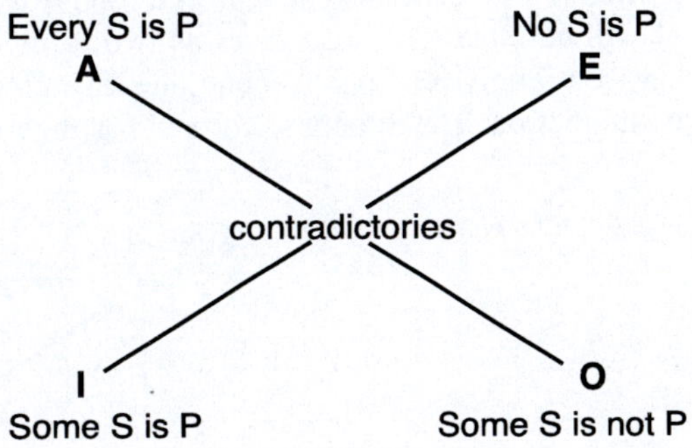

It must be noted here that even if the worlds represented by **PTSTF**, **SFPTF**, and **PFSTF** form the revised square, they belong to separate paradigms. Moreover, the truth conditions so reached have also followed a separate trajectory. We must disillusion ourselves from considering the above three paradigms as one (or identical) after obtaining an indistinguishable end result, i.e., the revised square.

In **STPT-World**, however, there is only one condition. Here, **A** and **I** will be true and **E** and **O** will be false. This gives us two results. First, there will be no relation of subalternation. Second, the contrariety and sub-contrariety will be replaced by contradiction. Here is its depiction (Fig. 5):

Fig. 5 STPT-World

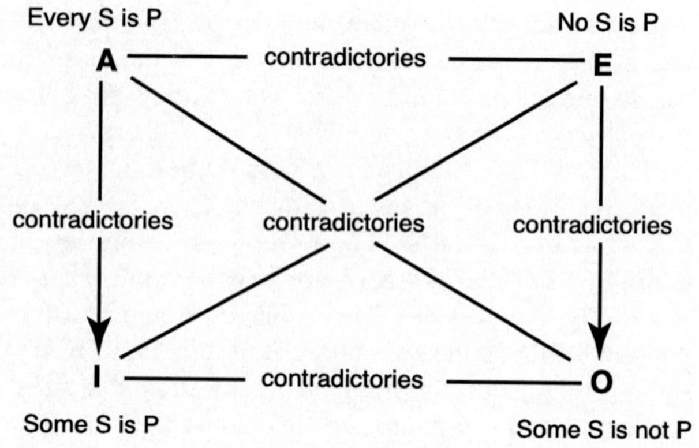

In **SFPF-World**, also there is one condition. Here, **A** and **E** will be true and **I** and **O** will be false. This also gives us two results. First, there will be contrariety and sub-contrariety relations. Second, the relation of subalternation will be replaced by contradiction. The Square so formed has the following depiction (Fig. 6):

Fig. 6 SFPF-World

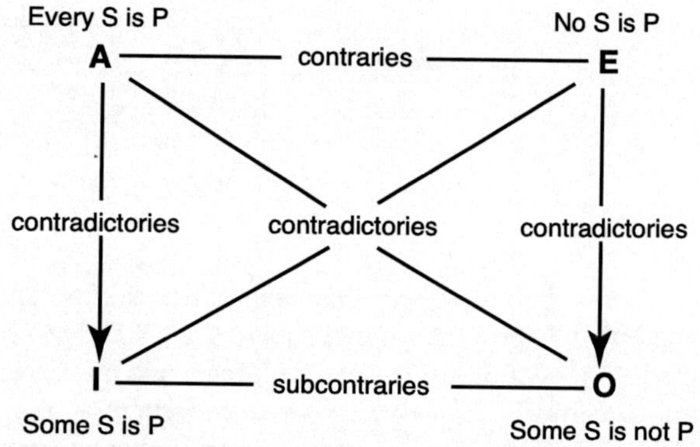

The reason behind carrying out the above exercise is not to show the formation of a variety of squares but rather to illustrate them as sprouting from different truth conditions, which must be considered as separate paradigms. Here, we can also see that none of them gave rise to the traditional square. One may be tempted to argue that if we consider the **STFPTF-World**, it will form the traditional square as well. **STFPTF** truth conditions (i.e., a combination of STPT, STPF, SFPT, and SFPF) will not form the traditional square but rather it will form the revised square, due to the presence of SFPT and SFPF conditions.

3.3 *Summary*

The point of this exercise is significant and relevant to this study. It is surprising to note that none of the truth conditions actually drew the traditional square. Kuhn remarks, “a lexicon which gives access to one set of possible worlds also bars access to others [14].” I will bring this issue later (in the next section). In this section, we found the following: First, the **STPTF-World** has contradictories and contradictories replacing the contraries and sub-contraries. Along with this, the relation of subalternation holds. Second, **PTSTF**, **SFPTF**, and **PFSTF Worlds** have contradictories only replacing the other three relations. Third, **STPT-World** has contradictories, contradictories replacing the contraries and sub-contraries and contradictories replacing subalterns as well. Fourth, **SFPF-World** has contradictories, contradictories replacing subalterns whereas the contraries and sub-contraries hold. I reiterate that the above six possible worlds (leaving **STFPTF** as it is a combination of all) are considered as separate paradigms, even though one may find certain similarities. The question, *'Why should we consider these as separate paradigms, when there are certain similarities?'* is answered in the following section, where we undertake to show these separate paradigms are incommensurable.

4 The Incommensurability of Squares

According to Kuhn, “if two theories [T_1 and T_2] are incommensurable, they must be stated in mutually untranslatable languages [13].” Incommensurabilty in a linguistic sense (though incorrectly) means incomparability. Precisely, if two theories T_1 and T_2 are incommensurble, it means that they do not have a common denominator and thus, they are impossible to measure. Incommensurability, like paradigm, is another outstanding thesis of Kuhn’s philosophy of science and a central tenet [7]. Kuhn distinguishes between “incommensurability” and “incomparability” and states “Most readers of my text have supposed that when I spoke of theories as incommensurable, I meant that they could not be compared. But “incommensurability” is a term borrowed from mathematics, and it there has no such implication. The hypotenuse of an isosceles right triangle is incommensurable with its side, but the two can be compared to any required degree of precision. What is lacking is not comparability but a unit of length in terms of which both can be measured directly and exactly [15].”

Kuhn has clarified and refined his notion of “incommensurabilty” in three stages [34]. It can be divided into early, middle, and later periods. He has further differentiated the notions of commensurability, comparability, communicability with sufficient rigor and argued that “incommensurable” does not mean that two or more theories are incomparable or impossible to communicate [13]. Briefly, they can be divided into two types [13–17, 34, 35]—Taxonomic incommensurability and Methodological incommensurability. The former refers to conceptual changes whereas the latter evaluates the epistemic values. In this section, I attempt to show that the traditional and revised squares are both taxonomically and methodologically incommensurable.

4.1 Taxonomic Incommensurability

In the last section, we have seen that none of the truth conditions (or **Worlds**) could endorse the traditional square. It must be understood that in the creation of possible **Worlds**, we took the modern symbolic interpretation into consideration. Any act of examining the equivalences of two theories (T_1 and T_2) requires translating their consequences into a neutral observation language [8]. The Boolean interpretation (1,0) or the symbolic representation (T/F) is not in complete harmony with the Aristotelian framework as presented in *De Interpretatione*. Thus, it not only fails to account for the traditional square but also shows the absence of a common denominator.

The squares are compared since they deal with propositions. However, whether these propositions can be classified under the same category is debated. The Squares are compared because (the relation of) "contradiction" is a shared opposition found in both of them. It must be understood that "contradiction" too comes in varying degrees. It is one thing to say "Not every S is P" contradicts "Every S is P" and it is another thing when we say "Some S is not P" contradicts "Every S is P." The squares are compared because logicians think that existential import is a common ground for examining resemblance. An affirmative proposition having a referent is not the same as Particular (Existential) propositions are having referents. Moreover, the absence of referent results in truth-value equals false cannot be compared with neither true nor false (or having undetermined truth-value) propositions. Therefore, Wreen opines that "the notion of existential import is itself confused, and should be banished from logical theory" [39].

In science, or any organized structure of knowledge accumulation, a seeker abreast her/himself with new terms and concepts after a breakthrough, invention, or revolution. S/he may not be in a position to make sense of such terms and concepts before. For example, wireless communication may not have been conceived in the thirteenth century B.C.E. even by the most brilliant minds of that time. Similarly, historians or ratiocinators can understand certain conceptions of the past by erasing and deleting conceptions in vogue, which may be the reason for unnecessary aberrations. This is akin to what Kuhn calls, incommensurability in terms of ineffability [17]. It is also stated that "the frequent insistence upon the fulfillment of requirements concerning such things as existence and universality has blinded us to the genuine restriction on the square of opposition" [4]. It must be clear that equating the two squares is a taxonomic failure, as it will be if we equate the proposed seven possible worlds.

4.2 Methodological Incommensurability

Two or more theories are methodically incommensurable if they do not have a standard measure, and such theories are compared by weighing historically developing values, not following fixed, definitive rules [35]. Consider the following questions and statements:

1. Is conversion by limitation acceptable?
2. Is DARAPTI a valid syllogism?
3. Why did Aristotle considered only three figures?
4. Why Aristotle called the first figure perfect?
5. Aristotle's logic is not equipped to deal with empty terms.
6. The traditional square can be saved with the help of presupposition.
7. The revised square supplants the traditional square.
8. The traditional square is only intuitively correct.

The above questions as well as statements are theory-laden. These questions were addressed several times and the statements were challenged in the last century by those, who awe their allegiance to Aristotle. However, as Kuhn points out that reality or truth is paradigm relative and paradigm dependent, moreover, scientists work in different worlds [17]. In such a scenario, what is proposed by the members of one world will fall flat on the deaf ears of members of another world. Neither their beliefs and principles remain the same nor do they speak the same language. Kuhn further remarks that "the claim that two theories are incommensurable is then the claim that there is no language, neutral, or otherwise, into which both theories, conceived as sets of sentences, can be translated without residue or loss [13]."

4.3 Summary

The traditional square of opposition is compared to the revised square based on its scope, acceptability, applicability, accuracy, and so on. These comparisons cannot have a unanimous acceptance or rejection. Aristotle's logic is not inconsistent but has clear-cut boundaries in its capacity to account for arguments and propositions both of the everyday and scientific language [22]. A system cannot be evaluated on the basis of the principles and norms of another system. A comparison of Aristotle's logic from the standpoint of modern formal logic is an invitation to both taxonomic and methodological incommensurability. Thus, the comparison between incommensurable squares is also unfounded and unwarranted. Kuhn himself admits, "In applying the term "incommensurability" to theories, I had intended only to insist that there was no common language within which both could be fully expressed and which could therefore be used in a point-by-point comparison between them [15]." In this section, we have argued that similarities between different systems do not necessarily permit commensurability. In the next section, we consider some cases where none of the relations present in any of the squares is applicable.

5 The Inapplicability of Squares

Until now, we have gathered that the relation of "contradiction" is common between squares. The seven worlds proposed in this paper, too, warrant its ubiquitousness. It not only acts as a link but is also a certain truth condition (or an opposition principle) in any paradigm. In this section, we strive to show that the opposition of "contradiction" does not hold in specific situation(s) or paradigm(s). For this, I devise the following four thought experiments.

5.1 Prof. Wascot vs. Prof. Palton

Imagine a class of undergraduates, studying Basic Physics, say, PHY101. Prof. Wascot is wave theorist, and Prof. Palton, who accepts the particle nature of light, are the course instructors. Some classes are taken by Prof. Wascot to energetically impart the fundamentals of wave theory, whereas other classes are taken by Prof. Palton to teach students the wonders of particle theory.

According to Prof. Wascot, the following truth conditions are acceptable:

(a) Every light is/are waves—True
(b) No light is/are waves—False
(c) Some light is/are waves—True
(d) Some light is/are not waves—False

However, according to Prof. Palton, the following truth conditions are acceptable:

(e) Every light is/are waves—False
(f) No light is/are waves—True
(g) Some light is/are waves—False
(h) Some light is/are not waves—True

It can be very well seen that the course instructors belong to rival paradigms and are incommensurable. However, this does not end the misery of students, as they have to go through all that, in a single semester. The students may not be confused whether light is a particle or wave, but they are concerned about who is setting the questions in the end semester examination. It is a standard practice that both instructors ask some questions. Mr. Sarry Horton, the nephew of the Dean, found out that Prof. Wascot sets all the questions on odd places, and the questions of the even place are by Prof. Palton. The objective section of the question paper looks like this:

Instruction Indicate which of the following statements are true or false by writing T or F. You will score one mark for every correct answer but will go on to lose a half mark for every incorrect response.

(1) Every light is/are waves
(2) No light is/are waves
(3) Some light is/are waves
(4) Some light is/are not waves

To score full marks in the above section, a student needs to answer all these questions in affirmation. Similarly, if all the questions on odd places are set by Prof. Palton and the even place's questions are by Prof. Wascot, then the answer to all these questions must have been a denial. In both these situations, we do away with the relation of "contradiction."

5.2 *Prof. Mind vs. Prof. Head*

A deductive logician, Prof. Mind, and an inductive logician Prof. Head are in the midst of a discussion.
Prof. Mind—*"Isn't it true that inductive arguments are invalid?"*
"No. They are either strong or weak", said, Prof. Head.
"Yes, they are invalid" said, Prof. Mind, *"as the truth of the premises never guarantee the truth of the conclusion."*
Prof. Head responds, *"Well, inductive arguments do not compete to be valid or invalid. Validity is not a parameter of inductive arguments. One cannot say an inductive argument is valid or invalid as it will be akin to commit a category mistake."*
Prof. Mind argues, *"Look, Prof. Head, ultimately, all we are talking about is nothing but arguments. It is high time you realize that inductive arguments are invalid first, then they are either strong or weak."* Mr. Sarry Horton from Prof. Wascot and Prof. Palton's class happens to pass by and is already planning what to answer in case this question comes in the end semester examination, and he knows who sets the question. Now, consider the following statements:

(a) All inductive arguments are invalid.
(b) No inductive arguments are invalid.
(c) Some inductive arguments are invalid.
(d) Some inductive arguments are not invalid.

They can all be true or false at the same time if we allow Prof. Mind and Prof. Head to take charge of odd and even cases, respectively and vice versa. Nevertheless, we do away with contradiction, in this case too.

5.3 *Tomato vs. Tomato*

New Jersey recognizes the tomato as the state vegetable, and Tennessee adopted the tomato as the official state fruit. Whether the rich tomato is a fruit or a vegetable is an age old question. However, it has a simple answer. Tomatoes are both. They are fruits as they develop from the flower of a tomato plant. The same goes for cucumbers, beans, peppers, pumpkins, peas, etc. So, tomatoes are fruits—is a botanical classification. However, tomatoes are widely used as vegetables. Thus, tomatoes being vegetable is a culinary classification. Moreover, in the United States of America, tomatoes are taxed as vegetables to endorse the social construct of everyday language and people's mindset.

I just forgot to mention that Mr. Sarry Horton's father is a Chef, his mother is a Botanist, his sister is a fun-loving individual who loves to laugh at the "tomato fiasco" and his grandfather was a Supreme Court judge. When I asked Sarry—*"What do you think ... "*, he stopped me in between, yelling *"Nothing holds!"*.

5.4 *HEisenberg vs. HeIsenberg*

A high school Physics class would tell us, Heisenberg uncertainty principle states that we cannot determine the position and momentum of a (sub-atomic) particle at the same time. However, we can separately determine them as such. When HEisenberg asked HeIsenberg (in Sarry's dream),

(i) The momentum of all particles can be determined—T/F?
(ii) The momentum of no particle can be determined—T/F?
(iii) The momentum of some particles can be determined—T/F?
(iv) The momentum of some particles cannot be determined—T/F?
Then, HeIsenberg answered (or re-questioned) HEisenberg (in Sarry's dream) as follows:
(v) The position of all particles can be determined—T/F?
(vi) The position of no particle can be determined—T/F?
(vii) The position of some particles can be determined—T/F?
(viii) The position of some particles cannot be determined—T/F?

The answers to the above questions depend on—"what we are trying to determine" rather than "what can be determined." This is one of the striking aspects of the difference between quantum physics, which is conceptually different from classical physics [6]. HEisenberg vs. HeIsenberg not only tell us that "contradiction" is not obtained in this paradigm but also that (i) and (iv) or (v) and (viii) is a "dialetheia" [28]. I do wonder whether any of these propositions from (i) to (viii) is/are "glut" (a sentence, which is both true and false) and/or "gap" (a sentence which is neither true nor false) at the same time. Nonetheless, I am not discussing this point any further, as it is beyond the scope of present work. However, this can be an avenue for further research.

5.5 *Summary*

We can devise and develop many more thought experiments like this. In such situations, we fail to rescue the relation of contradiction (or any relation of opposition what-so-ever), even if we know that these are separate paradigms, which are incommensurable. The PHY101 class is a paradigm, so is the logic in general, or Mr. Sarry Horton. HEisenberg vs. HeIsenberg is a double (or multiple) paradigm. Nevertheless, they all have one thing in common—the inability to accommodate the relation of "contradiction."

6 Conclusion

The history of squares or logic, in general, has seen various developments, which went on to question the established norms. Revisions, therefore, are an integral part of any system. Kuhn's thesis, in a nutshell, tells us that established norms are normal science. It has a puzzle-solving capacity. In the wake of new developments, challenges are posed, and there is a state of crisis due to the inability of normal science to address the threats of new questions. Rival paradigms compete with each other for supremacy. Once a paradigm can address the challenges posed, it is established as the new normal science. The past and the present paradigms are entirely different worlds. They are incommensurable since they are not only based on a different set of beliefs and principles but are also lacking a common ground. But, incommensurabilty is not the end of the road; it is instead a requirement to grow as "humanity did not begin its intellectual journey already possessing all the concepts and methodological tools that would ever be required, incommensurability becomes a requirement for progress [2]."

Squares too are no exception. Aristotle's ("assertoric") propositions, we consider today (i.e., "categorical" propositions as translated by Boëthius) are not (exactly) the same as they were in his era. Similarly, the paradigms of traditional and modern logicians are dissimilar and have no (exactly) equivalent explanations or notions to equate. Some examples at the cusp of two different worlds poses questions, which are intriguing to the residents of any world. We (may thus,) need a new and evolved vocabulary (including diagrams) to explain these concepts. The squares and the oppositions expressed are not handy in many cases. I reiterate the findings of this paper in the following three statements. First, the traditional and the revised square belong to two separate paradigms. Second, they are taxonomically and methodologically incommensurable. Third, none of the relations (including "contradiction") expressed in any of the squares holds in certain situations as shown by thought experiments. In addition to this, I disclaim that this paper (or any of its part) argues that the traditional and modern squares cannot be compared, as "even a complete incommensurability of two theories does not make them incomparable by various objective standards" [7]. The squares are comparable and so is its "contradiction."

Acknowledgments I am indebted to Shri Mata Vaishno Devi University Katra and its School of Philosophy and Culture, along with my co-supervisors, Dr. Anil Kumar Tewari and Dr. Madhu Mangal Chaturvedi for their academic and administrative support. I express my sincere gratitude to Prof. Jean-Yves Beziau and Prof. Ioannis Vandoulakis—the organizers of *Square 2018* at Crete, for providing me with an opportunity to present this work. I am honored to have a lively and informative discussion with Prof. Stephen Read. My thanks are due to the anonymous reviewers for their insightful comments and suggestions on an earlier draft.

References

1. J. Barnes, *The Complete Works of Aristotle*, Princeton University Press, New Jersey, 2014.
2. H. I. Brown, Incommensurability reconsidered, *Studies in History and Philosophy of Science* (2005), 149–169.
3. P. Corkum, Empty Negations and Existential Import in Aristotle, *Apeiron* (2018), 201–219.
4. G. Englebretsen, The Square of Opposition, *Notre Dame Journal of Formal Logic* (1976), 531–541.
5. P. Godfrey-Smith, *An Introduction to the Philosophy of Science: Theory and Reality* The University of Chicago Press, Chicago, (2003)
6. J. Hilgevoord and J. Uffink, The Uncertainty Principle: In *The Stanford Encyclopedia of Philosophy* (2017 Edition), E.D. Zalta (eds.) URL = https://plato.stanford.edu/archives/win2016/entries/qt-uncertainty/.
7. J. Hintikka, On the Incommensurability of Theories, *Philosophy of Science*. (1988), 25–38.
8. P. Hoyningen-Huene, *Reconstructing Scientific Revolutions. The Philosophy of Science of Thomas S. Kuhn* Chicago University Press, Chicago (1993)
9. D. Jacquette, Boole's Logic: In *Handbook of the History of Logic, Volume 4, British Logic in the Nineteenth Century* (2008), D.M. Gabbay and J. Woods (eds.)
10. D. Jacquette, Subalternation and existence presuppositions in an unconventionally formalized canonical square of opposition *Logica Universalis* (2016), 191–213.
11. J.N. Keynes, *Studies and Exercises in Formal Logic* Macmillan, London, 1887.
12. W. Kneale & M. Kneale, *The Development of Logic* Clarendon Pres, Oxford, 1962.
13. T. S. Kuhn, Commensurability, Comparability, Communicability: In *Proceedings of the Biennial Meeting of the Philosophy of Science Association* (1982), P.D. Asquith and T. Nickles (eds.)
14. T. S. Kuhn, Possible Worlds in History of Science: In *Possible Worlds in Humanities, Arts and Sciences* (1989), A. Sture (eds.)
15. T. S. Kuhn, Theory-Change as Structure-Change: Comments on the Sneed Formalism *Erkenntnis* (1975), 179–199.
16. T. S. Kuhn, *The Essential Tension* Chicago University Press, Chicago (1977)
17. T. S. Kuhn, *The Structure of Scientific Revolution* Chicago University Press, Chicago (1962)
18. D. Londey and C. Johanson, Apuleius and the Square of Opposition *Phronesis* (1984), 165–173.
19. D. Londey and C. Johanson, *The Works of Apuleius* E.J. Brill, London (1987)

20. M. Masterman, *The Nature of a Paradigm. In Criticism and the Growth of Knowledge* Cambridge University Press, Cambridge, 1970.
21. R. McKeon, *The Basic Works of Aristotle*, Ransom House, New York, 1941.
22. M. Mignucci, Aristotle on the Existential Import of Propositions *Phronesis* (2007), 121–138.
23. A. Moktefi and S.J. Shin, A History of Logic Diagrams: In *Handbook of the History of Logic, Volume 4, British Logic in the Nineteenth Century* (2008), D.M. Gabbay and J. Woods (eds.)
24. F.S. Northrop, An Internal Inconsistency in Aristotelian Logic, *The Monist* (1928), 191–210.
25. F.S. Northrop, A Reply Emphasizing the Existential Import of Propositions, *The Monist* (1929), 157–159.
26. T. Parsons, The Traditional Square of Opposition, *The Stanford Encyclopedia of Philosophy* (2017 Edition), E.N. Zalta (ed.), URL = https://plato.stanford.edu/entries/square/.
27. T. Parsons, Things that are right with the Traditional Square of Opposition, *Logica Universalis* (2008), 3–11.
28. G. Priest, Dialetheism, *The Stanford Encyclopedia of Philosophy* (2018 Edition), E.N. Zalta (ed.), URL = https://plato.stanford.edu/archives/fall2018/entries/dialetheism/
29. W.V.O. Quine, Peano as Logician, *History and Philosophy of Logic* (1987), 15–24.
30. S. Read, Aristotle and Łukasiewicz on Existential Import, *Journal of the American Philosophical Association* (2015), 535–544.
31. S. Read, Aristotle's Theory of the Assertoric Syllogism (2017) retrieved from URL = https://www.st-andrews.ac.uk/slr/The_Syllogism.pdf
32. S. Read, John Buridan's Theory of Consequence and His Octagons of Opposition: In *Around and Beyond the Square of Opposition. Studies in Universal Logic* (2012), J.Y. Béziau and D. Jacquette (eds.)
33. W.D. Ross, *Aristotle's Prior and Posterior Analytics*, Clarendon Press, Oxford, 1949.
34. H. Sankey, Kuhn's Changing Concept of Incommensurabilty, *British Journal of Philosophy of Science* (1993), 759–774.
35. H. Sankey and P. Hoyningen-Huene, *Incommensurability and Related Matters* Kluwer, Dordrecht (2001)
36. S.S. Sharma, Interpreting Squares of Opposition with the help of Diagrams: In *The Square of Opposition - A General Framework for Cognition* (2012), J.Y. Béziau and G. Payette (eds.)
37. P.F. Strawson, *Introduction to Logical Theory* Metheun, London (1952)
38. A.P. Uchenko, Aristotelian Logic and the Logic of Classes, *The Monist* (1929), 153–156.
39. M. Wreen, Existential Import, *Crítica: Revista Hispanoamericana de Filosofía,* (1984), 59–64
40. J. Wu, The Problem of Existential Import (From George Boole to P.F. Strawson), *Notre Dame Journal of Formal Logic* (1969), 415–424.

The Square of Opposition as a Framework for Stephen Langton's Theological Solutions

Marcin Trepczyński

Abstract In some texts of the prominent medieval thinker Stephen Langton (1150/55–1228), whose main theological works are being edited these days, it is possible to point out solutions based on the square of opposition. Although it is not clear whether he had such a structure in mind as a geometric representation, in his analyses devoted to God's will, he introduced from three to four options representing possible states of will connected by such relations as contradiction, contrariety and the relationships set up by the possible distributions of logical values. Regardless of whether he knew the square of opposition, it is argued that this was the framework of his theological solutions. The power of the square of opposition in theological consideration is also seen in the example of the problem of predestination and the problem of theodicy. Finally, the "square of will" based on Langton's analyses is further developed to a "hexagon of will".

Keywords Square of opposition · Stephen Langton · Predestination · Theodicy · Contradiction · Contrariety

Mathematics Subject Classification (2000) Primary 03B05; Secondary 03B80

1 Introduction

The aim of this chapter is to present how the medieval thinker Stephen Langton (1150/55–1228) used the square of opposition as a basis for his theological considerations and to underline the significance of this structure in his thought. At

This article was supported by the National Science Centre, Poland, under grant agreement No. 2015/18/E/HS1/00153.

M. Trepczyński (✉)
Faculty of Philosophy, University of Warsaw, Warsaw, Poland
e-mail: m.trepczynski@uw.edu.pl

J.-Y. Beziau, I. Vandoulakis (eds.), *The Exoteric Square of Opposition*, Studies in Universal Logic, https://doi.org/10.1007/978-3-030-90823-2_5

the same time, it is an occasion to analyse an interesting example of the application of the square and to further develop such an example to a hexagon,[1] as well as to take a deeper look at this interesting author whose works are today being brought to audiences from their manuscripts.

Stephen Langton was a very prominent person at the turn of the twelfth and thirteenth centuries. He was a very famous, appreciated and influential (cf. [7, 14]) master of theology at the University of Paris, where he lectured until 1206. He was called “Doctor Nominatissimus”, so “Excellent”, “Famous” or “Outstanding”. In 1206, he was appointed as cardinal by Pope Innocent III, and in 1207 (after 2 years of disputes around elections), the Pope ordained him archbishop of Canterbury, although—because of those tensions—he was only able to travel to England after 6 years, in 1213. Then, he became involved in a struggle of the barons against King John Lackland and—to bring both sides to peace—he proposed a draft of the future Magna Carta Libertatum (The Great Charter of Liberties) signed by the King in 1215, being the most important document issued in medieval ages in England (cf. [10, 15–26], [3, 832–835]). Let us note that he is the first person mentioned in the Great Charter by the King. It is also likely that he is an author of the famous sequence of Pentecost Come, Holy Spirit (*Veni Sancte Spiritus*) (cf. [10, 39–41]). What is more, he is believed to have divided the Bible into chapters in the way that it is used today.

Finally, the ongoing study shows that he was a brilliant and very original theologian, often conducting philosophical considerations and using many advanced logical tools, which we can find especially in his short *Summa*[2] and *Theological Questions*.[3] In the introduction to the partial edition of Langton’s *Summa*, S. Ebbesen remarks that “there is plenty of logic—semantic theory, theory of inference, and not least, use of the *eadem ratione* principle”. As regards this specific tool of logic mentioned at the end, Ebbesen explains that:

> This principle, which is ever-present though never explicitly formulated, states that if some proposition of inference, p, is true or valid, then q is so too unless it can be shown that q differs from p in some relevant respect. In particular, if the truth or validity of p is defended by appeal to some rule, it must be shown that the rule does not apply to q or that there is some other relevant difference if q is to be declared less acceptable than p. [15, 574–576]

The same can be said about Langton’s *Theological Questions*, where we find a great number of examples of reasoning introduced by the phrase “eadem ratione” or “pari ratione”, based on the mentioned principle.[4] Furthermore, there are many syllogisms and considerations in which Langton analyses possible arguments,

---

[1] As described by R. Blanché [5] and developed by J.-Y. Béziau [4] and many collaborators.

[2] It is an unfinished work, barely started; see the partial edition: [1].

[3] See the first book: Stephen Langton, *Quaestiones Theologiae*, Liber I, eds. R. Quinto and M. Bieniak, Oxford University Press, Oxford 2014 and two recently published volumes of the third book (hereinafter: QT) [11], [12], [13].

[4] Only in q. 1 of QT do we find six arguments based on this principle: p. 237, ll. 45–48, p. 238, ll. 63–67, ll. 79–83, p. 241, ll. 150–153, ll. 156–161, ll. 164–168.

pointing out premises and conclusions and for example showing that an argument (inference) is not valid, but a conclusion is true or that a conclusion is false, despite the fact that a premise is true or that the argument is invalid and something else should be inferred.[5] We should also add that an important basis for many of his solutions is semiotic theory, within which such relationships as *suppositio*, *significatio*, *consignificatio*, *copulatio*, *appellatio*, *notatio*, and *connotatio* are discerned and used in a very sophisticated manner (cf. [15, 579–583]).[6]

Unfortunately, probably due to his political and pastoral involvement, his works were not widely distributed in his time and were not well known (cf. [3, 811]), despite the fact that he influenced many important scholars and was a famous and recognised person. Furthermore, even today only some of his works are edited and published. One of his most important works, beside his short *Summa* and commentaries to the *Sentences* of Peter Lombard[7] as well as to the *Epistles* of Saint Paul, is the already mentioned collection of more than two hundred *Theological Questions*.[8] The first book of these questions was published in 2014 by Riccardo Quinto and Magdalena Bieniak at Oxford University Press. A complete critical edition of consequent questions is now being prepared as part of a large project at the University of Warsaw.[9] Let us underline that except for some selected questions edited in the twentieth century, the majority of Langton's questions existed only in manuscripts, whereas—in comparison to his *Summa* and commentaries—they contain definitely wider and deeper analyses of the discussed problems, also from the logical point of view. These special features of *Quaestiones Theologiae* provide much richer and more interesting material, also for logical investigations, and should be the main resource for studies of his thought.

One of Langton's theological questions published in 2014 is q. 17 on God's will. In fact, the editors identified three different texts of Langton on this topic and after a thorough analysis claimed that they were not three versions of the same question, but three separate questions, which were subsequently marked as: q. 17, q. 17*, and q. 17**. In the last two questions, Langton analyses examples of sentences referring to what God wants and what God does not want (respectively, "God wants him to sin"/"God doesn't want him to sin" and "God doesn't want him to be bad"), discerning contradictions and contrarieties. It will be shown that the framework

[5] From the many examples, let us recall for example the following two statements: *Non ualet prima argumentatio. Deberet enim inferre "(...)"* (QT, q. 1, p. 239, ll. 107–108); *Solutio. Prima uera, conclusio falsa. Set ex hac sequitur conclusio "(...)", set hec falsa; et hec uera...* (QT, q. 13b, p. 232, ll. 12–14).

[6] An example of such an utterance from q. 62c: *Set hoc uerbum iudicare' duas habet significationes prout copulat iudicium auctoritatis: (1) in una copulat pure iudicium auctoritatis nichil connotando, et secundum hoc dicitur de tota trinitate; set de hac significatione nichil ad presens; (2) in alia significatione copulat iudicium auctoritatis et connotat ministerium* [11, 247, ll. 138–144].

[7] See the edition [6].

[8] Cf. [10, 161–166] and cf. also the updated working catalogue of the questions published at the project website: http://langton.uw.edu.pl/theological-questions (access: 31.12.2021).

[9] See http://langton.uw.edu.pl (access: 31.12.2021).

of these considerations was the square of opposition. This presentation will be supplemented with Langton's remarks on this topic, formulated in the commentary to the *Sentences*, but there will be no links to his *Summa*, as *Doctor Nominatissumus* had not addressed those problems in this work.

2 Langton's Logical Structures in Questions on God's Will

The first passage of q. 17* referring to relationships based on the square of opposition can be found in Chap. 4, namely:

> *Dicimus quod utraque istarum est falsa: "deus uult istum peccare", "deus non uult istum peccare", ita quod hec uox non uult' sit una dictio; nec sunt contradictorie, immo utraque affirmatiua. Set si non' – uult' sint due dictiones, tunc uera est ultima. Quod dicitur "Nolo mortem peccatoris" exponi debet: idest "non uolo".* (QT, q. 17*, c. 4, resp., p. 376, v. 60–64)

Fig. 1 Langton's logical structure in q. 17*

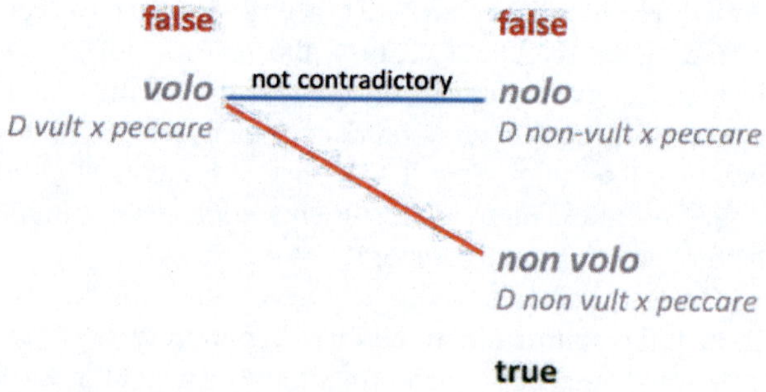

> We say that both of the following are false: "God wants him to sin", "God doesn't want him to sin" in a way that this phrase "doesn't want" is one word; they are also not contradictory, and they are both affirmative. But if "doesn't" – "want" are two words, then the latter [sentence] is true. When it is said "I don't-want (*nolo*) the sinner's death" should be exposed as "I don't want (*non volo*)".

At this stage, Langton states that the relationship between "God wants him to sin" and "God doesn't want him to sin" is not a contradiction. Furthermore, both sentences are false. He also adds that there is another sentence negating the first one and states that this sentence is true. Finally, let us see that these three sentences are based on three options of using Latin verbs which express one's will:

- *volo* (I want)
- *nolo* (I do-not-want or I want not)
- *non volo* (I do not want, which according to Langton's suggestions could be understood as it is not true that I want)

Hence, Langton offers here a complex relationship between three sentences with the possible distribution of logical values and information that two of those sentences (which are both affirmative) are not contradictory. The situation described by Langton is presented in Fig. 1.

Let us compare this first passage from q. 17* with a short passage on the same topic from q. 17**, Chap. 5:

> *Vnde hec duplex: "deus non uult istum esse malum", quia si hec non uult' sit oratio, uera est, cum eius contradictoria sit falsa; si est tertia persona huius uerbi nolo', falsa est, quia sequeretur quod deus uellet eius contrarium.* (QT, q. 17**, c. 5, resp., p. 380–381, v. 57–61)

> (...) From this, this [sentence] "God doesn't want him to be bad" has two meanings, because if this "doesn't want" is a sentence (is complex), it is true, as its contradictory is false; if it is the third person of the verb "to want-not", it is false, as it entails that God would want what is contrary to this.

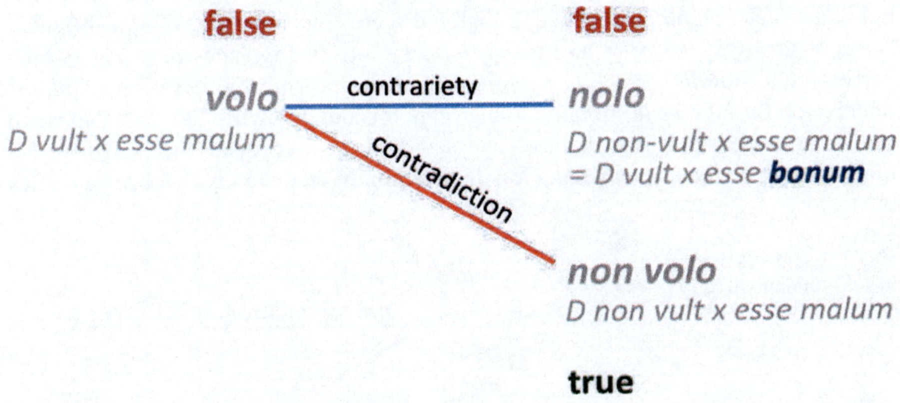

Fig. 2 Langton's logical structure in q. 17**

Although in this passage Langton analyses different sentences than in q. 17*, they are based on the same structure and have a very similar meaning. Just as in the previous example, Langton uses three predicates: *volo*, *non volo*, and *nolo* and offers a distribution of possible logical values of the sentences based on those three options. But this time he also precisely names the relationships between them. In the first case, he is talking about contradiction. And in the case of the sentence with the second meaning of *non vult*, he points out that then God would want something contrary, so in this situation one does not negate that God wants x to be bad, but one says that God wants x to be good (as bad and good are contrary terms). We should underline that Langton speaks here about the contrariety of terms and does not explicitly mention a contrariety between sentences. However, it seems that it is crucial to make a distinction by using different names for these different relations (contrariety and contradiction), and as these two contrary terms are used as arguments of the same formula "God wants x to be ... ", we can transpose it onto the relationship between the sentences. The situation given in q. 17** has been presented in Fig. 2.

Now, let us take into account that in q. 17* Langton continues his analysis. In the argument that opens Chap. 5 (v. 69–72), he shows that the relationship among *volo*, *nolo*, and *non volo* is the same as among "likes" (*placet*), "dislikes" (*discplicet*), and "doesn't like" (*non placet*). Hence, Langton states "when this sentence God dislikes sin' is true, so similarly this [sentence] Sin happens when God doesn't want it' and also this [sentence] God doesn't want him to sin' in a way that "doesn't want" is one word, which has been negated before".

In his answer to this argument (v. 73–80), he explains that "to like" (*placere*) can be understood in two ways: as "to want" (*velle*) taken strictly and as "to tolerate" (*approbare*). And similarly, "to dislike" (*displicere*) may be understood as "to not want" (*nolle*). According to this meaning, these two sentences "God doesn't want a sin to happen" (*deo nolente fit peccatum*) and "God dislikes this sin" (*deo displicet hoc peccatum*) are false. But the situation is different when we say that God does not like it (*deo non placet*). Hence, Langton discerns two meanings of *placet* (which refer, respectively, to *velle* and to *approbare*) and distinguishes *displicet* (which corresponds to *nolle*) and *non placet* (which results from the negation of *placet* understood through *velle*). Finally, Langton points out that when "to like" can stand for "to tolerate" and "not to like" can stand for "to reprobate" (*reprobare*). He underlines that when we say that God dislikes sin, in fact, we say that he reprobates

Fig. 3 Langton's developed logical structure in q. 17*

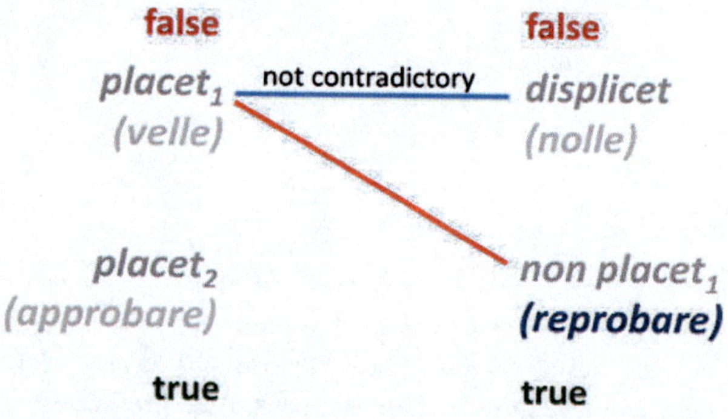

Fig. 4 Langton's merged logical structure (qq. 17* and 17**)

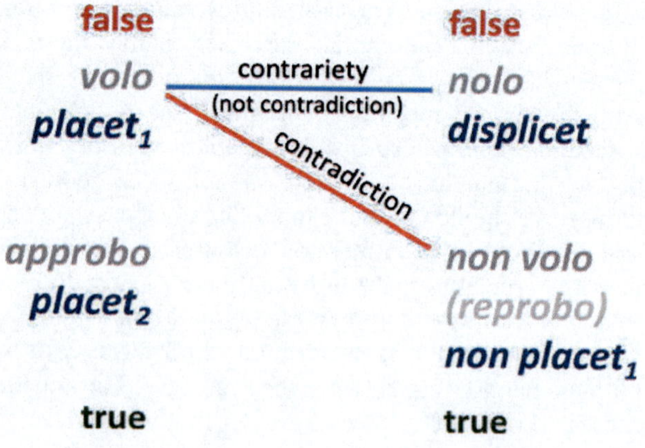

it, because he hates it, so he does not want it, but not in an "active" way, as when we understand "dislikes" as "one word".

Thanks to this substitution, Langton obtains a similar structure, isomorphic to that presented in Fig. 1, with some defined relationship and with the distribution of possible logical values. This structure is shown in Fig. 3.

In order to collect all the information delivered by Langton about such a structure, we can merge the figures based on considerations from q. 17* and q. 17** and thus Figs. 2 and 3. We obtain the picture presented in Fig. 4:

Finally, we can add that in the *Commentary to the Sentences*,[10] Langton discerns two meanings of *non vult*, using an example similar to the one presented in q. 17** and referring to the concept of reprobation, as in q. 17*. He briefly analyses here a short reasoning: "God doesn't want x to be bad. Hence: God wants the contrary of this". Langton shows that if we treat *non vult* as two words, then "God wants x to be good" is false, as x can be reprobated. But when we treat it as one word based on the form *nolo* in the third person, "as we would say: *nult*", then the reasoning presented works.

3 Towards the Square

We can find four options in Langton's considerations, which can be presented as the four corners of the square of opposition and which properly meet the standards of the four conventional corners of A, E, I, and O (cf. [4, 5–6]). Within this square are two relations described and named correctly, namely, contrariety (between A and E corners) and contradiction (between A and O corners).

What is more, *Doctor Nominatissimus* provides a possible distribution of logical values, which specifies the relation between the corners I and O: they "may be both true", so in fact it can be identified as a subcontrariety. Furthermore, this distribution shows that it is possible that the relationship between A and I and the relationship between E and O are subalternations, and when we take into account the content of these corners, it indeed turns out that the sentences corresponding to A and I, such as "God wants x" and "God tolerates x" (or "it is not the case that God wants not x"), remain in the relationship of a subalternation and similarly in the case of the sentences "God wants not x" and "it is not that God wants x", corresponding to E and O.

Therefore, it seems that Langton delivered all the information sufficient to reconstruct his square of opposition in a complete version, which can be presented as Fig. 5.

[10] Stephen Langton, Commentarius in Sententias, lib. I, d. 45, c. 7, n. 411: "Hoc non valet: Deus non vult hoc; ergo vult eius contrarium. Secundum quod sunt due dictiones: non vult, quia non vult istum esse malum. Tamen hec falsa: vult istum esse bonum. Sit, quod reprobus sit. Set secundum quod est una dictio huius verbi nolo tertia persona, acsi dicatur nult, bene sequitur" (Der Sentenzkommentar..., p. 63).

However, a question may be posed: why has Langton not named other relations? He could not only have named the I–O relationship as a subcontrariety, but he could also have explicitly mentioned that there was a contradiction between E and I (so between *nolo*-corner and *approbo*-corner) and the subalternation between the pairs A–I and E–O. He could do this, as he was well educated in logic and familiar with such concepts.

Fig. 5 Langton's completed square (based on qq. 17* and 17**)

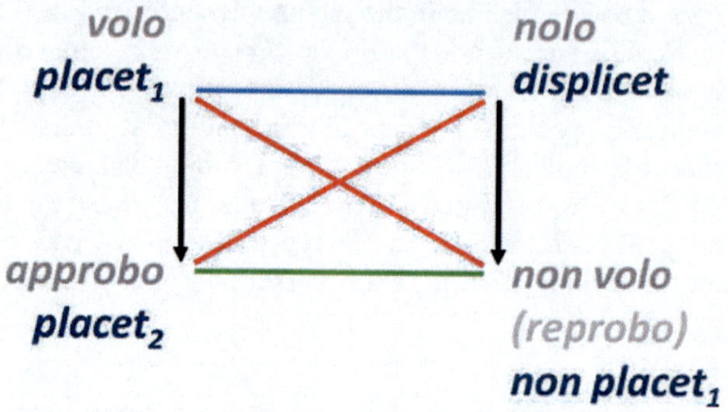

We cannot be sure about the reasons of this "negligence", but there is a hypothesis that can explain it. In q. 17**, there is a passage referring to a different topic, but containing the following useful hint:

> Si autem queratur pro quo fiat suppositio cum dicitur "quod deus permittat hoc etc.", potius logica est questio quam theologica, nec spectat ad theologum hoc inquirere, set tantum modum loquendi et causam dicti esplanare (QT, q. 17**, c. 4, resp., p. 380, v. 51–52).

It means that within his *Theological Questions*, Langton is not interested in addressing logical problems directly. When he perceives something as rather a logical question than a theological one, he stops, as it should not be expected that a theologian will investigate such topics deeply. Langton needs logic to the extent that it is useful in explaining some theological problems. So, he limits the logical considerations to the extent that they suffice as a tool for the theological discussion. Therefore, we can guess that he was not interested in presenting a complete structure of the four mentioned options, but just sketched those relationships which he needed to explain the difference between the similar notions.

However, let us remember that Langton mentioned all the substantial elements needed to complete such a square. And what is more, a deep understanding of those elements is needed to provide such subtle solutions as he did. This is why, in my opinion, the square was a logical framework on which he based his considerations.

4 The Significance for Theology

The square discerning four possible states of will mentioned above (let us call it "the square of will") is very powerful in discussions on several important theological problems.

The examples used by Langton in questions on God's will: "God doesn't want x to sin" or "God doesn't want x to be bad", as well as his explicit reference to the theological concept of reprobation, indicate that the square is useful to explain the problem of predestination. Langton himself addressed this issue in several theological questions transmitted in various versions (q. 13b, q. 13c, q. 13*, q. 14, q. 15a, q. 15b, q. 15*a, q. 15*c), and in some of these texts, he discussed several problems concerning reprobation. However, we should underline that he had not considered it in respect to God's will, as he had in q. 17* and q. 17**, which should not be very surprising if we take into account that in the most common theological textbook of his time, which was also the basic material for his work, namely the *Sentences* of Peter Lombard, predestination and reprobation are considered in terms of God's prescience.[11] A similar situation can be found in William of Auxerre's *Summa aurea* (which followed some of Langton's ideas), despite the fact that William quoted St. Augustine's "will-laden" definition.[12] It was St. Thomas Aquinas who finally combined the concept of reprobation with God's will again, saying in the question on predestination that "reprobation includes the will to permit a person to fall into sin".[13] Nevertheless, as it was said, Langton set up a distinct link between will and reprobation in two questions devoted to God's will (in q. 17* directly and in q. 17** indirectly). And in the examples from *Doctor Nominatissimus*, we can see that the square of will makes it easy to understand two key statements supported by some medieval theologians and generally by Catholic theology:

1. When someone is reprobated by God, it does not mean that God wills for this person to be condemned (this would be the *nolo* corner—E), but it is not true that God wills to save this person (*non volo* corner—O). It just means that this person did not obtain enough grace to be saved.
2. The theory of double predestination is excluded. According to this theory, either the *volo* corner (A) or *nolo* corner (E) must be applied to each person, so if it is not true that God wants someone to be saved, then it means that God wills to

[11] Peter Lombard, *Quattuor Libri Sententiarum*, lib. I, dist. XL, cap. 2: "reprobatio e converso intelligenda est praescientia iniquitatis quorundam et praeparatio damnationis eorundem" [9, 286].

[12] William of Auxerre, *Summa aurea*, lib. I, tr. IX, c. 3, q. 2: "Diffinit enim A u g u s t i n u s super Epistolam ad Romanos reprobationem hoc modo: reprobare est nolle misereri, et nolle misereri est non apponere gratiam in presenti et dampnare in futuro" [8, 184].

[13] Thomas Aquinas, *Summa theologiae*, I pars, q. 23, a. 3, resp. [14]).

have someone condemned. To understand this theory, one has to understand the difference between the situation of the *nolo* corner (E) and the situation of the *non vult* corner (O).

The second problem (both theological and philosophical) to which the square of will based on Langton's analyses can be successfully applied is the problem of theodicy, which is absent in the considerations of *Doctor Nominatissimus*, but which—despite being called this in modern times[14]—has been present in Catholic theology since late antiquity, thanks to St. Augustine.[15] In discussions devoted to this topic, it is very important to include both following possibilities: (1) that God tolerates some evil, although he has no active will this evil to happen and (2) and that God does not want some evil, but at the same time he has no active will this evil not to happen (because then this evil simply could not happen). This is crucial for the Catholic solution according to which God never wants any evil to happen but may tolerate it (according to St. Augustine, in order to draw from that evil some good). This situation is represented by two bottom corners taken together (I and O), namely *approbo* and *non volo* (*reprobo*) corners. The square of will shows clearly that (Fig. 6):

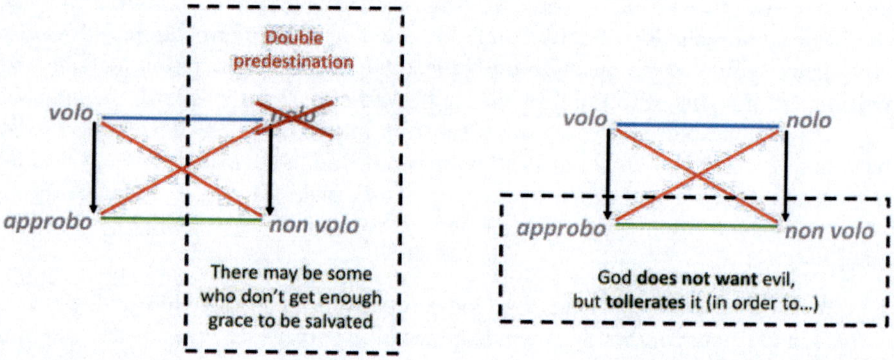

Fig. 6 The square of will used to discuss, respectively, the problem of predestination and the problem of theodicy

[14] The term "theodicy" was introduced by G.W. Leibniz in *Essais de thodice sur la bont de Dieu, la libert de l'homme et l'origine du mal*, published in 1710.

[15] Cf. St. Augustine, *Enchiridion ad Laurentium liber unus (de fide, spe et caritate liber unus)*, III, 11: "For the Omnipotent God, whom even the heathen acknowledge as the Supreme Power over all, would not allow any evil in his works, unless in his omnipotence and goodness, as the Supreme Good, he is able to bring forth good out of evil" (St. Augustine, *Handbook on Faith, Hope, and Love*, transl. A.C. Outler, Grand Rapids, MI, Christian Classics Ethereal Library, generated online 2019, p. 7 [2]).

1. *approbo* situation (I) is different from *volo* situation (A) and similarly that *non volo* (*reprobo*) situation (O) differs from *nolo* (E).
2. *approbo* situation (I) and *non volo* (*reprobo*) situation (O) can be both true at the same time, whereas *volo* (A) and *nolo* (E) situations cannot be true but can be false at the same time.

The use of the square of will not only helps to visualise the differences between those situations and these specific relationships among them but also assures that such theological solutions are based on logically correct conceptual structures. Furthermore, the four options, represented by four corners of the square, linked together by specific relationships are in fact necessary to solve both mentioned problems. This is because without discerning those options one will not understand that negation of the state of will does not mean another state of will with some opposite object, but rather absence of will.

5 The Hexagon of Will

Finally, as a kind of supplement, it is worth pointing out that it is possible to develop the square of will to obtain the hexagon which includes additional options for some positive state of will, so having some active will (let us call it "voluntas mea excitat"), and of a negative one, when there is no act of will (let us call it "voluntas mea non excitat"). An example of such hexagon has been presented in Fig. 7.

The hexagon clearly presents two triangles of contrarieties. The one where two contrary states of active will (vertices A and E), so $V(x, s)$ and $V(x, \neg s)$, are opposed to the situation in which there is no act of will (Y): $\neg V(x, \neg s) \wedge \neg V(x, s)$, and

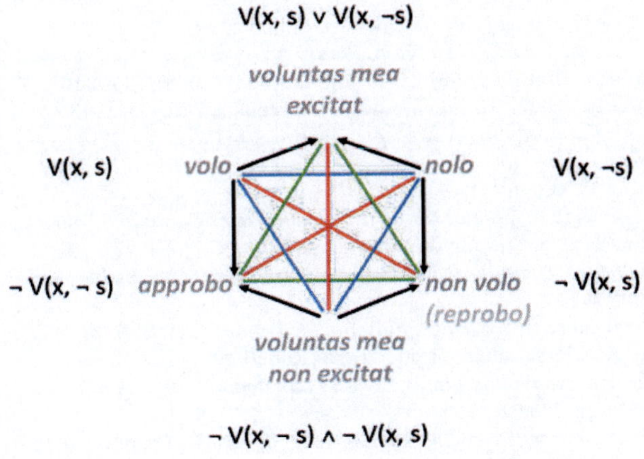

Fig. 7 The hexagon of will. For V(x, s)—**x** wants **s** to happen, where x is a person and s is a state of affairs

another one where two contrary states in which there is no active will (I and O): $\neg V(x, \neg s)$ and $\neg V(x, s)$ are opposed to the situation of some act of will (U): $V(x, s) \vee V(x, \neg s)$.

6 Conclusions

To conclude, let us briefly collect the most important remarks:

1. Stephen Langton had not drawn any square and had not pointed out explicitly all the square relations.
2. He mentioned necessary elements to easily complete the square.
3. We cannot guess whether he had in mind some square as a geometric representation. However, he considered a structure with four arguments combined by logical relations.
4. No matter what Langton imagined, the square of opposition was a framework for his theological solution, as such logical structure is necessary to formulate answers that he delivered.
5. The square is very useful or even necessary as a framework to solve some of theological problems like those connected with predestination or theodicy.

References

1. *A Partial Edition of Stephen Langton's Summa and Questiones with Parallels from Andrew Sunesen's Hexaemeron*, eds. S. Ebbesen and L. B. Mortensen, Cahiers de l'Institut du Moyen Age Grec et Latin, XLIX, 1985.
2. Augustine, *Handbook on Faith, Hope, and Love*, transl. A. C. Outler, Grand Rapids, MI, Christian Classics Ethereal Library, generated online 2019.
3. Baldwin J. W., *Master Stephen Langton, Future Archbishop of Canterbury: The Paris Schools and Magna Carta*, English Historical Review **123(503)** (2008), 811–846.
4. Béziau J.-Y., *The Power of Hexagon*, Logica Universalis **6** (2012), 1–43
5. Blanché R., *Sur l'opposition des concepts*, Theoria **19** (1953), 89–130.
6. *Der Sentenzkommentar des Kardinals Stephan Langton*, ed. A.M. Landgraf, Aschendorff Verlag, Münster 1952.
7. Lacombe G., Smalley B., *Studies on the Commentaries of Cardinal Stephen Langton*, Archives d'Histoire Doctrinale et Littraire du Moyen Age **5–6** (1931).
8. *Magistri Guillelmi Altissiodorensis Summa aurea*, ed. J. Ribaillier et al., Grottaferrata (Roma)—Paris, 1980.
9. *Magistri Petri Lombardi Parisiensis Episcopi Sententiae in IV libris distinctae*, ed. Collegii S. Bonaventurae Ad Claras Aquas, Grottaferrata (Roma) 1971.
10. Quinto R., *"Doctor Nominatissimus". Stefano Langton (1228) e la tradizione delle sue opere*, Aschendorff Verlag, Münster 1994.
11. *Stephen Langton, Quaestiones Theologiae, Liber I*, eds. R. Quinto and M. Bieniak, Oxford University Press, Oxford 2014.
12. *Stephen Langton, Quaestiones Theologiae, Liber III.1*, eds. M. Bieniak and W. Wciórka, Oxford University Press, Oxford 2021.

13. *Stephen Langton, Quaestiones Theologiae, Liber III.2*, eds. M. Bieniak, M. Trepczyński and W. Wciórka, Oxford University Press, Oxford 2022.
14. *Summa theologiae—The Summa Theologi of St. Thomas Aquinas*, Second and Revised Edition, 1920, transl. Fathers of the English Dominican Province, online ed. K. Knight 2017, URL: http://www.newadvent.org/summa
15. Valente L., *Logique et thologie trinitaire chez Étienne Langton: res, ens, suppositio communis et propositio duplex*, in: *Étienne Langton, prdicateur, bibliste, théologien*, Brepols, 2010.

The Limits of the Square: Hegel's Opposition to Diagrams in Its Historical Context

Valentin Pluder

Abstract The square of opposition hardly appears in German texts on logic from the early to mid-nineteenth century. This cannot be due to a lack of awareness of the square, for although it only appears occasionally in works from this period, these rare appearances present highly elaborate variations of it or show its historical development. But this is, almost without exception, the case only in works about the history of logic (Biese, Prantl, Rabus, Ueberweg) or in school textbooks (Fischer, Gockel, Jäger, Troxler, Lindner, Waitz). This might seem like a lot of references, but in fact these works only represent a small minority of the numerous logics that were published in German during this period. The absence of the square should not be taken as a sign of critical attitudes towards the logical relations represented by it; the opposite is more likely. As part of the Aristotelian heritage of traditional logic, the content of the square may have been considered so obvious that there was no need for further illustration. Logic was believed to have been perfected or nearly perfected for 2000 years, after all. Diagrams that use circles, rather than the square, can be found more often. However, this was probably not because they showed new logical content but because the practice of showing old content using circles was quite new at this time. The lack of interest in diagrams among the more orthodox logicians was not counterbalanced by an increased use by the non-orthodox ones. For example, neither circles nor squares are to be found in the works of Hegel. But Hegel at least discusses, through the lens of his *Science of Logic*, the limits to the use of diagrams in the context of the relations traditionally represented by the square of opposition. This paper aims to clarify the arguments that draw an opposition between Hegel's logic and diagrams like the square, in the light of his place within nineteenth-century German logic.

Keywords Square of opposition · Hegel · Nineteenth-century logic

V. Pluder (✉)
University of Siegen, Siegen, Germany
e-mail: valentin.pluder@uni-siegen.de

J.-Y. Beziau, I. Vandoulakis (eds.), *The Exoteric Square of Opposition*, Studies in Universal Logic, https://doi.org/10.1007/978-3-030-90823-2_6

Mathematics Subject Classification (2000) Primary; secondary

1 Introduction

The subject of this paper is the square of opposition in nineteenth-century German works on logic, with a particular emphasis on Hegel's attitude towards diagrams like the square. More precisely, the period between 1810 and 1870 will be explored. The 1870s marked the start of a period of great changes in logic, culminating in the publication of Frege's *Begriffsschrift* [6] in 1879. It is therefore quite clear why 1870 has been picked as an end point. Beginning with 1810 is rather more arbitrary. The transcript of Kant's *Lectures on Logic* was published in 1800 [17], and this would also have made a good starting point given the text's influence on ideas of logic. Hegel's *Wissenschaft der Logik* (*Science of Logic*), a work that contrasts sharply with Kant's relatively traditional theory of logic, was published a decade later: the first volume in 1812 [9], followed by the second volume in 1816 [10] and the revised and expanded second edition of the first volume in 1831 [11]. My investigation starts with the second decade of the nineteenth century because it is focused on Hegel's approach to diagrams. The aim is to establish why Hegel refused to use diagrams in his logic; given the constraints of space, I do not seek here to explain or justify his logic as a whole. Furthermore, the square of opposition is considered primarily with regard to its diagrammatic form and only secondarily with regard to its content.

The paper starts by examining the historical context. In this initial section, I firstly ask where exactly the square can be found between 1810 and 1870. As it turns out, the square is relatively rare during this period, and so – secondly – I consider why this might be the case. The following section focuses on Hegel. After a short examination of Hegelianism, Hegel's concept of the concept is briefly outlined. This is followed by a presentation of his critique of diagrams in his *Science of Logic*.

2 The Historical Context

2.1 *Where to Find the Square?*

A good place to start might be to look at some numbers. Between 1810 and 1870, at least 420 books on logic (and probably more) were published in German. This includes multiple editions but excludes Latin texts on logic written by German-speaking writers and, of course, all journal articles and suchlike. Out of these at least 420 books, I have only searched 120, so the following numbers are based only on this sample. The square can be found in 19 books by 11 authors. Nine of these books are works on the history of logic ([2], p.107; [21], pp.654, 677, 692, 694, 697; [22], p.271; [23], pp.14, 45, 417; [24], pp.22, 23, 208, 258; [26], pp.287, 305; [32], pp.136, 164; [33], pp.160, 161; [34], p.175) and seven are schoolbooks ([5], p.98; [8], p.63; [16], p.47; [19], p.92; [25], p.61; [31], p.285; [35], p.56). There are few surprises in the way the square is presented in the schoolbooks. The names of the relations between the categorical propositions are mainly kept in Latin, except for in the textbook by Troxler, which is in German (Figs. 1 and 2).

In most cases, the categorical propositions are arranged in the square as shown in the diagrams above. Only Trendelenburg, in his 1836 schoolbook on logic, swapped round the positions of the E and O propositions as a result of his study of Aristotle ([30], p.51). He was followed in this by Drobisch, who changed his square accordingly in the second edition of his *New Presentation of Logic*; in his first edition, also published in 1836, he had the square arranged the standard way ([3], p.40) (Fig. 3).

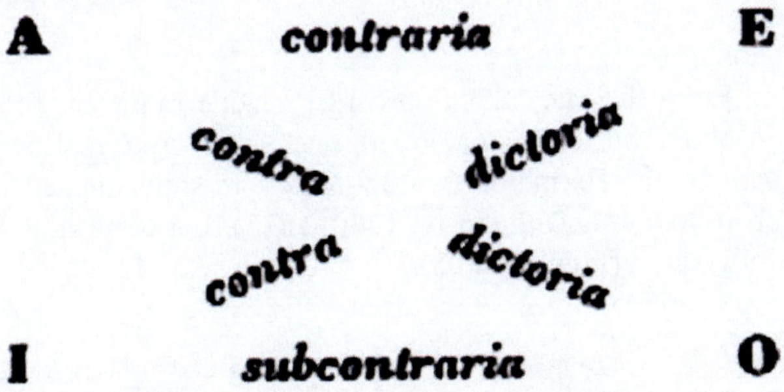

Fig. 1 Latin square from a schoolbook by F. Fischer, 1838 ([5], p.98). (Figure gratefully taken from the scans provided by the Münchner DigitalisierungsZentrum (www.digitale-sammlungen.de))

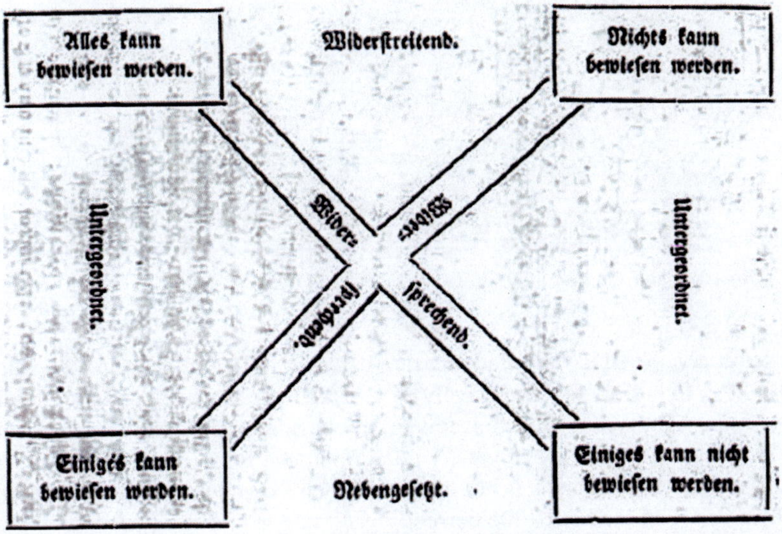

Fig. 2 German square from a schoolbook by I.P.V. Troxler, 1829 ([31], p.285). (Figure gratefully taken from the scans provided by the Münchner DigitalisierungsZentrum (www.digitale-sammlungen.de))

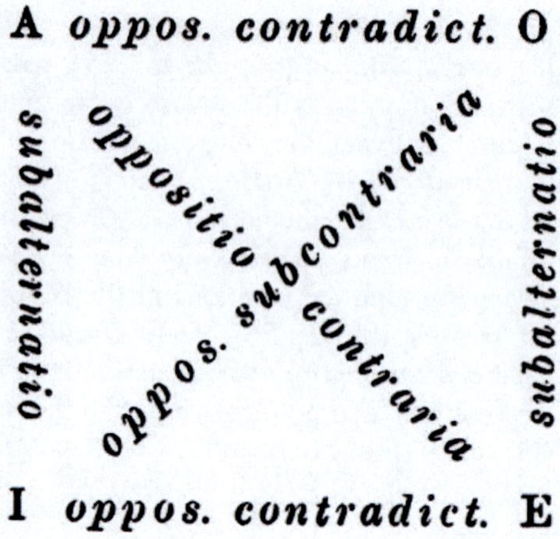

Fig. 3 Square with changed corner by M.W. Drobisch, 1851 ([4], p.76) . (Figure gratefully taken from the scans provided by the Münchner DigitalisierungsZentrum (www.digitale-sammlungen.de))

From the late 1850s onwards, a wide range of different squares can be found in books on the history of logic: from early Greek versions – for example, from Ammonius Hermeiou (c. 435–517) – to sophisticated scholastic versions, like one from Johannis Dullaert (c. 1480–1513) representing different kinds of hypothetical judgement (Figs. 4 and 5).

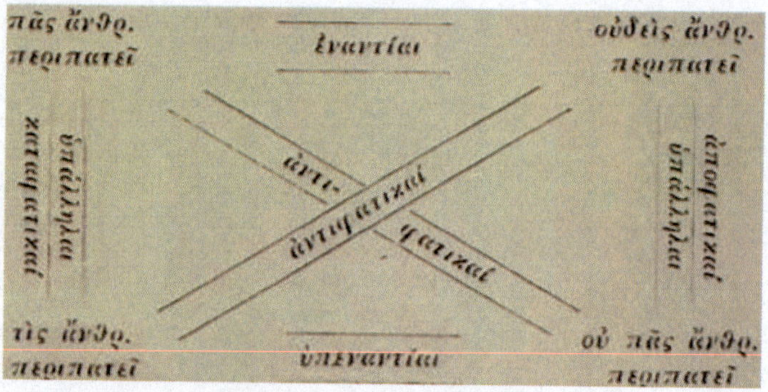

Fig. 4 Greek square from Prantl's *History of Logic*, 1855 ([21], p.654)

In summary, the square is relatively rare between 1810 and 1870 in the examined sample. It is found almost exclusively in reproductive works, like schoolbooks or books on the history of logic. It seems to be considered dead wood, since nobody works with the square or investigates its diagrammatic form. Even the change of the E and O corners by Drobisch is not motivated by his own ideas about the square but only by a desire to read Aristotle correctly.

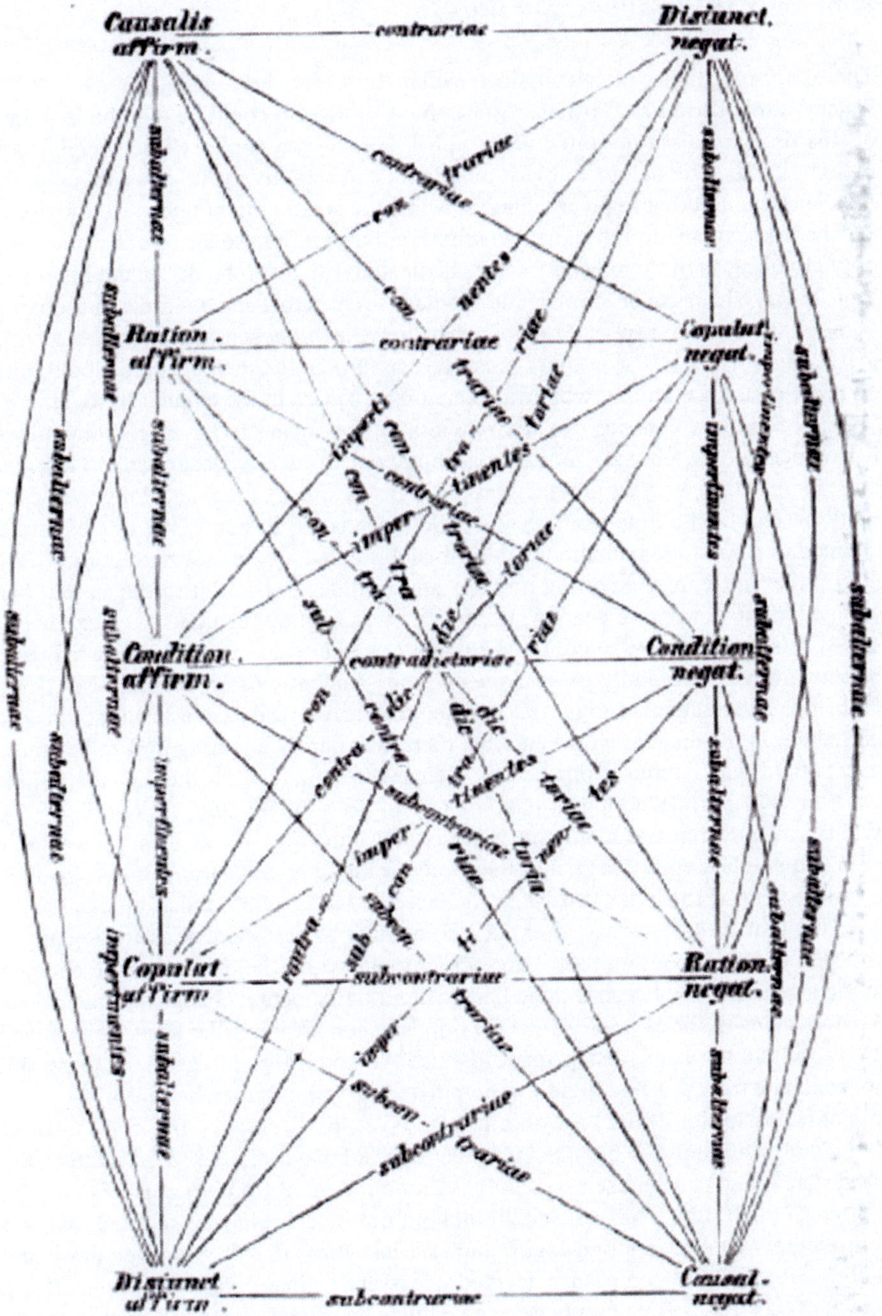

Fig. 5 Square for hypothetical judgements from Prantl's *History of Logic*, 1870 ([24], p.258). (Figure gratefully taken from the scans provided by the Münchner DigitalisierungsZentrum (www.digitale-sammlungen.de))

2.2 *Why Is the Square So Rare?*

There are, of course, many possible explanations for the relative rareness of the square in this period of German philosophy. One reason could be that the logicians of this time were simply unfamiliar with it. Or, on the contrary, they might have thought it was too obvious to merit mentioning. Alternatively, it is possible that the logicians thought that logic and diagrams like the square did not go well together.

The first explanation that they were unfamiliar with the square can be ruled out, not only because of the presence of sophisticated squares in books on the history of logic but also because of the very rudimentary squares that can be found in footnotes without any further explanations of what they are supposed to illustrate e.g. [2]., p.107, or [30], p.51. The authors must have assumed that their readers would have no problem understanding what was meant by a square in the context of logic.

The second explanation for the rareness of the square is far more convincing: it was deemed so obvious that no one bothered to take a closer look. This idea corresponds to a very common view of logic in the German-speaking world up until the end of the nineteenth century. Schopenhauer, to name just one example, claimed that Aristotle had already described logic to an 'extent of perfection' ([28], p.357) that left barely anything more to add in order to bring it to the state it had attained in the nineteenth century, when logic was 'rightly regarded as an exclusive, self-subsisting, self-contained, finished, and perfectly safe branch of knowledge, to be scientifically treated by itself alone and independently of everything else' ([29], §9/p.46). The belief that logic is a science that had already been brought close to perfection in the ancient world and that there was hardly anything new to be found in this field was famously phrased by Kant, who remarked on the fact that 'since the time of Aristotle it [logic] has not had to go a single step backwards [. . .]. What is further remarkable about logic is that until now it has also been unable to take a single step forward, and therefore seems to all appearance to be finished and complete' ([18], p.Bviii/p.106). Against this background, most logicians must have thought it unnecessary to explain something as self-evident as the square of opposition. Of course, there were exceptions to the widely held view that no changes in logic were needed because there had been no fundamental changes in logic since Aristotle. Hegel, for example, drew the opposite conclusion. Like Kant, he believed that logic had not taken a step forward since Aristotle. But unlike Kant, Hegel did not conclude from this that logic was completed early on. Rather, he saw all the more urgent necessity for a 'total reworking' of it ([12], p.31; cf. [11], pp.35–37). Hegel thought new life must be breathed into the 'ossified material', the 'dead matter', the 'devastated land' of the useless, empty wisdom inherited from scholasticism ([12], p.507; cf. [10], p.5). This resuscitation could not be achieved by supplementing a 'pure logic' with an 'applied logic' and still less through 'all the psychology and anthropology that is commonly deemed necessary to interpolate into logic' ([12], p.676; cf. [10], p.179). Such phenomena, which are often to be found in nineteenth-century German logic (e.g. in Fries's system), are for Hegel the symptom of the crisis and not its cure:

> A just published and most up-to-date adaptation of this science, Fries's *System of Logic* [7], goes back to its anthropological foundations. The shallowness of the representation or opinion on which it is based, in and of itself, and of the execution, dispenses me from the trouble of taking any notice of this insignificant publication. ([12], p.31; cf. [9], p.23)

One might think that someone like Hegel, who was willing to rethink all logic without being led astray by psychology and anthropology, might have found a new interest in diagrams like the square. But that proved not to be the case since, like other philosophers of this time, Hegel was convinced that diagrams and logic do not go well together. This belief could be explained by the idea that there is a fundamental difference between sensibility or spatial perception and thinking. Since logic is about thinking and diagrams are about spatial perceptions, mixing them together, and thus ignoring this difference, would obscure understanding of the genuine nature of thinking. This thought can be found in Kant too, when he explains: 'Hence we distinguish the science of the rules of sensibility in general, i.e., aesthetic, from the science of the rules of understanding in general, i.e., logic' ([18], pp.A52, B76/p.194). When he speaks of the difference between understanding and sensibility, Kant of course does not intend to keep them separated: 'Only from their unification can cognition arise' (ibid.). Therefore, it is not surprising that diagrams (though not, admittedly, squares) can be found in Kant's logic, at least if one trusts the Jäsche edition of Kant's *Lectures on Logic* ([17], pp.160, 168). Matters are different with Hegel.

3 Hegel and Hegelianism

In Hegel's writings on logic, there are no diagrams to be found. Even in the notes made by the students who attended his lectures on logic, there is only one small Euler-like diagram ([14], p.608) and one square ([13], p.8). But this square is just meant to illustrate how the train of thought in Hegel's *Science of Logic* proceeds. Clearly, Hegel was no friend of diagrams. So what is the point of investigating his attitudes towards them? The answer to this question is somewhat dialectic in its own right: if the more traditional logics, which tend to be affirmative of the square, have nothing to say, maybe the critiques of the square have something interesting or even something new to add to the topic. They do discuss the subject at least.

3.1 Hegelianism

The hostility towards diagrams representing certain crucial aspects of thinking has a long tradition within Hegelianism. One example from the twentieth century is Theodor Litt, who repeatedly claims in his *Individual and Community* that spatial perception is unable to illustrate the relation between the individual and the community adequately, because 'although the individual lives in the whole, none the less the whole lives in the individual' ([20], p.155; cf. ibid., pp.7–8, 117, 157,

239). Meanwhile, to take an example from the nineteenth century, Karl Rosenkranz explicitly expresses his scepticism towards the square of opposition, writing in his *Science of the Logical Idea* that:

> In the logics this square is usually presented with emphatic solemnity as if it were the altar in the holy of holies of the goddess of wisdom. But it seems like the square has contributed little to the essence of the logical insight. The reason for that is presumedly that thinking becomes rigid in such schematism and that the results that arise from such superficial reflections are too modest. ([27], pp.130–131)

Rosenkranz believes on the one hand that essential parts of thinking are not rigid and can therefore not be captured by schematisms and on the other that the aspects of thinking that can be captured by schematisms like the square are trivial. Both ideas are in line with the above-mentioned assumptions: many nineteenth-century logicians deemed the square too obvious to merit exploring, and some logicians were convinced that logic, or at least the interesting parts of logic, and diagrams do not go well together.

3.2 Hegel

3.2.1 Hegel's Concept of the Concept

Hegel writes about figures, forms, lines and circles in his *Science of Logic* when he finally presents his concept of the 'concept' at the beginning of the third and last book. The line of reasoning in his work is therefore already rather advanced at this point, and the conceptions he uses are quite complex. Nonetheless, it is necessary to outline roughly what Hegel means when he talks about the 'concept'. First of all, it is important to know that Hegel uses 'concept' (*Begriff*) as a technical term to refer to the structure and form of performance of all thinking and truly understood being. Since Hegel's use of the German term *Begriff* is idiosyncratic, it has been translated in several different ways: 'comprehension', which seems to be closer to the German *Begriff*, or 'logos', in a more or less Heraclitan sense, or 'notion'. In the following, *Begriff* will be translated as 'concept', simply because the translation of the *Science of Logic* that is used in this paper keeps it that way. To use 'logos' as a translation would of course have the advantage that the performative character of the structure that *Begriff* is meant to name would come to the fore.

Hegel's concept includes three moments or determinations: the universal (*Allgemeinheit*), the particular (*Besonderheit*) and the singular (*Einzelheit*), and of course these are technical terms as well. The three moments of the concept are closely interwoven with each other. The universal, for example, is supposed not to be abstract but a 'concrete universal'. It is established not by abandoning specific determinations but by embracing their differences as such. That is why it can only be properly conceived in relation to the particular and the singular. The singular is likewise not understood as a private or isolated element in its own right. It is a distinct part of a specific universal, which in turn is established as specific by the differences of its parts. All three moments of the concept are thus interdependent. At the same time, they are necessarily opposed to one another because their differences define what they mean.

Although Hegel clearly disapproves of diagrams, it is very common and sometimes even helpful to illustrate the one concept with its three opposing determinations by means of a triangle [15] (Fig. 6).

This triangle could be understood as representing the A and E corners. So one might be tempted to add the negations in the form of a complementary triangle, so as to form a hexagon (Fig. 7).

But the result would be very hard to interpret. Even though the square of opposition can be discerned within this diagram, it is – regardless of the semantic content of 'Singular' and '~Singular' – unclear how the new corners fit in, let alone how the new relations should be understood. The triangle of the singular, the universal and the particular is neither a triangle of contrariety nor a triangle of subcontrariety and even less a triangle of subalternity (cf. [1]) (Figs. 8 and 9).

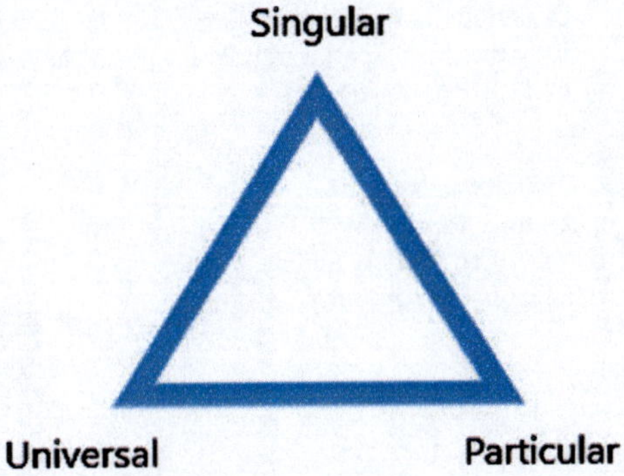

Fig. 6 The determinations of Hegel's concept, illustrated by a triangle

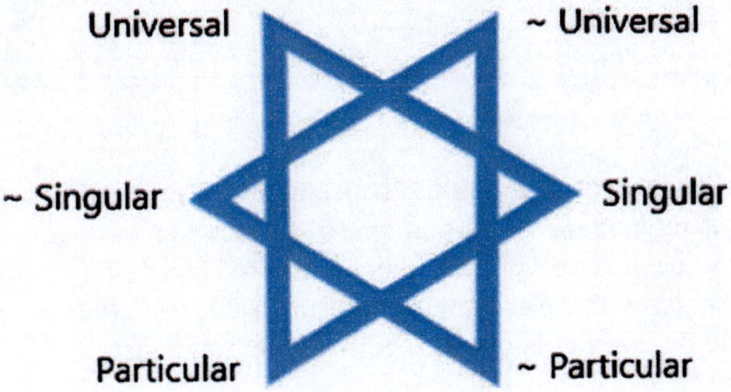

Fig. 7 A hexagon formed by the concept's triangle and its negation

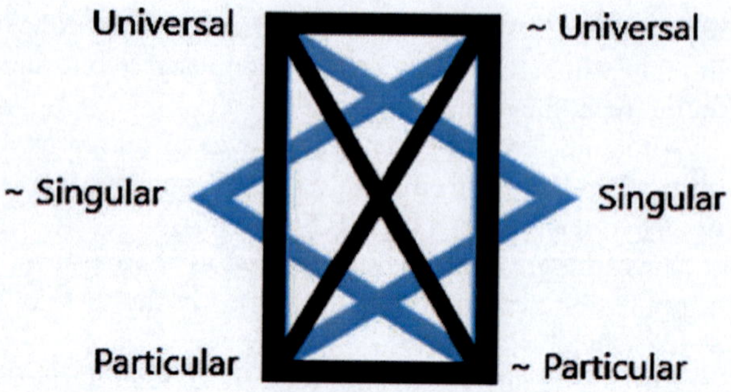

Fig. 8 The square within the concept's hexagon

In fact, the relations between the three determinations of the concept do not form a triangle at all, because no diagram could possibly be a proper approach to what Hegel has in mind when he talks about his concept:

> Since the human being has in language a means of designation that is appropriate to reason, it is otiose to look for a less perfect means of representation to bother oneself with. It is essentially only spirit that can grasp the concept as concept [. . .]. It is futile to want to fix it by means of spatial figures [. . .] for the sake of the *outer eye* [. . .]. ([12], pp.545–546; cf. [10], pp.48–49)

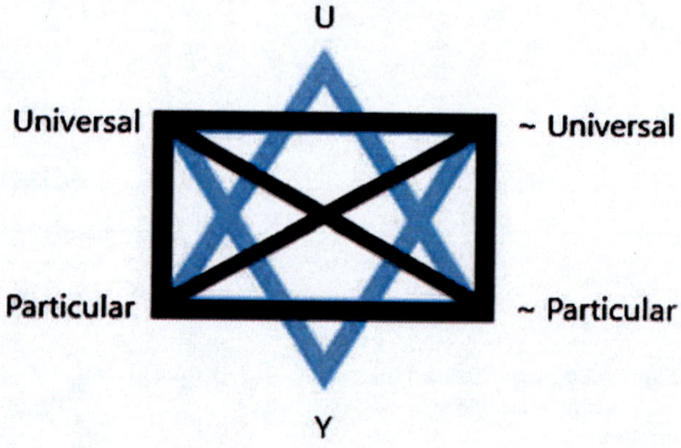

Fig. 9 The hexagon of oppositions formed by a triangle of contrariety and a triangle of subcontrariety

Hegel points out that the concept can be understood adequately only by the spirit, using language, and that spatial figures like diagrams are unsuitable for that task. Since the concept has a pivotal role in Hegel's logic, it is safe to assume that Hegel shares with several other logicians of his time the opinion that logic (or at least his logic) does not go well together with diagrams.

3.2.2 Hegel's Critique of Diagrams in Logic

But what is the reason for Hegel's objection to diagrams in his logic? In his words:

> It is characteristic of objects of this kind [lines, figures, numbers, etc.], as contrasted with the determinations of the concept, that they are mutually *external*, that they have a *fixed* determination. Now when concepts are made to conform to such signs, they cease to be concepts. Their determinations are not inert things [*Totliegendes*], like numbers and lines whose connections lie outside them; they are living movements; the distinguished determinateness of the one side is immediately also internal to the other side; what would be a complete contradiction for numbers and lines is essential to the nature of the concept. ([12], pp.544–545; cf. [10], p.47)

According to this passage, 'lines and figures' are, firstly, mutually external. Regarding the square of opposition, this makes sense. Neither the corners nor the different lines overlap each other. There are good reasons for this: for example, the square is meant to visualise the difference between the contrary and the contradictory relation. It is more difficult to understand how Hegel's critique or diagnosis can be applied to, say, Euler diagrams. In the case of one set that completely includes another set, it seems obvious that two circles do overlap. But that is not actually what the diagram shows. It could equally be that the outer circle ends at the edge of the inner circle, as if the area of the outer circle had a hole in it covered exactly by the smaller circle. That the outer circle continues beneath the inner one is only one possible interpretation of the diagram. This leads to Hegel's second point: the connections between the elements that form a diagram are made not by the forms themselves but by their interpretation. The diagonals in the square of opposition do not show the contradictory relation of the corners they connect. When Drobisch exchanges the E and the O corners (see Fig. 3), he simply interprets the diagonals as no longer signifying contradictory relations but rather contrary or subcontrary ones. Hegel thirdly points out that the determinations in diagrams are fixed. At least in his day, they did not move or morph. This means they do not change according to the relations they are in but simply always stay what they are.

In contrast to the 'lines and figures', the determinations of Hegel's concept are not isolated. They are in some manner internally connected. They react to each other and change relative to their counterparts, more like living beings than dead things (*Totliegendes*). So one might get the impression that Hegel's concept is some swirling, undifferentiated oneness. That is not what Hegel intends. His concept of the concept unites the differences of the determinations as well as the unity of the determinations. It is the unity of the oppositions or the identity of identity and non-identity.

That means Hegel's concept includes aspects that are clearly distinguished from each other. Therefore, it would be misleading to think Hegel means to overcome all differences by making them fade away in some blurry union. Moreover, the aspects of the concept, which are well differentiated from each other, could also be shown by a diagram: 'The determinations of the concept, universality, particularity, and singularity, certainly are, like lines or the letters of algebra, *diverse*; and they are also *opposed* and allow, therefore, the signs of *plus* and *minus*' ([12], p.544; cf.

[10], p.47). But focusing on this side alone would be misleading, because the other side is that:

> The determinations of the concept [. . .] themselves and especially their connections [. . .] are in their essential nature entirely different from [. . .] lines and their connections, from the equality and diversity of magnitudes, the *plus* and *minus*, or the superimposition of lines, or the joining of them in angles and the resulting disposition of space that they enclose. ([12], p.544; cf. [10], p.47)

This might seem plainly inconsistent. But there is quite a simple configuration that gives at least a hint of what Hegel has in mind with his unity of oppositions. The basic idea is that of opposites that mutually constitute each other. That means the one side is exactly what the other side is not, and neither side can be conceived without its complement. Both sides are necessarily connected to each other, because otherwise they would cease to be what they are. For this reason, they are one. And for the same reason, they are necessarily opposed, because their difference from each other defines what they are. That is why in this case 'the distinguished determinateness of the one side is immediately also internal to the other side'. Very roughly speaking, this is also the way the moments of Hegel's concept are related to each other. The determinateness can be understood as semantic content that derives from the relation between the opposite sides. Indeed, one of the main criticisms that Hegel makes of traditional logic, with its notations, lines, figures and calculus, is that it is only formal and therefore meaningless, whereas he claims that logic has to include content as well.

The possibilities of a visual illustration of such strictly contradictory relations are in fact limited. One might object that there is one obvious example that disproves this. A taijitu or 'yin–yang' symbol seems to show exactly this relation, for the white part seems to be formed by the black part and vice versa (Fig. 10).

Fig. 10 The taijitu or 'yin–yang' symbol

Fig. 11 The black form is partly independent of the white form

But a closer look reveals that one part can be changed without any effect on the other. The white and black parts determine each other only where they border on each other (Fig. 11).

Because there is no possible image where the black part completely encompasses the white part and the white part completely encompasses the black part at the same time – that would be a 'universe' – there are limits to visual illustration. These limits can also be shown by removing the aspects of the image that are determined by the opposition of the black and white parts. If the taijitu is not a symbol but showed the strictly contradictory relation that Hegel presumably has in mind, nothing should remain. But in fact, there would still be a circle because the outside of the image is independent of the internal relation of the black and white parts.

Of course, there might be ways to illustrate the impossibility of illustration by using graphics that work with these very limits, like a Penrose triangle, but these effects are probably hard to implement in diagrams. And of course there are non-Euclidian geometries that come closer to Hegel's concept than he might have thought possible, for example, projective geometries that were developed purely synthetically and understand figures without any reference to 'external determinations' like coordinates or formulae. But, again, a specific conception of geometry is probably hard to operationalise within diagrams.

Returning to the square of opposition, it might be pointed out that it displays relations which are not meant to be strictly contradictory. Therefore, it should not be a problem if they are illustrated by 'lines and figures' that lie apart from each other. The lines that show the contrary and contradictory relations do not even cross each other. While one could argue that the relation between the identical and contradictory relations should show their mutual determination, because it is a contradictory relation in itself, the relation between the contrary and contradictory relations is surely not a contradictory one. Nonetheless, Hegel stresses that the diagram is still misleading because the contrary relation depends on the contradictory one:

> They [the contrary and contradictory relation] are viewed as two particular *species*, each fixed for itself and indifferent towards the other, without any thought being given to the dialectic and the inner nothingness of these differences, as if that which is *contrary* would not equally have to be determined as *contradictory*. ([12], p.543; cf. [10], p.46)

Hegel does not believe every contrary relation is actually a contradictory one. Rather, he wishes to express that without contradictory relations there is no contrary relation.

A closer look at the contrary relation might indicate what Hegel has in mind. Within the contrary relation, two cases can be distinguished. Case 1, if one side is true, the other is determined to be false. Case 2, if one side is false, the other is indeterminate; it can be true or false. In regard to these two cases, two contradictory aspects can be recognised within the contrary relation: firstly of course in case 1 and secondly between case 1 and 2, because if case 1 is not the case, then case 2 has to be and vice versa. So the contrary relation reveals itself not to be a relation of its own quality that would back up its isolation from the contradictory relation in the square. It is only a lack of determination, as can be seen in case 2, that differentiates the contrary from the contradictory relation.

This does not mean that Hegel considers it pointless to determine a relation as contrary. The long and complex line of thought in his *Science of Logic* is intended to show that the categories of logic are interrelated and context-sensitive. They are not isolated like rocks but answer to each other and change their determination in the light of the relations they participate in. That this conception does not lead to arbitrariness and chaos but can be understood rationally, without losing the different determinations logic provides, is the other central purpose of Hegel's logic. This should be kept in mind when he says that within his concept of the concept, the determinations are not separated but rather exist in a state of permanent interchange:

> The universal has proved itself to be not only the identical, but at the same time the diverse or *contrary* as against the particular and the singular, and then also to be opposed to them, or *contradictory*; but in this opposition it is identical with them, and it is their true ground in which they are sublated. The same applies to particularity and singularity, which are likewise the totality of the determinations of reflection. ([12], p.543; cf. [10], p.46)

Even though it is questionable whether this makes any sense at all, it is clear why Hegel can hardly agree to illustrate his concept using a diagram:

> It is, therefore, entirely inappropriate, in order to grasp such an inner totality, to want to apply [. . .] spatial relations in which the terms fall apart; such relations are rather the last and the worst medium that could be used. ([12], p.545; cf. [10], p.48)

But one might doubt whether there is any appropriate way to 'grasp such an inner totality' at all, since the propositions Hegel uses to describe it reveal contradictions. This is, of course, what one would expect from Hegel. However, this is also helpful to understand why Hegel's logic is not fixed in the way that (on his view) the determinations in diagrams are. In fact, contradictions appear systematically within Hegel's logic. Every single proposition provokes the formulation of an opposed proposition, due to the link between the semantic determination of a proposition and the negative relation to its opposite (*omnis determinatio est negatio* ([12], p.87;

cf. [11], p.101)). But just as systematically, these contradictions are solved, not by levelling down the opposed determinations but by embedding these determinations into a more complex and richer context where they are properly related to each other and no longer appear as contradictions. This more complex context is the ground in which the determinations of the former contradiction are sublated. But this more complex context will reveal itself to be contradictory as well, and so the logic proceeds from the simplest notion of pure being to the very complex notion of Hegel's concept and beyond that until finally the whole system has evolved. While in Hegel's logic a single proposition has the power to overcome its isolated fixation and reveal its true nature as part of a wider context, diagrams and traditional logic do not provoke such a development. In Hegel's view, their 'terms fall apart' and are fixed in formal isolation from each other.

4 Conclusion

Hegel does not wholly object to the use of diagrams. As mentioned above, they can illustrate the aspects of difference within the concept. But at the same time, he makes perfectly clear that in his eyes these aspects should not be mistaken for the whole structure of thinking or logic that is the concept in its entirety. To understand these aspects of difference within the concept adequately, they must be embedded in their wider context, and in this context, their differences no longer appear isolated and independent from each other but become relative and fluid.

Broadening the perspective and bringing the nineteenth century back into view, one could say that the aspects of thinking that can be shown using the diagrams of this time correspond to the traditional and back then largely unquestioned logic of the time. The study of the context and conditions in which this traditional logic works could be called 'metalogic'. Since it was common among nineteenth-century German-speaking logicians to think that logic was already completed and perfected, these logicians often laid emphasis on metalogical questions of this sort. This leads to another answer to the question of why the square was so rare at this time: even if the square is doubtlessly useful to illustrate traditional logic, it may be deemed useless for reflecting on metalogical issues. This was the view taken by, for example, Hegel and his school.

But even if it is true that diagrams are limited in their ability to display certain kinds of relations, it does not necessarily render them useless in these cases. Because once understood, the limit itself might be operationalised and express determinations. Or, conversely, by defining limits, one always also to a certain extent defines what lies beyond those limits. That too is a very Hegelian thought.

Acknowledgements I would like to express my gratitude to the audience of the sixth World Congress on The Square of Opposition that took place in Crete in 2018 and to the anonymous reviewers for comments that contributed to the improvement of this paper. I would also like to thank Jens Lemanski who supported me in writing this paper.

References

1. Beziau, J.-Y.: The power of the hexagon. In: Beziau, J.-Y. (ed.): Special Double Issue on the Hexagon of Opposition. Logica Univers. **6** (1–2) (2012)
2. Biese, F.: Die Philosophie des Aristoteles, vol. 1: Logik und Metaphysik. G. Reimer, Berlin (1835)
3. Drobisch, M.W.: Neue Darstellung der Logik. L. Voss, Leipzig (1836)
4. Drobisch, M.W.: Neue Darstellung der Logik, second totally revised edition. L. Voss, Leipzig (1851)
5. Fischer, F.: Lehrbuch der Logik. J.B. Metzler'sche Buchhandlung, Stuttgart (1838)
6. Frege, G.: Begriffsschrift. L. Nebert, Halle a.S. (1879)
7. Fries, J.F.: System der Logik. Mohr and Zimmer, Heidelberg (1811)
8. Gockel, Ch.F.: Propädeutische Logik und Hodegetik. Ch.Th. Groos, Karlsruhe (1839)
9. Hegel, G.W.F.: Wissenschaft der Logik, vol. 1: Die objektive Logik. In: Gesammelte Werke, vol. 11. Meiner, Hamburg (1978)
10. Hegel, G.W.F.: Wissenschaft der Logik, vol. 2: Die subjektive Logik. In: Gesammelte Werke, vol. 12. Meiner, Hamburg (1981)
11. Hegel, G.W.F.: Wissenschaft der Logik, part 1: Die objektive Logik, vol. 1: Die Lehre vom Sein. In: Gesammelte Werke, vol. 21. Meiner, Hamburg (1985)
12. Hegel, G.W.F.: The Science of Logic (1812, 1816), translated and edited by G. Di Giovanni. Cambridge University Press, Cambridge (2010)
13. Hegel, G.W.F.: Vorlesungen über die Wissenschaft der Logik I. In: Gesammelte Werke, vol. 23,1. Meiner, Hamburg (2013)
14. Hegel, G.W.F.: Vorlesungen über die Wissenschaft der Logik III. In: Gesammelte Werke, vol. 23,3. Meiner, Hamburg (2017)
15. hegel.net/en/e1311.htm#tree
16. Jäger, J.R.: Handbuch der Logik. J.G. Heubner, Wien (1839)
17. Kant, I.: Logik. Ein Handbuch zu Vorlesungen, edited by G.B. Jäsche. F. Nicolovius, Königsberg (1800)
18. Kant, I.: Critique of Pure Reason, translated and edited by P. Guyer and A. Wood. Cambridge University Press, Cambridge (1998)
19. Lindner, G.: Lehrbuch der formalen Logik. C. Gerold's Sohn, Wien (1867)
20. Litt, Th.: Individuum und Gemeinschaft. Grundlegung der Kulturphilosophie, second totally revised edition. B.G. Teubner, Leipzig, Berlin (1924)
21. Prantl, C.: Geschichte der Logik im Abendlande, vol. 1. S. Hirzel, Leipzig (1855)
22. Prantl, C.: Geschichte der Logik im Abendlande, vol. 2. S. Hirzel, Leipzig (1861)
23. Prantl, C.: Geschichte der Logik im Abendlande, vol. 3. S. Hirzel, Leipzig (1867)
24. Prantl, C.: Geschichte der Logik im Abendlande, vol. 4. S. Hirzel, Leipzig (1870)
25. Rabus, L.: Lehrbuch der Logik. A. Deichert, Erlangen (1863)
26. Rabus, L.: Logik und Metaphysik, part 1: Erkenntnislehre, Geschichte der Logik, System der Logik. A. Deichert, Erlangen (1868)
27. Rosenkranz, K.: Wissenschaft der logischen Idee, vol. 2: Logik und Ideenlehre. Gebrüder Bornträger, Königsberg (1859)
28. Schopenhauer, A.: Philosophische Vorlesungen. Theorie des Erkennens (1819–1822). In: A. Schopenhauers sämtliche Werke, vol. 9, edited by P. Deussen. R. Piper and Co, München (1913)
29. Schopenhauer, A.: The World as Will and Representation, vol. 1 (1819), translated by E.F.J. Payne. Dover Publications, New York (1966)
30. Trendelenburg, F.A.: Elementa Logices Aristotelicae. In Usum Scholarum. G. Bethge, Berlin (1836)
31. Troxler, I.P.V.: Logik, part 1. Cotta'sche Buchhandlung, Stuttgart, Tübingen (1829)
32. Ueberweg, F.: System der Logik und Geschichte der logischen Lehren. A. Marcus, Bonn (1957)

33. Ueberweg, F.: System der Logik und Geschichte der logischen Lehren, second revised edition. A. Marcus, Bonn (1965)
34. Ueberweg, F.: System der Logik und Geschichte der logischen Lehren, third expanded and improved edition. A. Marcus, Bonn (1968)
35. Waitz, F.H.W.: Die Hauptlehren der Logik. Hennings and Hopf, Erfurt (1840)

Augustus De Morgan's Unpublished Octagon of Opposition

Anna-Sophie Heinemann and Lorenz Demey

Abstract The British logician Augustus De Morgan (1806–1871) sought to unify the traditional syllogistics with the new algebraic logic that he and George Boole (1815–1864) were developing. Although there are hardly any diagrams or figures in De Morgan's published writings, in his unpublished manuscripts, one can find various attempts to draft certain figures of opposition, which are evidently meant to fit the relations between the propositions of De Morgan's extended syllogistics. In this paper, we present some archival findings and discuss their contemporary relevance. We will focus on one of De Morgan's unpublished diagrams, which occurs several times throughout his manuscripts. Based on a detailed analysis of this octagon, in combination with De Morgan's unpublished notes as well as his published materials, we argue that this diagram belongs to the type of so-called KJ octagons. Throughout the twentieth century, this type of diagram has been studied quite extensively in philosophical logic, and in recent years even in computer science. Historically speaking, it has long been assumed that this type of octagon was studied for the first time by John Neville Keynes and William E. Johnson around the turn of the twentieth century. However, the manuscript findings presented in this paper clearly show that this type of octagon was already known by De Morgan, some five decades earlier than Keynes and Johnson.

Keywords Augustus De Morgan · Syllogistics · Logical notation · Logical geometry · Square of opposition · Keynes-Johnson octagon

Mathematics Subject Classification (2000) Primary 01A55; Secondary 03A05, 03B65, 68T30

A.-S. Heinemann
Universität Paderborn, Paderborn, Germany
e-mail: annasoph@mail.uni-paderborn.de

L. Demey (✉)
Center for Logic and Philosophy of Science, KU Leuven, Leuven, Belgium
e-mail: lorenz.demey@kuleuven.be

J.-Y. Beziau, I. Vandoulakis (eds.), *The Exoteric Square of Opposition*, Studies in Universal Logic, https://doi.org/10.1007/978-3-030-90823-2_7

1 De Morgan, Symbols, Syllogistics, and Diagrams?

Augustus De Morgan[1] does not belong to those historical characters who are very well-known today. Although every contemporary logician is familiar with the propositional laws regarding conjunction, disjunction, and negation that have come to bear his name,[2] relatively little is known about De Morgan himself. He is notoriously referred to in the historiography of algebraic logic as a more traditionally minded contemporary to George Boole and described as "not a clear-thinking philosopher" who "worked largely within the syllogistic tradition" [18, p. 27].[3] His involvement in the debates over the so-called quantification of the predicate is habitually mentioned for the very, maybe the only, reason that there is a reference to it in the preface to Boole's *Mathematical Analysis of Logic*, published in 1847 [1, p. 1].[4] His own contributions to the logical literature of his times are usually not discussed very extensively.[5]

As a Cambridge-trained mathematician, De Morgan thought of logic as a set of operations defined to be performed upon symbols, rather than upon intellectual

[1] During most of his adult life (1828–1831 and 1836–1866), De Morgan served as professor of mathematics at London University and University College of London, respectively. He had studied and taken his BA degree at the University of Cambridge, but refrained from proceeding to the MA degree for reasons of religious discrimination. De Morgan was born on 27 June, 1806, at Maduras, Madras, India. He was admitted to Trinity College at the University of Cambridge on 1 February, 1823. He took the chair of mathematics at the newly founded London University in 1828. In 1864, he became the first president of the Mathematical Society. De Morgan died on 18 March, 1871, in London (information to be found in the records of the University of Cambridge, available on the Cambridge Alumni Database: http://venn.lib.cam.ac.uk/, last access 26 August 2019).

[2] In classical propositional logic, De Morgan's laws state that $\neg(p \vee q)$ is equivalent to $\neg p \wedge \neg q$ and that $\neg(p \wedge q)$ is equivalent to $\neg p \vee \neg q$. Despite their name, these laws were already known by scholastic authors such as William of Ockham, Walter Burley, and Paul of Pergula [2]. De Morgan himself stated them differently, viz., in terms of classes instead of propositions: "The contrary of an aggregate is the compound of the contraries of the aggregants [...]. The contrary of a compound is the aggregate of the contraries of the components" [14, p. 40]. His presentation corresponds roughly to $\overline{(X \cup Y)} = \left(\overline{X} \cap \overline{Y}\right)$ and $\overline{(X \cap Y)} = \left(\overline{X} \cup \overline{Y}\right)$.

[3] As early as 1918, C. I. Lewis also stated that De Morgan's "methods and symbolism ally him rather more with his predecessors than with Boole and those who follow" and that De Morgan's articles "are ill-arranged and interspersed with inapposite discussion" [25, p. 38]. Both Lewis and Grattan-Guinness do admit that there are some noteworthy novelties in De Morgan's logic. Lewis credits him with 14 pages (pp. 37–51), Grattan-Guinness with 12 pages (pp. 25–37). However, the general impression of De Morgan's contributions seems to have adjusted to the quotes just given.

[4] A popular narrative has it that De Morgan's *Formal Logic* [9] came off the press on the very same day as Boole's *Mathematical Analysis of Logic* [1]. An overview of the debate between De Morgan and Sir William Hamilton which Boole refers to is to be found in [20, pp. xxi–xxiv]. For an extended discussion, see [21].

[5] There are of course exceptions. For example, in the early 1990s, De Morgan's logic has been a subject in Maria Panteki's [27, pp. 407–492]. De Morgan's logic of relations has been dealt with in Daniel D. Merrill's [26]. More than half of [21] is devoted to De Morgan.

contents.[6] In other words, he tended to conceive of logic as inherently symbolical–but, one should assume, not as diagrammatic.[7] When it came to relations between propositions, for example, De Morgan usually spoke of them in terms of implication instead of illustrating them by figures of opposition such as the classical square of opposition.[8] Indeed, there are hardly any diagrams or figures in De Morgan's published writings.[9]

Given De Morgan's apparent reluctance to make use of logical diagrams, it is all the more surprising to learn that among De Morgan's unpublished manuscripts, there remain the documents of various attempts to draft certain figures of opposition which are evidently meant to fit the relations between the propositions of De Morgan's extended syllogistics. These documents suggest that the reason why De Morgan did not publicly invoke such schemata is not that he rejected them altogether but, rather, that he seemed to have remained unclear about how to extend the classical square of opposition in order to be applicable to his version of syllogistics. Some of these findings and their interpretations will be the subject of the present paper.

In particular, we will argue that one of De Morgan's unpublished diagrams, which occurs several times throughout his manuscripts, is an example of the type of so-called KJ octagons. Throughout the twentieth century, this type of diagram has been studied quite extensively in philosophical logic, and in recent years even in computer science. Consequently, its logical and diagrammatic properties are now well-understood. Historically speaking, it has long been assumed that this type of octagon was studied for the first time by John Neville Keynes and William E. Johnson around the turn of the twentieth century (1894–1921); hence the term "KJ

6 His approach culminated particularly in his theory of the "abstract copula" and attempts at a logic of relations as prepared in his second paper in the "syllogism series" [10, pp. 107–116] and elaborated on in the fourth [13]; see [15] for a modern edition. For De Morgan's logic of relations, see [26].

7 With the exception of rudiments of "pictorially" motivated choices of symbols, such as an enclosing parenthesis – the remnant of a circle – to signify a term's being quantified universally (cf. Sect. 3.1).

8 De Morgan himself did not use the term "implication" but rather spoke of "affirmation" and "containment." Even in his very first little textbook on logic [9], which was published in 1839 and kept quite close to the traditional syllogistics of his times, he stated that "Every *A* is *B*" "affirms and contains" the particular affirmative "Some *A* is *B*," while it "denies" the universal negative "No *A* is *B*" as well as the particular negative "Some *A* is not *B*." "No *A* is *B*" in turn "affirms and contains" the particular negative "Some *A* is not *B*" but "denies" the universal affirmative "Every *A* is *B*" as well as the particular affirmative "Some *A* is *B*." Again, "Some *A* is *B*" in turn "does not contradict" the universal affirmative "Every *A* is *B*" nor the particular negative "Some *A* is not *B*," but it "denies" the universal negative "No *A* is *B*." Finally, the particular negative "Some *A* is not *B*" "does not contradict" the universal negative "No *A* is *B*" nor the particular affirmative "Some *A* is *B*," but it "denies" the universal affirmative "Every *A* is *B*" [9, p. 7].

9 With the exception of De Morgan's "zodiac" of syllogisms, printed in his late *Syllabus of a Proposed System of Logic* [14, p. 21]. It is a circular arrangement of symbolical expressions for triplets of propositions, and it is meant to illustrate the relations between syllogisms, differing in strengthened or weakened premises.

octagon." However, the manuscript findings presented in this paper clearly show that this type of octagon was already known by De Morgan, some five decades earlier than Keynes and Johnson.

The paper is organized as follows. In Sect. 2 we present the manuscript findings and provide a transcription of the relevant sections of De Morgan's handwritten notes. In Sect. 3 we discuss some key aspects of De Morgan's logic, which will help us to make sense of the diagrams found in his manuscripts. With these prerequisites in place, we then show in Sect. 4 that De Morgan's octagon belongs to the type of KJ octagons, and briefly discuss the significance of this result for the historiography of logical diagrams. Finally, Sect. 5 wraps things up and suggests some questions for future research.

2 The Senate House Library (SHL) Diagrams

At this point, the manuscript findings will be presented, and a transcription of De Morgan's note will be given. This section is to be followed by an exposition of some core pieces of De Morgan's logic and notation, in order to facilitate subsequent interpretive steps.

2.1 Provenance and Context

The drawings to be discussed in the present paper are held by the Senate House Library (SHL) of London.[10] Hence in the following, they will be referred to as "the SHL diagrams" or "the SHL figures."

The Senate House Library itself was founded upon De Morgan's private library, purchased and presented to the University of London by Samuel Jones Loyd, 1st Baron Overstone,[11] after De Morgan's death in 1871. The De Morgan Library, as it is called today, is still one of the special collections held by the Senate House. It

[10] Large portions of De Morgan's multitudinous manuscripts are held by other archives, such as the manuscript collections at the British Library, the special collections of University College, the Department of Manuscripts and University Archives at the Cambridge University Library, or the special collections at the Bodleian Library of Oxford University. A list of institutions holding manuscripts of De Morgan's can be found at https://discovery.nationalarchives.gov.uk/details/c/F49779 (last access 26 August 2019). During the preparation of the present paper, it has not been possible to systematically search those stocks for similar figures.

[11] As De Morgan himself, Samuel Jones Loyd was an alumnus of Trinity College at the University of Cambridge. Born in Manchester on 25 September 1796, he enrolled on 18 September 1813 and earned his first degree in 1818, the second in 1822. Having been employed in several other functions, he succeeded his father as head of Jones Loyd and Co. Bankers, London, and became a well-respected authority on matters of finance (information to be found in the records of the University of Cambridge, available on the Cambridge Alumni Database: http://venn.lib.cam.ac.uk/, last access 26 August 2019).

comprises around 3800 items, printed between 1474 and 1870, most of which cover mathematical topics. Some of them are interleaved and annotated by De Morgan himself.

Several of the annotated exemplars of books from De Morgan's library have been allotted to a collection of unpublished material by or relating to De Morgan, kept in the Library's Archives and Manuscripts Department. Besides the De Morgan family papers and correspondence of private or of scientific character, this collection includes papers such as lecture notes and notebooks as well as drafts and annotated copies of De Morgan's own works. The diagrams to be discussed are partly among the papers and correspondence files (MS.775), partly among the annotated copies of De Morgan's works (MS.776).[12] To be more precise, some are contained in a notebook (MS.775/355) containing addenda to De Morgan's *Formal Logic* [9], which extended on De Morgan's first "syllogism" paper, "On the structure of the syllogism" [8]. Others are to be found in De Morgan's own copy of the same book, annotated and interleaved with notes, letters, and newspaper cuttings (MS.776/1). It is reasonable to assume that De Morgan's printed copy of his *Formal Logic* [9] has been held by the Senate House Library ever since De Morgan's private collection was presented to the Library after his death in 1871. The provenance of the handwritten notebook (MS.775/355), however, is not so clear. According to the archivists, it was almost certainly deposited at the Senate House Library in the twentieth century.[13]

While the additions to the printed copy (MS.776/1) are from 1846 through 1859 (one letter maybe later), the entries to the notebook (MS.775/355) date around 1850–1853. It seems that the notebook (MS.775/355) in particular may relate to a projected second edition of *Formal Logic*. Indeed, it seems reasonable to assume that De Morgan wanted to prepare a second edition of *Formal Logic* around the beginning of the 1850s. By this time, he had completed his second paper in the "syllogism" series [10], which was to improve his extensions of syllogistics as proposed in the first edition of *Formal Logic* and the first "syllogism" paper [8] from 1847. This hypothesis becomes even more plausible if one takes into account that the figures bear inscriptions to express propositional forms invoking the notational system of his second "syllogism" paper [10].

2.2 *The Manuscript Materials*

Although the De Morgan papers contain several more drawings of logical diagrams, the present exposition will focus on three figures only. The reason is that all three of them contain a similar octagonal figure. As this figure seems to be a recurring theme

[12] For MS.775 and MS.776, there exists a typewritten "interim handlist," which sums up the contents of the files. It is available on https://archives.libraries.london.ac.uk/resources/MS775.pdf (last access 26 August 2019).

[13] We thank Richard Temple, archivist at the Senate House Library, for this information.

in De Morgan's diagrammatic explorations, it will be worthwhile to concentrate on it.[14]

2.2.1 The Figures

The manuscripts to be referred to are MS.775/355-36, MS.776/1-90b, and MS.776/1-94b, as reproduced in Figs. 1, 2, and 3. The materials have officially been digitized by the Senate House Library document supply service with the permissions to publish the images for purposes of research in an international venue.

A comparison of the octagons in MS.775/355-36 (Fig. 1) and MS.776/1-90b (Fig. 2) reveals that both indeed contain the same figure as to the inscriptions of the corners. Reading clockwise and indicating the corners by lowercase letters (*a*) to (*h*), MS.775/355-36 (Fig. 1) gives the following order of inscriptions:

(a)	) · (
(b)	) ·)
(c)	))
(d)	()
(e)	(·)
(f)	(· (
(g)	((
(h)	)(

The octagon in MS.776/1-90b (Fig. 2) displays the same arrangement. The octagon contained in MS.776/1-94b (Fig. 3), however, seems to have been rotated 180°. Consequently, there is an interchange of (*a*) and (*e*), i.e., the uppermost and lowermost corners of the octagon. Similarly, (*b*) and (*f*) are interchanged, (*c*) and (*g*) are interchanged, and finally, (*d*) and (*h*) are interchanged.

While MS.775/355-36 (Fig. 1) and MS.776/1-90b (Fig. 2) agree in the ordering of corners, they also contain the same kinds of lines drawn between them, such as thick lines and thin lines, thick dashed lines, thin dashed lines, and others. In MS.776/1-94b (Fig. 3), which displays the rotated figure, most of the thinner lines differ from the other drawings graphically.

From De Morgan's notes, it is quite clear that the inscriptions at the corners of the figures are meant as symbolic expressions for propositional forms, and the lines between them are meant to represent their interrelations of opposition. Thus, MS.775/355-36 (Fig. 1) bears an explanation of the drawing:

[14] The octagonal figure occurs at least once more in De Morgan's manuscripts held by the Senate House Library, viz., at MS.776/1-94a. That diagram is not reproduced here, for reasons of copyright, but it is very similar to the one at MS. 776/1-94b (Fig. 3), modulo a 180° rotation.

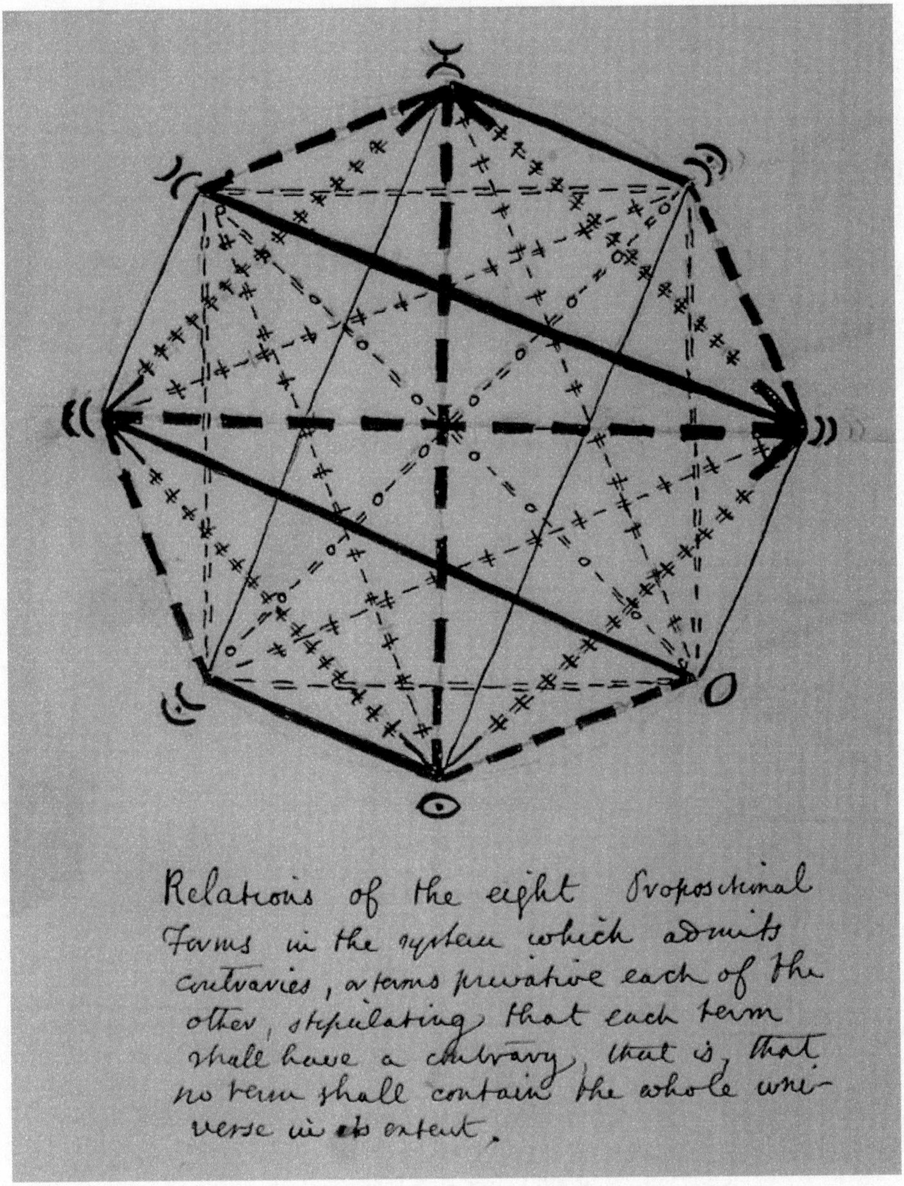

Fig. 1 De Morgan's MS.775/355-36

"Relations of the eight propositional Forms in the system which admits contraries, or terms privative each of the other, stipulating that each term shall have a contrary, that is, that no term shall contain the whole universe in its extent."

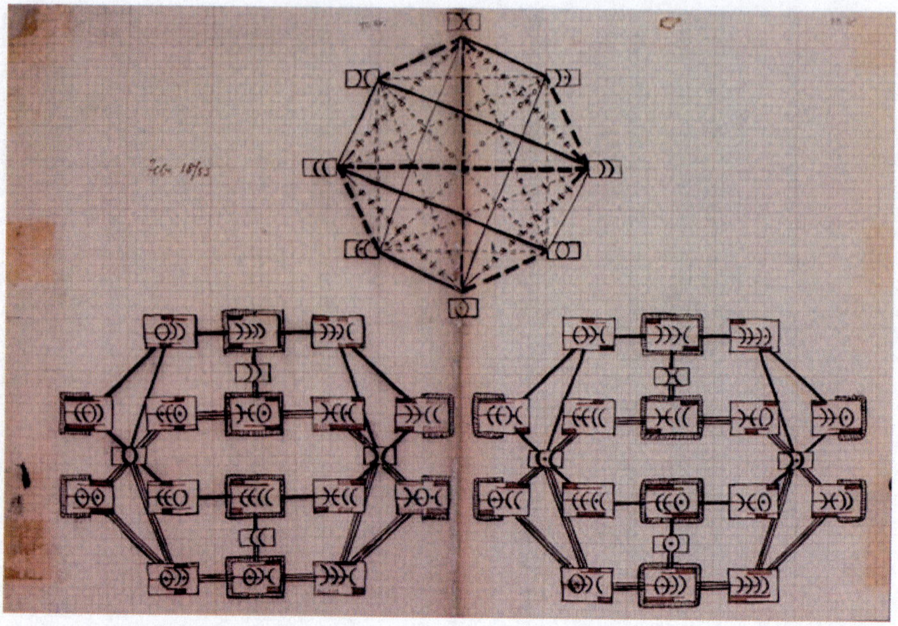

Fig. 2 De Morgan's MS.776/1-90b

2.2.2 The Comments

De Morgan commented on his drawings more extensively. His handwritten comments are to be found on two sheets, namely, MS.775/355-28 and MS.775/355-37 to MS. 775/355-38.

Judging from the kinds of lines contained in the explanatory remarks, MS.775/355-28 (Fig. 4) refers to MS.776/1-94b (Fig. 3). By contrast, MS.775/355-37 and MS.775/355-38 (Figs. 5 and 6) comment on MS.775/355-36 (Fig. 1) and may also be applied to MS.776/1-90b (Fig. 2).

MS.775/355-28 (Fig. 4) contains an explanation of the inscriptions at the corners of the figures, which is also presented in Table 1. Table 2 lists De Morgan's explanations of the lines drawn between the poles specified above. As MS.775/355-37 and MS.775/355-38 (Figs. 5 and 6) differ from MS.775/355-28 (Fig. 4) both in the graphics for some of the oppositions and in their ordering, the transcriptions have partly been rearranged to run parallel for both.

3 Reading Sense Into the Manuscript Materials

In order to read some more sense into De Morgan's drawings, it will be helpful to get acquainted with De Morgan's logical notation and with some core pieces of

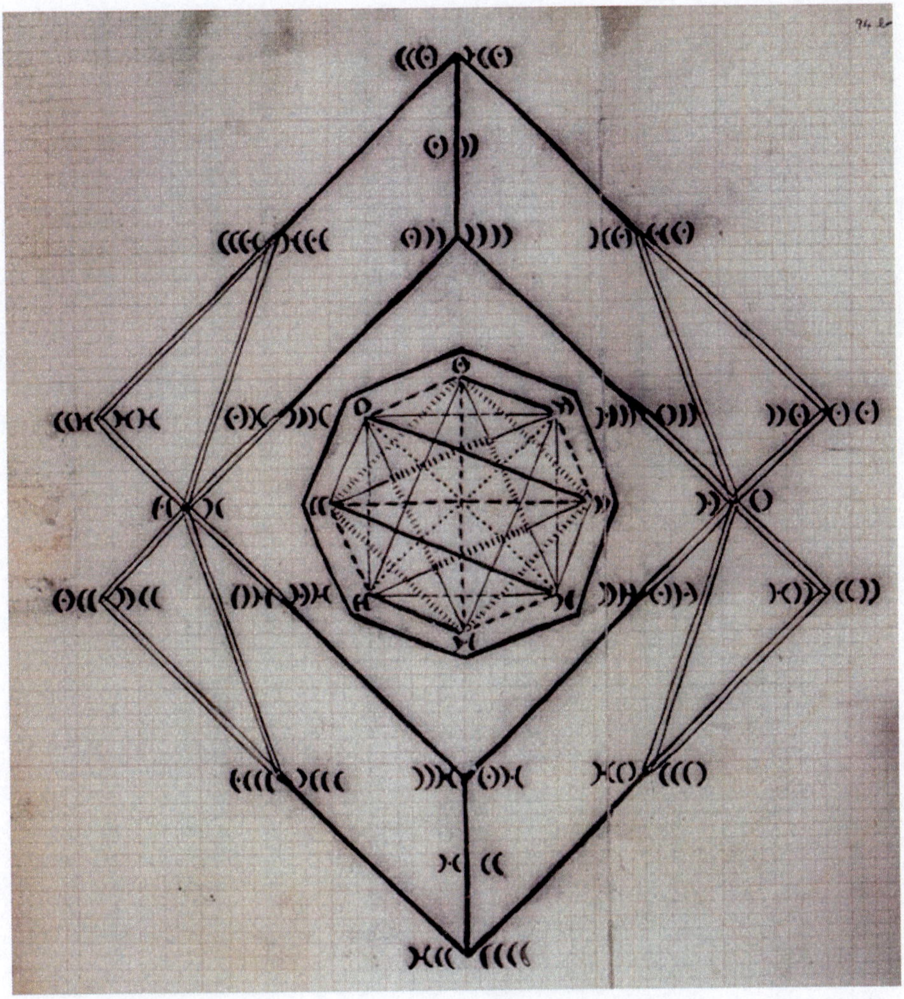

Fig. 3 De Morgan's MS.776/1-94b

Table 1 De Morgan's interpretations of the corners of the octagon in Fig. 3

))	*X*))*Y*	Every *X* is *Y*	Univ.
(· (	*X*(· (*Y*	Some *X*s are not *Y*s	Part.
((	*X*((*Y*	Every *Y* is *X*	Univ.
) ·)	*X*) ·)*Y*	Some *Y*s are not *X*s	Part.
) · (	*X*) · (*Y*	No *X* is *Y*	Univ.
()	*X*()*Y*	Some *X*s are *Y*s	Part.
(·)	*X*(·)*Y*	Everything is either *X* or *Y* or both	Univ.
)(	*X*)(*Y*	Some things are neither *X*s nor *Y*s	Part.

Table 2 Comparison of De Morgan's explanations of the lines in Fig. 3 and Figs. 1, 2

	MS.775/355-28 (Fig. 4)		MS.775/355-37 and MS.775/355-38 (Figs. 5 and 6)
	"The octagon in the middle has then 8 forms at its corners – Every form is joined to every other either by side or diagonal, and the joining lines, 28 in number, are marked in a manner which symbolized the connexion of the propositions.		"The eight forms are marked at the eight points of an octagon, which has, in sides and diagonals, 28 lines. Each line is marked in a way which symbolises the connexion of the propositions at the two ends, as follows
━━━━	joins a Univ[ersal] and a Part[icular] when the latter follows from the former and to the greatest possible extent. Thus)) ━)(means that when every *X* is *Y*, some things are neither *X*s nor *Y*s to the utmost, namely, *all* that are not *Y*.	━━━━	The propositions are a universal and a *consequent* particular the extension of the particular character being maximum, and independent of the extent of *X*. Thus in)) and)(, every *species* has the utmost amount of *subcompletion*, for no part of the species of *Y*, large or small, is in the contrary.
────	joins a Univ. and a part. when the extent of the character of the particular is indefinite. Thus)) ━() means that *X*))*Y* gives *X*()*Y*, the part of *Y* filled by *X* not being ascertained by the proposition.	────	The propositions are a universal and a consequent particular, the amount of extension of the particular character depending upon that of *X*. Thus in)) and (), the partience of *X* and *Y* depends on the extent of *X*.
━ ━ ━	joins a universal & particular or two universals, which together form a *complex* proposition: thus)) ━ ━ ━).) or *X*))*Y* and *X*).)*Y* give *X* a *subidentical* of *Y*.	━ ━ ━	The two prop.ns may or may not coexist, and when they coexist, they form one of the *complex propositions* used by the writer of a work of Formal Logic, whom may Aristotle confound. Thus)) and) ·) coexisting, say)o), give *X* his *subidentical* of *Y*.

(continued)

Table 2 (continued)

	MS.775/355-28 (Fig. 4)		MS.775/355-37 and MS.775/355-38 (Figs. 5 and 6)
....	joins two particulars of which neither, either, or both may be true together	$o - = o - =$	The propositions are two particulars of which neither [o] or one [–] or both [=] may be true
——- - - - -——	joins two particulars of which one must, both may, be true	$- = - = - =$	The propositions are two particulars which are one [–] or both [=] true. Thus () or) ·) must be true, *X* is either partient or subtotal of *Y*. and may be both
——‖‖‖‖‖‖——	joins a universal & a particular of which one must, both cannot be true	$- \neq - \neq - \neq$	The propositions are a contradictory (or as he w.m.A.c. says, *contrary*) particular and universal, one [–], not both [≠], true.
‖‖‖‖‖‖‖‖‖‖	joins two universals of which both cannot, and neither may, be true."	$\neq\neq \neq \neq \neq\neq$	The propositions are two universals which are not [≠] both true"

his logic and terminology. The following paragraphs will introduce the necessary prerequisites for successfully reading and interpreting the diagrams.

3.1 *Notation*

The inscriptions at the corners of the SHL diagrams are based on a notational system which De Morgan arrived at in his second paper from the "syllogism" series, published in 1850 [10]. He proposed it as a relevant refinement of his earlier notation.[15] As in his earlier writings, the uppercase letters *X*, *Y*, *Z* represent terms, while the lowercase letters *x*, *y*, *z* represent their "contraries" [8, p. 379]. De Morgan now applies parentheses as signs of quantity to each[16] of the term signs. An

[15] In 1847, De Morgan had made use of a notational system based on parentheses as signs of quantity and dots as signs of quality. However, as this earlier version is irrelevant to the reading of the inscriptions to the SHL diagrams, it may at this point be neglected. De Morgan's notational system of 1850 is outlined in [20, pp. xxvi–xxvii]. It should be noted that there is also an exposition of it in Lewis's [25, pp. 38–42]. For a more detailed discussion see [22].

[16] The earlier notation had not consistently applied a sign of quantification to each of the terms but basically to the subject terms only, which makes conversions hard to manipulate.

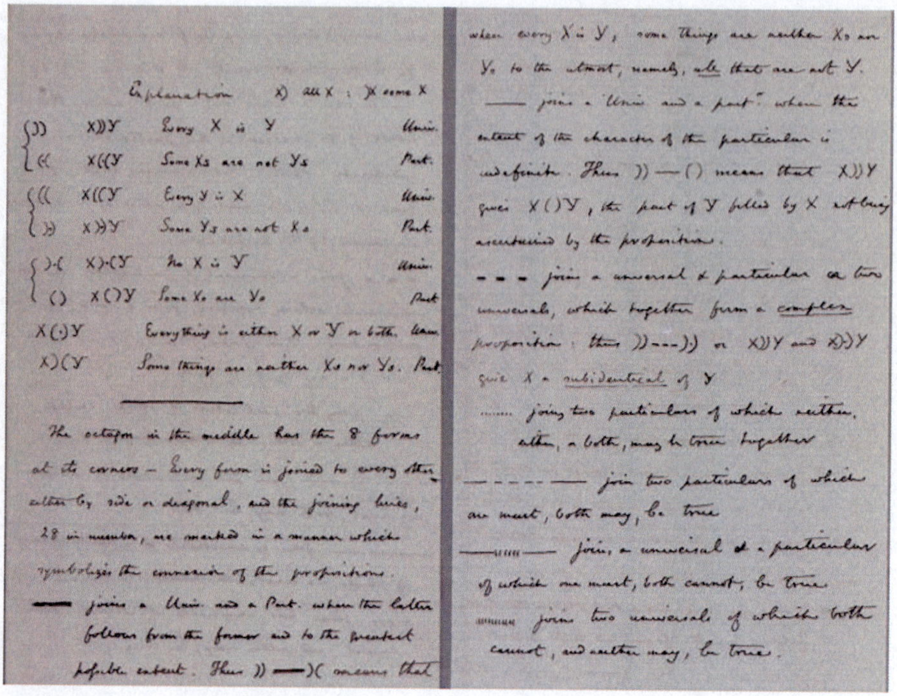

Explanation X) all X ;)X some X

))	X))Y	Every X is Y	Univ.
((	X(.(Y	Some Xs are not Ys	Part.
((	X((Y	Every Y is X	Univ.
))	X).)Y	Some Ys are not Xs	Part.
).(	X).(Y	No X is Y	Univ.
()	X()Y	Some Xs are Ys	Part.
	X(.)Y	Everything is either X or Y or both.	Univ.
	X)(Y	Some things are neither Xs nor Ys.	Part.

The octagon in the middle has the 8 forms at its corners – Every form is joined to every other either by side or diagonal, and the joining lines, 28 in number, are marked in a manner which symbolizes the connexion of the propositions.

—— joins a Univ. and a Part. when the latter follows from the former and to the greatest possible extent. Thus)) ——)(means that when every X is Y, some things are neither Xs nor Ys to the utmost, namely, all that are not Y.

—— joins a Univ. and a part. when the extent of the character of the particular is indefinite. Thus)) —— () means that X))Y gives X()Y, the part of Y filled by X not being ascertained by the proposition.

- - - joins a universal & particular or two universals, which together form a complex proposition: thus)) - - -).) or X))Y and X).)Y give X a subidentical of Y

.... joins two particulars of which neither, either, or both, may be true together

- - - - —— join two particulars of which one must, both may, be true

——''''''—— joins a universal & a particular of which one must, both cannot, be true.

'''''''' joins two universals of which both cannot, and neither may, be true.

Fig. 4 De Morgan's MS.775/355-28

"inclosing parenthesis, as in *X*) or (*X*" denotes that the term sign it relates to "enters universally" [10, p. 86]. An "excluding parenthesis, as in)*X* or *X*(" signifies that *X* "enters particularly" [10, p. 86]. The use of dots to indicate a proposition's quality is also systematized: "[A]n odd number [of dots], usually one, denote[s] negation or non-agreement" [10, pp. 86–87], while affirmation is signified by "an even number of dots, or none at all" [10, p. 86].[17] An advantage of this notational system is that complex expressions can be read in both directions: for instance, "*X*))*Y* and *Y*((*X* both denote that every *X* is *Y*" [10, p. 87]. The same goes for negative forms and for forms including contraries. For example, *X*). (*Y* is equivalent to *Y*). (*X*, and *X*))*y* is equivalent to *y*((*X* [10, p. 91].

In order to simplify his notational system even further, De Morgan chose to fix a certain order of reference relating to the terms: he fixed the subject to be *X* and the predicate to be *Y*. The theoretical justification for this notational convention is that any of the expressions that are admissible within De Morgan's system can be reduced to an equivalent expression which has *X* as its subject and *Y* as its predicate.

[17] At the time of De Morgan's second "syllogism" paper [10], De Morgan inscribed the dots at the bottom of the line. In later writings, such as [14], he raised them to the middle of the line. The latter version can be recognized in the SHL figures.

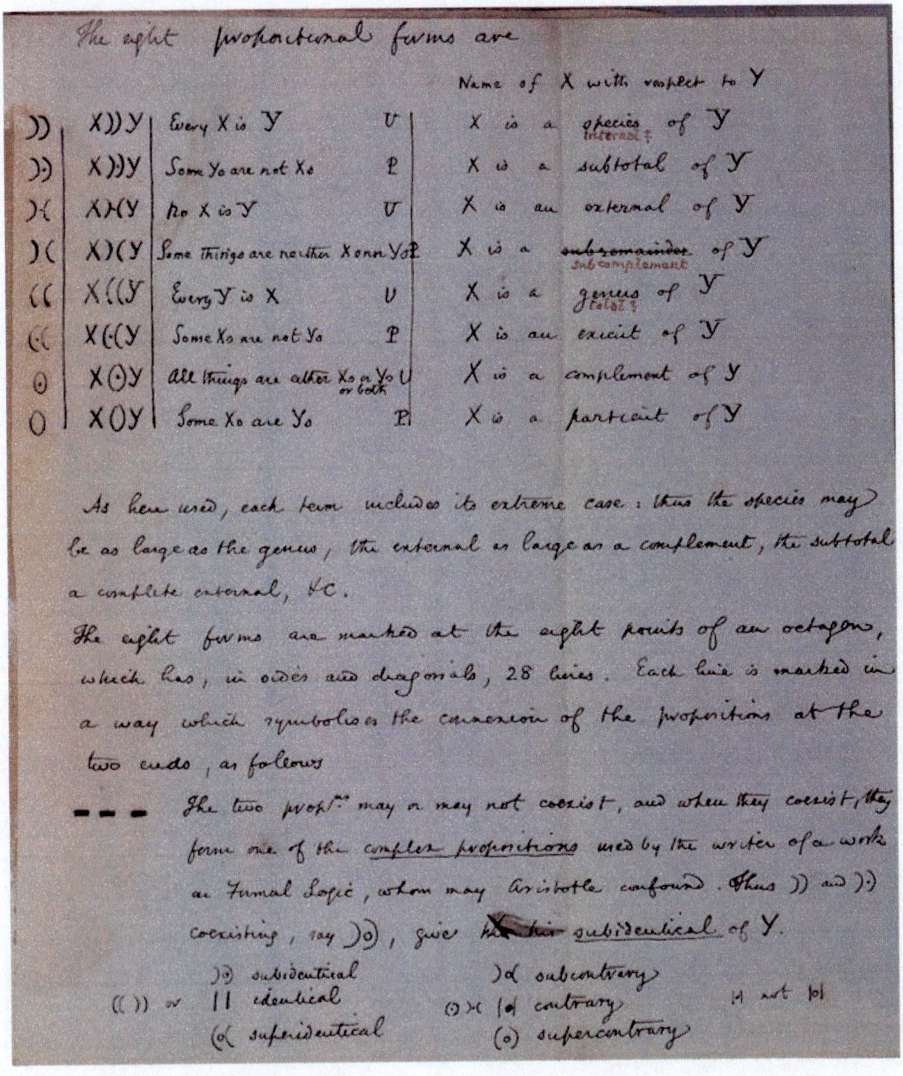

The eight propositional forms are

				Name of X with respect to Y
))	X))Y	Every X is Y	U	X is a species of Y (internal?)
)·)	X)·)Y	Some Ys are not Xs	P	X is a subtotal of Y
)·(	X)·(Y	No X is Y	U	X is an external of Y
)(	X)(Y	Some things are neither X nor Y	P	X is a ~~subremainder~~ subcomplement of Y
((	X((Y	Every Y is X	U	X is a genus of Y (total?)
(·(	X(·(Y	Some Xs are not Ys	P	X is an exient of Y
(·)	X(·)Y	All things are either Xs or Ys or both	U	X is a complement of Y
()	X()Y	Some Xs are Ys	P	X is a particent of Y

As here used, each term includes its extreme case: thus the species may be as large as the genus, the external as large as a complement, the subtotal a complete external, &c.

The eight forms are marked at the eight points of an octagon, which has, in sides and diagonals, 28 lines. Each line is marked in a way which symbolises the connexion of the propositions at the two ends, as follows

- - - The two propns may or may not coexist, and when they coexist, they form one of the complex propositions used by the writer of a work on Formal Logic, whom may Aristotle confound. Thus)) and)·) coexisting, say))·), give ~~X~~ subidentical of Y.

)·) subidentical —)(subcontrary
(()) or || identical — ()(|·| contrary — |·| not |o|
((superidentical — (o) supercontrary

Fig. 5 De Morgan's MS.775/355-37

For example, *x*)(*Y* can be transformed into the equivalent expression *X*(. (*Y*, and similarly, *X*)(*y* can be rewritten as the equivalent expression *X*).)*Y*. (For the full set of rules that enable these transformations, see Table 3, given in the following section.) In light of this convention, i.e., now that it has been determined that the subject term will always be *X* (rather than *x*, *Y* or *y*) and the predicate term will always be *Y* (rather than *y*, *X* or *x*), the notation can be simplified further. For example, it is now possible to leave out the terms altogether, and simply write)) without any danger of ambiguity:)) will unequivocally stand for *X*))*Y*, rather than

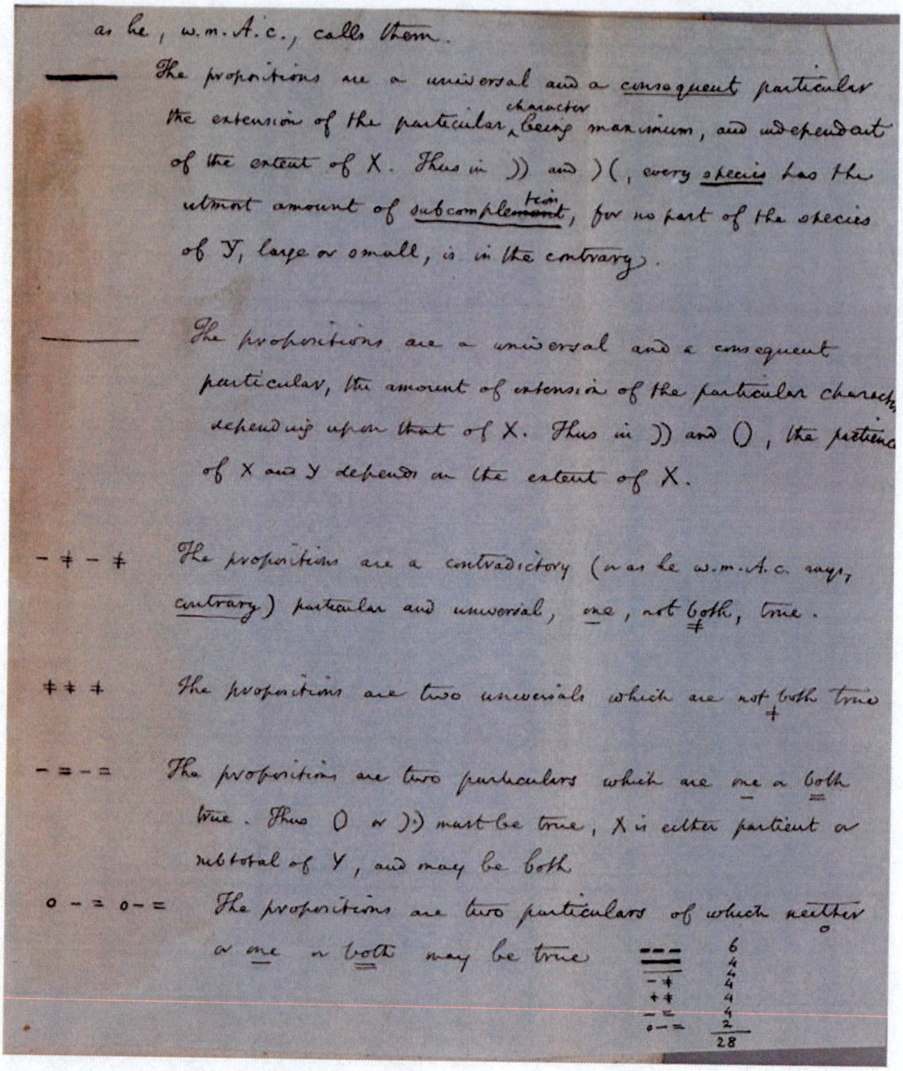
as he, w.m.A.c., calls them.

—— The propositions are a universal and a consequent particular the extension of the particular character being maximum, and independent of the extent of X. Thus in)) and)(, every species has the utmost amount of subcomplement term, for no part of the species of Y, large or small, is in the contrary.

—— The propositions are a universal and a consequent particular, the amount of extension of the particular charact. depending upon that of X. Thus in)) and (), the partience of X and Y depends on the extent of X.

-‡-‡ The propositions are a contradictory (or as he w.m.A.c. says, contrary) particular and universal, one, not both, true.

‡‡‡ The propositions are two universals which are not both true

-=-= The propositions are two particulars which are one or both true. Thus () or)) must be true, X is either partient or subtotal of Y, and may be both

o-=o-= The propositions are two particulars of which neither or one or both may be true

=== 6
—— 4
—— 4
-‡ 4
‡‡ 4
-= 4
o-= 2
28

Fig. 6 De Morgan's MS.775/355-38

for *x*))*Y*, for *X*))*y*, or for any other expression. It is exactly this simplified notation (with subject and predicate terms fixed to *X* and *Y*, respectively) which De Morgan uses in his diagrams (Figs. 1, 2, and 3) as well as his notes (Figs. 4, 5, and 6; also cf. Table 1).

3.2 Propositions

Generally speaking, a proposition contains two terms: subject and predicate. On De Morgan's account, each of these two terms has a "contrary": the subject is either *X* or *x*; the predicate is either *Y* or *y*. Furthermore, each of these two terms enters into the proposition either universally or particularly (as to notation, the bracket is either "enclosing" or "excluding," cf. Sect. 3.1). Finally, the copula that links both terms is either affirmative or negative (as to notation, either there is no dot or there is exactly one, cf. Sect. 3.1). These five binary parameters (*X* or *x* as subject; universal or particular subject; affirmative or negative copula; *Y* or *y* as predicate; universal or particular predicate) clearly yield $2^5 = 32$ propositions.

De Morgan's 32 propositions cluster together into eight groups of four pairwise equivalent propositions.[18] In each of these eight groups, any proposition may be equivalently converted into the three other ones of that group, by interchange of terms and rotation of parentheses, following the rule of transformation: "To use the contrary of a term, without altering the import of the proposition, alter the curvature of its parenthesis, and annex or withdraw a negative point" [10, p. 92]. By applying this rule of transformation, equivalent expressions are generated by restating one expression by reference to contraries of terms or contraries of contraries, respectively. De Morgan himself gave an incomplete table of the resulting equivalences [10, p. 91]. It can be completed as in Table 3. The verbal interpretations are De Morgan's own; it is noteworthy that except for their ordering,

[18] De Morgan himself rather thought of eight fundamental propositions in the following way: for any combination of subject and predicate, the choice may be made from the set of terms or from the set of their contrary terms, respectively. If both subject and predicate are replaced by their respective contraries, De Morgan spoke of "contranominals" [9, p. 62]. For example, "All *X*s are *Y*s" and "All Non-*X*s are Non-*Y*s" are contranominals [9, p. 62]. Starting from the four classical fundamental propositions and adding their four contranominals, De Morgan arrived at eight fundamental propositions instead of the traditional four; see [8, p. 381; 9, p. 62]. However, it should be noted that De Morgan thought the eight fundamental propositions eventually reduced to six propositional forms. De Morgan's distinction between fundamental propositions and propositional forms is somewhat different as to terminology: "When the subject and the predicate are of the same sort of quantity, both universal or both particular [i.e., respectively], the converse forms give the same proposition" [9, p. 58]. The reduction to six forms is arrived at by various approaches: according to De Morgan himself, certain equivalences may be established between propositions and contranominals, as stated on converted order of terms. In particular, he held "All Non-*X*s are Non-*Y*s" to be equivalent to "All *Y*s are *X*s," and "Some Non-*X*s are not Non-*Y*s" to be equivalent to "Some *Y*s are not *X*s"; see [8, p. 382; 9, p. 61]. But in this case, while the order of terms is converted, the form of the proposition remains constant: a classical *A* or a classical *O*, respectively. Hence *A* and *O* give rise to two distinct fundamental propositions, but as propositional forms, *A* and *O* each occur only once. However, according to De Morgan, the classical *E* and *I* each transform into two distinct propositional forms. The reason is that besides "No *X* is *Y*" and "Some *X*s are *Y*s," there are the contranominals "No Non-*X* is Non-*Y*" and "Some Non-*X*s are Non-*Y*s," which share the form of a negative universal negative and an affirmative particular, but cannot be reduced by conversion since the result would not match their import [8 , p. 382]. Hence there are only two forms that should be added to the classical four.

they match the ones from the manuscripts, as quoted in Table 2, Sect. 2.2.2. We add a column with transcriptions based on our own notation, which is based on the two propositional forms *all*(–, –) and *some*(–, –) and allows both the subject term and predicate term to be negated.

Table 3 De Morgan's 32 propositions and their interpretation

X))Y	= X). (y	= x((y	= x(.)Y	Every X is Y	*all*(X, Y)
x))y	= x). (Y	= X((Y	= X(.)y	Every Y is X	*all*(∼X, ∼Y)
X). (Y	= X))y	= x(.)y	= x((Y	No X is Y	*all*(X, ∼Y)
x). (y	= x))Y	= X(.)Y	= X((y	Everything is X or Y or both	*all*(∼X, Y)
X()Y	= X(. (y	= x)(y	= x).)Y	Some Xs are Ys	*some*(X, Y)
x()y	= x(. (Y	= X)(Y	= X).)y	Some things are neither Xs nor Ys	*some*(∼X, ∼Y)
X(. (Y	= X()y	= x).)y	= x)(Y	Some Xs are not Ys	*some*(X, ∼Y)
x(. (y	= x()Y	= X)(y	= X).)Y	Some Ys are not Xs	*some*(∼X, Y)

A difficulty with De Morgan's convention to fix the subject and predicate terms to *X* and *Y*, respectively, is that at least some of the resulting expressions are assigned a quite unnatural interpretation. For example, for the listings given in Tables 1 and 3, it is neither straightforward why *X*(.)*Y* should be read as "Everything is *X* or *Y* or both" nor why *X*)(*Y* corresponds to "Some things are neither *X*s nor *Y*s."[19] Indeed, De Morgan's verbal descriptions of his formulae are sometimes at variance. A slightly different version, which is still relatively close to the interpretations given in Tables 1 and 3, is to be found in his second "syllogism" paper [10, p. 101]:

All X are all Ys	)(
Some X s are some Ys	()
All X s are some Ys	))
Some X s are all Ys	((
No Xs are Ys	). (
Some Xs are not some Ys	(.)
No Xs are some Ys	).)
Some X s are no Ys	(. (

As a final alternative, De Morgan also offered a set of relational descriptions [10, p. 113]:

[19] If we drop the requirement that the subject and predicate terms be fixed to resp. *X* and *Y*, we can obtain more natural readings. For example, the expression *X*(.)*Y* can be transformed into the equivalent expression *x*))*Y*, which straightforwardly reads "Every Non-*X* is *Y*." Similarly, *X*)(*Y* transforms into *x*()*y*, which straightforwardly reads "Some Non-*X* is Non-*Y*."

X)(Y	Each X is related to all the Ys.
X(.)Y	Some Xs are not related to some of the Ys.
X))Y	Each X is related to one or more Ys
X(. (	Some Xs are not related to any Ys.
X((Y	Some Xs are (among them) related to all the Ys.
X).)Y	No X is related to some one or more Ys.
X). (Y	No X is related to any one Y.
X()Y	Some Xs are related to some one or more Ys.

3.3 *Relations Between Terms and Propositions*

Up to the times to which the manuscript materials are dated, De Morgan's modifications to syllogistic were based on an extensional understanding of mutually exclusive complements. In *Formal Logic*, for example, he made it quite clear that he proceeded on the assumption that "all the instances of a name are counted" [9, p. 71]. Stipulating that given this assumption there are extensionally (even numerically) "complemental" and "non-complemental" terms [9, p. 75], he pointed out that "[e]xclusion from one complement is inclusion in the other" [9, p. 75]. As the notion of a complement is obviously defined by mutual exclusion and joint exhaustion, it reminds of conditions that are classically associated with contradictory relations. However, De Morgan chose to speak of "contraries" [8, p. 379; 9, p. 60]. The notion of a "contrary" is the core piece of De Morgan's logic, with regard to terms as well as to propositions. In both respects, he redefined the relations holding between terms and between propositions. As his terminology is different from the traditional wording, the following paragraphs will introduce De Morgan's main ideas.

3.3.1 "Contraries," "Subcontraries," "Supercontraries"

De Morgan chose to speak of "contraries" of terms [8, p. 379; 9, p. 60] as well as of their "subcontraries" and "supercontraries" [9, p. 67]. Similarly, he spoke of "contrary," "subcontrary," and "supercontrary" propositions [9, p. 60, 148). The term "complement" extends over at least "contrary" *and* "subcontrary" (see [9], correction to p. 62, line 23 as indicated in the corrigenda to the table of contents for Ch. IV, p. ix; see also [9, p. 120; 12, p. 200]).

The contrary of a term X is another term Non-X, such that taken together, both "fill up" a given "universe of a proposition, or of a name" [8, p. 380]; also see [9, p. 55].[20] In other words, if "[b]y the *universe* (of a proposition) is meant the collection of all objects which are contemplated as objects about which assertion or denial may take place," then "every thing [within the universe] is either X or Non-X; nothing is both," as De Morgan later put it in 1860 [14, pp. 12–13]. If two terms are indeed

[20] As C. I. Lewis, for example, recognizes, the term "universe of discourse" was not so much coined by Boole but already by De Morgan [25, p. 37].

contraries, then "[t]here is nothing which is both *X* and *Y*, [and] there is nothing which is neither" [9, p. 67]. If, however, "*X* and *Y* are clear of each other, but [...] do not fill up the universe," then they should be called "subcontraries" [9, p. 67]. Finally, it is possible for two terms to "overfill" the universe, namely, in the case that "some things are both *X*s and *Y*s" [9, p. 67]. In this case, De Morgan speaks of "supercontraries" [9, p. 67].

In characterizing the relations holding between propositions, De Morgan continued and extended his specific use of the term "contrary." As before, he took the term "contrary" to mean "what logicians usually call *contradictory*" [9, p. 147]. Its meaning also covers relations between propositions because the truth and falsity of propositions are determined by whether a given term or its contrary applies [9, p. 147]. But again, De Morgan modified the traditional terminology even further: "[T]he propositions usually called *contraries*, 'Every *X* is *Y*' and 'No *X* is *Y*' [...] I shall call *subcontraries*: while those usually called *subcontraries* 'Some *X*s are *Y*s' and 'Some *X*s are not *Y*s' I shall call *supercontraries*" [9, p. 60]; also see [9, p. 148].

3.3.2 "Subidentical," "Subcomplement," "Subtotal"

The comments (presented and transcribed in Sect. 2.2.2) which De Morgan added to his drawings (presented in Sect. 2.2.1) quite obviously relate to further aspects of relations between propositions. However, they involve some specific terminology, which will be useful to consider in order to grasp the sense of De Morgan's comments.

The explanatory remarks in MS.775/355-28 (Fig. 4) are somewhat easier to read, while those in MS.775/355-37 (Fig. 5) and MS.775/355-38 (Fig. 6) involve a bit more of De Morgan's specific terminology. The terms "complex proposition" and "subidentical" occur in all comments. Additionally, MS.775/355-37 and MS.775/355-38 (Figs. 5 and 6) make use of the terms "consequent," "species," "subcomplement," "partient," and "subtotal." In order to draw a comparison between De Morgan's remarks from MS.775/355-28 (Fig. 4) on the one hand and those from MS.775/355-37 and MS.775/355-38 (Figs. 5 and 6) on the other, the relevant terms will now be given a quick explanation.

The notions of a "complex proposition" and of a "subidentical" were already introduced in *Formal Logic* [9]. A proposition is said to be complex if it includes explicit quantifications of both the subject and the predicate term, in the sense that it involves a conjunction of two assertions: one assertion that states the quantitative relation of the first term to the second, and another assertion that states the quantitative relation of the second term to the first. For example, "Every *X* is *Y* and every *Y* is *X*" is a complex proposition; "Every *X* is *Y* and some *Y*s are not *X*s" is another [9, p. 56]. According to De Morgan, if "[a]ny two of the eight forms [i.e., fundamental propositions]" are considered, "it is clear either that they cannot exist together or that one must exist when the other exists, or that one may exist either with or without the other" [9, p. 63]. Consequently, a "*complex proposition* is one which involves within itself the assertion or denial of each and all of the eight simple propositions" [9, p. 65].

The notion of being "identical" may be applied to pairs of terms. Identical terms are of the kind that "[w]here either can be applied, there can the other also" [9, p. 66]; such terms thus define the very same extension. For example, the terms "equilateral and equiangular are identical names," because "what figure soever has a right to either name, it has the same right to the other" [9, p. 66]. If, however, "there are more *Y*s than *X*s, and *X* stops short of a complete claim to identity with *Y*," then *X* is "subidentical" of *Y* [9, p. 67]. Correlatively, if "[e]very *Y* is *X*" but "there are more *X*s than *Y*s," then *X* is a "superidentical" of *Y* [9, p. 67].

However, not only terms but also propositions can be called identical, subidentical, or superidentical to one another. In this case, the notion of a "consequent" comes into play. For example, a proposition *P* is a subidentical of a proposition *Q* "if every case in which *P* is true be one in which *Q* is true, but so that *Q* is sometimes true when *P* is not" [9, p. 147]. In this case, *Q* "is usually mentioned as *essential* to *P*, and as a *necessary consequence* of it" [9, p. 147]. Therefore, De Morgan thought, "superidentical or identical" and "necessary consequent" are synonymous [9, p. 147].

The terms of "genus," "species," and "partient" may more readily be explained in the context of a set of descriptions summed up in De Morgan's *Syllabus of a Proposed System of Logic* [14]. Here, De Morgan states that a "class [not a term!] *X* be called a *species* of the class *Y*, and *Y* a *genus* of *X*," if "[e]very *X* is *Y*, *X*))*Y* or *Y*((*X*." Furthermore, if "some *X*s are *Y*s, *X*()*Y* or *Y*()*X*," then each of them is "called a *partient* of the other" [14, p. 52].

The terms "subcomplement" and "subtotal" are not often used in De Morgan's writings. However, there is at least one occurrence of the term "subtotal" in a small article on "Logical Phraseology," published in 1853 [11]. Here, the term "subtotal" occurs in De Morgan's attempt to paraphrase those particular propositions which would deny a universal indicating that *X* is a genus of *Y*, i.e., that every *Y* is *X*. A "subtotal" would be what is "[n]ot a genus, that is, not entirely filling up" [11, p. 30]. In other words, "*x* is a subtotal of *y*" if "Some *y*s are not *x*s" [11, p. 30].

The term "subcomplement" is hard to find in De Morgan's writings. Perhaps it is meant to signify a "subtotal" of a "complement," i.e., what is not a complement of a term in that it does not entirely fill up that term's contrary. In the article just quoted, however, De Morgan coins the terms "subremainder or subremnant" for this case [11, p. 30].

3.3.3 Aristotelian Relations

We will now gather De Morgan's pairs of opposites and compare them to the Aristotelian relations encoded in logical diagrams such as the square of opposition.

First of all, we have already seen that De Morgan's "contraries" correspond to what is traditionally called the contradictory relation. These propositions are such that "one must be true and one false, differ[ing] both in quantities and copula.

Thus *X*))*Y* and *X*(. (*Y* are contraries" [10, p. 92]. Hence, in order to obtain the contradictory of a proposition, we have to change both its quantities (notationally encoded by the enclosing/excluding brackets) and also its copula (notationally encoded by the presence/absence of a dot). In total, there are four applications of this rule (the first one was already given by De Morgan himself):

- *X*))*Y* and *X*(. (*Y* are contradictories, i.e., De Morgan's "contraries";

in our own notation, *all*(*X*, *Y*) and *some*(*X*, ∼*Y*) are contradictories.

- *X*((*Y* and *X*).)*Y* are contradictories, i.e., De Morgan's "contraries";

in our own notation, *all*(∼*X*, ∼*Y*) and *some*(∼*X*, *Y*) are contradictories.

- *X*). (*Y* and *X*()*Y* are contradictories, i.e., De Morgan's "contraries";

in our own notation, *all*(*X*, ∼*Y*) and *some*(*X*, *Y*) are contradictories.

- *X*(.)*Y* and *X*)(*Y* are contradictories, i.e., De Morgan's "contraries";

in our own notation, *all*(∼*X*, *Y*) and *some*(∼*X*, ∼*Y*) are contradictories.

Secondly, we have already seen that De Morgan's "subcontraries" and "supercontraries" correspond to what are traditionally called the contrary and subcontrary relations, respectively. De Morgan's rule for these is that "The alteration of one quantity, and the copula, turns a universal into another and inconsistent [i.e. contrary] universal, and a particular into another and a consistent [i.e., subcontrary] particular: as *X*))*Y* into *X*(.)*Y*, or *X*()*Y* into *X*).)*Y*" [10, pp. 92–93].[21] Hence, in order to obtain the contraries of a universal proposition, we have to change exactly one of its quantities (notationally encoded by the enclosing/excluding brackets) and also its copula (notationally encoded by the presence/absence of a dot). As the De Morgan's logic contains four universal propositions, there are in total four applications of this rule (the first one was already given by De Morgan himself):

- *X*))*Y* and *X*(.)*Y* are contraries, i.e., De Morgan's 'subcontraries';

in our own notation, *all*(*X*, *Y*) and *all*(∼*X*, *Y*) are contraries.

- *X*))*Y* and *X*). (*Y* are contraries, i.e., De Morgan's "subcontraries";

in our own notation, *all*(*X*, *Y*) and *all*(*X*, ∼*Y*) are contraries.

- *X*((*Y* and *X*). (*Y* are contraries, i.e., De Morgan's "subcontraries";

in our own notation, *all*(∼*X*, ∼*Y*) and *all*(*X*, ∼*Y*) are contraries.

- *X*((*Y* and *X*(.)*Y* are contraries, i.e., De Morgan's "subcontraries";

in our own notation, *all*(∼*X*, ∼*Y*) and *all*(∼*X*, *Y*) are contraries.

[21] The second example mentioned in this quotation literally reads: "*X*(.)*Y* into *X*).)*Y*" [10, p. 93]. However, that would not be a valid illustration of the rule it is supposed to illustrate, since the copula (the dot) has not been altered. The most charitable interpretation is to view this as a misprint for "*X*()*Y* into *X*).)*Y*," which *would* effectively be a valid illustration of this rule. This interpretation has also been adopted in the main text of this paper.

Furthermore, the same rule also specifies that in order to obtain the subcontraries of a particular proposition, we have to change exactly one of its quantities (notationally encoded by the enclosing/excluding brackets) and also its copula (notationally encoded by the presence/absence of a dot). As the De Morgan's logic contains four particular propositions, there are in total four applications of this rule (the second one was already given by De Morgan himself, but cf. Footnote 21):

- *X*()*Y* and *X*(. (*Y* are subcontraries, i.e., De Morgan's "supercontraries";

in our own notation, *some*(*X*, *Y*) and *some*(*X*, ∼ *Y*) are subcontraries.

- *X*()*Y* and *X*).)*Y* are subcontraries, i.e., De Morgan's "supercontraries";

in our own notation, *some*(*X*, *Y*) and *some*(∼*X*, *Y*) are subcontraries.

- *X*)(*Y* and *X*).)*Y* are subcontraries, i.e., De Morgan's "supercontraries";

in our own notation, *some*(∼*X*, ∼*Y*) and *some*(∼*X*, *Y*) are subcontraries.

- *X*)(*Y* and *X*(. (*Y* are subcontraries, i.e., De Morgan's "supercontraries";

in our own notation, *some*(∼*X*, ∼*Y*) and *some*(*X*, ∼*Y*) are subcontraries.

In his second "syllogism" paper, De Morgan also gives a rule for subalternations. The rule reads: "In a universal proposition, any *one* quantity may be altered, either from universal to particular, or from particular to universal; and the result is always a true deduction, though not an equivalent. Thus *X*))*Y* gives both *X*()*Y* and *X*)(*Y*" [10, p. 92]. Hence, in order to obtain the propositions that stand in subalternation to a universal proposition, we have to change exactly one of its quantities (notationally encoded by the enclosing/excluding brackets), while leaving its copula (notationally encoded by the presence/absence of a dot) unaltered. There are four universal propositions in De Morgan's system, each of which yields two subalternations. In total, we thus get eight subalternations (the first two were already given by De Morgan himself):

- Subalternations from *X*))*Y* to *X*()*Y* and to *X*)(*Y*;

i.e., subalternations from *all*(*X*, *Y*) to *some*(*X*, *Y*) and to *some*(∼ *X*, ∼ *Y*).

- Subalternations from *X*((*Y* to *X*)(*Y* and to *X*()*Y*;

i.e., subalternations from *all*(∼*X*, ∼*Y*) to *some*(∼*X*, ∼*Y*) and to *some*(*X*, *Y*).

- Subalternations from *X*). (*Y* to *X*(. (*Y* and to *X*).)*Y*;

i.e., subalternations from *all*(*X*, ∼*Y*) to *some*(*X*, ∼*Y*) and to *some*(∼*X*, *Y*).

- Subalternations from *X*(.)*Y* to *X*).)*Y* and to *X*(. (*Y*;

i.e., subalternations from *all*(∼*X*, *Y*) to *some*(∼*X*, *Y*) and to *some*(*X*, ∼*Y*).

Finally, we should consider the relation of unconnectedness or independence. Unconnectedness essentially amounts to the absence of any relation: two propositions are said to be unconnected if and only if they do not stand in any Aristotelian relation whatsoever. This is entirely in line with De Morgan's remark that such propositions are "perfectly indifferent" [10, p. 92]. De Morgan's rule of unconnectedness (which he calls "concomitance") reads as follows: "The *concomitants* of a universal, to which it is perfectly indifferent, differ from it in quantities, or in copula, not in both. Thus *X*))*Y* coexists either with *X*((*Y* or *X*).)*Y*" [10, p. 92]. Hence, in order to obtain the propositions that are unconnected to a universal proposition, we have to change either both of its quantities (notationally encoded by the enclosing/excluding brackets) or its copula (notationally encoded by the presence/absence of a dot). As De Morgan's rule contains four universal propositions, we get a total number of six applications of this rule for unconnected pairs (the first two were already given by De Morgan himself):

- *X*))*Y* and *X*((*Y* are unconnected;

i.e., *all*(*X*, *Y*) and *all*(∼*X*, ∼*Y*) are unconnected.

- *X*))*Y* and *X*).)*Y* are unconnected;

i.e., *all*(*X*, *Y*) and *some*(∼*X*, *Y*) are unconnected.

- *X*((*Y* and *X*(. (*Y* are unconnected;

i.e., *all*(∼*X*, ∼*Y*) and *some*(*X*, ∼*Y*) are unconnected.

- *X*). (*Y* and *X*(.)*Y* are unconnected;

i.e., *all*(*X*, ∼*Y*) and *all*(∼*X*, *Y*) are unconnected.

- *X*). (*Y* and *X*)(*Y* are unconnected;

i.e., *all*(*X*, ∼*Y*) and *some*(∼*X*, ∼*Y*) are unconnected.

- *X*(.)*Y* and *X*()*Y* are unconnected;

i.e., *all*(∼ *X*, *Y*) and *some*(*X*, *Y*) are unconnected.

These six unconnected pairs each involve at least one universal proposition. However, later on in his second "syllogism" paper, De Morgan also adduces additional considerations that "justify us in extending the general name of comcomitants [i.e. unconnectedness] to particulars in which both quantities differ" [10, p. 93]. Hence, in order to obtain the propositions that are unconnected to a particular proposition, we have to change both of its quantities (notationally encoded by the enclosing/excluding brackets), while leaving its copula (notationally encoded by the presence/absence of a dot) unaltered. This extension of the definition of unconnectedness yields two more unconnected pairs (which involve two particular propositions):

- *X*()*Y* and *X*)(*Y* are unconnected;

i.e., *some*(*X*, *Y*) and *some*(∼*X*, ∼*Y*) are unconnected.

- $X(.\ (Y \text{ and } X).\ )Y$ are unconnected;

i.e., *some*$(X, \sim Y)$ and *some*$(\sim X, Y)$ are unconnected.

Adding these two pairs to the six pairs that we already obtained above, we thus find that De Morgan's logic yields a total number of eight pairs of unconnected propositions.

To summarize, in this section, we have shown that De Morgan's rules yield a total number of four pairs of contradictory propositions, four pairs of contrary propositions, four pairs of subcontrary propositions, eight pairs of propositions in subalternation, and finally, eight pairs of unconnected propositions. Adding these all up yields the total number of $4 + 4 + 4 + 8 + 8 = 28$ pairs of propositions. This is exactly the number that we should expect: De Morgan's logic (his "system of contraries") contains 8 propositions, and an easy combinatorial calculation shows that this gives rise to $\binom{8}{2} = \frac{8 \times 7}{2} = 28$ pairs of propositions. In other words, we can be certain that the lists above are completely exhaustive: every single pair of propositions that exists within De Morgan's logic has now been "classified" as either contradictory, or contrary, or subcontrary, or in subalternation, or unconnected.

4 Logical Analysis of De Morgan's Octagon of Opposition

Having completed our discussion of De Morgan's logical system (as he presented it in his published materials and unpublished notes), we are now fully equipped to analyze the octagonal diagram that occurs several times throughout his manuscripts (cf. Figs. 1, 2, and 3). In order to facilitate this analysis, we will start with the diagram in Fig. 1 and add our own, modern notation for the propositions to the corners of the octagon (cf. Table 3 and Sect. 3.2). Furthermore, we will also emphasize the various lines between those corners, following the contemporary color convention for the Aristotelian relations (cf. Table 2 and Sect. 3.3), i.e., contradictions in red, contrarieties in blue, subcontrarieties in green, and subalternations as black arrows. (Note that unconnectedness is typically not visualized at all,[22] but De Morgan himself visualized unconnectedness by means of thick dashed lines.) The resulting diagram is shown as Fig. 7.

Before moving on, it should be emphasized that the diagram in Fig. 7 is only meant as a visual aid and thus contains exactly the same logical information as De Morgan's original diagram in Fig. 1. In particular, the visual elements that have been added in Fig. 7 do not carry any new logical information: we have seen in Sect. 3 that they all fit within De Morgan's overall system of logic, his notation for propositions, and his rules for the relations between propositions.

[22] And for good reasons, recall that unconnectedness is the absence of any Aristotelian relation, so this is naturally "visualized" by the absence of any colored lines in the diagram.

It should be clear that the octagon in Fig. 7 is an *Aristotelian diagram*, just like the traditional square of opposition and many others. After all, this octagon visualizes a number of propositions and the (Aristotelian) relations holding between those propositions. This is also emphasized in De Morgan's accompanying handwritten comments: "Relations of the eight propositional Forms in the system which admits contraries" (cf. the transcription given in Sect. 2.2.1). Nevertheless, the octagon in Fig. 7 does have some peculiar features. For example, contradiction is not represented by means of central symmetry: usually, the red lines that represent the contradiction relations are the diagonals of the diagram, but in Fig. 7, this is clearly

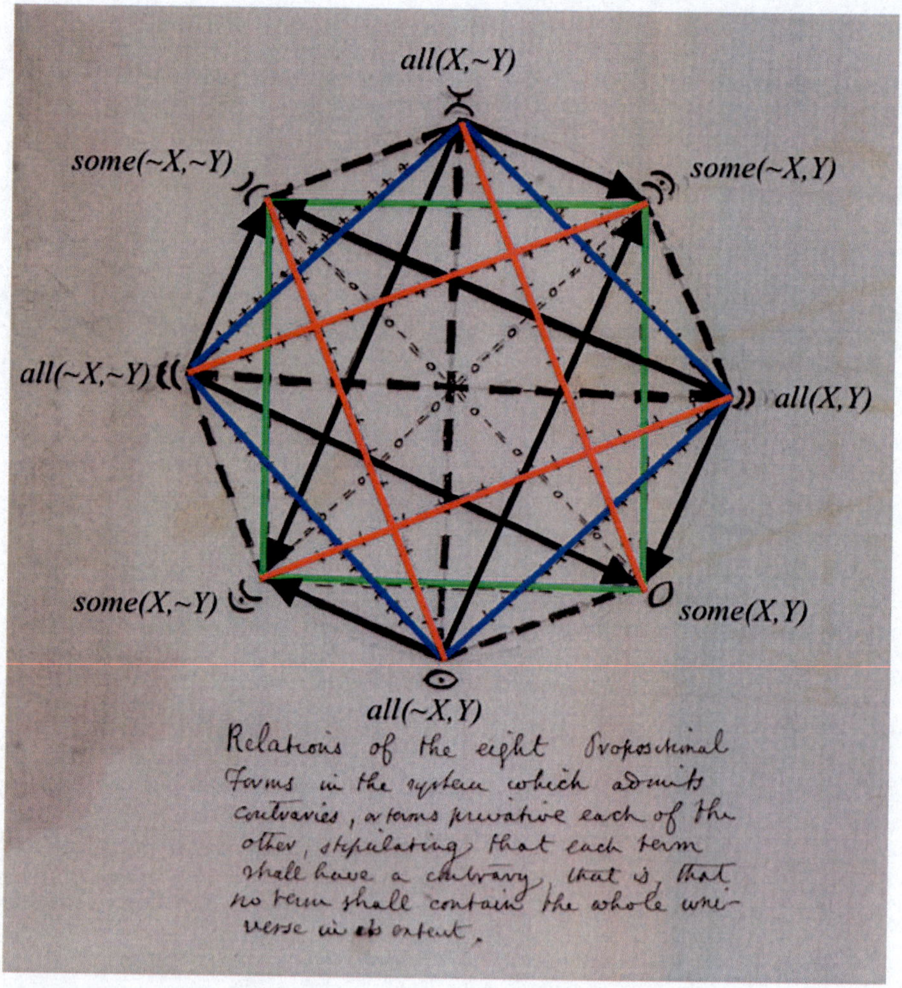

Fig. 7 De Morgan's MS.775/355-36, labelled according to our own notation and with the Aristotelian relations highlighted in color

not the case. However, this visual-diagrammatic peculiarity by no means disqualifies the octagon in Fig. 7 from being an Aristotelian diagram. Although they are not extremely common, there certainly exist other examples of Aristotelian diagrams which do not visualize the contradiction relation by means of central symmetry [3, 31]. The octagon in Fig. 7 is just a new example to be added onto that list.

Since De Morgan's unpublished octagon is an Aristotelian diagram, the following question naturally arises: which *type* of Aristotelian diagram is it precisely? After all, recent research in logical geometry has shown that there exist several types of Aristotelian diagrams, each with their own distinctive logical properties [7, 29]. For example, one can show that there exist 2 types of Aristotelian squares (including the traditional square of opposition), 5 types of Aristotelian hexagons, and 18 types of Aristotelian octagons. We will now show that De Morgan's unpublished octagon belongs to 1 of these 18 types, viz., the type of so-called KJ octagons (this terminology will be explained later).

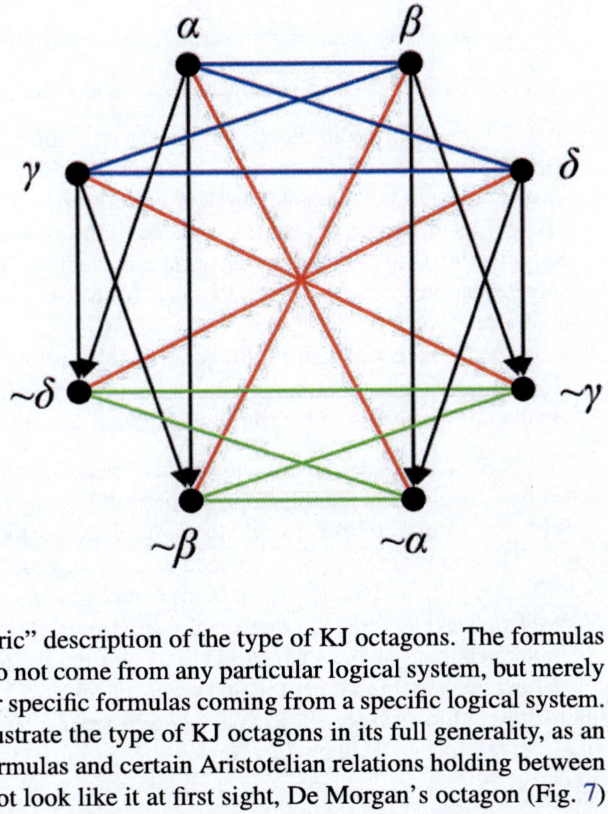

Fig. 8 Generic description of the type of KJ octagons

Figure 8 shows a "generic" description of the type of KJ octagons. The formulas α, β, etc. in this diagram do not come from any particular logical system, but merely serve as "placeholders" for specific formulas coming from a specific logical system. The goal of Fig. 8 is to illustrate the type of KJ octagons in its full generality, as an abstract pattern of eight formulas and certain Aristotelian relations holding between them. Although it might not look like it at first sight, De Morgan's octagon (Fig. 7) is a perfect instantiation of this abstract pattern (Fig. 8). We do need to transform the diagram quite a bit to see this. To make this precise, we define a bijective mapping f from the generic description in Fig. 8 to De Morgan's actual octagon in Fig. 7. The full definition of f is given below. For example, when we say that $f(\alpha) =$)), we mean that the "placeholder" α from the generic description is to be filled in with the specific formula)) from De Morgan's specific logical system, i.e., his system of contraries.

x	α	β	γ	δ	$\neg\alpha$	$\neg\beta$	$\neg\gamma$	$\neg\delta$
$f(x)$	))	). (	((	(.)	(. (	()	).)	)(

It can now be shown that this function f is an *Aristotelian isomorphism*, i.e., it perfectly preserves and reflects the Aristotelian relations. More formally, for all φ, ψ that occur in the octagon in Fig. 8 and for all Aristotelian relations R, the following holds:

φ and ψ stand in relation R (in the octagon in Fig. 8)

if and only if

$f(\varphi)$ and $f(\psi)$ stand in that same relation R (in the octagon in Fig. 7).

In order to illustrate this isomorphism, consider the following three examples:

- α and δ are contrary in the generic description of the type of KJ octagons (Fig. 8), and we have already seen in Sect. 3.3 that $f(\alpha)$ and $f(\delta)$, i.e.,)) and (.), are also contrary in De Morgan's system of contraries (Fig. 7).
- There is a subalternation from γ and $\neg\beta$ in the generic description of the type of KJ octagons (Fig. 8), and we have already seen in Sect. 3.3 that there is also a subalternation from $f(\gamma)$ to $f(\neg\beta)$, i.e., from ((to (), in De Morgan's system of contraries (Fig. 7).
- α and γ are unconnected in the generic description of the type of KJ octagons (Fig. 8), and we have already seen in Sect. 3.3 that $f(\alpha)$ and $f(\gamma)$, i.e.,)) and ((, are also unconnected in De Morgan's system of contraries (Fig. 7).

The existence of this Aristotelian isomorphism shows that De Morgan's unpublished octagon (Fig. 7) belongs to the type of KJ octagons (Fig. 8).

Over the course of the twentieth century, KJ octagons have gathered a rich history. Diagrams of this type have been used by authors such as Dopp [16], Thomas [32], Hacker [19], and Sauriol [28]. In more recent years, they have also been studied by philosophers and logicians such as Dekker [6] and García-Cruz [17], and even by computer scientists such as Ciucci, Dubois and Prade [4, 5]. Historically speaking, it has long been assumed that the first occurrences of KJ octagons were to be found in the third edition of John N. Keynes's *Studies and Exercises in Formal Logic* from 1894 [24][23] and in William E. Johnson's *Logic* from 1921 [23]. This is also the

[23] The first (1884) and second (1887) editions of Keynes's *Studies and Exercises in Formal Logic* contain a traditional square of opposition, but not yet a KJ octagon. The KJ octagon was included in the third edition (1894), and it again appears in the fourth (1906), which was the last major

reason why we now speak of the type of KJ octagons: "KJ" stands for "Keynes-Johnson." Ironically, however, this now turns out to be a misnomer. After all, in this section, we have shown that De Morgan's unpublished octagon (Fig. 7) from the 1850s also belongs to the type of KJ octagons. In other words, this type of diagram was already known by De Morgan, some five decades earlier than Keynes and Johnson.

5 Conclusion

In this paper we have uncovered some hidden gems from Augustus De Morgan's unpublished manuscripts. Although De Morgan's published writings hardly contain any diagrams or figures, his unpublished manuscripts contain several attempts to extend the traditional square of opposition to a larger diagram. In particular, one "octagon of opposition" occurs several times throughout his manuscripts. Based on a detailed analysis of this octagon, in combination with De Morgan's unpublished notes as well as his published materials, we have argued that this diagram belongs to the type of so-called KJ octagons. It has long been assumed that this type of octagon was studied for the first time by John Neville Keynes and William E. Johnson around the turn of the twentieth century. However, the manuscript findings presented in this paper clearly show that this type of octagon was already known by De Morgan, some five decades earlier than Keynes and Johnson.

These results naturally lead to a vast array of further questions. For example, on the historical side, one might wonder if Keynes and Johnson's octagon of opposition was in any way influenced by De Morgan's unpublished diagram. On the logical side, now that De Morgan's diagram has been identified as a KJ octagon, it would be interesting to systematically investigate its logical properties by means of the contemporary tools and techniques of logical geometry (e.g., bitstring semantics [30]). These topics will be addressed in future research.

Acknowledgments The archival research in preparation of the present paper would not have been possible without the help of Dr. Karen Attar, curator of the rare books collections at the Senate House Library and herself a specialist on De Morgan and his reading, as well as Richard Temple, archivist. We are also grateful to Charles Harrowell, the special collections administrator. Finally, thanks to two anonymous reviewers for their helpful suggestions. Lorenz Demey's research is funded through a postdoctoral fellowship of the Research Foundation – Flanders (FWO) and a research professorship (BOFZAP) at KU Leuven.

edition of this work. When giving his octagon for the first time, Keynes explicitly states: "For the octagon of opposition in the form in which it is here given I am indebted to Mr. Johnson" [24, p. 113].

References

1. G. Boole, *The Mathematical Analysis of Logic: Being an Essay towards a Calculus of Deductive Reasoning* (Macmillan, Barclay & Macmillan, Cambridge, 1847)
2. P. Boehner, Bemerkungen zur Geschichte der de Morganschen Gesetze in der Scholastik. Archiv für Philosophie **4**, 113-146 (1951)
3. K.-F. Chow, General patterns of opposition squares and 2n-gons, in *Around and Beyond the Square of Opposition*, ed. by J.-Y. Béziau, D. Jacquette (Springer, Basel, 2012), pp. 263-275
4. D. Ciucci, D. Dubois, H. Prade, Structures of opposition in fuzzy rough sets. Fundamenta Informaticae **142**, 1-19 (2015)
5. D. Ciucci, D. Dubois, H. Prade, Structures of opposition induced by relations. The Boolean and the gradual cases. *Annals of Mathematics and Artificial Intelligence* 76, 351-373 (2016)
6. P. Dekker, Not *only* Barbara. Journal of Logic, Language and Information **24**, 95-129 (2015)
7. L. Demey, Computing the maximal Boolean complexity of families of Aristotelian diagrams. Journal of Logic and Computation **28**, 1323-1339 (2018)
8. A. De Morgan, On the structure of the syllogism, and on the application of the theory of probabilities to questions of argument and authority. Transactions of the Cambridge Philosophical Society **8**(3), 379-408 (1847)
9. A. De Morgan, *Formal Logic, or: The Calculus of Inference, Necessary and Probable* (Taylor & Walton, London, 1847)
10. A. De Morgan, On the symbols of logic, the theory of the syllogism, and in particular of the copula, and the application of the theory of probabilities to some questions of evidence. Transactions of the Cambridge Philosophical Society **9**(4), 79-127 (1850)
11. A. De Morgan, Some suggestions in logical phraseology. Proceedings of the Philological Society **6**(129), 27-30 (1853)
12. A. De Morgan, On the syllogism, No. III and on logic in general. Transactions of the Cambridge Philosophical Society **10**(1/10), 173-230 (1858)
13. A. De Morgan, On the syllogism, No. IV, and on the logic of relations. Transactions of the Cambridge Philosophical Society **10**(2/7), 331-358 (1860)
14. A. De Morgan, *Syllabus of A Proposed System of Logic* (Walton and Maberly, London, 1860)
15. A. De Morgan, *On the Syllogism and Other Logical Writings*, ed. by P. Heath (Routledge & Kegan Paul, London, 1966)
16. J. Dopp, *Leçons de logique formelle* (Éditions de l'Institut Supérieur de Philosophie, Leuven, 1949)
17. J. D. García-Cruz, From the square to octahedra, in *The Square of Opposition: A Cornerstone of Thought*, ed. by J.-Y. Béziau, G. Basti (Springer, Cham, 2017), pp. 253-272
18. I. Grattan-Guinness, *The Search for Mathematical Roots 1870-1940* (Princeton University Press, Princeton and Oxford, 2000)
19. E. A. Hacker, The octagon of opposition. Notre Dame Journal of Formal Logic **16**, 352-353 (1975)
20. P. Heath, Editor's introduction, *On the Syllogism and Other Logical Writings by Augustus De Morgan*, ed. by P. Heath (Routledge & Kegan Paul, London, 1966), pp. vii-xxxi
21. A.-S. Heinemann, *Quantifikation des Prädikats und numerisch definiter Syllogismus* (Mentis, Münster, 2015)
22. A.-S. Heinemann, 'Horrent with mysterious spiculae'. Augustus De Morgan's logic notation of 1850 as a 'calculus of opposite relations'. History and Philosophy of Logic **39**, 29-52 (2017)
23. W. E. Johnson, *Logic. Part I* (Cambridge University Press, Cambridge, 1921)
24. J. N. Keynes, *Studies and Exercises in Formal Logic* (3rd edition) (MacMillan, London, 1894)
25. C. I. Lewis, *A Survey of Symbolic Logic* (University of California Press, Berkeley, 1918)
26. D. D. Merrill, *Augustus De Morgan and the Logic of Relations* (Springer, Dordrecht, 1990)
27. M. Panteki, *Relationships between Algebra, Differential Equations and Logic in England: 1800-1860* (Ph.D. Middlesex University, London, 1991)

28. P. Sauriol, La structure tétrahexaédrique du système complet des propositions catégoriques. Dialogue **15**, 479-501 (1976)
29. H. Smessaert, L. Demey, Logical geometries and information in the square of oppositions. Journal of Logic, Language and Information **23**, 526-565 (2014)
30. H. Smessaert, L. Demey, The unreasonable effectiveness of bitstrings in logical geometry, in *The Square of Opposition: A Cornerstone of Thought*, ed. by J.-Y. Béziau, G. Basti (Springer, Cham, 2017), pp. 197-214
31. C. Thiel, The twisted logical square, in *Calculemos ... Matemáticas y libertad: homenaje a Miguel Sánchez-Mazas*, ed. by J. Echeverría, J. de Lorenzo, L. Peña (Editorial Trotta, Madrid, 1996), pp. 119-226
32. I. Thomas, CS(n): an extension of CS. Dominican Studies **2**, 145-160 (1949).

A Bitstring Semantics for Calculus *CL*

Jens Lemanski and Fabien Schang

Abstract The aim of this chapter is to develop a semantics for Calculus *CL*. *CL* is a diagrammatic calculus based on a logic machine presented by Johann Christian Lange in 1714, which combines features of Euler-, Venn-type, tree diagrams, squares of oppositions etc. In this chapter, it is argued that a Boolean account of formal ontology in *CL* helps to deal with logical oppositions and inferences of extended syllogistics. The result is a combination of Lange's diagrams with an algebraic semantics of terms: Bit-*CL*, in which any ordered objects are identified by characteristic bitstrings. Then, a number of objections to Bit-*CL* are answered to, and the process of inference is explained in this new logical framework.

Keywords Calculus *CL* · Bitstring semantics · Logic diagrams · Diagrammatic reasoning · Extended syllogistics · Ontology

Mathematics Subject Classification (2010) Primary 03A99; Secondary 03G99, 00A66

1 Introduction

In 1714, Johann Christian Lange and his staff introduced a logic machine that serves to automate three faculties of human cognitive abilities: (1) the representation of knowledge by means of hierarchically ordered concepts or classes, (2) the judgement concerning relationships of hierarchically ordered knowledge, and (3) the inferential reasoning based on explicit or implicit knowledge (cf. [17, 91]). Lange understood his approach as a contribution to a 'logica universalis', since the

J. Lemanski (✉)
FernUniversität in Hagen, Institute for Philosophy LG I, Hagen, Germany
e-mail: jens.lemanski@fernuni-hagen.de

F. Schang
Universidade Federal de Goiás, Faculty of Philosophy, Goiânia (GO), Brazil

J.-Y. Beziau, I. Vandoulakis (eds.), *The Exoteric Square of Opposition*, Studies in Universal Logic, https://doi.org/10.1007/978-3-030-90823-2_8

logic machine should unite the advantages of all logic diagrams known at that time, in particular tree diagrams, Eulerian diagrams and the square of opposition.

Calculus *CL* is the name of a diagram type that is based on the principles of Lange's logic machine and has the form of a square or a cube. Since Lange himself speaks of a logical cube, the abbreviation *CL* stands for '*C*ubus *L*ogicus' or '*C*ubus *L*angianae' and thus recalls the origin of the calculus presented here. *CL* also combines many features of today's common logic diagrams, which are currently used in areas such as visual representation (cf. [7]), proof theory (cf. [1]) or Artificial Intelligence (cf. [12]). In these research areas, diagrams are regarded as equivalent to conventional notations in a linguistic or sentential form, since they also have their own syntax and semantics (cf. [30]). As with many other diagram types, *CL* also has the special feature that it can depict information that was not intended when the diagram was created (cf. [29]).

In recent years, there has been much progress in research on various types of diagrams, which build the basis of *CL*: tree diagrams have been used to clarify semantic problems in ontology engineering (cf. [14]), Euler diagrams have been combined with systems of natural deduction (cf. [22]), and a bitstring semantics has been developed for the square of opposition (cf. [8]), to name just a few examples. Since *CL* combines various elements of these diagram types into a new one, it is obvious that advances in other diagram types can also be applied to *CL*.

Moreover, research on *CL* is still in its infancy: In the last 2 years, *CL* has been studied first as a two-dimensional square. However, this square forms the basis of the three-dimensional cube that Lange had in mind. The researches on *CL* so far have gone in different directions: oppositional structures have been investigated (cf. [21]), analogy relations have been analyzed (cf. [2]), possible applications in area domains have been examined [18], Lange's role in AI history has been discussed (cf. [24]), *CL* has been presented as an extended syllogistics (cf. [20]) and currently it has been proven that *CL* can be used as a formal system that is sound and complete (cf. [15]).

In this chapter, we want to develop a bitstring semantics for *CL*, which helps to organize, display and verify information. Thus, we assume that the progress for the square of opposition can also be transferred to *CL*. We will see that there are differences between the use of a bitstring semantics in the square of opposition and in *CL*. Nevertheless, a bitstring semantics developed specifically for *CL* is very well suited to represent classes, display propositions and test inferences. We expect from this approach on the one hand to obtain a well-founded and suitable semantics, which offers easy verifiability of *CL* as a formal system, and a model for *CL* diagrams, which will eventually make arithmetic operations in *CL* possible with the help of Boolean algebra.

Section 2 presents the current research on bitstring semantics, which is particularly oriented towards the square of opposition. In Sect. 3, we will design the basic principles of *CL* using a specific bitstring semantics. Section 4 then shows how to test inferences in *CL* using a bitstring semantics. Finally, in Sect. 5, we will briefly summarize the results on bitstring semantics in *CL* and venture an outlook on future research in this area. It should be noted that in Sects. 3 and 4, we refer only to the basic principles of *CL* and to one possibility to use bitstrings semantics in *CL*. Furthermore, we use a simple *CL* diagram and use extended syllogisms with only few propositions.

2 Bitstring Semantics

It is well known that consequence is the central issue of formal logic, whereas opposition is a derived notion that can be explained in truth-conditional terms. For instance, two sentences are said to be contradictory to each other if, and only if, they can be neither true together nor false together. At the same time, the process of partition has to do with opposition and appeared as a relevant feature in the history of philosophy as well. Some cases in point are Socrates' definition by dichotomy (cf. [32]), the Pythagorean table of opposites [10] or Seneca's and Porphyry's trees for classifying categories of being (genera, species, accidental differences and the like) (cf. [11]). The latter two organized ontology into a range of increasingly higher order entities, whilst the former mentioned characterized the meaning of concepts by ramifying the discourse into positive and negative.

The present section wants to go back to the latter tradition by emphasizing upon the logic of *terms*, i.e. a set of values and operations upon the components of sentences. The result is an *extensional* logic of terms, where properties are the basic items of meaning and behave like extensions of structured objects instead of being viewed as *intensional* entities. Unlike the usual distinction between extensions and intensions, the following semantics will depict properties as extensionally greater than objects.

2.1 Identity Without Existence

The ontology of the twentieth century was shaped by the existential questions of the Quine-Carnap debate, and today many argue that Quine's main question of *On what there is* remained the dominant one (cf. [13, 34]). With Quine's existential question, two further questions that relate to formal ontology and formal semantics are connected: The first question is: how many objects can there be in an arbitrary world? The second question is: how to define an arbitrary object within such a world? Such interrelated questions also relate to the issue of individuation, and they can find a preliminary answer in Quine: the two necessary criteria for individuating any object are existence and identity. Only the latter will be retained in the following semantics, insofar as existence is not required to make sense of entities.

Indeed, the first criterion of existence assumes a primary metaphysical distinction between two main kinds of entities: those existing ones, or individuals, and the other ones. This distinction is deeply entrenched from Aristotle's metaphysics to the Frege–Russell tradition of first-order logic, as witnessed by the usual distinction between logically singular and general terms. In this sense, singular terms are those which can never be predicated of another one, contrary to general terms. This matches with Aristotle's definition of primary substances or Russell's proper names, that is, whatever cannot be the predicate in a categorical proposition and always occurs in the position of subject term [33, chap. V].

Our intention is neither to take a position on these questions of ontology nor to make any connection between Quine's criteria of identity and the previous philosophy of Lange. On the other hand, a way back to Lange's logical works may have the merit of calling into question some contemporary ontological presuppositions, including the significance of atomism in logic and the resulting logical atomism. Instead of positing such a metaphysical theory of preexisting objects, the following wants to account for the meaning of any entity uniquely in terms of properties. This brings us closer to recent trends in ontology (e.g. [26, 28]), which regard the question of ordered structures as more important than existential questions in the sense of the Quine–Carnap debate. However, from Quine we are borrowing the insight that ontological matters are relative to how speakers or experts take the world to be structured into related entities and operations on them. But unlike Quine, our coming semantics wants to take the issue of existence apart and construe a formal logic on a prior formal ontology of constructed or structured objects. These are not mere atoms, departing from the mainstream tradition of logical atomism inherited from Russell and Wittgenstein.

2.2 *Boolean Ontology*

Let us consider a fragment of discourse including 3 properties only: B for blue, R for round and S for soft. Then, the existential question of how many individuals there *are* that are blue, round or soft in the world is a secondary question from our perspective. Rather, the formal question of how many individuals there *can be* is to be answered to by a mere exhaustive combination of the available properties. One first object may be both blue, round and soft; a second one may be blue, round and not soft; a third one may blue, not round and soft; and the like, resulting in a total set of 8 kinds of objects. More generally, for any set $S = \{P_1, \dots, P_n\}$ of n given properties P, there is a total of 2^n kinds of possible objects which correspond to the single subsets of S. Of course, any two different objects may share the same properties in this partial domain, and the question about whether properties are sufficient to characterize any individual is debated in philosophy of language. In addition, imagine an all-comprehensive set of properties S^*, so that there could not exist any additional property $n + 1$ outside S^*. Then given that every object is to be characterized by an ordered set of these properties, none of the objects could have all or none of the properties of S^* for it would be nothing or everything, respectively. In other words, any object must satisfy at least one property and not satisfy at least one other property. At the same time, any two objects can be distinguished from each other by some of their characteristic properties whenever there is one property the one has that the other one does not.

2.3 Term Semantics

The following semantics is a systematization of earlier works (cf. [8, 27]), where bitstrings were introduced to deal with fragments of language centered upon specific expressions like binary sentences, modalities and the like. Unlike these pioneer works, this chapter wants to deal with any kind of meaningful information without restriction: concepts or individuals; besides that, any kind of information is intended to be codified by a sort of logical code—a characteristic bitstring, to be compared to the biological code of DNA and related to any other information irrespective of its content. Here is the main added contribution of this chapter, given that logical opposites are traditionally limited to structured formulas whose content is the same.

Let us consider now our intended *Bitstring Semantics* (thereafter, BS). It is an algebra of terms according to which objects are to be defined as ordered sets of properties. $BS = \langle L, D \rangle$ consists in a language L including a set of *properties* $P = \{P_1, \ldots, P_n\}$ and a set of operations on these properties (complement, join, meet, inclusion), together with a Boolean domain of valuation D including the two bits 1 and 0. The valuation function β assigns the value 1 and 0 to every element of P, so that every ordered set of bits $\langle \beta(P_1), \ldots, \beta(P_n) \rangle$ is a *bitstring* that denotes a corresponding *object*. Hence, every object x_i of BS corresponds to an ordered set of properties that are satisfied or not by this object. Taking again the above example of $m = 3$ properties, this yields an exhaustive set of 2^3 objects $x = \{a, b, c, d, e, f, g, h\}$ denoted by their characteristic bitstrings $\beta(x) = \langle \beta(P_1), \beta(P_2), \beta(P_3) \rangle$. Thus, $\beta(a) = 111, \beta(b) = 110, \beta(c) = 101, \beta(d) = 011, \beta(e) = 100, \beta(f) = 001, \beta(g) = 010$ and $\beta(h) = 000$.

A calculus of predications may also be made specified according to the relation any objects of a given ontology $x = \{x_1, \ldots, x_n\}$, thereby giving a Boolean expression to sets, relations and entities beside the language of classical logic. Let $\top$ and $\bot$ be constant values of bitstrings, such that $\top$ and $\bot$ are finite sequences $1\ldots1$ and $0\ldots0$ including only 1s-bits and 0-bits, respectively. Then, a set of four main operations $\{\cap, \cup, \subset, \overline{}\}$ may be performed into BS upon structured objects x, to be characterized by a *length* of n finitely ordered properties $\beta(x) = \langle \beta_1(x), \ldots, \beta_n(x) \rangle$ where every ordered valuation function $_i$ assigns a bit to the ith property P_i. Then, for any object variable $x_1, x_2 \in x, x_1, x_2$ may be turned into other objects by one unary term operator and three binary term operators.

Term negation (complement):

$\beta(\overline{x_1}) = \langle \overline{\beta_1(x_1)}, \ldots, \overline{\beta_n(x_1)} \rangle$

Example: if $\beta(x_1) = 1000$, then $\overline{\beta(x_1)} = 0111$.

Term conjunction (meet):

$\beta(x_1 \cap x_2) = \beta(x_1) \cap \beta(x_2) = \langle \beta_1(x_1) \cap \beta_1(x_2), \ldots, \beta_n(x_1) \cap \beta_n(x_2) \rangle$

Example: if $\beta(x_1) = 1110$ and if $\beta(x_2) = 0111$, then $\beta(x_1 \cap x_2) = 0110$.

Term disjunction (union):

$\beta(x_1 \cup x_2) = \beta(x_1) \cup \beta(x_2) = \langle \beta_1(x_1) \cup \beta_1(x_2), \ldots, \beta_n(x_1) \cup \beta_n(x_2) \rangle$

Example: if $\beta(x_1) = 1000$ and if $\beta(x_2) = 0001$, then $\beta(x_1 \cap x_2) = 1001$.

Term inclusion:

$\beta(x_1 \subset x_2) = \beta\overline{(x_1 \cap \overline{x_2})} = \beta(\overline{x_1} \cup x_2) = \langle \beta_1(\overline{x_1}) \cup \beta_1(x_2), \ldots, \beta_n(\overline{x_1}) \cup \beta_n(x_2) \rangle$

Example: if $\beta(x_1) = 1100$ and $\beta(x_2) = 0110$, then $\beta(x_1 \subset x_2) = 0011 \vee 0110 = 0111$.

Sentential formulas can be gathered from this term calculus by defining the truth-values True and False in terms of inclusion. Whilst the usual truth-conditional semantics sentential assigns basic values to sentences in terms of truth and falsity, BS is in position to account for the truth-value of predicative sentences by explaining the relation of subsumption set-theoretically. For example, let p be the closed sentential formula 'b and c are a'. That p is interpreted as true in the above domain can be shown by the fact that whatever b and c are is also b: a is blue and round and soft, whereas b and c are jointly blue; however, that b and c are neither round nor soft jointly entails that b and c are not a.

2.4 *Truth and Falsity*

Equipped with this term calculus, the sentential values of truth and falsity can now be defined according to what predications hold or not. More precisely, any predicative sentence can be evaluated in terms of inclusion such that, for any different objects x_1, x_2:

Truth and Falsity

The sentence 'x_1 is x_2' is *true* if, and only if, $\beta(x_1 \subset x_2) = \top$, i.e.

The sentence 'x_1 is x_2' is *false* if, and only if, $\beta(x_1 \subset x_2) \neq \top$.

In other words, 'x_1 is x_2' is true if and only if whatever is x_1 is also x_2, whereas 'x_1 is x_2' is false if and only if there is at least one x_1 that is not x_2.

Such a sentence seems to be absurd from the perspective of first-order logic, where both x_1 and x_2 are normally intended to be instantiated by individual constants. For if so, then how can any value of x_2 fall upon the range of predicate functions insofar as x_1 is also taken to be an individual? It turns out that, from our formal perspective of terms, x_1 and x_2 are any arbitrary terms that can occur in the positions of either subject or predicate terms; in other words, terms are not automatically considered as denoting individuals by merely being denoted with the same sort of letters, in the sense of not being predicates; rather, they are arbitrary objects which may be basic individuals or more abstract entities like predicates. For this reason, the formal symbolism of BS does not rely on the usual distinction between capital $(P, Q, R, \ldots)$ and lowercase letters $(a, b, c, \ldots)$. This also means that the ontological distinction between different kinds of entities and levels of abstraction will be expressed without specific letters, for want of any metaphysical hierarchy of beings in the language for BS. Rather, the difference between ontological levels will be expressed in terms of their characteristic bitstrings. Assuming that what makes a subject different from a predicate in any predicate statement is that the latter be more abstract than the former, the abstractness degree can be defined in *BS* as the number of properties a given object is in position to satisfy.

2.5 Hamming Measurements

In the following, but especially in Sect. 3, we will use Hamming measures to better describe bitstrings and their relations. We use especially *Hamming weight* and *Hamming distance* of bitstrings.

Hamming weight:

For any x, its Hamming weight $w(x)$ is the number of occurrences of 1-bits in $\beta(x)$.

Example: Let $\beta(x_1) = 1000$, and let $\beta(x_2) = 1110$, then $w(x_1) = 1$ and $w(x_2) = 3$.

Let $d(x_1, x_2)$ be the Hamming distance between any terms x_1 and x_2 into a given set of $2^n - 1$ terms, to be defined by the number of Boolean bits that differ in x_1 and x_2.

Hamming distance:

For any x_1, x_2, the Hamming distance $d(x_1, x_2)$ is the number of occurrences at which their corresponding bits $\beta_i(x_1), \beta_i(x_2)$ differ from each other.

$$\sum_{i=1}^{n-1} |\beta_i(x_1) - \beta_i(x_2)|$$

that is, $d(x_1, x_2) = |(\beta_1(x_1) - \beta_1(x_2)| + \ldots + |(\beta_n(x_1) - \beta_n(x_2)|$.

Example: Let $\beta(x_1) = 1000$, and let $\beta(x_2) = 1100$, then $d(x_1, x_2) = |(\beta_1(x_1) - \beta_1(x_2)|, \ldots, |(\beta_n(x_1) - \beta_n(x_2)| = 0 + 1 + 0 + 0 = 1$.

Hamming measurements will be detailed further in Sect. 3, due to their usefulness when dealing with bits and bitstrings. Furthermore, Hamming distance and Hamming weight may also contribute to the definition of logical oppositions.

For example, any x_1 and x_2 are said to be *contradictories* if and only if their Hamming distance is maximal, that is, $d(x_1, x_2) = n$. In the above case, $w(x_1) = 1$ and $n = 4$, so that the contradictory $cd(x_1)$ of x_1 is such that $d(x_1, x_2) = 4$, i.e. $\beta(cd(x_1)) = 0111$.

Likewise, the relation of *subalternation* can be defined as a case of subsumption or possible predication such that, if x_1 is subaltern to x_2, then $w(x_1) > w(x_2)$.

2.6 Boolean Quantifications

Finally, any pair of objects of our Boolean domain may be related to each other by using the traditional relations of *oppositions*. The special contribution of BS in this domain is to afford a Boolean calculus on categorical propositions, by accounting for these quantified statements in terms of range over bits. Let us take the basic, non-quantified sentence 'x_1 is x_2'. Then, the quantified accounts of the basic predication 'x_1 is x_2' yield the four usual Aristotelian propositions **a**, **e**, **i**, and **o**, such that:

a: 'Every x_1 is x_2' is true if, and only if, there is no x_1 that is not x_2, that is, $\overline{\beta(x_1 \cap \overline{x_2})} = \top$, i.e. $\beta(x_1 \cap \overline{x_2}) = \bot$.
e: 'No x_1 is x_2' is true if, and only if, there is no x_1 that is x_2, that is, $\overline{\beta(x_1 \cap x_2)} = \top$, i.e. $\beta(x_1 \cap x_2) = \bot$.
i: 'Some x_1 is x_2' is true if, and only if, not every x_1 is not x_2, that is, $\overline{\beta(x_1 \cap x_2)} \neq \top$, i.e. $\beta(x_1 \cap x_2) \neq \bot$.
o: 'Some x_1 is not x_2' is true if, and only if, not every x_1 is x_2, that is, $\overline{\beta(x_1 \cap \overline{x_2})} \neq \top$, i.e. $\beta(x_1 \cap \overline{x_2}) \neq \bot$.

It could be noted that the above quantified formulas express a number of relations between the terms x_1 and x_2; for this reason, the above definitions already include some relations of opposition insofar as the relations between x_1 and x_2 internalize the four Aristotelian oppositions between terms within one and the same proposition. More generally, there is a structural isomorphism between the oppositions between sentences of the forms 'x_1 is (not) x_2' and those between terms like x_1 and x_2. Bits are for x_1 and x_2 what models are for propositions p and q, that is, an ordered set of conditions satisfied by both kinds of relata. The relations of opposition can thus be characterized as follows.

Contrariety: $R = CT$

x_1 and x_2 are *contrary* to each other if, and only if, every property P_i that is satisfied in x_1 is not satisfied in x_2. That is, if $\beta_i(x_1) = 1$, then $\beta_i(x_2) = 0$.

Contradiction: $R = CD$

x_1 and x_2 are *contradictory* to each other if, and only if, every property P_i that is satisfied in x_1 is not satisfied in x_2 and conversely. That is, $\beta_i(x_1) = 1$ if, and only if, $\beta_i(x_2) = 0$.

Subcontrariety: $R = SCT$

x_1 and x_2 are *subcontrary* to each other if, and only if every property P_i that is not satisfied in x_1 is satisfied in x_2 and conversely. That is, if $\beta_i(x_1) = 0$, then $\beta_i(x_2) = 1$.

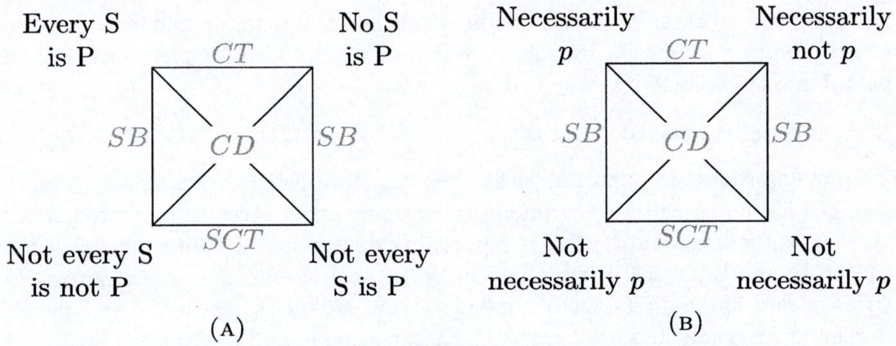

Fig. 1 Square of Oppositions: **(a)** Categorical propositions. **(b)** Modal propositions

Subalternation: $R = SB$

x_2 is *subaltern* to x_1 if, and only if, every property P_i that is satisfied in x_1 is also satisfied in x_2. That is, if $\beta_i(x_1) = 1$, then $\beta_i(x_2) = 1$.

All these logical properties may be illustrated into the famous square of oppositions or Aristotelian square, irrespective of the kind of formulas in use. As illustrated in Fig. 1, these were restricted to categorical and modal propositions, in Aristotle's *Organon* and in the Aristotelian commentaries (cf. [6]).

It is worthwhile noting that such square relates logical entities like categorical or modal propositions to each other, but rarely ontological entities like singular terms or properties.[1] This means that our bitstring semantics goes farther into the Aristotelian square by relating any kind of meaningful entities, beyond the sole cases of propositions. This extension already occurred in the recent history of logic, when the authors like Sesmat or Blanché showed that logical oppositions apply beyond categorical and modal propositions to define a large range of concepts including binary connectives (and, or etc.), ordering relations (greater than, lesser than etc.) and the like (cf. [16]).

[1] An analysis of objects in terms of oppositions has been made in [9]. This method could be applied to the ontology presented in Sect. 4.

Now, our Boolean approach to semantics means that whatever can be characterized by bitstrings may occur in such a logical square, including the sole Ss and Ps. Let x_1 be any such term that can be defined by a set of 4 properties. Then, a logical square for x_1 may be illustrated as given in Fig. 2.

It should be clear that such a term like x_1 does not have only one characteristic square: its Boolean structure helps to determine the cardinality of its contraries, subcontraries and subalterns (or superalterns), whilst every term has one and only one contradictory.

And finally, a semantics for quantified predicates may also be afforded into BS with formulas like 'Every x_1 is every x_2', 'Every x_1 is every x_2', 'Some x_1 is every x_2' or 'Some x_1 is some x_2'. Let '$\ldots x_1$ is---x_2' be such an expression including quantifiers on both subject terms and predicate terms. Assuming that the extension of the predicate term x_2 is broader than that of x_1, the above expressions may be parsed into nested quantifiers such that:

'$\ldots x_1$ is---x_2' is satisfied such that, for $\ldots x_1$ and---x_2, $\beta(x_1 \text{ is } x_2) = \top$.

In syllogistics, the oppositions, as they are represented in the square (Figs. 1 and 2), play an essential role: among other things, they serve to prove imperfect syllogisms for their validity. This is especially essential for the indirect modes (such as Baroco and Bocardo). We will see in Sect. 3 that the meaning of the arrows in CL is related to the directions of the lines in the square of opposition. In contrast to traditional syllogistics, the arrows in CL can also be used to prove inferences of extended syllogistics such as '$\ldots x_1$ is---x_2'.

2.7 Objections and Answers

A number of objections may be raised upon the foundations of BS, however.

A first objection is about the usual distinction between material and formal truth. How can it be sustained in the light of this semantics, given that any true sentence is defined there in terms of $\top$, whereas the latter symbol normally expresses formal truth? A way to overcome this problem is to introduce *dispositional* predicates, by distinction with real predicates or predicates actualized in the real world. Thus, any object x of a given domain is to be defined as an ordered set of properties that x can possess. Although this account may appear undesirable by introducing modal expressions, one can also interpret what an object can possess in the light of updated empirical data. For example, any number *cannot* be colored by virtue of what numbers are supposed to be. Hence, numbers never satisfy properties having to do with colors in any possible domain of interpretation, whilst a mere pencil may be blue or not in a given domain of interpretation.

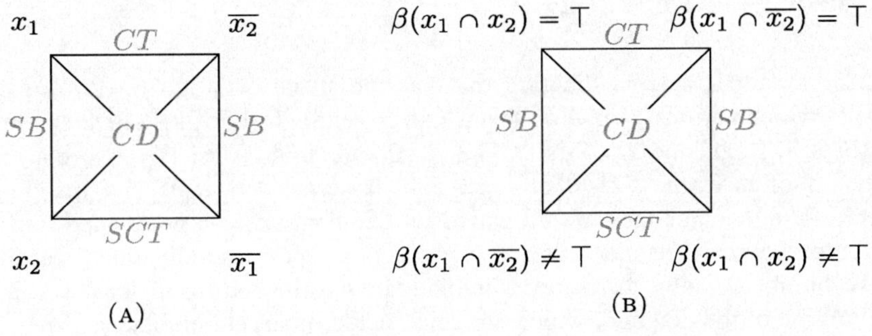

Fig. 2 Square of Oppositions: **(a)** Oppositions between individuals. **(b)** Oppositions between atomic propositions

A second objection is about the process of individuation. What are the necessary and sufficient conditions for an object to be individuated in BS, according to the length of its characteristic bitstring? By virtue of the Leibnizian law of identity, any x is individuated *only if* its characteristic bitstring makes it different from all the other objects in a given domain of interpretation. At the same time, *if* any such x_1 is to be characterized, then it is made different from any other object x_2 whose characteristic bitstring is not that of x_1. Another issue is whether any object can always be characterized in concrete situations of natural language. But just as logicians need not investigate what makes a sentence materially true or false outside their logical system, BS need not explain how many properties a given object needs in order to be characterized completely.

A third and last objection is about the concept of identity in itself. How can a given object x be defined in BS without circularity, given that any set of properties $\beta_1(x_1) \ldots \beta_n(x_1)$ characterizing bitstrings already assumes identity by stating that $\beta_i(x_1) \neq \beta_j(x_1)$ for any compound properties $\beta_i(x_1), \beta_j(x_1)$ of $\beta(x)$. BS states Leibniz's law of identity, by redefining identity in terms of bitstrings. Thus,

$x_1 = x_2$ if, and only if, $\beta(x_1) = \beta(x_2)$.

Nevertheless, the criterion of existence disappears from *BS* since such a logic of terms does not take quantifiers as relevant parts of its language. So just as Quine said, *BS* assumes that no entity may hold without identity, but unlike Quine, identity does not assume existence in the sense that such a sentence as $x = x$ holds whether x is said to exist in the real world or not. This is, by passing, a way to avoid the difficulty of existential import (cf. [5]) by assigning truth to sentences without assuming no ontological commitment.

Once the Boolean semantics BS is set up, let us see how it can be implemented into Lange's Calculus *CL*. Then, we will talk about the advantages of combining such diagrammatical and algebraic devices in logic.

3 Basic Principles of *CL*

A *CL* diagram is a logic diagram that has the shape of a square (2D) or cube (3D) to which various geometric forms can be applied. (For illustrative examples, see Figs. 3, 4, 5, and 6.) First of all, a *CL* diagram consists of several rows of boxes of different sizes that represent classes/sets/concepts and the objects /members/individuals they are grounded in. *CL* diagrams can be divided into two valid types, namely, regular and irregular diagram types. Both diagram types differ in the number of objects/members/individuals (in the following 'basics') in the bottom row of the diagram, which are constructed or structured objects. The more basics a *CL* diagram has, the more rows with different larger classes/sets/concepts (in the following 'classes') it has in the upper levels.

A regular *CL* diagram is based on 2^n basics. Any *CL* diagram that does not consist of 2^n basics is considered irregular. The number of basics also determines the size of the *CL* diagram: the more basics there are, the more classes the *CL* diagram contains and the larger it is. *CL* diagrams can be enlarged at will.

Further geometric forms can be applied to the various boxes, which then represent logical relations within the diagram. Up to now, arrows, lines, tensors and shadings were used in *CL* to represent propositions, logical connectives or general content information for propositional calculus or predicate logic. Many more elements are conceivable.

In the following, we use arrows and describe the basics and the classes in a regular 2D *CL* diagram. So, we only use a few possible functions of the logic diagram in order to show how *BS* works in *CL*.

3.1 Basics

Basics are represented by solid boxes, i.e. □, in the bottom row of a *CL* diagram. Unlike classes (see Sect. 3.2), basics do not contain dotted lines. At first sight, (1) basics can be interpreted as having some similarities to Aristotle's primary substances, Boethius's species specialissima (or individua) or maybe to Russell's proper names (cf. Sect. 2.1) and (2) the bottom row, including a series of basics (□, □, ...), can be interpreted as having some similarities to what has been called a 'fundamental layer' [4] or a '(partial) ultimate ontological basis' [23]. But these are only aids to understand what basics can be. For unlike these interpretations and debates mentioned above, *CL* does not use logic to argue what an ontology is that represents the world, but what logical assumptions we can make when we choose an ontology that can represent the world (cf. also Sect. 4.2).

Another given aid of understanding basics is *BS* as described in Sect. 2: each basic can be defined by a bitstring, the length of which is determined by the number of basics in the *CL* diagram. The bitstring of a basic always has a Hamming weight of 1, where the position of 1 in the bitstring determines the position of the basic's solid box in relation to all other basics in the *CL* diagram.

Basics can be defined in *BS* not by their mere existence, but by a property (cf. Sect. 2.1). In the variant of the *BS* presented here, it is the property that they are located at a certain position within the diagram. More precisely, each digit of a bitstring denotes one property: since each *n*th digit of a bitstring corresponds to the *n*th position of a basics in a *CL* diagram, each digit can be understood as an answer to the question whether this basic is meant or not. The position of a basic in a *CL* diagram is represented by a solid box in the lowest row. Diagrammatically speaking, each basic has its own solid box; in terms of *BS*, each basic has its own bitstring that functions as a unique identifier, thereby distinguishing each basic from all others into an ordered sequence from left to right or from right to left (cf. the DNA example in Sect. 2.3).

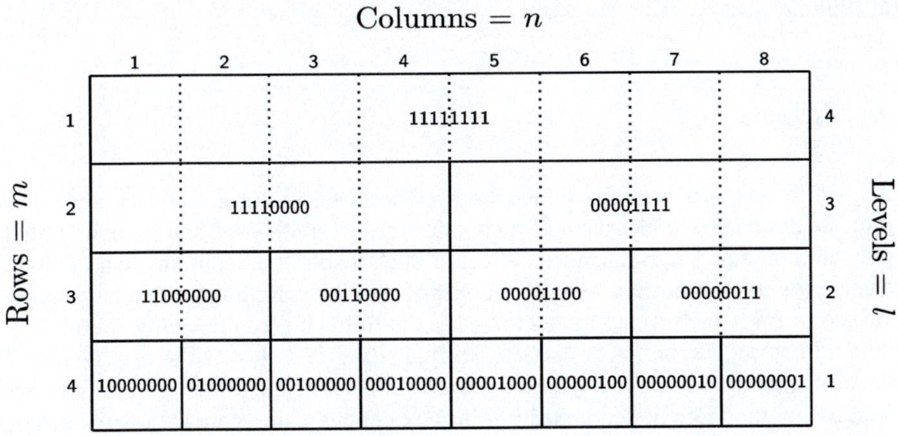

Fig. 3 A regular CL^8 diagram (4 × 8 matrix)

Let a regular *CL* diagram be a $m \times n$ matrix a_{ij} with m for horizontal rows and n for vertical columns (cf. [31, sect. 1.3]), such that

$$a_{ij} = \begin{bmatrix} a_{11} & a_{12} & \cdots & a_{1n} \\ a_{21} & a_{22} & \cdots & a_{2n} \\ \vdots & \vdots & \ddots & \vdots \\ a_{m1} & a_{m2} & \cdots & a_{mn} \end{bmatrix}.$$

A bitstring with a Hamming weight of 1 that begins with 1 at the first bit position denotes the basic at the solid box at the outer left edge of the lowest row, a_{m1}; a bitstring with the same Hamming weight, which is flagged 1 at the last bit position, is located at the outer right edge, a_{mn}. The Hamming distance between two basics is 2: to generate a basic's bitstring to the right of a_{m1}, i.e. a_{m2}, the first bit of a_{m1}

must be converted from 1 to 0 and the second bit must be converted from 0 to 1. To generate a basic bitstring to the left of a_{mn}, the last bit of a_{mn} must be converted from 1 to 0 and the penultimate 0 of the bitstring must be converted from 0 to 1. All basics with $w = 1$ and $d = 2$ denote solid boxes in the *CL* diagram located between a_{m1} and a_{mn}. A bitstring with $w = 0$ is outside the *CL* diagram, so that basics are all ways of forming minimal Hamming weights in *CL*.

Example In a regular *CL* diagram with $2^3 = 8$ basics, each bitstring has a length of $n = 3$, so that each following diagram will be an instance of a CL^{2^n} diagram. In the CL^8 diagram or 4×8 matrix of Fig. 3, there are eight solid boxes denoting basics each with a Hamming weight of 1, i.e. 10000000, 01000000, 00100000, ..., 00000001. The basic at the lower left corner is in $CL^8 = a_{41} = 10000000$, and the basic at the lower right corner is in $CL^8 = a_{48} = 00000001$. The bitstring 01000000 is read from the left, at the second position, i.e. a_{42}. The bitstring 00001000 is read from the left, at the fifth position, i.e. a_{45}.

3.2 Classes

Any solid box that contains dotted lines is not a basic but a class. Dotted lines indicate dotted boxes, but most of their sides are covered by the solid box in which they are contained, e.g.  and  $\cong$ 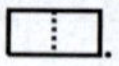. Such a solid box contains two or more basics, which are mirrored by dotted boxes within the solid box of the class. In the variant of *BS*, which we apply here for *CL*, the following applies: only solid boxes hold a bitstring, i.e. basics or classes. Each bitstring of a class is the *disjunction* of the bitstrings of two or more basics. For each class, $w < 1$. The higher the Hamming weight, i.e. the more dotted boxes a solid box contains, the more disjuncted basics the bitstring of a class denotes.

In the following, we distinguish between rows (m) and levels (l). Rows are horizontal series of dotted or solid boxes within the matrix. Levels denote the horizontal series of classes with the help of solid boxes. All solid boxes with the same Hamming weight form a level of classes. The higher the Hamming weight of a box, the higher the level in the *CL* diagram.

Basics have a Hamming weight of 1 and are therefore located within the first level seen from the bottom-up. In the level above there are only bitstrings with $w = 2$ in which two basics are disjuncted. In the third row above, there are bitstrings with $w = 4$ etc. More generally, at any level ($l\alpha$), there are only bitstrings with a Hamming weight of $2^{\alpha-1}$. If there is only one solid box in the highest row of the matrix ($m = 1$), then it has the maximum Hamming weight corresponding to the bitstring length.

Example In a regular CL^8 diagram (Fig. 3), the classes in the second level series ($l = 2$) consist of bitstrings with a Hamming weight of 2: 11000000, 00110000, 00001100, 00000011. In the CL^8 diagram, the classes with

$w = 4$ form the third level series ($l = 3$): 11110000, 00001111 and the top class ($l = 4$) has the maximum Hamming weight, i.e. 11111111.

For each higher class of the CL^8 diagram *top-down* applies in relation to all other classes:

($l = 4$): {11111111} = {11110000 ∨ 00001111}
($l = 3$): {11110000 = (11000000 ∨ 00110000), 00001111 =
(00001100 ∨ 00000011)}
($l = 2$): {11000000 = (10000000 ∨ 01000000), 00110000 =
(00100000 ∨ 00010000), 00001100 =
(00001000 ∨ 00000100), 00000011 =
(00000010 ∨ 00000001)}

Bottom-up applies to all classes within a row or level:

($l = 1$): {10000000, 01000000, 00100000, 00010000, 00001000,
00000100, 00000010, 00000001}
($l = 2$): {11000000, 00110000, 00001100, 00000011}
($l = 3$): {11110000, 00001111}

3.3 *Arrows*

So far we have only examined the ontology of *CL*, and we have presented in the given examples the regular order of basics and classes according to the JEPD (Jointly Exhaustive and Pairwise Disjoint) principle (cf. [14]). *CL* has this advantage in comparison to ordinary ontologies, however, that it intuitively offers a spatial structure to represent relations between solid boxes (basics or classes) in the form of propositions. Arrows can be used to make the information explicit, which are only given implicit in the ontology. It will be shown that the arrows provide more information about relations than is needed for the construction of the *CL* diagram (cf. Sects. 3.1–3.2).

The relations between two boxes are always displayed with the help of straight arrows. With these straight arrows, the shortest connection between two solid boxes is searched, and all other possible connections with straight arrows (between dotted boxes inside the solid ones) should be considered. The arrow shaft shows the first term of a proposition and the arrowhead the second term. The arrows have a different meaning depending on the direction. Similar to the square of opposition (cf. Sect. 2.6), one can generally say that vertical arrows are positive, and horizontal and transversal arrows are negative. In detail, the following applies:

(1) If arrows go bottom-up from all dotted boxes of a lower solid box to some dotted boxes of a higher solid box, all lower dotted boxes are included in the higher box.

Example: The proposition '11000000↑↑11111111', i.e. 'All 11000000 is some 11111111', can be represented by two arrows leading from the lower boxes a_{31}, a_{32} to the boxes a_{11}, a_{12}.

(2) If arrows go bottom-up from some dotted boxes of a lower solid box to some dotted boxes of a higher solid box, some lower dotted boxes are included in the higher box.
Example: The proposition '11000000↑↓11111111', meaning 'Some 11000000 is some 11111111', could be represented by an arrow going from the lower box a_{31} to the box a_{11}.
NB!: (2) is a partial expression of (1) since between 11000000 and 11111111 a further arrow between a_{32} and a_{12} can be drawn.

(3) If arrows go top-down from some dotted boxes of a higher solid box to dotted boxes of a lower solid box, this means that some higher boxes correspond to some lower boxes.
Example: The proposition '11110000↑↓00110000', meaning 'Some 11110000 is some 00110000', can be represented by two arrows going from the higher box a_{23}, a_{24} to the boxes a_{33}, a_{34}.

(4) If arrows go horizontally from one box to another, the proposition is completely negative.
Example: The proposition '11110000⇄00001111', i.e. 'No 11110000 is any 00001111', can be displayed with one arrow going from the boxes $a_{21}, a_{22}, a_{23}, a_{24}$ to the boxes $a_{25}, a_{26}, a_{27}, a_{28}$.

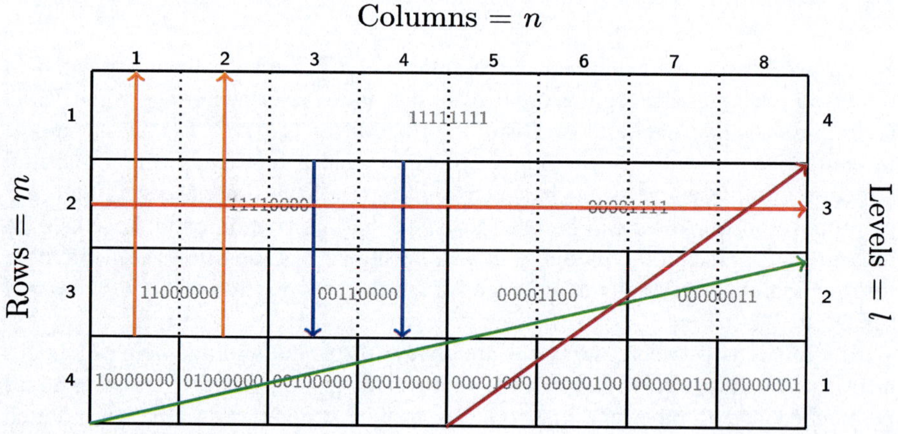

Fig. 4 A CL^8 diagram with arrows

(5) If an arrow crosses transversally from one box to another and no vertical arrow can be drawn between these two boxes, the proposition is completely negative.
Example: The proposition '10000000⤢00000011', i.e. 'No 10000000 is any 00000011', can be represented by an arrow leading from the box a_{41} to the boxes a_{37}, a_{38}.

NB! ‘10000000⤢00000011’ = ‘10000000⇄00000011’ because ‘10000000↑↓ 00000011’ is impossible (cf. also below (6)).

(6) If an arrow crosses transversally from one box to another and a vertical arrow can be drawn between these two boxes, the proposition is partially negative.
Example: The proposition ‘00001000⤢00001111’, i.e. ‘Some 00001000 is not some 00001111’, will be represented by an arrow leading from the box a_{45} to the boxes a_{27}, a_{28}.
NB! ‘00001000⤢ 00001111’ ≠ ‘00001000⇄00001111’ because ‘00001000↑↓ 00001111’ is possible.

It must be pointed out that we have six different interpretations of arrows, although there are only four different arrow directions, corresponding to the Aristotelian propositions (cf. Sect. 2.6). But this is not untypical for logic diagrams (cf. [3, Sect. 3.2.1]). The reason is that either one can draw less arrows than possible (1 and 2) or the meaning of an arrow depends on the possibility of implicit arrows between given boxes (5 and 6). So, we can either make less explicit than implicitly present (1 and 2) or consider more implicitly than should be made explicit (5 and 6). In any case, we can already see from the bitstrings which relations, expressed by arrows, are possible at all. In general, the following principles apply to the six types of arrows:

(1) Each 1-position of the first bitstring must also be a 1-position of the second bitstring, and the second bitstring must have a higher Hamming weight than the first. (All 1-bits are made explicit by arrows.)
(2) Each 1-position of the first bitstring must also be a 1-position of the second bitstring, and the second bitstring must have a higher Hamming weight than the first. (Only some 1-bits are made explicit by arrows.)
(3) Some 1-position of the first bitstring must also be a 1-position of the second bitstring, and the first bitstring must have a higher Hamming weight than the second one.
(4) Each 1-position of a bitstring must be a 0-position in the second bitstring. Both bitstrings must have the same Hamming weight.
(5) Each 1-position of a bitstring must be a 0-position in the other bitstring and vice versa. Both bitstrings must have the same Hamming weight.
(6) Each 1-position of one bitstring must be a 0-position in the other bitstring and vice versa. Both bitstrings cannot have the same Hamming weight.

4 Testing Inferences with *CL*[4]

CL can be used for very different purposes and for different logics. Lange has already used his logic machine for an extended syllogistic with numeric definite predicates as well as for propositional logic and modal logic [19]. Regardless of which logic is used, *CL* can be applied to test the validity of inferences, supplement incomplete inferences or extrapolate additional information from given inferences. The following subsections are limited to extended syllogisms (cf. also [20]), and *CL* is used only for the purpose of testing inferences consisting of two premises and one conclusion. To show how this works, we use four examples each to demonstrate the validity and invalidity of inferences.

In order to save space, we will work with the smallest possible *CL* diagram, in which all forms of proposition shown in Sect. 2 can be applied. This smallest possible *CL* diagram is an $m \times 4$ matrix with a bitstring length of 4 and is called CL^4. The following color assignment is suitable for the project of testing inferences and the following rule is set up.

Color assignment:

The first premise is displayed in blue, the second premise in green and the conclusion in red.

Rule (R):

Three propositions form a valid inference if **(RI)** all three propositions can be displayed by arrows in the *CL* diagram and **(RII)** two arrow ends meet each in three different solid boxes.

By using **R**, we now have a possibility to prove the validity of given inferences. If three propositions satisfy the **RI** subrule, they can be interpreted as true judgements about the ontology of the *CL* diagram. If they also fulfil subrule **RII**, the three arrows form a valid inference pattern in *CL*.

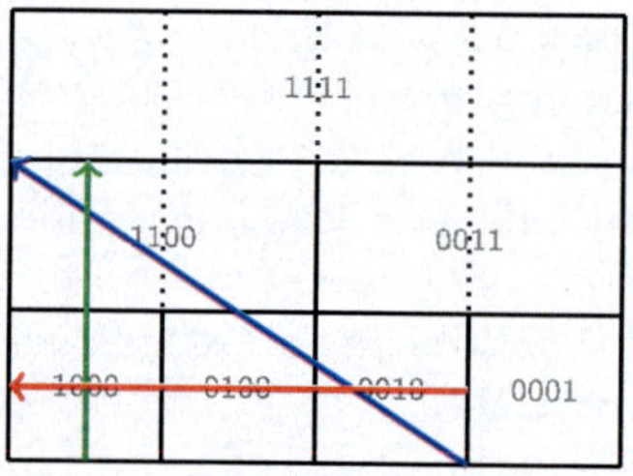

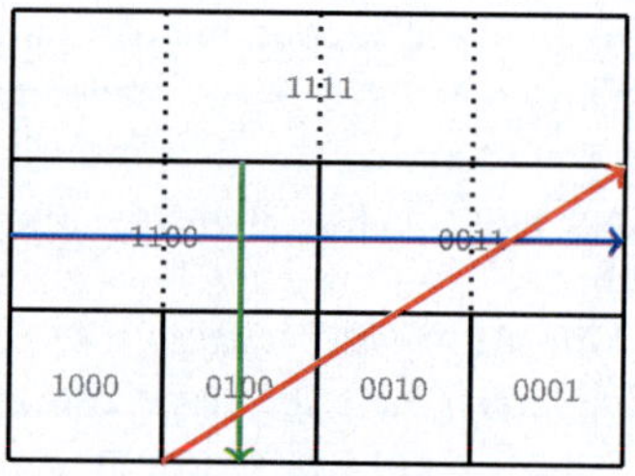

Fig. 5 Examples: **(a)** Ex1. **(b)** Ex2

In this way, on the one hand, the truth and falsehood discussed in Sect. 2.4 can be examined diagrammatically and, on the other hand, the technique for proving inferences mentioned at the end of Sect. 2.6 can both be replaced and applied to inferences of extended syllogistics.

4.1 Examples

We now take the following four inferences as given and check with CL^4 whether these inferences are valid or not.

Ex1 (Fig. 5A): 0010⤢1100 and 1000⇈1100, therefore 0010⇄1000.
Ex2 (Fig. 5B): 1100⇄0011 and 1100↑↓0100, therefore 0100⤢0011.
Ex3 (Fig. 6A): 0011⇈1111 and 0010⇈0011, therefore 0010⇈1111.
Ex4 (Fig. 6B): 0011⤢0010 and 0011↑↓1111, therefore 1111⤢0010.

Each example corresponds to one of the CL^4 diagrams in Figs. 5 and 6. In each diagram, it can be seen that all three corresponding propositions of Ex1–4 can be displayed. The rule **(RI)** is therefore fulfilled in all examples. Since in three different solid boxes of each diagram two arrow ends meet, the rule **(RII)** for Ex1–4 is also fulfilled.

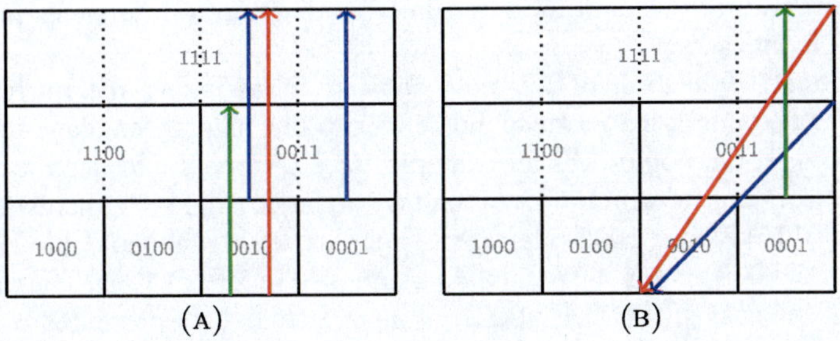

Fig. 6 Examples: **(a)** Ex3. **(b)** Ex4

Let us take a look at the following four inferences for comparison. All four examples are a repetition of Ex1–4, but one proposition was altered in each case. For convenience, the changed proposition has not been color-marked. Therefore Ex5–8 are only partially identical with Figs. 5 and 6.

Ex5: 0010⤢1100 and 1000⇄1100, therefore 0010⇄1000.
Ex6: 1100⇈0011 and 1100↑↓0100, therefore 0100⤢0011.
Ex7: 0011⇈1111 and 0011⇈1111, therefore 0010⇈1111.
Ex8: 1000⇈1100 and 0011↑↓1111, therefore 1111⤢0010.

All four given examples, Ex5–8, show invalid inferences: in Ex5, the second premise cannot be represented in a *CL* diagram, since it is required to draw a horizontal arrow between 1000 and 1100, which is not possible. 1000 and 1100 are on different levels of the diagram and cannot represent a universal negative statement since 1100 is 1000 ∨ 0100. Also, the first premise of Ex6 is false, because 1100 and 0011 are on the same level and both exclude each other. A universal positive proposition with bottom-up arrows between 1100 and 0011 is therefore impossible to display in CL^4. Ex5 and Ex6 therefore both do not fulfil the requirements of **RI**.

In Ex7 and Ex8, all premises can be represented in CL^4, but they do not fulfil **RII**: in Ex7, premises 1 and 2 are identical, so that in box 0010, the arrow shaft of the conclusion remains unconnected, but in box 1111 three arrowheads meet. In Ex8, the first premise has no connection to the other remaining arrows, so that there is no valid inference here either.

4.2 Application

In Sect. 4.1, we have shown how to test inferences with *CL* using four valid (Ex1–4) and four invalid examples (Ex5–8). We started from bitstring inferences which were given and constructed a CL^4 diagram for each in the second step in order to prove whether the inferences are valid or not. Of course, one can also first draw a *CL* diagram with arrows and then translate the diagram in inferences including a bitstring semantics.

A practical application of *CL* now consists of assigning the bitstring to an ontology or taxonomy that corresponds either to our natural language usage or to specific domain ontologies. As an example for CL^4, one could use the following ontology of organisms which we will call `OrganOnt`: 1111 = Organisms, 1100 = Animals, 0011 = Plants, 1000 = terrestrial animals, 0100 = aquatic animals, 0010 = land plants and 0001 = aquatic plants. As mentioned in Sect. 3.1, one can understand the 1-bit of the basics, 1000, ..., 0001, as a property to be represented by a certain box within the *CL* diagram. Similar to Sects. 2.2–2.3, however, the 1-bit of the basics can also be interpreted as referring to objects which can only fulfil one of the following properties: living on land, living in water, growing on land and growing in water.

If we translate our first four examples into a natural language, they would read in terms of an extended syllogistic (cf. [25]) as follows:

Ex1*: No land plant is any animal. All land animals are some animals. No land plant is any land animal.

Ex2*: No animal is any plant. Some animals are all aquatic animals. So no aquatic animal is any plant.

Ex3*: All plants are some organisms. All land plants are some plants. So all land plants are also some organisms.

Ex4*: Some plants are not land plants. Some plants are some organisms. So some organisms are not land plants.

If one also translates Ex5–8 into a natural language ontology, one quickly finds out why they cannot be valid. We will therefore not discuss these invalid examples by using `OrganOnt`.

Of the valid examples, Ex1*–4*, only Ex4* is probably worth discussing, since it does not necessarily have to be accepted in the normal language form. At least in natural language, it is unclear what the quantified subject of the second premise is dealing with. Furthermore, the proposition would also be valid if one uses the universal quantifier 'All'. Regardless of which quantification of the subject is used in the second premise, the inference would be much clearer in everyday language if it indicates that all three judgements are about the same subject, i.e. 'Some plants are not land plants. But these plants are also organisms. So some organisms are not land plants'.

In the *CL* diagram and in the bitstrings, however, we see immediately that all three propositions of Ex4* are about the same subject: the arrow shafts, i.e. the diagrammatic subjects, are all in the last dotted box of the diagram, and the bitstrings of the negative propositions (first premise and conclusion) show a Hamming distance to the fourth bit. Thus, the *CL* diagram and the corresponding bitstrings indicate implicit information, namely that all three propositions of Ex4 or Ex4* deal with the box $a_{34} = 0001$, which is not explicitly mentioned in one of the propositions. Aquatic plants in `OrganOnt` are thus the truthmaker for all three propositions in Ex4 or Ex4*. In other words, Ex4 or Ex4* is valid, since they are a shortening of the valid inference '0011⇄ 0010 and 0001↑↓1111, therefore 1111⤢0010'. This information gain can be interpreted diagrammatically as a free ride (cf. [29]).

5 Conclusion and Outlook

A combination of the diagrammatic calculus *CL* with a bitstring semantics has been endorsed in this chapter, the result of which is the so-called Bit-*CL*. It purports to augment the explanatory virtues of Lange's cubes with those of a Boolean set of bitstrings, in such a way that logic diagrams can be both justified in terms of bitstrings and naturally extended to the logical relations of oppositions.

After explaining its basic principles, the aim of such a non-standard semantics was to revisit the foundations of formal ontology through the key concept of bit. Thus, the latter has been applied for characterizing the special class of categorical propositions, as a way to qualify both the precondition of existence for identifying objects and the extensional definition of classes in terms of their elements. Some expected objections have been presented and replied to, although some other ones could have been exposed as well.

To concentrate on the detailed discussion of bitstring semantics and its application using the property of the location or position of the basics, we have not introduced a new *CL* system here but explained a diagrammatic version of extended syllogistics with the help of bitstrings. Future work will thus be left to review the formal system of this syllogistics and to discuss how to use Boolean algebra to compute inferences in *CL*.

Acknowledgments We would like to express our gratitude to the audience of the 6th World Congress on the Square of Opposition in Crete 2018 and to the journal's anonymous reviewers for comments that contributed to the improvement of this chapter.

References

1. G. Allwein, J. Barwise, eds., *Logical Reasoning with Diagrams*. Oxford Studies In Logic And Computation Series, 1996.
2. Barbot, N., Miclet, L., Prade, H., Gilles, R. (2019), A New Perspective on Analogical Proportions, *in* Kern-Isberner, G., Ognjanovi, Z., ed. 'Symbolic and Quantitative Approaches to Reasoning with Uncertainty'. ECSQARU 2019. LNCS, vol. 11726, Springer, Cham, 163–174.
3. P. Bernhard, *Euler-Diagramme: Zur Morphologie einer Repräsentationsform in der Logik*. Paderborn: mentis 2001.
4. R. Cameron, *Turtles all the Way Down: Regress, Priority and Fundamentality*, The Philosophical Quarterly **58** (2008), 1–14.
5. S. Chatti, F. Schang, *The Cube, the Square and the Problem of Existential Import*, History and Philosophy of Logic **34**:2 (2013), 101–132.
6. M. Correia, *The Proto-Exposition of Aristotelian Categorical Logic*, in: J.Y. Béziau, G. Basti (eds) The Square of Opposition: A Cornerstone of Thought. Cham: Birkhuser 2017, 21–34.
7. F. Dau, A. Fisch, *Conceptual Spider Diagrams*. Eklund P., Haemmerl O. (eds) Conceptual Structures: Knowledge Visualization and Reasoning. ICCS 2008. Lecture Notes in Computer Science, vol. **5113**. Springer, Berlin, Heidelberg 2008, 104–118.
8. L. Demey, H. Smessaert, *Combinatorial Bitstring Semantics for Arbitrary Logical Fragments*. Journal of Philosophical Logic **47**:2 (2018), 325–363.
9. L. Demey, *From Euler Diagrams in Schopenhauer to Aristotelian Diagrams in Logical Geometry*, in J. Lemanski (ed.) Language, Logic, and Mathematics in Schopenhauer. Basel: Birkhäuser, 2020, 181–207.
10. O. Goldin, *The Pythagorean Table of Opposites, Symbolic Classification, and Aristotle*, Science in Context **28**, 2015, 171–193.
11. I. Hacking, *Trees of Logic, Trees of Porphyry*, J. Heilbron (ed.) Advancements of learning. Firenze, L.S. Olschki, 2007. p. 219–261.
12. S. Cave, M. Jamnik, J. Hernandez-Orallo, *Artificial intelligence is Growing up Fast: What's Next for Thinking Machines?*, Research Horizons **35** (2018), 26–27.
13. T. Hofweber, *A Puzzle about Ontology*, Noûs **39** (2005), 256–283.
14. L. Jansen, *Classifications*, in Munn, K., Smith, B. (ed.) Applied Ontology: An Introduction, Ontos, Heusenstamm 2008, 159–173.
15. L. Jansen, J. Lemanski, *Calculus CL as a Formal System*, in A.-V. Pietarinen, P. Chapman, L. Bosveld-de Smet, V. Giardino, J. Corter, S. Linker (eds.) Diagrammatic Representation and Inference, 11th International Conference, Diagrams 2020, Tallinn, Estonia, August 24-28, 2020, Proceedings 2020. Cham: Springer 2020, 445–460.
16. D. Jaspers, P.A.M. Seuren, *The Square of Opposition in Catholic Hands: A Chapter in the History of 20th-Century Logic*, Logique et Analyse **233** (2016), 1–35.
17. J.C. Lange, *Inventvm Novvm Quadrati Logici Vniversalis*, Giessen (Gissae-Hassorum), Müller 1714.
18. J. Lemanski, *Calculus CL—From Baroque Logic to Artificial Intelligence*, Logique & Analyse **249–250** (2020), 109–127.
19. J. Lemanski, *Euler-Type Diagrams and the Quantification of the Predicate*, Journal of Philosophical Logic **49**:2, 2020, p. 401–416.
20. J. Lemanski, *Extended Syllogistics in Calculus CL*, in: Daniele Chiffi, M. Carrara, C. De Florio (eds.) Proceedings of Assertion and Proof 2019, Lecce. Special Issue of Journal of Applied Logics - IfCoLoG Journal of Logics and their Applications **8**:2 (2021), 557–577.

21. J. Lemanski, *Oppositional Geometry in the Diagrammatic Calculus CL*, South American Journal of Logic **3**:2 (2017), 517–531.
22. K. Mineshima, M. Okada, R. Takemura, *A Diagrammatic Inference System with Euler Circles*, Journal of Logic, Language and Information **21**:3 (2012), 365–391.
23. A. Paseau, *Defining Ultimate Ontological Basis and the Fundamental Layer*, The Philosophical Quarterly **60**:238 (2010), 169–175.
24. Prade, H., Marquis, P., Papini, O. (2020), *Elements for a History of Artificial Intelligence*, in P. Marquis. O. Papini, H. Prade (eds.) A Guided Tour of Artificial Intelligence Research. Bd. 1: Knowledge Representation, Reasoning and Learning, Springer, 1–43.
25. I. Pratt-Hartmann, *The Hamiltonian Syllogistic*, Journal of Logic, Language and Information **20** (2011), 445–474.
26. J. Schaffer, *On What Grounds What*, in D. Manley, D. J. Chalmers, R. Wasserman (eds.), Metametaphysics: New Essays on the Foundations of Ontology. Oxford University Press, 2009, 347–383.
27. F. Schang, *Abstract Logic of Oppositions*. Logic and Logical Philosophy **21** (2012), 415–438.
28. F. Correia, B. Schnieder: *Grounding: An Opinionated Introduction*, in F. Correia and B. Schnieder (eds.), Metaphysical Grounding: Understanding the Structure of Reality. Cambridge University Press, 2012, 1–37.
29. A. Shimojima, *Semantic Properties of Diagrams and Their Cognitive Potentials*. Stanford, CSLI Publications, 2015.
30. S.-J. Shin, *The Logical Status of Diagrams*, Cambridge University Press, 1994.
31. G. Strang, *Introduction to Linear Algebra*, 5th ed., Wellesley-Cambridge Press, 2016.
32. J.J. Vlasits, *Platonic Division and the Origins of Aristotelian Logic*, UC Berkeley 2017, ProQuest ID: https://doi.org/Vlasits_berkeley_0028E_17182. Merritt ID: https://doi.org/ark:/13030/m5zm0ckf.
33. M.V. Wedin, *Aristotle's Theory of Substance: The Categories and Metaphysics Zeta*, Oxford: Oxford Univ. Press, 2004.
34. S. Yablo, *Does Ontology Rest on a Mistake?*, Proceedings of the Aristotelian Society **72** (1998), 229–261.

Logical Diagrams, Visualization Criteria, and Boolean Algebras

Roland Bolz

Abstract This paper considers logical diagrams as a method for visualizing information concerning logical/linguistic/conceptual systems. I introduce four criteria for assessing visualization: (1) completeness, (2) correctness, (3) lack of distortion, and (4) legibility. Next, I present well-known families of diagrams, based on the geometrical figures of (a) the hexagon, and (b) the tetrakis hexahedron. These two families of diagrams are generally regarded as exemplars in the logical geometry literature. To understand better why they succeed so well at visualizing logical information, they are presented as visualizations of complete (finite) Boolean algebras. This also establishes the connection between the combinatorial concept of partition and the logical concept of opposition (i.e., contradiction, contrariness, and subcontrariness). Finally, the paper suggests that the two geometrical figures in question are part of a larger family of polytopes with deep connections to Boolean algebras.

Keywords Oppositional geometry · Logical geometry · Logical hexagon · Logical tetrakis hexahedron · Opposition · Partition · Boolean algebra · Visualization · Diagram

Mathematics Subject Classification (2000) 03G05 · 03E02 · 03B05 · 52B11

R. Bolz (✉)
Humboldt University Berlin, Berlin, Germany
e-mail: rolandbolz@gmail.com

J.-Y. Beziau, I. Vandoulakis (eds.), *The Exoteric Square of Opposition*, Studies in Universal Logic, https://doi.org/10.1007/978-3-030-90823-2_9

1 Introduction[1]

The purpose of logical diagrams such as the famous square or hexagon of opposition is to *visualize* information about a system of logical expressions, concepts, or objects. In that sense, the use of diagrams in logic is very much like the use of bar charts, boxplots, etc. in statistics. A diagram is not used to generate new knowledge (to derive new facts) but to *communicate* a set of available data effectively and without visual clutter. As in statistics, such visualizations should be assessed according to four criteria:

1. *Completeness* – are all the relevant data shown by the diagram?
2. *Correctness* – are the data displayed in the diagram correct?
3. *Absence of distortion* – does the diagram contain as few as possible visual elements that may be misinterpreted by its reader?
4. *Legibility* – how easy is it to read the data off the diagram?

Since it is easy to create diagrams which are incomplete and incorrect and/or distort the data, there exists an independent subject of inquiry which studies rules and guidelines for creating felicitous diagrams. The potential for error warrants separate methodological inquiry.[2]

Another way to phrase the above is as follows: if logical diagrams such as the square and hexagon are to become part of a *shared language* among logicians and linguists, we will need transparent criteria concerning *successful* visualization. The purpose of this paper is to contribute to developing these criteria by presenting two classical diagrams which perform surprisingly well in this regard. To establish this, I strongly emphasize the connection between logical diagrams and small (i.e., finite) *Boolean algebras*. It appears that this is the correct mathematical basis on which to assess visualizations. This perspective also results in a relative emphasis on the connection between logical geometry and the fundamental concepts of combinatorics (subset, set partition, set permutation).

---

[1] I would like to thank Alessio Moretti for stimulating discussions during the 2018 conference on the square of opposition. This paper was also discussed in the logic colloquium of Prof. Dr. Karl-Georg Niebergall at the Humboldt-University of Berlin. Finally, I would like to thank two anonymous referees for their valuable comments.

[2] Logical diagrams of the square, hexagon, etc. kind have only rarely been explicitly studied as a *method* for visualization of (independently existing) logical subject matter. It seems that the labels "oppositional geometry" (Moretti, [8]) and "logical geometry" (Smessaert & Demey, www.logicalgeometry.org) are somewhat misleading with regard to this matter. What is at stake here is neither a division of geometry which focuses on logic or opposition nor a form of logic which is geometrical in its approach, as if geometrical methods are being used to solve logical problems or the other way around. The study of visualization techniques in statistics is called *data visualization* and not *statistical geometry* (or *geometrical statistics*). There is no reason not to treat the use of diagrams in logic as a form of data visualization also. The concept of *visualization* is really most apt here, because the emphasis is on the nontrivial character of the process of turning nonvisual data (in this case, logical data) into visual data (i.e., images). That said, whenever I refer to the existing literature, I use the label "oppositional/logical geometry."

The paper has four parts. The first part describes different facets of visualizing logical structures in diagrams with the goal of separating geometrical considerations from abstract-structural ones. The following two sections describe the hexagon and the tetrakis hexahedron in detail. The uniqueness of the description lies in the fact that the paper treats both diagrams in relation to the Boolean algebras on $P(\{a, b, c\})$ and $P(\{a, b, c, d\})$.[3] I believe that this allows one to describe the connection between abstract logical (and combinatorial) structure and geometrical structure in a more systematic and simple fashion than has been done so far. In the final section, I briefly focus on the idea that certain diagrams can be "contained in" other diagrams.

The subject treated in this paper has been studied most intensely by Alessio Moretti in his thesis and subsequent papers [7, 8] and by Hans Smessaert and Lorenz Demey [3–6, 9, 10] in their many publications on logical geometry. My main aim is to simplify what is already known. The aforementioned authors have contributed tremendously to the classification and understanding of existing diagrams. Here, I foreground the question "what makes a really good diagram?" against the larger classificatory effort. And I show how the usual concepts of contrariness and subcontrariness can be elegantly described using the combinatorics of set partitions. So far, no texts have been published which contain systematic criteria for assessing diagrams, so I aim to fill this gap by making a number of observations concerning diagrams which, according to scholarly consensus, perform *very well.*

2 Visualizing Logical Structure

A typical visualization of logical data in a diagram involves the following aspects/steps:

1. The context is usually (a fragment of) some *natural* or *artificial language* in which certain expressions occur among others. Usually, a small number of expressions is chosen, because of the strong *structural relations* between them. The goal is then to exhibit the relations between these expressions.
2. The isolated logical expressions may first be regarded as a nonspatial relational structure. The structure consists of a domain (the expressions) and relations between them, such as implication and opposition (contradiction, contrariety, etc.) or other "Aristotelean relations" (Smessaert & Demey). It is important to see that this abstract structure is not yet the diagram but simply a finite relational structure. Depending on what expressions and relations are lifted from the larger linguistic/logical context at 1., one may get a canonical type of structure or a more nonstandard type of structure.

[3] $P(x)$ is the power set of x.

3. Next, a geometric entity (again, not yet a diagram) is chosen which fits the abstract structure from 2. For example, if the structure contains five expressions, then the geometric entity should be a polytope (shape) with five vertices. Since the diagram will be printed on paper, the most typical geometric figures are two-dimensional polygons or projections of three-dimensional polyhedra.
4. Finally, a diagram is designed which is based on this geometric entity. Typically, the vertices are labeled with the chosen expressions, and the edges are given different colors or are turned into arrows. Perhaps further lines (e.g., diagonals) are added to indicate further relations from 2.

A quick survey of the literature shows that most authors have focused on steps 1 and 4. Typically, the step from choosing expressions to presenting a diagram is direct, without any extended reflection on the abstract structures and the features of the purely geometrical objects behind the diagrams. The goal of this paper is to show that a detailed consideration of the link between certain algebraic structures and certain polytopes greatly aids our understanding of successful visualization. With regard to the geometric side of logical geometry, I emphasize a combinatorial interpretation over a logical one.

In the following sections of the paper, I present diagrams based on the hexagon and the tetrakis hexahedron, focusing on the link between the abstract structural dimension and the geometric dimension (points 2 and 3 of the above list). These two types of diagrams are the most complete up to a certain level of complexity. It will become apparent that these correspond to the Boolean algebras on the power sets of three- and four-element sets, respectively. Finally, and perhaps most interestingly, I make a number of observations with regard to the lack of distortion in these diagrams. Above all, it is striking that many of the visual (i.e., geometric) features of these diagrams correspond to relevant logical (or combinatorial) features of the abstract structures. I introduce several connections that seem to have escaped notice so far (especially in relation to set partitions and set permutations). As such, these two (families of) diagrams are exemplars of effective visualization in the sense of the four criteria completeness, correctness, absence of distortion, and legibility.

3 The Logical Hexagon

An important starting point for the recent renaissance in the study of logical diagrams descending from Aristotle's square is the discovery of a family of hexagon diagrams (see [1] and [2]). The latter are usually presented as natural completions or extension of square diagrams. Let us now examine two such hexagon diagrams in order to describe their common structure.

The following figure shows the completion of a square to a hexagon (Fig. 1):

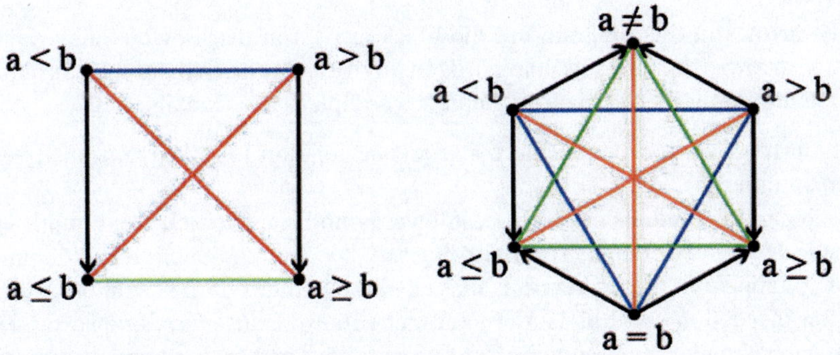

Fig. 1 The completion of a square diagram to a hexagon diagram

Here is a quick reminder of the meaning of the arrows and colored lines:

- A black arrow signifies logical consequence.
- A red line (contradictories) signifies that the two expressions:
  - Cannot both be true.
  - Cannot both be false.
- A blue line (contraries) signifies that the two expressions:
  - Cannot both be true.
  - Can both be false.
- A green line (subcontraries) signifies that the two expressions:
  - Can both be true.
  - Cannot both be false.

The diagram displays six expressions with four different relations between them. This much is manifest. However, is this abstract structure, with the given relations, a familiar mathematical structure – something canonical? The answer is that it is the finite Boolean algebra of eight elements with two of its eight elements (*True* and *False*) omitted. To see this, consider the following Hasse diagram (Fig. 2):

Fig. 2 Hasse diagram for Fig. 1

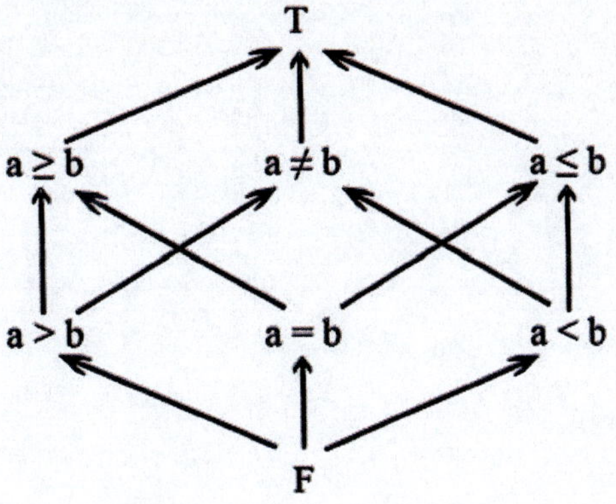

The arrows in this diagram are the black arrows in the previous diagram (plus "trivial" arrows from and to T and F). Transitivity and reflexivity of the consequence relation are implicit. The diagram depicts a complete Boolean algebra:

- The arrows form a poset (the consequence relation is reflexive, transitive, and antisymmetric).
- Each pair of elements has a greatest lower bound and a least upper bound. This is not immediately obvious from the diagram, because conjunction and disjunction do not appear in the expressions that label the vertices of the hexagon. However, upon inspection and this is a crucial fact about the structure, one surmises that for each conjunction/disjunction of two formulas, one finds a logically equivalent expression already in the structure. For example, $a \leq b$ is logically equivalent to $a < b \vee a = b$. The usual approach is to treat the formulas as representatives for equivalence classes of formulas (where equivalence is logical equivalence).
- The structure has a top and a bottom (T and F, respectively).
- Disjunction and conjunction are distributive (an assumption about conjunction and disjunction in the source theory, which we take to be classical).
- Each element a has a complement b, such that $a \wedge b$ (or its equivalent) is F and $a \vee b$ (or its equivalent) is T (this corresponds to the relation of contradictoriness).

The generality of the Boolean background of hexagon diagrams can be illustrated by way of another (more familiar) example, this time using the classical quantifier hexagon (Figs. 3 and 4):

Fig. 3 Quantifier hexagon

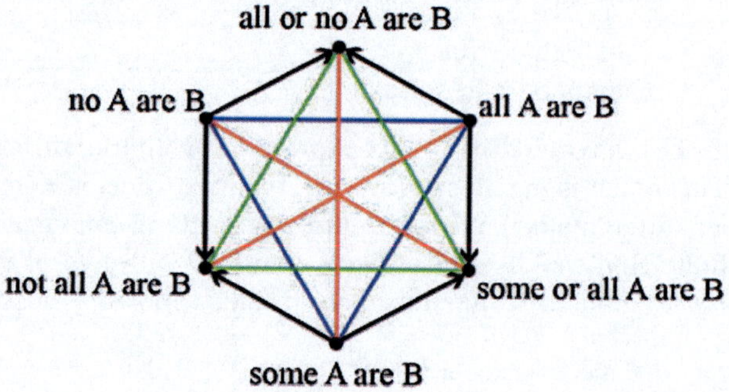

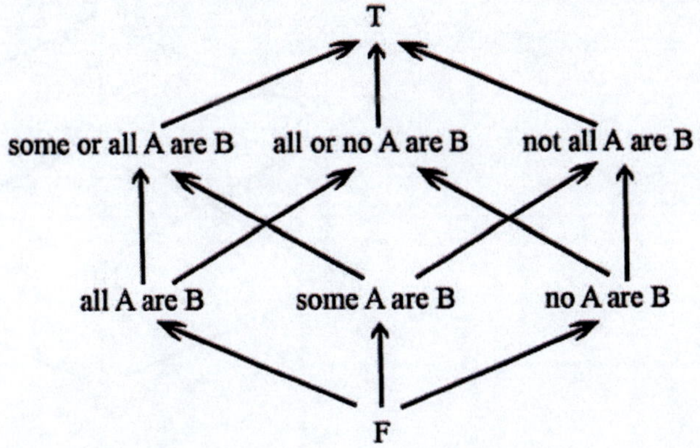

Fig. 4 Hasse diagram for quantifiers

Any hexagon in which the arrows and lines correctly display the aforementioned logical relations can be regarded as a Boolean algebra of this type. Hence, it will be useful for our purposes to choose a canonical representation of this structure. The natural candidate for this purpose is the inclusion relation on the powerset of a three-element set. Here are the Hasse diagram and the hexagon (with and without universal and empty sets):

Fig. 5 Canonical hexagon for {a, b, c}

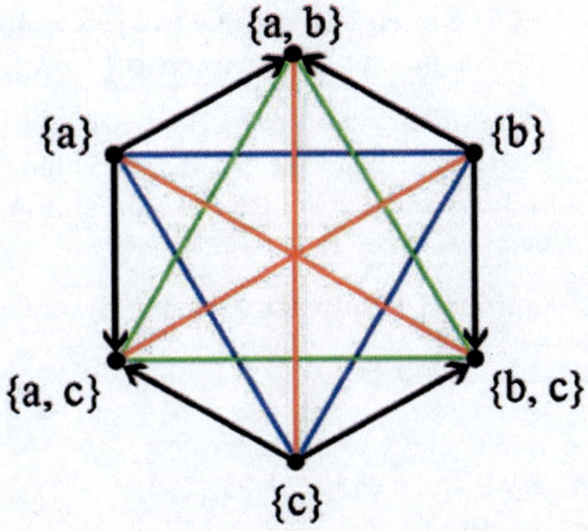

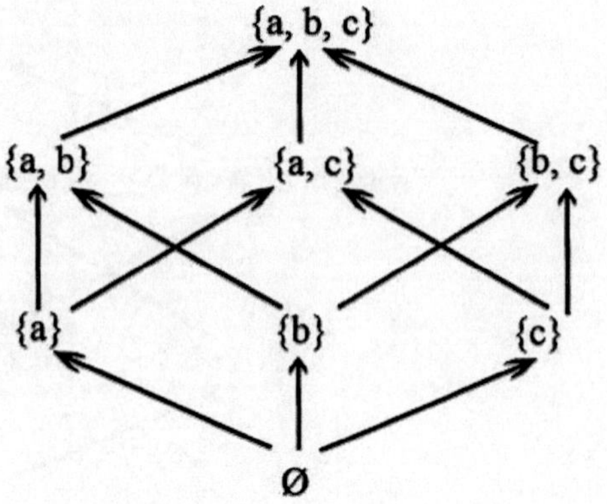

Fig. 6 Hasse diagram for {a, b, c}

Focusing on the hexagon now, the arrows and colored lines receive a different but related interpretation:

- A black arrow means inclusion.
- A red line means that the two sets are complements relative to $\{a, b, c\}$.
- A blue line means that:
  - The intersection of the two sets is empty.
  - The union of the two sets is not $\{a, b, c\}$.
- A green line means that:
  - The intersection of the two sets is non-empty
  - The union of the two sets is $\{a, b, c\}$.

A set theorist will point out that inclusion and complementation (black and red) are canonical concepts but that the blue and green lines signify relations that are a bit nonstandard. We can give the following definitions for these relations for the general set-theoretical case[4]:

Definition 1 (Contrariness) x and y are *contrary in* z iff:

- $x \subseteq z$ and $y \subseteq z$
- $x \neq y$
- $x \neq \varnothing$ and $y \neq \varnothing$
- $x \neq z$ and $y \neq z$
- $x \cap y = \varnothing$
- $x \cup y \neq z$

Definition 2 (Subcontrariness) x and y are *subcontrary in* z iff:

- $x \subseteq z$ and $y \subseteq z$
- $x \neq y$

[4] These definitions are not minimal; it is more expedient to simply list the relevant features here.

- $x \neq \varnothing$ and $y \neq \varnothing$
- $x \neq z$ and $y \neq z$
- $x \cap y \neq \varnothing$
- $x \cup y = z$

Although these definitions adequately translate the logical concepts of contrariness and subcontrariness into a set-theoretical context, I will propose below that these be replaced with the more fine-grained (and well-understood) concept of set partition.

The canonical eight-element Boolean algebra on $P(\{a, b, c\})$ (Figs. 5 and 6) adequately represents the abstract structure behind the other hexagonal diagrams.[5] Complementation, closure under conjunction/disjunction, etc. are now easy to read off the diagram. This constitutes a relative advantage over other proposals for canonical diagrams/structures (e.g., Smessaert and Demey's *bitstring* approach in [4]). We may now focus on the fit between this structure and the *geometric* figure of the regular hexagon. What follows are three neat convergences between logic (consequence and opposition), the combinatorics of $P(\{a, b, c\})$, and the geometrical features of the regular hexagon.

3.1 Inclusion and Logical Consequence

The relation of logical consequence is modeled by set inclusion. This much is obvious from the fact that the arrows in Figs. 1 and 5 are the same. Geometrically, the regular hexagon is a good fit because each of its six edges (all of the same length, suggesting no difference to the reader) can be made to stand for exactly one of the inclusions. Of course, this means that *edges* are replaced by *arrows* pointing in the right direction (Fig. 7).

Fig. 7 The inclusion arrows form a regular hexagon

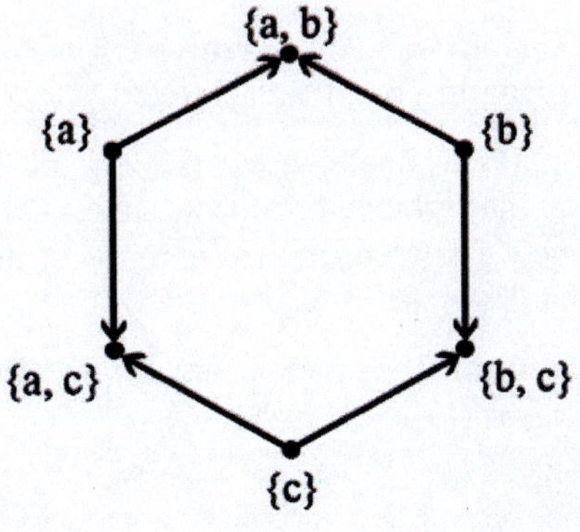

[5] All other Aristotelean relations (Smessaert & Demey) can be defined inside it.

3.2 Partitions and Opposition

The concept of contradictoriness in logic neatly fits set complementation relative to $\{a, b, c\}$ (the red lines of Fig. 5). However, the original logical hexagon contains two more relations: contrariness and subcontrariness (blue and green lines) (Definitions 1 and 2). It will be instrumental in the next section on the tetrakis hexahedron to define *all relations of opposition* uniformly in relation to the more fundamental concept of *set partition*. On this basis, the deep connection between the algebraic structure and the polytope becomes easier to grasp.

Definition 3 (Partition) A *partition* of a set X is a set Y of non-empty subsets of X such that X is the disjoint union of Y.

The partitions of $\{a, b, c\}$ are:

- $\{\{a\}, \{b\}, \{c\}\}$
- $\{\{a, b\},\{c\}\}$
- $\{\{a, c\}, \{b\}\}$
- $\{\{a\}, \{b, c\}\}$
- $\{\{a, b, c\}\}$

In total, $\{a, b, c\}$ has one ternary partition, three binary partitions, and one unary partition. It is easy to see that the red lines in the hexagon (complementation/contradictoriness) correspond to the binary partitions. The blue triangle (contrariness) *as a whole* corresponds to the single ternary partition. Hence, one can reframe the relation of contrariness between (nonidentical) elements of $P(\{a, b, c\})$ in terms of the single ternary partition of $\{a, b, c\}$, which is $\{\{a\}, \{b\}, \{c\}\}$:

Lemma 1 Two elements $x, y \in P(\{a, b, c\})$ are *contrary in* $\{a, b, c\}$ iff $x \neq y$ and $x, y \in \{\{a\}, \{b\}, \{c\}\}$.

This follows from Definition 1.

To get subcontrariness, the following simple construction does the trick:

Definition 4 (Co-partition) The *co-partition* Z of a partition Y of a set X is the set of all the complements (relative to X) of all the elements of Y.

For example, the co-partition of $\{\{a\}, \{b\}, \{c\}\}$ is simply $\{\{b, c\}, \{a, c\}, \{a, b\}\}$.[6]

The relation of subcontrariness between (nonidentical) elements of $P(\{a, b, c\})$ is now easily understood in terms of the single ternary co-partition of $\{a, b, c\}$, which is $\{\{b, c\}, \{a, c\}, \{a, b\}\}$.

Lemma 2 Two elements $x, y \in P(\{a, b, c\})$ are *subcontrary in* $\{a, b, c\}$ iff $x \neq y$ and $x, y \in \{\{b, c\}, \{a, c\}, \{a, b\}\}$.

[6] Only co-partitions of ternary and quaternary partitions are of interest in this paper. Co-partitions of binary partitions are simply those binary partitions themselves.

This follows from Definition 2. For example, $\{a, b\}$ and $\{b, c\}$ are subcontraries in $\{a, b, c\}$ because they are both members of $\{\{b, c\}, \{a, c\}, \{a, b\}\}$, which is the co-partition of $\{\{a\}, \{b\}, \{c\}\}$, the single ternary partition of $\{a, b, c\}$.

All this shows that the three varieties of opposition can be defined in reference to the partitions of $\{a, b, c\}$, structure that comes "for free" with our canonical representation.

I want to stress the subtle difference between *relations* of contrariness and *partitions*. In $P(\{a, b, c\})$, these concepts are not interestingly different, which should be clear from the above. However, when one increases the complexity of the Boolean algebra (taking a basis of four, not three, elements), the number of ternary partitions will increase from one to six. At that point, the contrariness of two sets will come to mean that they are members of *some* partition (ternary or higher). Hence, it is beneficial to consider the concept of partition as more fundamental (also, more fine-grained) than the concept of contrariness.[7] The same goes for the dual concept of subcontrariness/co-partition. This will become clearer in the next section, which contains a concrete example of a structure where these distinctions start to matter.

Turning to the regular hexagon as a geometric figure, one notices that the red lines are the three mirror symmetries (where the axis is a line through two vertices of the figure). This is a first observation regarding the link between mirror symmetry and partitions. I submit the following analogy between the partition structure and the symmetries of the hexagon:

- The ternary partition can be identified with the figure itself (the motivation for this will become clear in the next section).
- The binary partitions can be identified with the segments that are mirror symmetries of the figure (where the axis is a line through two vertices of the hexagon).[8]
- The unary partition can be identified with the point at the center of the hexagon (the central symmetry of the figure).

One also observes that the central point *lies on* all the symmetry segments, which in turn *lie on* the figure itself.

Definition 5 (Refinement of a partition) Let the sets A and B both be partitions of the set X. Then A is a *refinement* of B iff every element of A is a subset of an element of B.

One can order the partitions of $\{a, b, c\}$ by refinement, resulting in the following refinement lattice:

---

[7] It seems that the concept of contrariness arose in the context of trichotomy (as opposed to contradiction for dichotomy). Two choices of a trichotomy are contraries since there is a third option (the law of excluded middle does not hold here). But when the complexity is increased (tetrachotomy, pentachotomy), the number of concepts increases and so does the number of conceptual partitions. In that case, the full set of complete conceptual partitions will be of more interest than the mere existence of a contrariness relation between two concepts.

[8] Without this restriction, the hexagon has three more mirror symmetries.

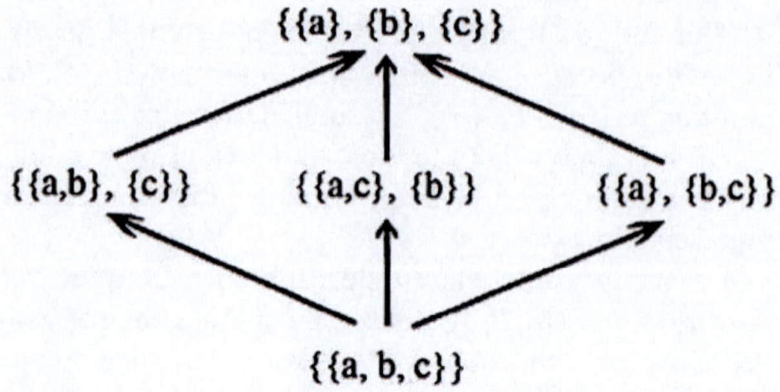

Fig. 8 Refinement Lattice on {a, b, c}

Turning once more to a comparison between the hexagon and the abstract structure, one observes the following: let A' be the part of the hexagon (i.e., the point, segment, or figure) assigned to the partition A. Then the following correspondence holds between the above refinement lattice and our hexagon: A is a refinement of B iff B' lies on A' (the reader should compare Figs. 5 and 8).

3.3 Permutations

There is a final analogy between the Boolean algebra on $P(\{a, b, c\})$ and the regular hexagon. The relevance of this connection is perhaps the least understood of the three connections that are explored here. However, the simplicity of the matter is so apparent that I cannot fail to mention it.

Any finite set with *n* elements has *n!* permutations. The permutations of $\{a, b, c\}$ appear as complete paths through the inclusion lattice (exemplified in Fig. 9).

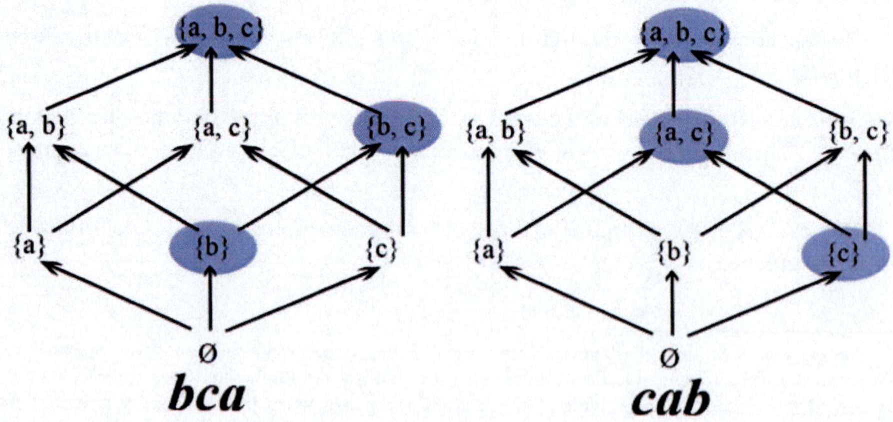

Fig. 9 Permutations as complete paths through the poset

The set $\{a, b, c\}$ has six permutations. These appear in the hexagon once a single point in the center of the hexagon is added (Fig. 10).

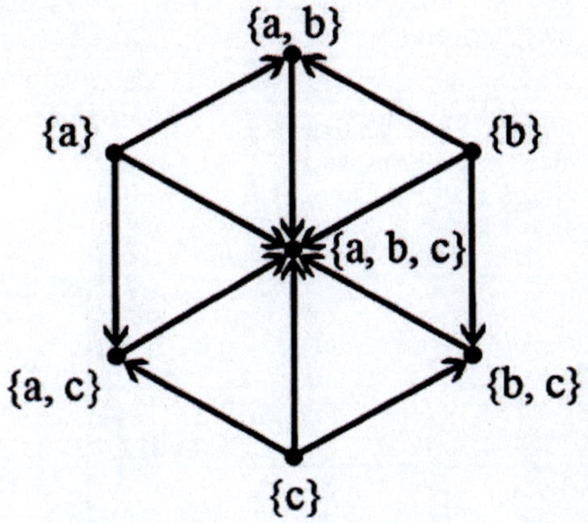

Fig. 10 The permutations appear as six regular triangles

The figure shows how the hexagon decomposes into six equal triangles, each corresponding to a unique complete path through the lattice: a permutation.

The relevance of permutations to oppositional/logical geometry lies in the connection to *modal graphs* (see Chap. 11 of Moretti's [8]).

In this chapter I described an elaborate analogy between the Boolean algebra on $P(\{a, b, c\})$ and the logical hexagon in terms of the combinatorial (set-theoretic) concepts of subsets, partitions, and permutations. In what looks like an amazing pattern, the same analogy exists between $P(\{a, b, c, d\})$ and the three-dimensional figure of the tetrakis hexahedron.

4 The Logical Tetrakis Hexahedron

Logical diagrams based on the tetrakis hexahedron have been described, among others, by Régis Pellissier, Alessio Moretti, and Hans Smessaert and Lorenz Demey.[9] Although these authors seem well aware of the Boolean background of

[9] Pellissier and Moretti [8] speak of a "logical tetraicosahedron," emphasizing the fact that the figure has 24 faces. Smessaert and Demey [3, 5, 9] present the figure as a "rhombic dodecahedron," which amounts to collapsing the 24 faces into 12. Given that the 24 faces of the tetrakis hexahedron correspond to 24 permutations (combinatorics), i.e., 24 entailment paths (logic), it seems that collapsing them into 12 is to suppress valuable information. In other words, the rhombic dodecahedron seems to sacrifice completeness for no apparent reason, except perhaps that the data in question has not been recognized as relevant. Furthermore, by collapsing 24 triangular faces into 12 rhombic faces, the diagram also *risks distorting* the data, since the reader may ask what the meaning of these rhombic faces is – a question without answer. This specifically geometric feature has no reading in terms of the logical structure.

this diagram, I believe that the full extent of the connection has not been grasped and expressed fully.[10] This section of the paper works toward filling this gap.

It seems natural to orient ourselves using the following diagram, which sets the Boolean algebra on $P(\{a, b, c, d\})$ (omitting the universal and empty set) to a tetrakis hexahedron (Fig. 11):

Fig. 11 Tetrakis hexahedron diagram for inclusion on $P(\{a, b, c, d\})$

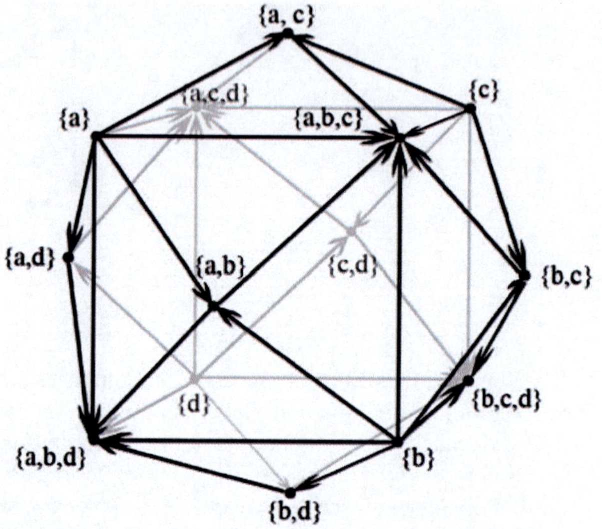

This diagram is a good basis for exploring the match between the algebraic structure and the geometric figure (polyhedron) without getting distracted by specific logical or linguistic content. It should be observed that the diagram is already quite busy as is, without the relations of opposition drawn in (this concerns our criterion of *legibility*).

[10] The connection between Boolean algebras and logical diagrams has been studied most deeply by Smessaert and Demey [4, 6, 9, 10]. However, in their work, the elements of the Boolean algebra appear as *bitstrings*. Of course, such Boolean algebras are isomorphic to Boolean algebras on power sets. However, it seems that the connection between the inclusion structure and the partition and permutation structure becomes much simpler to grasp when one builds the Boolean algebras around power sets. Most importantly, the deeper connection between the Boolean structure and the opposition concepts (contradiction, contrariness, subcontrariness) remains below the surface using the bitstring approach – they appear as more or less independent aspects of the theory. Using power sets, one gets a partition structure for free which aids our understanding of opposition tremendously. All the relevant logical structure is simply already in the power set structure and can hence be defined on its basis. Also, the bitstring approach has the disadvantage of obscuring the relations between this subject and existing work in combinatorics, which is almost always done in a set-theoretic setting.

4.1 Inclusion and Logical Consequence

The vertices of the diagram are labeled by the 14 subsets of $\{a, b, c, d\}$ (excluding, as with the hexagon, the universal set and the empty set). Furthermore, each of the 36 edges of the tetrakis hexagon corresponds to one subset inclusion in $P(\{a, b, c, d\})$. All the vertices are equidistant from the center, suggesting no hierarchy.

In $P(\{a, b, c, d\})$ without the universal and empty sets, there are two types of proper inclusion:

- $A \subset B$ where B has exactly one element more than A (call this a 1-inclusion)
- $A \subset B$ where B has exactly two element more than A (call this a 2-inclusion)

The edges of the tetrakis hexahedron come in two lengths, which reflect this difference (see Demey & Smessaert [5]). The longer edges are the two-inclusions, the shorter ones the one-inclusions.

As has been pointed out by Smessaert, it is sensible to consider a variation of the diagram in which the universal and empty set are given a vertex in the center of the figure.

4.2 Partitions and Opposition

In this section it will become apparent that a characterization of opposition in terms of set partitions pays off.

The following diagram (also a hexagon, but by coincidence) gives the 15 partitions of a four-element set and the refinement relation on them, with the finer partitions always above the coarser ones:

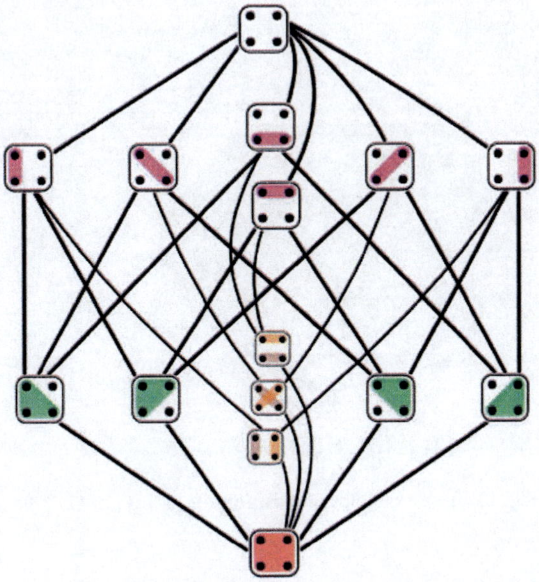

Fig. 12 Partitions of a four-element set ordered by refinement. (Source: Wikipedia – "Partitions of a Set")

The figure shows that $\{a, b, c, d\}$ has:

- 1 quaternary partition (white)
- 6 ternary partitions (purple)
- 7 binary partitions (green for 3–1 and orange for 2–2)
- 1 unary partition (red)

Furthermore, the refinement relation on these 15 partitions is nontrivial, as opposed to the refinement relation on the 5 partitions of $\{a, b, c\}$ (compare Figs. 8 and 12).

Surprisingly, these partitions and their refinement relation find a natural place in the tetrakis hexahedron as canonical diagram for $P(\{a, b, c, d\})$:

- The *binary partitions* are relations of complementation/contradiction between two subsets of $\{a, b, c, d\}$. In the diagram, these always appear at opposite sides of the center of the polyhedron. Hence, each binary partition is represented by a segment which is one of seven line symmetries (rotation) of the tetrakis hexahedron.
- The *ternary partitions* are sets of three elements. The three corresponding vertices always determine a plane. As it turns out, this plane is always a unique plane through the center of the tetrakis hexahedron, more precisely, one of six mirror symmetries of the tetrakis hexahedron in a plane.
- The *quaternary partition* is a set of four elements. These four vertices do not determine a plane; they determine a sphere in which the tetrakis hexahedron is inscribed. Hence, one can identify the entire polyhedron with this partition.
- The *unary partition* can be identified with the point at the center of the tetrakis hexahedron.

The following diagram highlights examples of ternary and binary partitions:

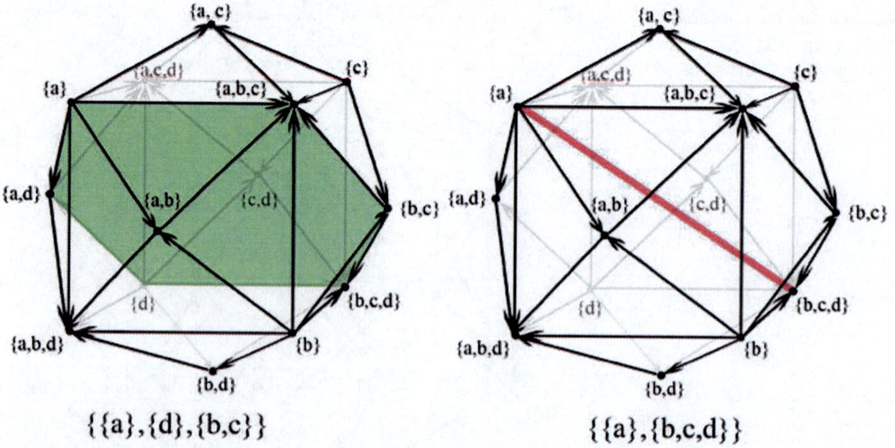

Fig. 13 Two examples of partitions of {a, b, c, d} "in" the tetrakis hexahedron

The above description amounts to a mapping between the partitions of $\{a, b, c, d\}$ and certain vertices, segments, planar figures, and shapes in the tetrakis hexahedron. Interestingly, as before, the refinement relation translates to *geometrical incidence*. This is most meaningful for the relations between the seven binary and six ternary partitions. A binary partition A corresponds to a segment A', and a ternary partition B corresponds to a plane B'. We now observe that B is a refinement of A iff A' lies on B' (see Fig. 13 for an example). But here, unlike with the hexagon, there are examples where such a refinement relation does not occur between some partitions A and B. In that case, the segment B' *does not* lie on the plane A'. The reader can verify these claims by comparing Figs. 11 and 12.

The above establishes a fit between the partition structure of $\{a, b, c, d\}$ and the symmetries of the tetrakis hexahedron. It remains to be asked how the concept of partition connects to the more familiar concepts of contrariness and subcontrariness. The way has already been paved for this in the section on the hexagon.

When two (logical, linguistic) expressions labeling edges of the tetrakis hexahedron are contraries, then they cannot both be true but may (in some models) both be false. The set-theoretic equivalent of this was given in Definition 1.

Lemma 3 Two sets x and y are contrary in $\{a, b, c, d\}$ iff $x \neq y$, and there exists some z which is a ternary or quaternary partition of $\{a, b, c, d\}$ such that $x \in z$ and $y \in z$.

This follows directly from our definitions. The partition concept is more fine-grained (and combinatorially canonical).

There are two reasons to focus on partitions and not only on relations of contrariness once the degree of complexity exceeds the ternary framework of $P(\{a, b, c\})$:

1. As has been shown for the hexagon and the tetrakis hexahedron, there is a correspondence between the partition lattice and the various symmetries of the polytope. The systematic nature of this relation is only visible if partitions are considered. In the tetrakis hexahedron, it was important to see all six ternary partitions and not just a single overall binary relation of contrariety. It seems to the author that the concept of partition is one of the keys to understanding the link between the algebraic structures and the geometric polytopes at the heart of oppositional/logical geometry.
2. As an analytic tool for conceptual, linguistic, or logical analysis, the degree of resolution offered by our approach is desirable. The mere fact that two expressions are contraries does not indicate whether or not there is an underlying trichotomy, tetrachotomy, or n-chotomy. A complete oppositional analysis of a conceptual system, however, should be able to present *all* possible dichotomies, trichotomies, tetrachotomies, etc.[11] The focus on partitions leaves this option

[11] Similarly, to know one's way around a certain conceptual apparatus means to be aware of the underlying dichotomies, trichotomies, etc. Usually, when one grasps that two concepts are contrary, one also grasps the third (and fourth, etc.) remaining option of the tri- or tetrachotomy. Cognitively speaking, contrariness is founded in n-chotomy, not the other way around.

open; it is the right level of abstraction. To reduce the existence of a tetrachotomy plus six trichotomies into a single relation of contrariness is to simplify the data at the outset.[12]

Subcontrariness can be treated as in the section on the hexagon. Each of the six ternary partitions and the one quaternary partition of $\{a, b, c, d\}$ has a unique co-partition. Each of these is interesting in its own right as the dual of the partition. The connection between subcontrary sets and co-partitions is established in the following lemma:

Lemma 4 Two sets x and y are subcontrary in $\{a, b, c, d\}$ iff $x \neq y$, and there exists some z which is a ternary or quaternary partition of $\{a, b, c, d\}$ and some w such that w is the co-partition of z and such that $x \in w$ and $y \in w$.

Again, it is more informative to know in which co-partitions the expressions A and B figure than to merely know that they are subcontraries.

So much for contrariness and subcontrariness.

From the point of view of *visualization*, we face the following difficulty. The structure that has been the focus of this section contains six ternary partitions, one quaternary partition, six ternary co-partitions, and one quaternary co-partition. Obviously, the diagram will no longer be legible (our fourth criterium for successful visualization) if all 14 of these relations are combined into a single diagram using different colors – a problem we did not face with $P(\{a, b, c\})$ because it was much simpler. It may very well be more informative to give these trichotomies and this tetrachotomy in seven additional diagrams, each based on a partition and its co-partition.[13] On the other hand, the fact that the tetrakis hexahedron implicitly conveys the partition structure via its symmetries is reassuring.

The following diagram plus a short instruction for the reader is an attempt at balancing between the aim of legibility and completeness:

How to read this diagram: Black arrows signify inclusion/logical consequence. The four blue points display a single tetrachotomy (a partition into 4 distinct concepts) and the four green points display the co-partition of this tetrachotomy. Any set of six points which form a plane symmetry of the figure indicates both a trichotomy (2 blue vertices and 1 black vertex at the other side of the center) and its dual co-partition (2 green vertices and 1 black vertex at the other side of the center). Finally, any pair of points at opposite ends of the center display a dichotomy (Fig. 14).

[12] Again, it seems the concept of contrariness (subcontrariness) emerges in the context of trichotomy, where there is no real distinction between the contrariness relation and the partition structure, as I noted in the section on the hexagon.

[13] These are essentially Moretti's "logical bi-simplexes" from his [8].

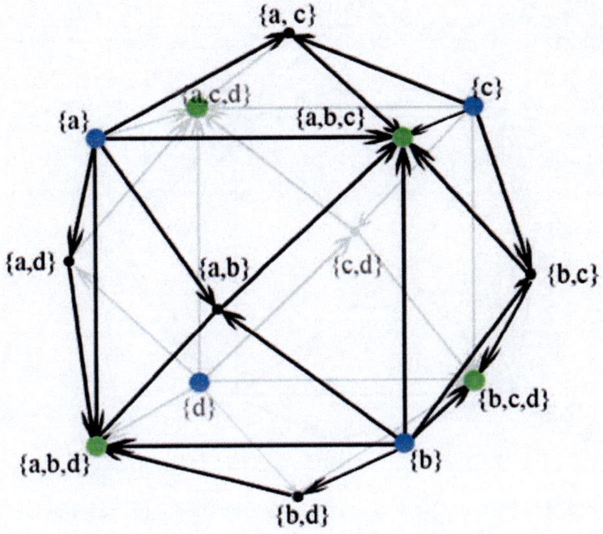

Fig. 14 A very informative diagram

This may perhaps be simplified further by dropping all mention of the co-partitions. At this level of complexity, the partitions (tetrachotomies, trichotomies, and dichotomies) are probably of primary interest. The co-partitions appear to be conceptually and cognitively derivative.

Presented with these instructions, the diagram nonetheless succeeds at conveying quite a lot of information. The biggest difficulty is probably that it presupposes that the reader is able to visualize the tetrakis hexadron in three dimensions in order to identify its symmetries. Nonetheless, it can be regarded as an extreme case, where completeness, correctness, and lack of distortion are fairly optimal but where legibility is perhaps already somewhat compromised. Presenting a combination of several diagrams, with the tetrakis hexahedron functioning as "global" diagram (integrating the other diagrams, so to speak), may be a good compromise.

4.3 Permutations

In the section on the hexagon, it was pointed out that the permutations of the set have a natural geometric interpretation. We briefly indicate how the same holds true for the tetrakis hexahedron.

$\{a, b, c, d\}$ has 24 permutations. In the geometrical solid, they appear as the 24 faces of the tetrakis hexahedron. To see this, it is helpful to add a vertex at the center of the solid (Fig. 15).

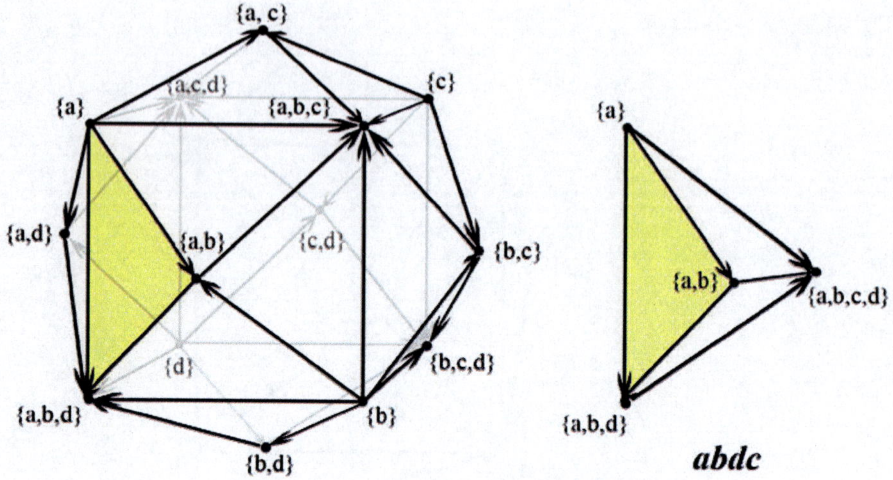

Fig. 15 Permutations as unique faces/tetrahedra in the tetrakis hexahedron

Each of the 24 permutations then appears as a tetrahedron (from the "outside," one sees the 24 faces, i.e., triangles). Each of these tetrahedra has the same shape. There are two path orientations (clockwise and counterclockwise).

Whenever two such tetrahedra share an edge, they introduce some character at the same stage. For example, *abcd* shares an edge with *bcad* because they both add *d* in the fourth stage. Whenever they share a face (i.e., whenever they are adjacent), they introduce two characters at the same stages. It seems that many more of the relations between the permutations of $\{a, b, c, d\}$ are reflected in the geometry of the tetrakis hexahedron and its constituent 24 tetrahedra. For example, the *abcd* face is at the opposite side of the *dcba* face. These connections cannot be explored fully here.[14]

5 Hexagons inside the Tetrakis Hexahedron?

There has been some discussion on the subject of smaller logical diagrams "inside" bigger diagrams (on the subject of nonstandard hexagons, see Moretti's [7]). In the following section, I attempt to simplify some of that work in the light of my own conceptual apparatus. Readers unfamiliar with these debates may find the following discussion overly technical, especially since my conclusion is that

[14] Both our canonical hexagon and our canonical tetrakis hexahedron are geometrically dual polytopes to Cayley graphs created using the so-called permutohedra on $\{1, 2, 3\}$ (a hexagon) and $\{1, 2, 3, 4\}$ (a truncated octahedron).

these inquiries are poorly motivated, given how well-understood the canonical combinatorial superstructures are.

The simplest example of a diagram "in" a diagram is a square inside a hexagon. The hexagon was first discovered as a *completion* of the square, so it is only to be expected that a square (three in fact) appears "inside" the hexagon (see Fig. 1). Especially when the complexity of the diagram is high (e.g., when it is based on the tetrakis hexahedron), the number of smaller diagrams to be "discovered" inside the larger diagram grows quickly. Several authors have considered it important to classify these diagrams inside diagrams exhaustively. Let us briefly see what the focus on Boolean algebras can bring to the discussion. It appears that these questions can be answered using simple combinatorics, before any visualization effort.

Since both the simple and the more complex diagram are based on a poset or Boolean algebra structure, it is only to be expected that any structural parthood relation that exists between the diagrams should already exist between the purely relational structures. There is no need to study these connections at the level of the visualization – it will be much more straightforward to study (and classify) them as mappings between canonical algebraic structures. To support this claim, let us use the example of hexagons in the tetrakis hexahedron. It seems to me that this entire subject reduces to the classification of structure-preserving mappings between $P(\{a, b, c\})$ and $P(\{a, b, c, d\})$, since these mappings pick out exactly the "hexagons in the tetrakis hexahedron." The different types of hexagons (e.g., "weak" and "strong") in the tetrakis hexahedron appear as maps with different degrees of structure-preservation. It becomes straightforward to specify what constitutes the weakness or strength of such a substructure of the larger structure.

$P(\{a, b, c\})$ and $P(\{a, b, c, d\})$ are lattices with 8 and 16 elements, respectively. It is fruitful to examine *injective* functions $f : P(\{a, b, c\}) \longrightarrow P(\{a, b, c, d\})$. These pick exactly 8 of the 16 elements of $P(\{a, b, c, d\})$. A second condition that should be imposed is:

Preservation of inclusion if $x \subseteq y$ in $P(\{a, b, c\})$, then also $f(x) \subseteq f(y)$ in $P(\{a, b, c, d\})$.

This guarantees that the function f picks out a set of eight elements with an inclusion relation matching the six arrows of the hexagon diagram. The image of such a function thus "lifts" an eight-element poset from the more complex structure of $P(\{a, b, c, d\})$.

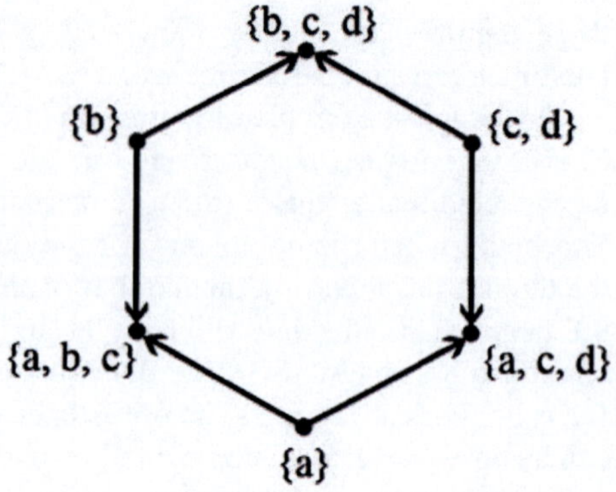

Fig. 16 Arrow hexagon based on the image of f

Table 1 Example: inclusion-preserving mapping f

Argument	Value
$\varnothing$	$\varnothing$
$\{a\}$	$\{b\}$
$\{b\}$	$\{c, d\}$
$\{c\}$	$\{a\}$
$\{a, b\}$	$\{b, c, d\}$
$\{a, c\}$	$\{a, b, c\}$
$\{b, c\}$	$\{a, c, d\}$
$\{a, b, c\}$	$\{a, b, c, d\}$

Table 1 and Fig. 16 give an example of such a mapping which picks out a hexagon "in" $P(\{a, b, c, d\})$. Because of the condition that the function preserves inclusion, the image of the function can always be "set" to a hexagon arrow diagram. It should however be noted that the structure portrayed in Fig. 16 is not closed under complements, unions, and intersections. So the condition of preserving inclusion is too weak to pick out interesting (important) substructures.

It is useful here to identify the functions of the above type with the substructures – each function of the specified kind picks out one substructure of the desired kind (a commonplace identification in category theory).[15] Any hexagon diagram based on the image of such a function f is what Moretti has called an "arrow hexagon" [7] (e.g., my Fig. 16). In other words, it has the canonical six arrows, but the canonical opposition relations (and closure under conjunction and disjunction) are not necessarily present (no restrictions on these functions have been introduced yet).

[15] If one requires a one-to-one correspondence between functions and substructures, define equivalence on the functions as follows: two functions $f, g : P(\{a, b, c\}) \longrightarrow P(\{a, b, c, d\})$ are *equivalent* iff there exists an automorphism $h : P(\{a, b, c\}) \longrightarrow P(\{a, b, c\})$ such that $g \circ h = f$ and $f \circ h^{-1} = g$. Assuming this notion of equivalence, there is a one-to-one relation between equivalence classes of injective inclusion-preserving functions $f : P(\{a, b, c\}) \longrightarrow P(\{a, b, c, d\})$ and substructures of the desired kind. For the purposes of this paper, it suffices to characterize the substructure as some map from $P(\{a, b, c\})$.

The substructure picked out by such functions is generally a weaker type of poset structure than the Boolean algebra exhibited in the canonical logical hexagon of Fig. 5 (not all the axioms of a Boolean algebra are true in it). It is not very clear what the information value of this wider class of hexagonal *diagrams* is (qua visualization). The fact that the corresponding structures (six nodes and six arrows) can be arranged hexagon-wise does not guarantee that the resulting diagrams are relevant for our visualization efforts. Also, the fact that there are many of them is not surprising in and of itself.

Nonetheless, among the many arrow hexagons in the tetrakis hexahedron, there are a number of so-called strong hexagons. Informally, these are the arrow hexagons such that the canonical opposition relations hold between their six corners (in the larger structure!) and which are closed under conjunction and disjunction. One can characterize strong hexagons in the tetrakis hexahedron by imposing two further conditions of the injective inclusion-preserving functions $f : P(\{a, b, c\}) \longrightarrow P(\{a, b, c, d\})$.

Condition 1 (preservation of complements) If X and Y are set complements relative to $\{a, b, c\}$ in $P(\{a, b, c\})$, then $f(X)$ and $f(Y)$ must be set complements relative to $\{a, b, c, d\}$ in $P(\{a, b, c, d\})$.

It follows from this that the top and bottom of $P(\{a, b, c\})$ are mapped onto the top and bottom of $P(\{a, b, c, d\})$.

Condition 2 (preservation of conjunction/disjunction) If $X = Y \cap Z$, then $f(X) = f(Y) \cap f(Z)$ and if $X = Y \cup Z$ then $f(X) = f(Y) \cup f(Z)$.

It follows from Conditions 1 and 2 that the image of f is itself a Boolean algebra (Fig. 17). The following figures and table present a function f' which meets these conditions. The image of this function is rendered as a hexagon in Fig. 18, which is indeed a strong hexagon in the tetrakis hexahedron (Table 2).

There are only six such strong hexagons based on images of such functions "in" the tetrakis hexahedron, one for each of the six ternary partitions of $\{a, b, c, d\}$ – *each strong hexagon is simply a ternary partition together with its co-partition*, complements at opposite sides of the hexagon. As was established earlier, they can be grasped in the geometric figure as mirror symmetries in a plane. Any diagram based on such a function f is a strong hexagon "in" the tetrakis hexahedron. The definitions that have been given here are generalizable to other types of inclusion between families of diagrams.

The classification of the weak hexagons "in" the tetrakis hexahedron may be undertaken as follows. Drop Conditions 1 and 2 concerning the preservation complements, unions, and intersections. Observe the following: the Hasse diagram on $P(\{a, b, c\})$ (Fig. 6) has four levels (the empty set, the singletons, the two-element sets, and the three-element set at the top). $P(\{a, b, c, d\})$ has five in total. Since f is injective and preserves inclusion, the image of f must contain elements from at least four of the five levels of $P(\{a, b, c, d\})$. The following definition will be useful:

Definition 6 *f preserves top and bottom* iff $f(\varnothing) = \varnothing$ and $f(\{a, b, c\}) = \{a, b, c, d\}$.

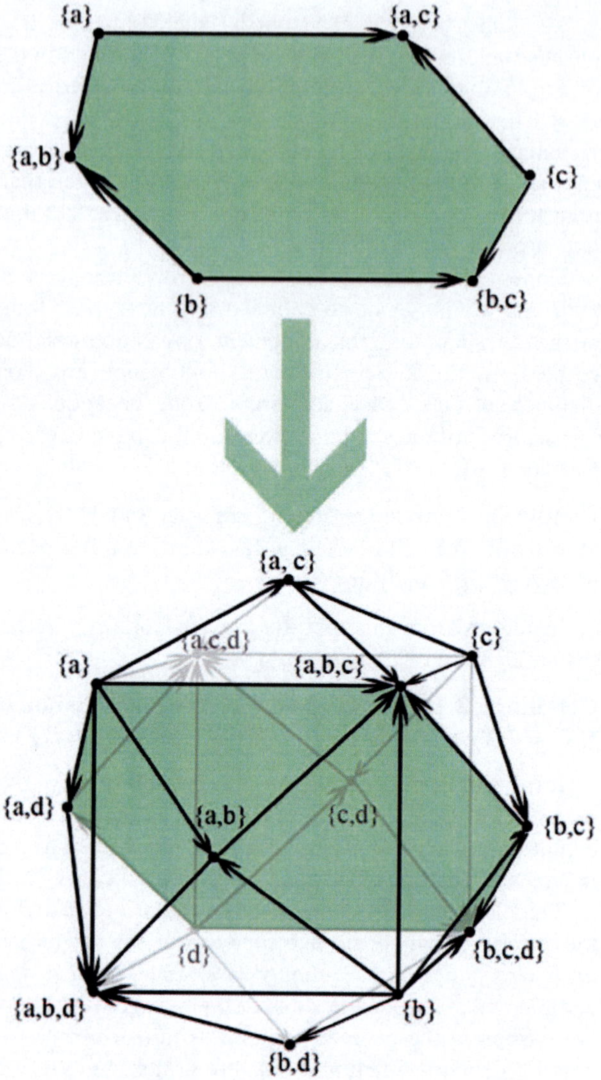

Fig. 17 Boolean-preserving function f'

One may now classify the weak hexagons in the tetrakis hexahedron on the basis of the following distinction regarding f:

– f does *not* preserve top and bottom. Since the image of f contains elements from at least four of the five levels of $P(\{a, b, c, d\})$, it follows that it preserves *either* the top *or* the bottom.

  • Suppose that f preserves the bottom, i.e., that $f(\varnothing) = \varnothing$. That means that f maps the top of $P(\{a, b, c\})$ to the fourth level of $P(\{a, b, c, d\})$. In other words, it maps it to some three-element set X. But since f preserves inclusion, the

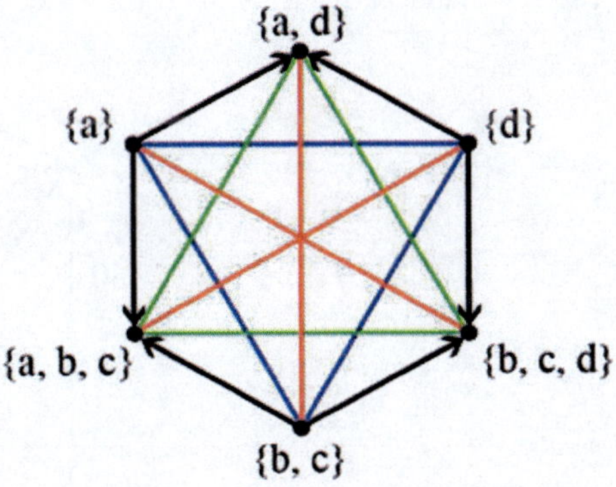

Fig. 18 Hexagon of the image of f'

Table 2 Function f'

Argument	Value
$\varnothing$	$\varnothing$
$\{\boldsymbol{a}\}$	$\{a\}$
$\{\boldsymbol{b}\}$	$\{d\}$
$\{\boldsymbol{c}\}$	$\{b, c\}$
$\{\boldsymbol{a}, \boldsymbol{b}\}$	$\{a, d\}$
$\{\boldsymbol{a}, \boldsymbol{c}\}$	$\{a, b, c\}$
$\{\boldsymbol{b}, \boldsymbol{c}\}$	$\{b, c, d\}$
$\{\boldsymbol{a}, \boldsymbol{b}, \boldsymbol{c}\}$	$\{a, b, c, d\}$

entire image of f is determined by this choice of X, since there are only three two-element sets and only three one-element sets in $P(\{a, b, c, d\})$ which are subsets of X. There are four such structures to be found in $P(\{a, b, c, d\})$, since there are four three-element subsets of $\{a, b, c, d\}$. Each of these structures is a complete Boolean algebra, but their internal relations of opposition do not hold true for the larger BA $P(\{a, b, c, d\})$. In other words, $P(\{a, b, c\})$, $P(\{a, b, d\})$, $P(\{a, c, d\})$, and $P(\{b, c, d\})$ are *substructures* of $P(\{a, b, c, d\})$.

- Suppose that f preserves the top, i.e., that $f(\{a, b, c\}) = \{a, b, c, d\}$. That means that f maps the bottom of $P(\{a, b, c\})$ to the first level of $P(\{a, b, c, d\})$. In other words, it maps it to some one-element set X. But since f preserves inclusion, the entire image of f is determined by the choice of X, since there are only three two-element sets and only three three-element sets in $P(\{a, b, c, d\})$ such that X is a subset of them. There are four such arrow hexagons in $P(\{a, b, c, d\})$, since there are four one-element subsets of $\{a, b, c, d\}$.
- It is interesting to note that these two types of structure appear as hexagons *on the surface* of the tetrakis hexahedron (not as mirror symmetries). Figure 19 shows a hexagon based on $P(\{a, b, c\})$ on the face of the tetrakis hexahedron, where the vertex $\{a, b, c\}$ is the *center* of the hexagon (sticking out, as it were). Incidentally, the tetrakis hexahedron can be assembled by taking four such shapes.

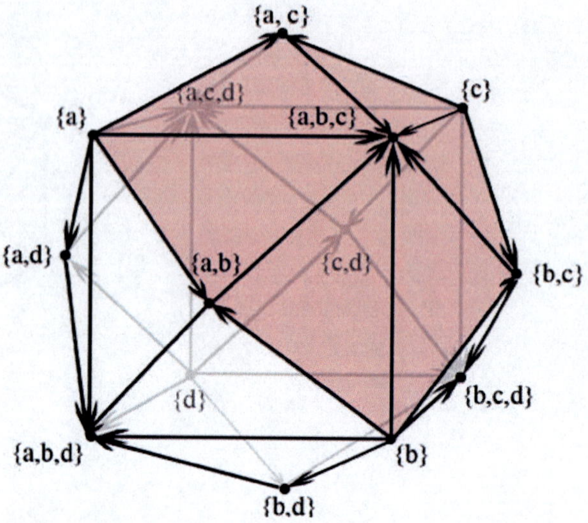

Fig. 19 A hexagon on the face of the tetrakis hexahedron

– f preserves top and bottom but fails to preserve complements and/or conjunction/disjunction. The internal unity/completeness of these structures is weak – there is no good reason to isolate them from the larger structure in a separate diagram. Moretti has classified these from a geometric point of view in his [7]. His classification can be captured in terms of restrictions on the mappings between the structures alone (there is nothing inherently geometric in the matter); one merely needs to specify that the structures fail to be closed under some combination of complements, unions, and intersections. Is it really relevant to meticulously count and classify all these possible degrees of Boolean incompleteness? I do not believe that these substructures play a very important role in the project of visualizing logical data – perhaps only in the negative sense that it is good to know of their existence so that we can avoid mistaking diagrams of such partial structures with diagrams of complete (i.e., Boolean) structures. That said, if one were to completely spell out the classification, it would certainly be more straightforward at the level of structure-mappings than at the level of the geometric objects.

Figure 20 shows that the difference between partitions and the wider contrariness/subcontrariness relation is relevant if we want to differentiate between weak hexagons (i.e., which depict a non-Boolean poset) and strong hexagons.

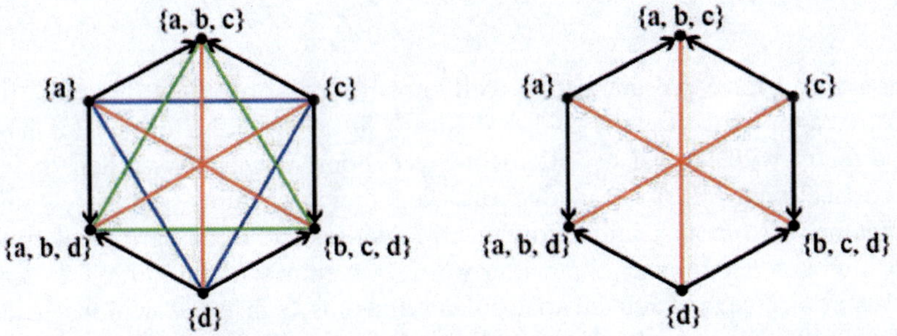

Fig. 20 **(a)** and **(b)**: Relevance of the difference between partitions and contrariness/subcontrariness

Figure 20 a and b depict the same weak hexagon inside the tetrakis hexahedron for $P(\{a, b, c, d\})$. It has been customary in the literature to indicate a relation of contrariness (our Definition 1) between two expressions with a blue line and a relation of subcontrariness (our Definition 2) with a green line. This would yield Fig. 20a, which is visually indistinguishable from a canonical hexagon (Fig. 5).But Fig. 20a is quite nonstandard, since it is not closed under conjunction and disjunction (union and intersection) vis-à-vis the larger structure $P(\{a, b, c, d\})$. We see that $\{d\} \neq \{a, b, d\} \cap \{b, c, d\}$ and also that $\{a, b, c\} \neq \{a\} \cup \{b\}$. But according to the conventional approach, this figure is entirely correct, even if it violates the (implicit but important) expectation of closure under conjunction and disjunction (modulo logical equivalence) – which is certainly part of the larger ideal of completeness.

To remedy this, I suggest to *prioritize complete partitions* over the relation of contrariness. This results in the following rule:

Rule (partitions over contrariness) *A blue (green) triangle may only appear if the three nodes form a complete partition (co-partition).*

According to this convention, which I hereby submit to the scrutiny of my colleagues, Fig. 20b is to be regarded as correct and Fig. 20a as incorrect. From the perspective of visual communication, such a rule is motivated by the fact that a triangle suggests a *complete* ternary relationship. It will be desirable to introduce normative rules which prevent us from drawing misleading/incomplete diagrams in the first place. I have argued that this is a good step from the jungle to the desert.

6 Conclusion

In this paper I have presented two well-known types of logical diagrams (based on the hexagon and the tetrakis hexahedron) in a new light. Instead of labeling them directly with logical or linguistic expressions, it was shown that they have deep connections with Boolean algebras – algebraic structures which are known to be fundamental to both combinatorics and logic. Instead of using logical concepts (opposition and entailment), the matter was reframed in terms of three fundamental combinatorial concepts related to Boolean algebras (subsets, partitions, and permutations).[16] I have shown that the diagrams in question are based on geometric objects (polytopes) which happen to display a lot of fundamental facts about the subsets, partitions, and permutations of the Boolean algebra in the relations between their vertices, facets, and symmetries. In fact, the connection between the Boolean algebras and the polytopes is so striking that I am tempted to call them "Boolean polytopes."

A few remarks on our four criteria for successful visualization:

It seems that the *completeness* of logical diagrams can be approached fruitfully via Boolean closure. The completion of incomplete logical diagrams often relies on the use of conjunction and disjunction (modulo logical equivalence). There are strong reasons to believe that "complete" diagrams are simply depictions of small Boolean algebras (at least when the source language/theory is classical!). The search thus becomes for the best geometrical objects to represent the structure of these Boolean algebras. I have made the case that the hexagon and the tetrakis hexahedron are exemplars for logical geometry exactly because they represent the Boolean algebras (and a lot of combinatorial structure!) on $P(\{a, b, c\})$ and $P(\{a, b, c, d\})$ elegantly.

Although no *incorrect* diagrams have been studied here, it is clear that a systematic view on the underlying structures and their geometric portrayal can only be of help here.

Distortion appears when the geometric features of the diagram suggest information about the structure that is not within the intended scope of communication. In this paper two diagrams have been presented with the goal of showing that many of their prominent geometric features correspond to something relevant about the logical structure. Importantly, any diagram whatsoever suggests completeness. An incomplete diagram thus also distorts the data, insofar as it is easily taken by a reader to be a complete representation of the system. Hence, the question of canonical types

[16] The choice of these three concepts is not arbitrary. I have been guided by the following fact: in category-theoretic treatments of set theory, the connection between subsets, partitions, and permutations is quite manifest. *Subsets* of a set X appear as equivalence classes of monic arrows into X. *Partitions* appear as equivalence classes of epic arrows out of X. "Epic arrow" and "monic arrow" are strictly dual concepts. Finally, *permutations* appear as automorphisms (endomorphisms that are isomorphisms). In the category of sets, isomorphisms are arrows that are epic and monic. Hence, the connection between subsets, partitions, and permutations appears at a deep and conceptual level.

of completeness becomes very important. In this paper I have developed the idea that Boolean closure is an obvious candidate for completeness. Nonetheless, this limitation would greatly restrict the range of acceptable diagrams. In practice, only very few diagrams from the literature are Boolean complete. Perhaps more nuanced criteria may be developed.

Legibility is the most relative of our criteria, especially since the *language* of logical diagrams is just being developed. Here I have merely suggested that legibility emerges, among others, by balancing the careful choice of geometric features with sound instructions to the reader. Finally, legibility emerges above all by standardizing visualization practices. We have a long way to go in this direction.

The following final remark pertains to possible generalizations of the pattern at the heart of this paper.

So far, a correspondence has been established between the following:

- $P(\{a, b, c\})$ and the hexagon
- $P(\{a, b, c, d\})$ and the tetrakis hexahedron

It is natural to inquire whether this pattern is more general. It is easy to see that $P(\{a, b\})$ can be set to a one-dimensional segment. Among these three polytopes, one observes that each increase of complexity amounts to an increase by one dimension. Hence, if the pattern continues as expected, $P(\{a, b, c, d, e\})$ (with a heptachotomy as basis) will correspond to a four-dimensional polytope inscribed in a four-dimensional sphere. As such, one should not expect such figures to be of great help in the project of visualizing logical relations, since humans are mostly unable to visualize higher-dimensional objects. On the other hand, they might be of independent mathematical interest as a series of Boolean polytopes of dimension n. I postpone the matter to a future paper.

References

1. Béziau, J.-Y. (2012). The Power of the Hexagon. *Logica Universalis*, *6*(1–2), 1–43.
2. Béziau, J.-Y., & Jacquette, D. (Eds.). (2012). *Around and Beyond the Square of Opposition*. Birkhäuser.
3. Demey, L., & Smessaert, H. (2018). Geometric and Cognitive Differences between Logical Diagrams for the Boolean Algebra B4. *Annals of Mathematics and Artificial Intelligence*, *83*(2), 185–208.
4. Demey, L., & Smessaert, H. (2018). Combinatorial Bitstring Semantics for Arbitrary Logical Fragments. *Journal of Philosophical Logic*, *47*(2), 325–363.
5. Demey, L., & Smessaert, H. (2017). Logical and Geometrical Distance in Polyhedral Aristotelian Diagrams in Knowledge Representation. *Symmetry*, *9*, 204.
6. Demey, L., & Smessaert, H. (2014). The Relationship Between Aristotelean and Hasse Diagrams. In T. Dwyer, H. Purchase, & A. Delaney (Eds.), *Diagrammatic Representation and Inference. Lecture Notes in Artificial Intelligence* (pp. 213–227). Heidelberg: Springer.
7. Moretti, A. (2015). Arrow-Hexagons. In A. Koslow & A. Buchsbaum (Eds.), *The Road to Universal Logic: Festschrift for the 50th Birthday of Jean-Yves Béziau: Volume II* (pp. 417–488). Birkhäuser.

8. Moretti, A. (2009). *The Geometry of Logical Opposition*. University of Neuchâtel. Retrieved from https://doc.rero.ch/record/12712/files/Th_MorettiA.pdf
9. Smessaert, H., & Demey, L. (2016). Visualising the Boolean Algebra B4 in 3D. In M. Jamnik, Y. Uesaka, & S. Elzer Schwartz (Eds.), *Diagrammatic Representation and Inference. Lecture Notes in Artificial Intelligence* (pp. 289–292). Heidelberg: Springer.
10. H. Smessaert and L. Demey, "The Unreasonable Effectiveness of Bitstrings in Logical Geometry" in J-Y Beziau and G. Basti. *The Square of Opposition: A Cornerstone of Thought:* Springer 2017.

Turnstile Figures of Opposition

Jean-Yves Beziau

Abstract We present many figures of opposition (triangles and hexagons) for simple and double turnstiles. We start with one-sided turnstiles, corresponding to sets of tautologies, and then we go to double-sided turnstiles corresponding to consequence relations. In both cases, we consider proof-theoretic (with the simple turnstile) and model-theoretic (with the double turnstile) figures. By so doing, we discuss various central aspects of notations and conceptualizations of modern logic.

Keywords Square of opposition · Turnstile · Tautology · Truth · Proof · Consequence · Model

Mathematics Subject Classification (2000) Primary: 03-01 Secondary 03A05; 03B22;03B45;03F99: 03 C99

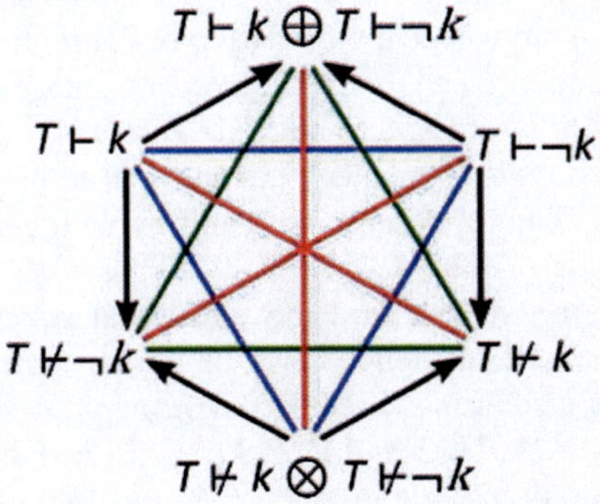

J.-Y. Beziau (✉)
Federal University of Rio de Janeiro, Rio de Janeiro, Brazil
e-mail: jyb@ufrj.br

J.-Y. Beziau, I. Vandoulakis (eds.), *The Exoteric Square of Opposition*, Studies in Universal Logic, https://doi.org/10.1007/978-3-030-90823-2_10

1 The Hexagon of Opposition and the Turnstile

The hexagon of opposition was introduced by Robert Blanché.[1] It is an improvement and/or reconstruction of the famous square of opposition. Figure 1 is a picture of it:

Fig. 1 Hexagon of opposition

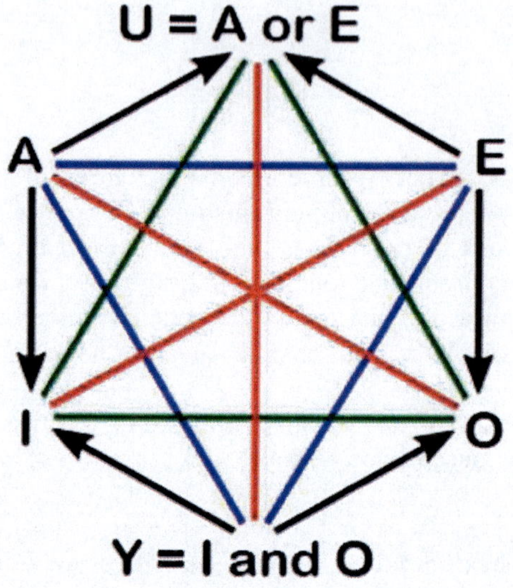

We have the same four relations as in the square: the black arrow is the relation of subalternation, and in red we have the relation of contradiction, while in blue the relation of contrariety and in green the relation of subcontrariety. We recall the basic definitions: two propositions are said to be *contradictory* iff they cannot be true and cannot be false together, *contrary* iff they cannot be true but can be false together, and *subcontrary* iff they cannot be false but can be true together. *Subalternation* is an implication.

In the above hexagon, we can find the traditional square of opposition with corners A, E, I, O. Blanché introduced two additional corners that he named U and Y and which are defined as indicated. In the hexagon, we can see two additional squares of opposition – Y, A, U, O and E, Y, I, A – as well as a contrariety triangle in blue and a subcontrariety triangle in green.

The hexagon of opposition has been applied to many topics ranging from deontic notions to the theory of colors, through music, economy, and quantum physics

[1] His main book on the topic is [1], but his first works were published in the 1950s, and at this time other people had similar ideas (for details, see [2]).

(cf. [3, 4]), and other papers in the many volumes of collected papers [5–8] and special issues [9–12] which have been published since the revival of the square (cf. [13, 14]) and the *First World Congress on the Square of Opposition* in Montreux in 2007. It can also be applied to the theory of opposition itself (see [15]).

Here we will apply it to logical notions. This paper is a follow-up of my paper "The Metalogical Hexagon of Opposition" [16]. It is also related with the talk "Beyond Truth and Proof" I gave in Tübingen at the workshop *Consequence and Paradox Between Truth and Proof (March 2–3, 2017)* and the tutorial I gave at UNILOG'2018 (*World Congress and School on Universal Logic*) in Vichy in June 2018: "The Adventures of the Turnstile."

The present paper connects two aspects of symbolism: diagrammatic symbolism and the use of nonalphabetical signs. In logic, among the first category, the square is the most famous representative followed by Venn diagrams. Among the second category, we have in particular the connectives "$\vee$, $\neg$, $\wedge$, $\rightarrow$" and the quantifiers "$\forall$, $\exists$." But probably the most famous one is "$\vdash$."

This symbol was introduced by Frege (1879, cf. [17]) with a specific meaning that we will not discuss here (see, e.g., [18]). It is nowadays used with another meaning which is not always clear. The aim of this paper is to clarify the contemporary meaning(s) of "$\vdash$," using the theory of opposition, in particular triangles of contrariety and hexagons of opposition.

Doing that we will deal with the sister symbol "$\models$," which is called the *double turnstile*, by contrast to "$\vdash$," called *simple turnstile*. "$\vdash$" is also called *Frege's stroke*, but we will not use here this terminology, because on the one hand we are not dealing with the original meaning given to it by Frege and on the other hand the turnstile terminology is nice because it allows to use the same word "turnstile" to qualify two different connected notions. It would make no sense to talk about Frege's *simple stroke* and *Frege's double stroke*, since Frege did not introduce "$\models$" (this symbol was introduced in the 1950s).

2 Tautological Figures of Opposition

2.1 Two Pretty Different Contrariety Triangles

There is the dichotomy between truth and falsity that we find in particular in classical propositional logic. We can go beyond this dichotomy by adding a third value or more values. This in particular is what Łukasiewicz [19] did, inspired by Aristotle. If we have three-values, we have then the following triangle of contrariety, depicted in Fig. 2:

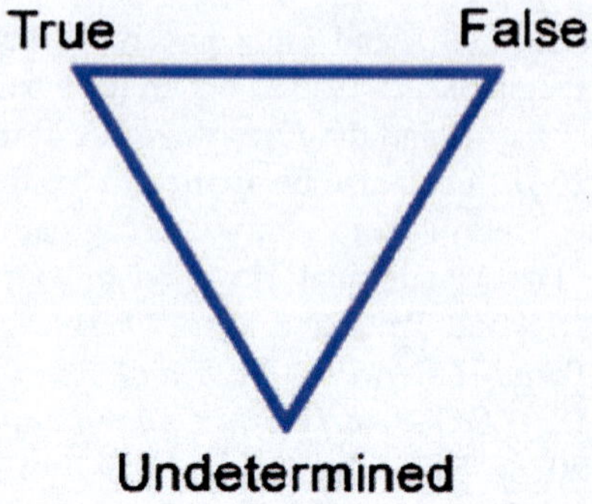

Fig. 2 Three-valued triangle of contrariety

On the other hand, there is a distinction which is at another level. This is the distinction between truth and logical truth, promoted in particular by Wittgenstein [20] putting forward the notion of tautology. This leads to a subtler triangle of contrariety:

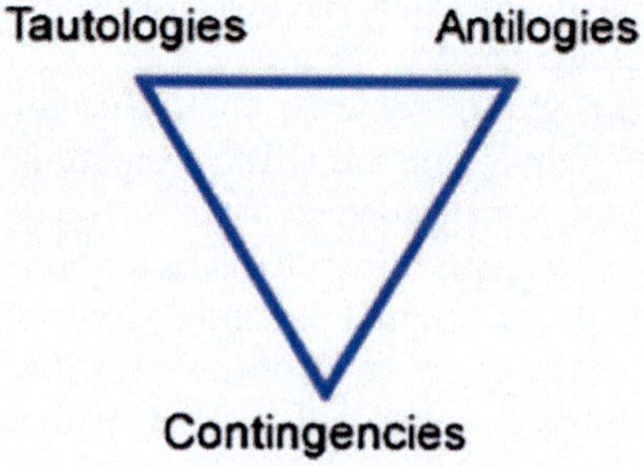

Fig. 3 Tautological triangle of contrariety

Wittgenstein didn't use the words "antilogies" and "contingencies." This terminological choice is explained in [15, 16, 21]. But although we are using a different terminology, we are presenting here the same trichotomy as in the *Tractatus*: a tautology is always true, an antilogy is always false, and a contingency can be true and can be false.

2.2 *Symbolic Representation of the Tautological Triangle*

Wittgenstein presented the trichotomy among propositions using a framework which is nearly identical to the one used nowadays for the semantics of classical propositional logic based on valuations that he calls "truth-possibilities" (*Tractatus* 4.3). We can present it in the following table:

We have called what is on the right column "mathematical definition" to emphasize that it is not just symbolism. In modern logic, mathematical tools, objects, and concepts are used. Here, for example, the numbers 0 and 1 (not only the notations "0" and "1") are used, as well as the notions of function and equality.[2]

In modern mathematical logic, the expression "$\forall v\ v(p) = 1$" is also written "$\models p$." The latter can be seen as an abbreviation of the former; it is indeed shorter. "$\forall$" can also be seen as an abbreviation of "All." This is the first letter of this word put upside down (notation introduced by Gentzen, following the same idea as for the sign of the existential quantifier "$\exists$" introduced by Peano). But mathematical writing is not only a question of abbreviation. There is the idea to use signs which are not completely arbitrary that have a serious symbolic aspect in the true sense of the word (see [24] and [25]). The sign "$\vdash$" was introduced by Frege with a real symbolic dimension expressing an important distinction through perpendicularity. "$\models$" is a symbol directly inspired by "$\vdash$." The similar graphic design of the two signs expresses the connection between their meanings, and the difference of meaning is expressed by doubling the horizontal line. This is nicely reflected in natural language by the expressions *simple turnstile* and *double turnstile*. Natural language is useful in particular when talking.

After having established a correspondence between "$\forall v\ v(p) = 1$" and "$\models p$," how can we go further on, rewriting the other mathematical definitions using the double turnstile? There is no "direct" way to do that. The best we can do with "$\forall v\ v(p) = 0$" is to write it as "$\models \neg p$" considering the definition of classical negation according to which $v(p) = 0$ iff $v(\neg p) = 1$.

It is even less straightforward to express contingency with the double turnstile. We have to use the symbol "$\not\models$," which uses a negation at the metalevel. According to that, "$\not\models p$" means $\exists v\ v(p) = 0$ (we are not putting quotes here, because we are not talking about this symbolic formula but about its meaning: p is false according to one valuation). "$\not\models p$" is the syntactic (metalevel syntax) negation of "$\models p$," which itself means $\forall v\ v(p) = 1$. Here we have to be very careful because there is a mix between logic and metalogic. $\exists v\ v(p) = 0$ is the negation of $\forall v\ v(p) = 1$ at the metalogical level (again we don't use quotes here because we are not talking about "$\exists v\ v(p)= 0$" and "$\forall v\ v(p)= 1$," but about their meanings).

The ambiguity is that the symbols "$\exists$" and "$\forall$" are generally used as symbols for quantifiers in first-order logic, at a logical level, not at a metalogical level. Here we are using them at a metalogical level. One may think that the metatheory of propositional logic (classical or not) can be carried out in first-order logic. This is true up to a certain point. But it is not necessarily obvious, details have to be checked, and someone may defend another point of view. Here we stay neutral. If we use the symbols "$\exists$" and "$\forall$," it is rather a question of abbreviation. There are no other symbols standardly used for that, like in the case of implication where we can make the distinction between implication and meta-implication, respectively,

[2] Wittgenstein uses "F" and "W," not "0" and "1." In general, his framework is not explicitly mathematical, although he uses the notion of function, following Frege and Russell. About 0 and 1 as truth-values, the notion of truth-function, etc.; see [22] and [23].

using the symbols "⟶" and "⟹".[3] Likewise, in the case of conjunction and meta-conjunction, we can make the distinction using the symbols "∧" and "&."

Using the latter symbol, we can rewrite the mathematical definition of contingency as "$\exists v\ v(p)=1\ \&\ \exists v\ v(p)=0$," which in turn we can express using the double turnstile: "$\nvDash p\ \&\ \nvDash \neg p$." At the end, using also negation at the metalevel for the second part of the mathematical definition of contingency, we have the Table 1:

Table 1 Trichotomy of propositions in propositional logic using bivaluations

TERMINOLOGY	MATHEMATICAL DEFINITION
Tautology	$\forall v\ v(p) = 1$
Antilogy	$\forall v\ v(p) = 0$
Contingency	$\exists v\ v(p) = 1$ and $\exists v\ v(p) = 0$

Base on the right column, we can represent the triangle of contrariety of Fig. 3 in the following manner:

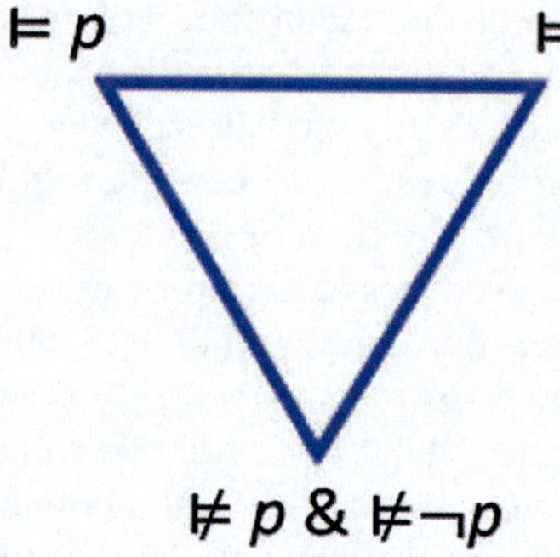

Fig. 4 Double turnstile tautological triangle of contrariety

Nowadays there is a clear distinction between "⊨" and "⊢," the latter being used in proof theory (also called syntax) by contrast to the former used in model theory (also called semantics). For classical propositional logic, the bridge was established by Emil Post in a paper published in 1921 [26], the same year of the publication of Wittgenstein's *Tractatus*,[4] and in 1930 by Kurt Gödel for first-order logic [27]. We have therefore the Table 2:

Table 2 Trichotomy of propositions in propositional logic using double turnstyle

TERMINOLOGY	MATHEMATICAL DEFINITION	DOUBLE TURNSTILE
Tautology	$\forall v\ v(p) = 1$	$\vDash p$
Antilogy	$\forall v\ v(p) = 0$	$\vDash \neg p$
Contingency	$\exists v\ v(p) = 1\ \&\ \exists v\ v(p) = 0$	$\nvDash p\ \&\ \nvDash \neg p$

[3] The difference between the two levels is expressed here by doubling the horizontal line. For the turnstile, the doubling of the horizontal line is not used in this sense.

[4] Post was using only "⊢." As we said, "⊨" was introduced in the 1950s. Wittgenstein was using none of these symbols, and he rejected Frege's stroke (cf. *Tractatus* 4.442).

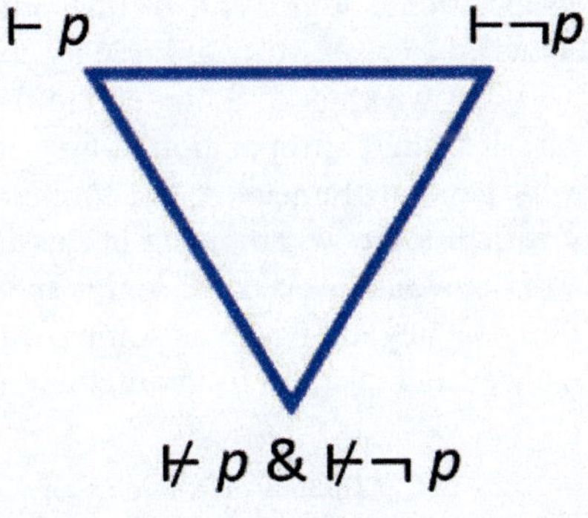

Fig. 5 Simple turnstile tautological triangle of contrariety

Accordingly, the triangle of Fig. 4. is equivalent to the following one:

Using the simple and double turnstiles, we are able to explicitly make the distinction between proof theory and model theory, truth and proof, syntax and semantics, a distinction which is not clearly made at the level of natural language. For example, the word "tautology" is not clearly attached to one of the side of the dichotomy. One may say that it is not important because of the completeness theorem. But in fact the distinction is important. If we don't make the distinction, the completeness theorem has no meaning. Moreover, if we have a general perspective, being interested not only in classical propositional logic but in many other systems of logic, there are some cases where the completeness theorem does not hold.

The two triangles of Figs. 4 and 5 clearly show the general structure of a triangle of contrariety. The bottom corner is the conjunction of the (metalogical) negations of the two top corners. The position of the decorations of the corners is mostly irrelevant, contrary to the spirit of the traditional theory of opposition with the labels "A," "E," "I," "O" for the corners of the square, which are moreover connected to a special version of the square, i.e., the original square of categorical propositions.

We have put on the top left corner the notion of tautology and the corresponding notations: "$\models p$" and "$\vdash p$." The reason to do so is because it is the most famous notion. This is also the reason why we have called these triangles "tautological triangles." Other words are used for the notion of tautology, for example, *logical truth*, but *tautology* is more striking. It is usual to call a figure of opposition (a triangle, a square, a hexagon) by the name of one of its corners, e.g., the analogical hexagon [28]. Another option is to use the name of the family of notions involved in the figure, i.e., the deontic hexagon. Here we could have used the expression "metalogical triangle" as we did in [16]. The reason not to do that here is that we want to make explicit the distinction between two metalogical figures, the one corresponding to logics as sets of tautologies and the one corresponding to logics as consequence relations, in which both are metalogical.

Another distinction is between propositional logic and first-order logic. Our two triangles of Figs. 4 and 5 can be seen from both perspectives. But on the one hand in our tables we have given only the mathematical definitions corresponding to propositional logic, and on the other hand the use of the letter "*p*" is generally attached to propositional logic by contrast to first-order logic.

What we want to do here is to make a uniform presentation for all variations of classical logic: propositional, first-order, second-order, etc. We are also aiming at a very general framework not limited to classical logic. The only typical feature of classical logic we are using is classical negation. Therefore, our framework applies to extensions of classical logics such as modal logics or nonclassical logics with a classical negation such as some paraconsistent logics.

We can make the following new version of Table 3:

Table 3 Trichotomy of propositions in propositional logic using simple turnstyle

TERMINOLOGY	DOUBLE TURNSTILE	SIMPLE TURNSTILE
Tautology	$\vDash p$	$\vdash p$
Antilogy	$\vDash \neg p$	$\vdash \neg p$
Contingency	$\nvDash p \,\&\, \nvDash \neg p$	$\nvdash p \,\&\, \nvdash \neg p$

When we write "$m \vDash k$" we are using the double turnstile in a different way as on the right column. It is a bit ambiguous but is based on a link between the two: the meaning of "$\vDash k$" is defined by $\forall M\ M \vDash k$. Generally "$M \vDash k$" is used only in first-order logic but Chang and Keisler in their famous book *Model Theory* [29] also use this notation for propositional logic. However they don't use the letters "*M*" and "*k*", but they use another notation.

On the left side we are using the letter "*M*" using a graphism different from the one of the letter "*k*" to emphasize that these are different kinds of entities. We use the 13th letter of the alphabet because it is the initial letter of the word "model." This word can be used in any context because it does not specify the internal nature of the thing, whether it is a bivaluation, a first-order structure, or a possible world, but only its function. It is a bit ambiguous because if "$M \vDash k$" can be read without problem as "*M* is a model of *k*," on the other hand "$M \nvDash k$" is read as "*M* is not a model of *k*," which is a bit paradoxical, because we have a model which is not a model! But this makes sense if we consider that *M* is a model of other formulas, here, for example, of $\neg k$.

Instead of writing "$M \vDash k$" and "$M \nvDash k$" we could have, respectively, written "v(*M* ; *k*)=1" and "v (*M* ; *k*)=0," but it would have been a bit cumbersome. Anyway "$M \vDash k$" is usually read as "*k* is true in *M*" and "$M \nvDash k$" as "*k* is false in *M*." What is important to stress is that a principle of bivalence is used at the metalogical level whether we are dealing with a non-classical logic and/or a first-order logic.

Why using the letter "*k*"? We want to avoid to use the letter "*p*" which is too much connected to propositional logic, so we chose the 11th letter of the alphabet which is quite neutral. We could have chosen the letter "*f*," considering that it is the first letter of the word "formula." This word is quite neutral and is used to talk either of formulas of propositional logic or first-order logic. But this word has an ambiguous

Table 4 Trichotomy of propositions from the point of view of model theory

TERMINOLOGY	MATHEMATICAL DEFINITION	DOUBLE TURNSTILE
Tautology	$\forall M\ M \vDash k$	$\vDash k$
Antilogy	$\forall M\ M \nvDash k$	$\vDash \neg k$
Contingency	$\exists M\ M \vDash k\ \&\ \exists M\ M \nvDash k$	$\nvDash k\ \&\ \nvDash \neg k$

meaning: it is also used for any symbolic expression (not necessarily connected to logic). It is important to emphasize the nature of the object we are dealing with and to which the three categories tautology, antilogy, and contingency apply. These are propositions, whether specified as formulas of a propositional language or another formal language. So we will keep using the word "proposition," but we prefer to use "k" than "p" to avoid the reader to immediately think that we are dealing only with propositional logic.

We have then the two following diagrams:

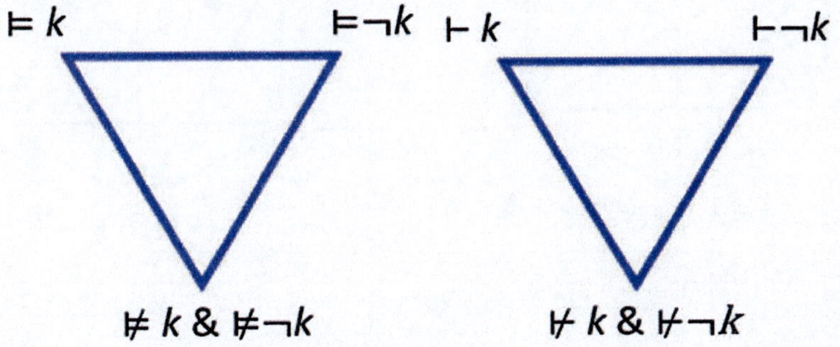

Fig. 6 Turnstile tautological triangles of contrariety

Based on Table 4, the left diagram can be designed as follows without using negation at the logical level:

Fig. 7 Model-theoretic triangle of contrariety

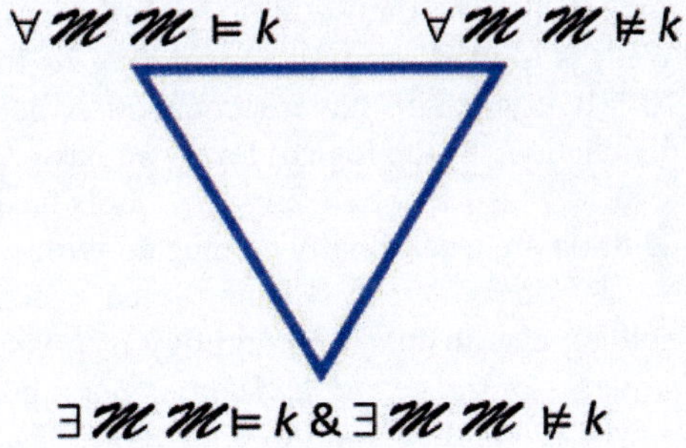

It is interesting because we have then a diagram not limited to logics with a classical negation, such as positive logic. This triangle could be called a "turnstile triangle" since the (double) turnstile is used, but not with the same meaning as in Fig. 4.

2.3 *Turnstile Tautological Hexagons*

Let's now apply the structure of the hexagon. We have then the following diagrams:

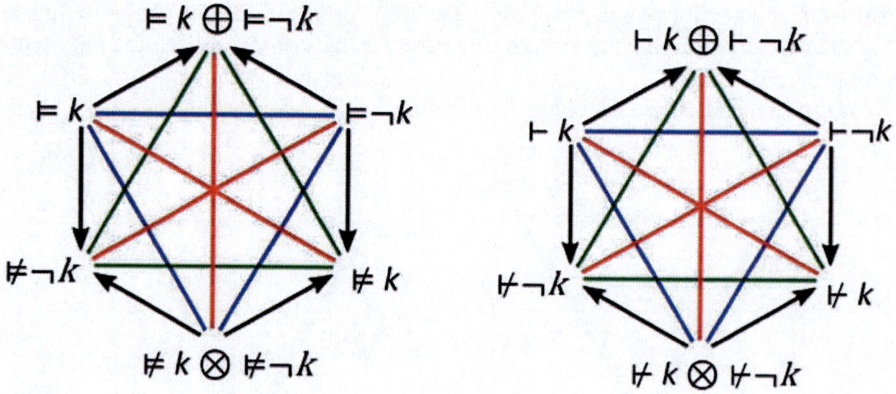

Fig. 8 Turnstile tautological hexagons

These hexagons have been generated using the logical structure of this figure of opposition. The three corners of the green triangles of subcontrariety are the (metalogical) negations of the three corners of the blue triangles of contrariety. We are using the symbol "⊕" to denote metalogical disjunction. And we have replaced "&" by "⊗" for metalogical conjunction. This is not only purely esthetical. A good notation has to be designed considering the general context, in relation with other notations. For example, the symbol for the empty set "∅" (introduced by André Weil) is a good notation considering its link with the symbol for the number zero "0." It is good to have a connection between the symbols for conjunction and disjunction. At the logical level, we have "∧, ∨" and that's nice. At the metalogical level, we also chose here two symbols having a connection (and multiplication and addition are traditionally connected with conjunction and disjunction).

"⊭ *k*" can literally be interpreted as follows: *k* is not a tautology, which means nothing else than *k* is an antilogy or *k* is a contingency, as clearly depicted by the structure of the hexagon. There is not a positive terminology for this situation, and maybe it could be good to create one. The same happens with the two other cases: the contradictory opposite of antilogy and the contradictory opposite of contingency.

3 Hexagons of Opposition for Consequence Relations

At some point a logic started to be considered as a consequence relation rather than a set of tautologies. The origin of this framework can be traced back to Tarski when in Poland. He put forward on the one hand the notion of consequence operator [30] and on the other hand the notion of logical consequence [31]. In both cases we have a binary setting: a formula is consequence of a set of formulas. These two notions studied by Tarski are not defined in the same way and he didn't use the same terminology for them.

Tarski at this time was using neither "⊢" nor "⊨." Nowadays it is common to use these symbols as binary relations in the following way:

Table 5 Proof-theoretic and model-theoretic consequence relations

SYMBOLISM	READING	MEANING
$T \vdash k$	k is a proof-theoretic consequence of T	There is a proof of k from T
$T \models k$	k is a model-theoretic consequence of T	All models of T are models of k

In continuity with what we have said in the previous section, what is on the right of the simple or double turnstile, we will call it a *proposition* and denoted it by "*k*." On the left side we have what is generally called a *theory*,[5] a set of formulas, or, to use our present language, a set of propositions. We use a capital letter for a theory to emphasize the difference between the size: multiplicity vs. oneness. Multiplicity on the right of the turnstile has also been considered (cf. [32]) but we will not deal here with this issue.

In the case of both turnstiles, the tautological framework can be seen as a particular limit case of the consequential framework, the case where the theory is the empty set: $\varnothing \vdash k$ and $\varnothing \models k$.

Symbolically we have then the two following consequential turnstile hexagons of which the two hexagons of Fig. 8 are limit cases:

[5] In Poland during the 1930's, the word "theory" was used in a different way: for what is nowadays called a "closed theory," a theory such that any formula which is a consequence of the theory is in the theory.

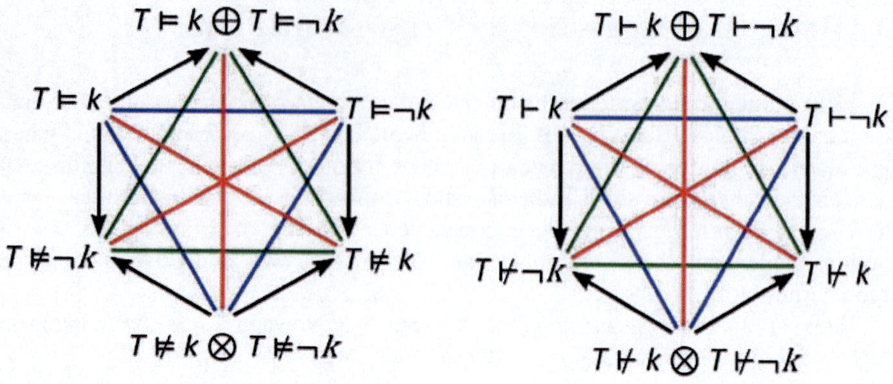

Fig. 9 Turnstile consequential hexagons

These hexagons perfectly depict the six possibilities we have for a relation of a proposition relatively to a theory, either from a model-theoretic point of view (on the left) or from a proof-theoretic point of view (on the right). The completeness theorem can be interpreted as the matching of these two hexagons.

These six positions do not always exist. For example, if we have a *complete theory*, the bottom position does not exist. The definition of a complete theory is given by the top position. A famous case of incomplete theory is Peano Arithmetic, *PA*. Gödel [33] has shown that there is a proposition g, inspired by the liar paradox, such that $PA \nvdash g$ and $PA \nvdash \neg g$. Sometimes such a proposition is called an *undecidable* proposition, but a better terminology is *independent.*[6]

Let's see what kind of names we can give to the other positions. We can design the following hexagon:

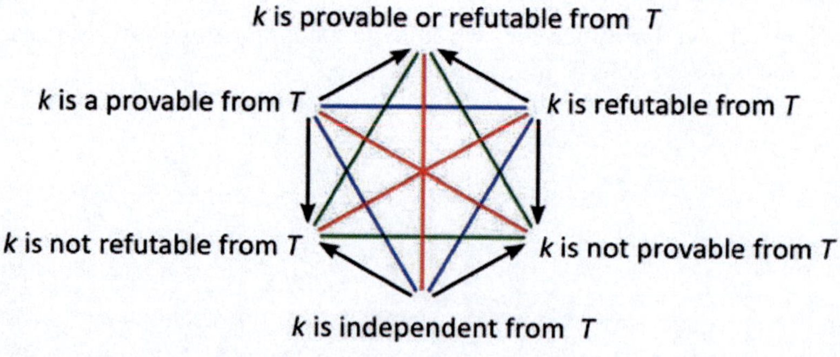

Fig. 10 Terminological proof-theoretic hexagon

[6] A theory can be incomplete and decidable, a famous case is the empty theory of classical propositional logic, and an atomic formula is independent from ∅ but ∅ is decidable.

We clearly have some positive terms for the three corners of the contrariety triangle. It is no clear that we can find some non-ambiguous terminology for the three other corners. But note that in this figure we have avoided to use negation at the logical level, so it can apply to any logical system.

Now let's turn to the model-theoretic hexagon. We have the following Table 6:

Table 6 Model-theoretic consequential hexagon

	MATHEMATICAL DEFINITION	DOUBLE TURNSTILE
A	$\forall M\, M \vDash T \Longrightarrow M \vDash k$	$T \vDash k$
E	$\forall M\, M \vDash T \Longrightarrow M \nvDash k$	$T \vDash \neg k$
Y	$(\exists M\, M \vDash T \otimes M \nvDash k) \otimes (\exists M\, M \vDash T \otimes M \vDash k)$	$T \nvDash k \otimes T \nvDash \neg k$
I	$\exists M\, M \vDash T \otimes M \vDash k$	$T \nvDash \neg k$
O	$\exists M\, M \vDash T \otimes M \nvDash k$	$T \nvDash k$
U	$\forall M\, M \vDash T \Longrightarrow M \vDash k \oplus \forall M\, M \vDash T \Longrightarrow M \nvDash k$	$T \vDash k \oplus T \vDash \neg k$

This allows us to have a consequential hexagon with the use of negation only at the metalevel similarly to the triangle presented in Fig. 7:

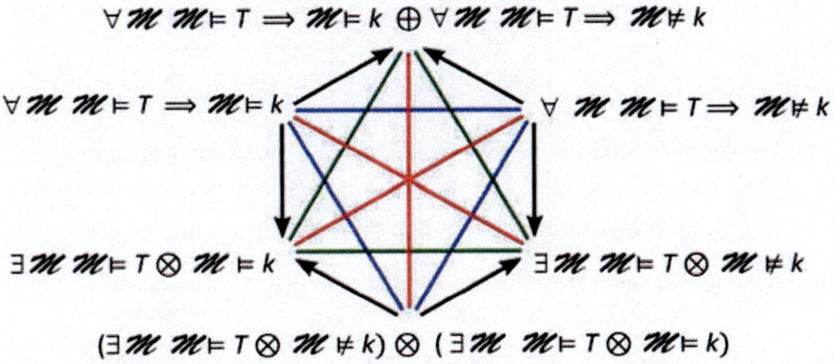

Fig. 11 Model-theoretic hexagon

If we want to use a truth terminology, we can interpret "$T \vDash k$" as "k is true in T" or "k is true according to T," for example, "$2 + 2 = 4$" is true according to Peano Arithmetic. And we can interpret "$T \vDash \neg k$" as "k is false in T" or "k is false according to T". For example, $2 + 2 \neq 4$ is false according to Peano Arithmetic. Up to now, no problems. From this point of view, the Y corner of the hexagon can be interpreted as neither true nor false in T (or according to T), but there is no straightforward terminology to summarize this in one word. And we can interpret the other corners of the hexagon in a pure negative way. We then have the situation as depicted in Fig. 12.

A very important point is that "k is true in T" is not equivalent here to "k is not false in T." Symbolically, $T \vDash k$ is not equivalent to $T \nvDash \neg k$. At the level of symbolism, it is interesting because we see that we have a logical negation and a metalogical negation, and the two together do not lead to affirmation.

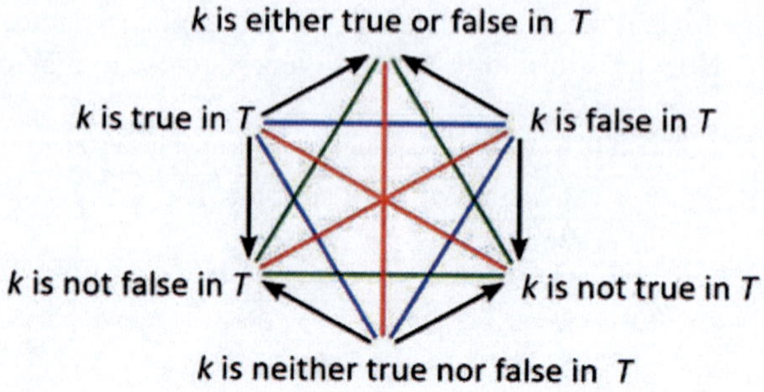

Fig. 12 Terminological model-theoretic hexagon: truth version

We may want to eliminate truth, and then we can a configuration as depicted in Fig. 13.

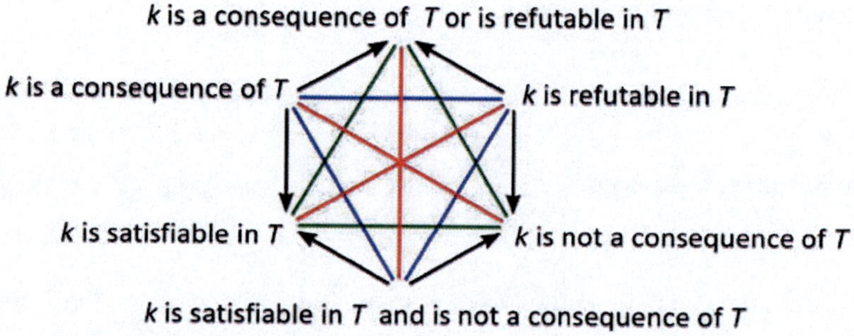

Fig. 13 Terminological model-theoretic hexagon: no truth version

The word "satisfiable" is clearly from model-theory, but generally it is not used in this way: we say that a formula is satisfiable in a model, not in a theory. The terminology "satisfiable" is quite natural for the I-corner when T is empty, i.e., in the case of the tautological model-theoretic hexagon. Then we say that a formula is satisfiable *tout court*.

On the E-corner we have put "refutable" which is rather from proof theory. We have used "in" rather than "from" to have a similar expression as with satisfiability and different from the proof-theoretic hexagon of Fig. 10.

From the point of view of model theory, it would make more sense to put refutable in the O-corner, where we have put "is not a consequence." Again, this is natural in the empty case, when we say refutable *tout court*. This is the reason why in our previous paper [16] we put refutable in the O-corner forming a nice subcontrariety pair with satisfiable rather than a contradictory pair as in Fig. 13. The problem we are facing here is that in proof theory it makes also sense to put it in the E-corner.

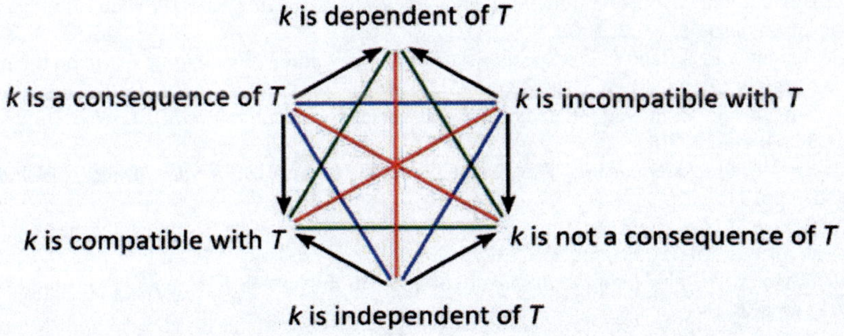

Fig. 14 Terminological consequential hexagon

To finish, let us present a new terminological decoration of the six corners of the consequential hexagon, depicted in Fig. 14 :

As in Figs. 10, 11, 12, and 13, we have avoided to use negation at the logical level, so this figure applies to any logical system. Moreover, the advantage of the terminology of this diagram is that it can be used both for proof theory and for model theory. The terminologies "(in)compatible" and "(in)dependent" are not usually univocally tight to one of these fields. This advantage turns of course into a defect if we want to emphasize one of the two specific fields.

Considering the turnstile symbolism, the two fields are clearly distinguished by the simple turnstile "⊢" and the double turnstile "⊨." There is not a symbol which is unambiguously used to deal with an abstract situation which is beyond proof and truth, although in recent years the tendency has been to use the simple turnstile for such a situation.

Acknowledgments Thanks to the participants of SQUARE'2018 and to Lloyd Humberstone for useful comments on a previous version of this paper.

Bibliography

1. R.Blanché, *Structures intellectuelles. Essai sur l'organisation systématique des concepts*, Vrin, Paris, 1966.
2. J.-Y.Beziau, "The power of the hexagon", *Logica Universalis*,**6** (2012), pp. 1-43.
3. D. Jaspers, "Logic and Colour", *Logica Universalis*,**6** (2012), pp. 227-248.
4. L.Magnani, "The violence hexagon", *Logica Universalis*,**10** (2016), pp. 359-371.
5. J.-Y.Beziau and D.Jacquette (eds), *Around and Beyond the Square of Opposition*, Birkhäuser, Basel, 2012.
6. J.-Y.Beziau and G.Payette (eds), *The Square of Opposition - a General Framework for Cognition*, Peter Lang, Bern, 2012.
7. J.-Y.Beziau and S.Geogiorgiakis (eds), *New dimensions of the square of opposition*, Philosophia, Munich, 2017.
8. J.-Y.Beziau and G.Basti (eds), *The Square of Opposition – a Cornerstone of Thought*, Birkhäuser, Basel, 2017.

9. J.-Y.Beziau and G.Payette (eds), Special issue of *Logica Universalis* on the square of opposition, issue 1, vol.**2**, 2008
10. J.-Y.Beziau, and S.Read (eds), Special issue of *History and Philosophy of Logic* on the history of the square of opposition, issue 4 vol. **35**, 2014.
11. J.-Y.Beziau and R.Giovagnoli (eds), Special issue *The Vatican Square*, *Logica Universalis*, issues 2-3, vol.**10**, 2016.
12. J.-Y.Beziau and J.Lemanski, "The Cretan Square", *Logica Universalis*, Issue 1, Volume 14 (2020), pp. 1-5.
13. J.-Y.Beziau, "New light on the square of oppositions and its nameless corner", *Logical Investigations*, **10**, (2003), pp. 218-232.
14. J.-Y.Beziau, "The new rising of the square of opposition", in J.-Y.Beziau and D.Jacquette (eds), 2012, pp.6-24.
15. J.-Y.Beziau "Disentangling Contradiction from Contrariety via Incompatibility", *Logica Universalis*, **10** (2016), pp. 157-170.
16. J.-Y.Beziau, "The metalogical hexagon of opposition", *Argumentos*, **10** (2013), pp. 111-122.
17. G.Frege, *Begriffsschrift, eine der arithmetischen nachgebildete Formelsprache des reinen Denkens*, L.Nebert, Halle, 1879.
18. F.Rombout, *Frege, Russell and Wittgenstein on the judgment stroke*, MD, University of Amsterdam, 2011.
19. J.Łukasiewicz, "O logice trójwartosciowej", *Ruch Filozoficny*, **5** (1920), pp. 170-171.
20. L.Wittgenstein, "Logisch-Philosophische Abhandlung" (Later published as *Tractatus Logico-Philosophicus*), *Annalen der Naturphilosophie*, **14** (1921), pp. 185-262.
21. J.-Y.Beziau, "Possibility, Contingency and the Hexagon of Modalities", *South American Journal of Logic*, **3** (2017).
22. J.-Y.Beziau, "Truth as a mathematical object", *Principia,* **14** (2010), pp. 31–46.
23. J.-Y.Beziau, "History of truth-values", in D.M.Gabbay, F.J.Pelletier and J.Woods (eds), *Handbook of the History of Logic , Vol. 11 - Logic: a history of its central concepts*, Elsevier, Amsterdam, 2012, pp.233-305.
24. M.Serfati, *La révolution symbolique. La constitution de l'écriture symbolique mathématique*, Petra, Paris, 2005.
25. J.-Y.Beziau, "La puissance du symbole" in *La pointure du symbole*, Paris, Petra, 2014, pp. 9-34.
26. E.Post, "Introduction to a general theory of elementary propositions", in *American Journal of Mathematics*, **13** (1921), 163–185, 1921.
27. K.Gödel, "Die Vollständigkeit der Axiome des logischen Funktionenkalküls". *Monatshefte für Mathematik und Physik,* **37** (1930), pp. 349–360.
28. J.-Y.Beziau, "An Analogical Hexagon", *International Journal of Approximate Reasoning* , **94** (2018), pp. 1–17
29. C.C.Chang and H.J.Keisler, *Model theory*, North-Holland, Amsterdam, 1973.
30. A.Tarski, "Remarques sur les notions fondamentales de la méthodologie des mathématiques", *Annales de la Société Polonaise de Mathématique*, **7** (1928), pp. 270-273.
31. A.Tarski, 1936, "O pojciu wynikania logicznego", *Przegląd Filozoficzny*, **39** (1936), pp. 58-68.
32. D.J.Shoesmith and T.J.Smiley, 1978, Multiple-conclusion logic, *Cambridge University Press*, Cambridge, 1978.
33. K.Gödel, "Über formal unentscheidbare Sätze der *Principia Mathematica* und verwandter Systeme, I", *Monatshefte für Mathematik und Physik*, **38** (1931), pp. 173–98.

The Naturalness of Jacques Lacan's Logic

Hubert Martin Schüler

Abstract In his works in the field of psychoanalysis, especially in his *Seminar XX – Encore* and in *L'Étourdit,* Jacques Lacan (1901–1981) developed a logic of incompleteness by rejecting universal negative propositions in the sense of *the traditional (quantificational) square of opposition.* By using the formula $\sim(\forall x)\Phi x$, entitled *pas-tout,* Lacan made a connection between psychoanalysis and formal logic. Recently, among logicians, a debate has been triggered that makes it possible to leave the depth of psychoanalytic theoretical content. This chapter compares Lacan's *four formulas of sexuation* with the *traditional (quantificational) square of opposition* and tries to explain the naturalness of Lacan's logic. Furthermore, it deals with Guy Le Gaufey's *Lacanian logical square* in order to compare it to the traditional square. Finally, it becomes clear that in Lacan's logic, a new contribution to the question of *naturalness* in logic can be found.

Keywords Square of Opposition · Psychoanalysis · Paraconsistent logic

Mathematics Subject Classification (2000) Primary 03B53 · Secondary 03B65

1 Introduction

Beyond our trivial knowledge that nothing is permanent, or – in more philosophical words – nothing is universal, for Jacques Lacan this issue is expressed in his well-known and much-cited cryptical statement that there is no sexual relationship. In the 1960s and 1970s (cf. Guy Le Gaufey [15], p. 2–3, as well as in William J. Urban [16], p. 167), Lacan presented his *four formulas of sexuation* which should support the thesis that "Il n'y pas de rapport sexuel" (cf. [2] – Alain Badiou and Barbara Cassin titled their book with this thesis.). These four formulas, which are based

H. M. Schüler (✉)
University of Hagen, Hagen, Germany

J.-Y. Beziau, I. Vandoulakis (eds.), *The Exoteric Square of Opposition*, Studies in Universal Logic, https://doi.org/10.1007/978-3-030-90823-2_11

on predicate-logic, raise many questions, which, however, culminate in one *main question:*

(mq) What is the difference between Lacan's logic and the traditional logic?

Regarding (mq) and the procedure of the paper discussed in this chapter, I will not focus on the psychoanalytic background, but rather on the formal-logical part. Furthermore, I will only mention the formalized propositions of *the four formulas of sexuation* and make a few introductory remarks (Sect. 2) so that a comparison with the traditional logic becomes possible (Sects. 3, 4, and 5). Sections 3 and 4 prepare the answer of (mq) which is finally given in Sect. 5. Section 3 prepares the following argument: Lacan's formulas are compared with *the traditional (quantificational) square of opposition.* This comparison leads to something that is very reminiscent of Jacques Brundschwig's *maximal particular* (Sect. 4). Section 5 deals with Guy Le Gaufey's *Lacanian logical square* in order to ask (mq): What is the difference between Lacan's logic and the traditional logic? At this point, it is important to mention that Newton da Costa and Jorge Forbes have already identified three basic ways of applying logic to psychoanalysis ([21]). With an answer to (mq), however, in this chapter, I would like to transfer to the question of naturalness (Sect. 6).

2 The *Four Formulas of Sexuation*

Regarding Lacan's *four formulas of sexuation,* within this section, I will argue the following: After introducing these four formulas, I will give a few introductory remarks on the variable x, the predicate Φ, and both of them in connection (Φx). Finally, based on two quotes of Lacan (Q1 and Q2), I will focus on the aforementioned four formulas regarding their quantification.

According to Lacan, "the absence of the sexual relationship" ([12], p. 69) or the proposition that "there is no sexual relationship" ([11], p. 5) is supported by *the four formulas of sexuation:*

$(\exists x) \sim \Phi x$ $(\forall x)\ \Phi x$	$\sim(\exists x) \sim \Phi x$ $\sim(\forall x)\ \Phi x$

Fig. 1 The four formulas of sexuation. (Redrawn from Hubert Martin Schüler [4], p. 73)

Figure 1 shows a redrawn part of Lacan's tabular given in [12], p. 73. For better readability, I have slightly changed the notation.

At first, we should focus on the variable x in Fig. 1. According to Russell Grigg, there is a fundamental difference between Lacan's use of x and the common use of the variable in predicate logic.

> [I]n Lacan's formulas the variable, x, ranges only over those things that fall under Φ [...], whereas the formulas of the predicate calculus are formulated in such a way that the variables, x, y, z, etc., range over everything. ([8], p. 54)

In other words, Lacan's x does not range over everything, it denotes "speaking being[s]" ([8], p. 54). According to Grigg, Lacan's formula $(\forall x)\ \Phi x$ corresponds to the formula $(\forall x)\ (Gx \rightarrow Hx)$ and $(\exists x)\ \Phi x$ to $(\exists x)\ (Gx \wedge Hx)$ of predicate calculus (cf. [8], p. 53).

But what is the predicate of the predicator Φ? For Lacan, Φ signifies nothing else than the phallus, or more precisely, phallic enjoyment (cf. [12], p. 73–81). However, according to Ellie Ragland, Lacan refers – contrary to Sigmund Freud – with the meaning of the phallus *not* to the penis (cf. [18], p. 2). Or, as Lacan says in *Seminar XX:*

> I designate Φ as the phallus insofar as I indicate that it is the signifier that has no signified, the one that is based, in the case of man, on phallic jouissance. ([12], p. 81)

Finally, the connection between x and Φ, i.e., Φx, can be identified as Lacan's "phallic function" ([12], p. 73). He uses the concept *function* based on Gottlob Frege, of course with the mentioned restriction of the blank space, the variable x. Furthermore, it is important that Lacan's phallic enjoyment, or "phallic jouissance" ([12], p. 81), means a special kind of enjoyment which is a characterization or structuring of the castration complex, i.e., a structure of lack. Thus, Φx can be interpreted as "x (a sexual not specified individual) is structured by the castration complex."

But how are *the four formulas of sexuation* to be understood regarding their *quantification?* Lacan refers with his neologism *sexuation* to the process of assignment in a particular gender. Each speaking being assigns either to the left or to the right column of Fig. 1. The left column corresponds to man, the right one to woman.

First, let us focus on the left column, i.e., the two formulas of sexuation of man. The following quote explains the meaning of the left column in detail:

> (Q1) We'll start with the four propositional formulas at the top of the table, two of which lie to the left, the other two to the right. Every speaking being situates itself on one side or the other. On the left – the lower line – $(\forall x)\ \Phi x$ – indicates that it is through the phallic function that man as whole acquires his inscription [. . .], with the proviso that this function is limited due to the existence of an x by which the function Φx is negated [..]: $(\exists x) \sim\Phi x$. That is what is known as the father function – whereby we find, via negation, the proposition $\sim\Phi x$, which grounds the operativity [..] of what makes up for the sexual relationship with castration, insofar as that relationship is in no way inscribable. The whole here is thus based on the exception posited as the end-point [..], that is, on that which altogether negates Φx. ([12], p. 79–80)

(Q1) not only confirms the inscription of every speaking being in one column mentioned above but also gives an explanation of $(\exists x) \sim \Phi x$ and $(\forall x)\ \Phi x$. The following results:

upper left: $(\exists x) \sim\Phi x$:

"There is (at least) one x which is not structured by the castration complex."

lower left: $(\forall x)\ \Phi x$:

"All x are structured by the castration complex."

However, a similar structure in vertical direction between the upper and the lower formula can also be found on the right side of the table (cf. Fig. 1), the column of woman:

> (Q2) On the other side, you have the inscription of the woman portion of speaking beings. Any speaking being whatsoever, as is expressly formulated in Freudian theory, whether provided with the attributes of masculinity – attributes that remain to be determined – or not, is allowed to inscribe itself in this part. If it inscribes itself there, it will not allow for any universality – it will be a not-whole, insofar as it has the choice of positing itself in Φx or of not being there. ([12], p. 79–80)

According to Fig. 1 and (Q2) we can say:

upper right: ~(∃x) ~Φx:

"There is no x which is not structured by the castration complex."

lower right: ~(∀x) Φx:

"Not-all *(pas-tout)*x are structured by the castration complex."

Concerning Lacan's so-called *pas-tout,* there are many translations that can be found in the literature. However, these translations already concern (mq) which will be discussed in the next sections.

3 Lacan's Formulas in Comparison to *the Traditional Square of Opposition*

It is well known that there are many illustrations of *the traditional square of opposition* in the literature on logic. For our concerns, it seems helpful to expand the quantificational square with formulas, such as the following:

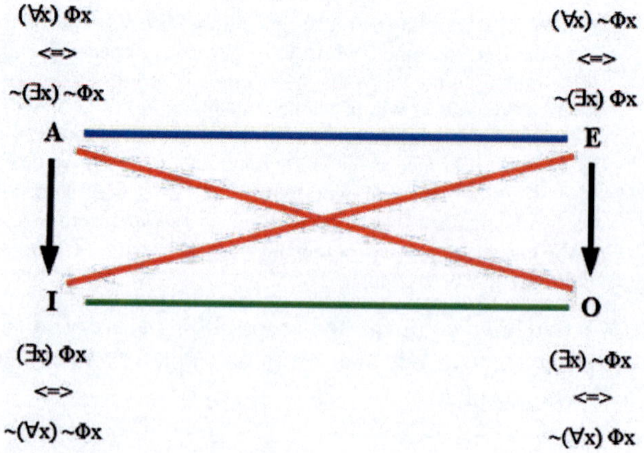

Fig. 2 The traditional (quantificational) square of opposition

Without going deeper into the various ways of drawing the square, there are basically four logical relationships, as we can see in Fig. 2. Figure 2 is based on the way of presentation introduced by Jean-Yves Béziau, cf. [5]: (1) contrary relationship between A and E (the blue line in Fig. 2); (2) contradictory relationship between A and O, and between E and I (the red line in Fig. 2); (3) subcontrary relationship between I and O (the green line in Fig. 2); and (4) implicatory relationship (subalternation) between A and I (A implies I), as well as between E and O (E implies O) (the black arrows in Fig. 2). In addition, each corner contains an equivalent formula, so that every proposition can be formalized with a ∀-quantifier *or* with an ∃-quantifier. For this chapter, I also use the following abbreviations: universal affirmative = UA, universal negative = UN, particular affirmative = PA, and particular negative = PN. However, I will *only* use these abbreviations if I refer to *the traditional (quantificational) square of opposition* (cf. Fig. 2).

Lacan's formulas obviously correspond only to the A- and O-corner: If we assign the four formulas of Lacan to the traditional square, then the A-corner corresponds to (lower left) $(\forall x)\ \Phi x$ <=> (upper right) ~ $(\exists x)$ ~ Φx, and the O-corner corresponds to (upper left) $(\exists x)$ ~ Φx <=> (lower right) ~ $(\forall x)\ \Phi x$. But this also means that in the formulas of Lacan, those formulas of the traditional square are not included, which have their place at the E- and I-corners: ~$(\exists x)\ \Phi x$ <=> $(\forall x)$ ~ Φx and $(\exists x)\ \Phi x$ <=> ~ $(\forall x)$ ~ Φx. Thus, in the logic of Lacan, there is no universalizable nothing as well as its contradictory, i.e., a particular something which affirms *not* a negation of a predicate. However, what that means has to be clarified in Sect. 5. Furthermore, *the four formulas of sexuation* describe only relationships of contradictions and equivalences.

Thus, if we recall (mq), a first superficial difference to the traditional logic comes up: First, *the four formulas of sexuation* describe no logical meanings of universal-negative (UN) and particular-affirmative (PA) propositions which do not affirm a negation of a predicate. Second, they contain only relationships of contradictions and equivalences. This fact is already reminiscent of another logical square, i.e., *the square of the maximal particular* by Jacques Brunschwig.

4 A "Source" of Lacan's Logic?

So far, Sect. 3 summarized the results of the comparison between Lacan's *four formulas of sexuation* and *the traditional (quantificational) square of opposition* and led to the assumption that Lacan refers to *the maximal particular.* This assumption is also confirmed by Grigg and Jacques-Alain Miller. Both identify the *maximal particular* – specially pursued by Jacques Brunschwig – "as the source of the pas-tout" ([8], p. 64).

Brunschwig's argument begins with the recognition of a fundamental problem in Aristotelian logic, i.e., the problem concerning the corners of particular propositions within *the traditional (quantificational) square of opposition.* Béziau examines this problem by dividing it into the I- and O-corner problems and proposes, with a reference to Robert Blanché, *the hexagon of opposition* (cf. [5]). For our concerns, it is sufficient to focus on the problems of the O-corner.

> (OP1) there is no primitive name of natural languages for the notion located at the O-corner.
>
> (OP2) tentative names used for the O-corner do not correspond to the meaning of the O-corner.
>
> (OP3) the notion located at the O-corner does not correspond to a natural notion of our thought. ([5], p. 8)

As said, Blanché and Béziau solve these problems by extending the traditional square. However, Brunschwig "argues that Aristotle came to realize that he had initially been misled by the working of natural language, and that this led to an internal problem" ([8], p. 62). With the introduction of *the minimal particular*, Aristotle already excluded *the maximum particular* as an important aspect of natural language. Grigg emphasizes that Brunschwig does not want to characterize the Aristotelian logic as being inconsistent, but he would notice that "it is one in which certain intuitions implicit in natural language have been disallowed specially in relation to particular statements" ([8], p. 62). Grigg summarizes that natural language, in the sense of *a usual meaning of the particular,* finds no place in *the traditional square of opposition.* This, so Grigg, can simply be shown by "three mutually inconsistent propositions" ([8], p. 62). (For better readability, I translate Grigg's propositions into predicate logic formulas):

> Three inconsistent axioms of the usual meaning of the particular (cf. Grigg [8], as well as Brunschwig [4]):
>
> a) $(\forall x)\ \Phi x \leftrightarrow \sim(\exists x)\sim\Phi x$
> b) $(\forall x)\ \Phi x \rightarrow (\exists x)\Phi x$
> c) $(\exists x)\Phi x \leftrightarrow (\exists x)\sim\Phi x$

Axiom a) and axiom b) are familiar to *the traditional (quantificational) square of opposition* (cf. Fig. 2): a) corresponds to the A-corner (with its equivalent proposition) and b) to the relationship between the A- and I-corners. But not axiom c): The two particular propositions reject the subcontrary relationship between the I- and O-corners of the quantificational square and claim to imply each other.

However, axiom c) can be explained with a simple example: Just imagine that we sit under an apple tree in the garden and someone in the house asks whether the tree carries red apples. So we look into the dense branches of the tree and though we did not take a close look at each apple, meaning we do not know if all apples are red, our answer might be: "Some apples are red.": $(\exists x)\ \Phi x$. But what do we actually mean by that? Usually we would not object to the person in the house if he or she understood: "Not all apples are red.": $\sim(\forall x)\ \Phi x$, which means: "Some apples are not red.": $(\exists x) \sim \Phi x$. This is what is meant with axiom c).

The three axioms are inconsistent because if it is true that *all* apples are red ($(\forall x)\ \Phi x$), it is also true that *some* apples are red ($(\exists x)\ \Phi x$) (cf. axiom b)). But according to axiom c) $(\exists x)\ \Phi x$ is equivalent to $(\exists x) \sim \Phi x$. So if someone claims that all apples are red ($(\forall x)\ \Phi x$), then he also claims that some apples are not red $(\exists x) \sim \Phi x$. This inconsistency can only be solved by discarding one of the three axioms.

By rejecting axiom c), *the traditional (quantificational) square of opposition* occurs; by rejecting axiom b), Brunschwig's *square of the maximal particular* comes up:

Fig. 3 Brunschwig's square of the maximal particular. (Redrawn from Hubert Martin Schüler [4], p. 7)

As we can see in Fig. 3, according to axiom c), Brunschwig draws an equivalency between the I- and O-corners. The diagonals were not affected by the three axioms, so they remain (as in the traditional square) contradictions. By rejecting axiom b), the implication becomes also a contradiction. Therefore, there is also a contradiction between the E- and O-corners. Otherwise, the inconsistency would be introduced once again. The consequence is that there is also an equivalency between the A- and E-corners, because both of them contradict both particular propositions which are equivalent.

Thus, why do we speak about a *maximum particular?* The answer is quite easy. According to Grigg, Brunschwig calls the particular corners of the traditional square the *minimal particular* because the truth of "Some apples are red" which means "At least one apple is red" does not exclude the possible truth of its antecedent "All apples are red." In contrast to the *minimal particular*, a *maximum particular* refers to the issue that the truth of "Some apples are red" and "Not all apples are red" contradicts every truth of a universal proposition, no matter whether affirmative or negative. So, if we say "Some apples are red," we mean "At least and at most some apples are red" (cf. [8], p. 63).

In comparison with Lacan's *four formulas of sexuation*, we have seen that they only refer to the formulas of the A- and O-corners of *the traditional (quantificational) square of opposition* (cf. Sect. 3). The consequence was that we are only dealing with contradictions and equivalences, just like Brunschwig's *square of the maximal particular.* The difference being that for Brunschwig – in contradiction to Lacan – the universal-negative (UN) and the particular-affirmative (PA) are included. The question of how Brunschwig interprets the UN and the PA in his square must be excluded here. However, for him as well as for Lacan, there is a basic consequence for the relationship between the particular and the universal level: Whether affirmative or negative, they are contradictions. In relation to Lacan's *four formulas of sexuation,* however, this led to confusion among the commentators.

Grigg – for example – interprets Badiou in agreement with himself and concludes that Lacan's ~ $(\forall x)\ \Phi x$ has "to be taken: no woman comes entirely under the phallic function" ([8], p. 57). Guy Le Gaufey, on the other hand, contradicts this conclusion:

> Here one must get rid of the idea that they would not entirely satisfy it (i.e. that a woman would never, as such, be would entirely taken up into the phallic function, etc. etc.) and that this would be the reason why Lacan would mark them with his not-all. ([15], p. 29)

This confusion manifests in the fact that Lacan assigns supposedly equivalent formulas different meanings. Why does he write $(\forall x)\ \Phi x$ if he could also write $\sim(\exists x) \sim \Phi x$? Why does he write $\sim(\forall x)\ \Phi x$ if he could also write $(\exists x) \sim \Phi x$? So, if we are not dealing with equivalents in terms of *the traditional (quantificational) square of opposition,* what are the relationships between these formulas? But this question affects the meaning of the quantifiers. In summary, there is a different meaning in Lacan's logic concerning its basic elements: the quantifiers, the negation (contradiction), and the equivalence (implication). But this reiterates (mq): What is the difference between Lacan's logic and the traditional logic?

5 Le Gaufey's *Lacanian Logical Square*

As we have seen in the last section, quantifiers, negation (contradiction), and equivalence (implication) have different meanings in Lacan than in the *traditional (quantificational) square of opposition* (cf. Fig. 2). In this section, these aspects will be pointed out, and my intention is to answer (mq).

Le Gaufey argues that there is nothing wrong by interpreting Lacan's *four formulas of sexuation* as a logical square (cf. [15], p. 32). Urban agrees with Le Gaufey and calls it *the Lacanian logical square* (cf. [16], p. 169).

Fig. 4 Le Gaufey's Lacanian logical square. (Redrawn from Hubert Martin Schüler, cf. [15], p. 32)

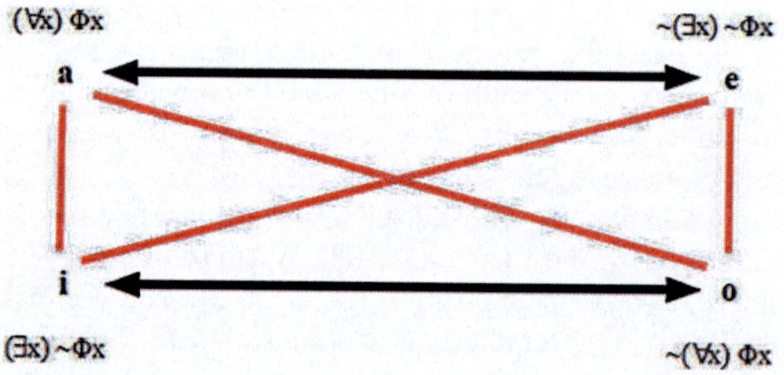

As we can see in Fig. 4, Le Gaufey adopts the logical relationships of the square of Brunschwig while replacing the formulas of the E- and O-corners with $\sim(\exists x) \sim \Phi x$ and $\sim (\forall x)\ \Phi x$ (cf. Sect. 3 and Fig. 3).

However, *the Lacanian logical square* raises questions that basically culminate in (mq). We are well advised to clarify this question, among other reasons because we have seen that we need to understand the logical relationships given in the square in different ways. If we want to understand the fundamental difference between Lacan's logic (i.e., *the Lacanian logical square* – cf. Fig. 4) and the traditional logic (i.e., *the traditional (quantificational) square of opposition* – cf. Fig. 2), it seems practical to form and follow a path of questions concerning *the four formulas of sexuation* in the sense of *the Lacanian logical square* (cf. Fig. 4). Although many paths lead to the same goal, I suggest the following questions:

1. Why do the positions of *the four formulas of sexuation* change as soon as they are put into a logical square?
2. What is the difference between Lacan's logic and the traditional logic, concerning the following:

   (a) The a-corner
   (b) The negation (contradiction)
   (c) The quantifiers: the ∀-quantifier and the ∃-quantifier
   (d) The o-corner
   (e) The equivalence (implication)
   (f) The i-corner
   (g) The e-corner

For the sake of readability, I will use lower case letters for the corners of *the Lacanian logical square* (as in Fig. 4: a, e, i, o) and uppercase letters for *the traditional (quantificational) square of opposition* (as in Fig. 2: A, E, I, O).

5.1 *Why Do the Positions of the* **Four Formulas of Sexuation** *change as Soon as They Are Put into a Logical Square?*

Comparing Le Gaufey's *Lacanian logical square* (cf. Fig. 4) with Lacans four formulas (cf. Fig. 1), it is obvious that the two formulas on the left ($(\forall x)\ \Phi x$ and $(\exists x) \sim \Phi x$) have switched positions. However, this problem is merely a bogus problem. Le Gaufey reverses these positions simply to improve readability. He is concerned with removing his square from the others as little as possible in an intention to make comparisons better. Urban goes a step further and exchanges the whole "man-column" with the "woman-column." He justifies his exchange with the claim that the formulas of woman are more primary. According to Urban, one would usually read from left to right, and thus, we should start with these formulas (cf. [16], p. 170). To what extent one can say that the formulas of woman are to be considered more primary cannot be discussed at this point yet. However, in comparison with other logical squares, the square of Le Gaufey seems – at least at this point – more workable.

5.2 *What Is the Difference Between Lacan's Logic and the Traditional Logic?*

(a) The difference between the a-corner (cf. Fig. 4) and the A-corner (cf. Fig. 2): First, it immediately becomes clear that the equivalent relationship between the formulas of the A-corner cannot be asserted in the a-corner. In Sect. 5.2.e, this issue will be explored. For now, we can only repeatedly say that in the a-corner, the relationship between $(\forall x)\ \Phi x$ and $\sim(\exists x) \sim \Phi x$ has to be understood in a different way than the traditional equivalence. Otherwise, there is no difference between them, at least so far. In both squares, the a- or A-corner means $(\forall x)\ \Phi x$ ("All x are Φx.", or in the sense of Lacan, and of course in considering the definition of *x* in Lacan (cf. Sect. 2), "All x are structured by the castration complex.").

(b) The different meaning of negation (contradiction): The more interesting point is that after starting our path of questions with the UA (cf. Sect. 5.2.a – $(\forall x)\ \Phi x$), according to Le Gaufey, Lacan establishes his further formulas only via *redoubled negation* (cf. [15], p. 34). Thus, there is a difference between the logical operation of negation (contradiction) in *the traditional (quantificational) square of opposition* and *the Lacanian logical square.* What is this difference? According to Slavoj Žižek, this important logical operation in Lacan is reminiscent of Georg Wilhelm Friedrich Hegel's *double negation* or *the negation of the negation.* Žižek points out that in *Seminar XX,* Lacan gives a "new definition" of Hegel's *negation of the negation* ([17], p. 86). According to Hegel, "the negation of the negation is something positive" ([9], p. 108) ("(S)o einfach die Einsicht ist, (. . .) daß die Negation der Negation Positives ist (. . .).). But Lacan's *redoubled negation* does not return to any kind of positivity. Of course, the special feature in Hegel's *double negation* is that the positivity unceasingly changes. This is not the point. Apart from the fact that Hegel would not only disagree to logical formalizations, the difference between his *double negation* and *redoubled negation* is that the latter does not mean that one can conclude by the first negation of $(\forall x)\ \Phi x$ to $\sim(\forall x)\ \Phi x$, while the second negation of $\sim(\forall x)\ \Phi x$ leads again to $(\forall x)\ \Phi x$. In this case, $(\forall x)\ \Phi x$ would then imply $(\exists x)\ \Phi x$ (cf. *the traditional (quantificational) square of opposition* – Fig. 2). However, in Lacan, "two quite different forms of negation" ([11], p. 17) are intertwined. If we negate $(\forall x)\ \Phi x$, will we conclude $\sim(\forall x)\ \Phi x$ (the o-corner) or $(\exists x) \sim \Phi x$ (the i-corner) – or both? In the sense of *the traditional (quantificational) square,* $\sim(\forall x)\ \Phi x$ and $(\exists x) \sim \Phi x$ would be equivalent (cf. Fig. 2) – not in the square of Le Gaufey, at least in that traditional meaning of equivalence. According to him, the "diagonal" negation leads to $\sim(\forall x)\ \Phi x$, which means that "[o]ur *not-all* is discordance" ([11], p. 17). On the other hand, the "vertical" negation of $(\forall x)\ \Phi x$ leads to $(\exists x) \sim \Phi x$. However, the i-corner is not discordance, it is "foreclosure" ([11], p. 17). Lacan explains the difference:

> But what is foreclosure? Assuredly it is to be placed in a different register to that of discordance. It is to be placed at the point at which we have written the term described as *function.* Here is formulated the importance of the said *(du dire). The only foreclosure is of the said,* of this something that exists – existence being already promoted to what assuredly (. . .) we have to give it as a status – that (..) something can be said or *not.* ([11], p. 17)

In summary: The interwined "*two types* of contradiction" ([15], p. 42) of $(\forall x)\ \Phi x$ lead on the one hand to $\sim(\forall x)\ \Phi x$ (negation as discordance) which means not that there *is* or *there exists a something.* That is the reason why Lacan places $\sim(\forall x)$ ("not all x") instead of $(\exists x)$ ("there is a x"). On the other hand, the second type of contradiction as foreclosure leads to existence: $(\exists x) \sim \Phi x$.

In the structure of Le Gaufey's *Lacanian logical square,* a problem arises: The two types of contradiction distinguish between existence and non-existence and combine the two in the sense of equivalence. Thus, there are equivalences between the a- and e-corners, as well as between the i- and o-corners. However, these equivalences are not to be understood in terms of equivalent relationships like in *the traditional (quantificational) square of opposition.* Insofar, the question of how *the Lacanian logical square* does not break down shifts first to the place of the different meaning of equivalent relationship and second to the question of the meaning of existence (cf. Sect. 5.2.e).

Furthermore, it is not only the mentioned aspect that strains *the Lacanian logical square.* For we are dealing with two different sets, which in turn pass on the question to the Lacanian meaning of equivalence. Le Gaufey describes it like this:

> It nevertheless remains that, in every case, when we write that any element whatsoever belongs to a determined set, we posit such a set as existing. No ∀ without the set that it is supposed to cover. And if no set... ([15], p. 16)

What Le Gaufey addresses here is nothing else than Lacan's "weld" in the midst of set-theory, formalized in his logical square by the *redoubled negation* on both horizontal planes as equivalence. Since this section only aims to explain the negation, it is necessary to stop at the mere naming of the problem and refer to Sect. 5.2.e. Again: *Whether and, if so, how existence in the whole Lacanian logical square is to be understood, ultimately depends on the question of equivalence.* So far, we have only dealt *in isolation* with the left and the right sides of the square.

Regardless of the unresolved problems, it makes more sense to pursue the mentioned structured path. We first started in the upper left corner of *the Lacanian logical square* and then turned to the negation in the diagonal, the o-corner follows. But before that, one has to consider the effects of the *redoubled negation* on the meaning of quantifiers themselves.

(c) The different meaning of the quantifiers: the ∀-quantifier and ∃-quantifier: Basically, the difference or the change of the meaning of the quantifiers by the *redoubled negation* already became clear in the last section. While in *the traditional (quantificational) square of opposition* the negation of the formula $(\forall x)\ \Phi x$ or its equivalent formula $\sim (\exists x) \sim \Phi x$ simply leads to $\sim (\forall x)\ \Phi x$ or $(\exists x) \sim \Phi x$ (cf. Fig. 2), the *redoubled negation* in Le Gaufey's *Lacanian logical square* affirms a different meaning of equivalence in distinguishing existence on the left ($(\forall x)\ \Phi x$ and $(\exists x) \sim \Phi x$) from non-existence on the right ($\sim(\exists x) \sim \Phi x$ and $\sim (\forall x)\ \Phi x$) columns of the square (cf. Fig. 4). As already mentioned above, this refers to the topic of set-theory and will be discussed in the context of the question of equivalent relationships (cf. Sect. 5.2.e). However, the difference between the meaning of the Lacanian quantifiers and the traditional use is not that the ∀- quantifier or ∃- quantifier would change their fundamental meaning. They change their meaning if they get negated: $\sim\forall x$ and $\sim \exists x$ no longer claim that any x exists, on the contrary:

> Starting from the negation brought to bear on all women, Lacan concludes to the inexistence of the woman as a strictly symbolic entity, and by there alone there vanishes the possibility of writing a relationship between an entity possessing a set of values to be covered (men) and another which does not possess such a range of Fregian values (women). ([15], p. 22–23)

This quote of Le Gaufey offers two interesting aspects: a first aspect for Lacan's argument that there is no sexual relationship. For how should these two essentially different values relate? Thus, there is no sexual relationship. Second, does this mean that the "man-column" correspond to existence, while the "woman-column" do not? Again, this issue is addressed.

(d) The difference between the o-corner (cf. Fig. 4) and the O-corner (cf. Fig. 2): Here, several aspects are simultaneously pressing for clarification – but one after another: Summarized, there are three logical relationships concerning the o-corner ($\sim(\forall x)\ \Phi x$): First, the two contradictions: the negation of the a-corner ($(\forall x)\ \Phi x$) and the negation of the e-corner ($\sim(\exists x) \sim \Phi x$), and second, the equivalence to the i-corner ($(\exists x) \sim \Phi x$). First, concerning the two contradictions: On the position of the o-corner, we do not say that there exists at least one which is not, which is equivalent to the proposition that not all are so, in the sense of existence. On the one hand, we negate the existence of $(\forall x)\ \Phi x$ in the sense of discordance, and on the other hand, the non-existence of $\sim(\exists x) \sim \Phi$x in the sense of foreclosure. The latter negation is perhaps the more interesting because of its contradiction of non-existence. Or, like Žižek perhaps would say: That which does not exist anyway is negated so that it exists even less than before. With this logical operation, Lacan brings that psychoanalytic structural moment into the whole context of his logic what he calls the real. Second, the question of equivalence is again asked at this point. Just like the a-corner, the o-corner does not contain the equivalent formula expressed with the quantifier of existence. The Lacanian change of meaning concerning the ∀- and ∃- quantifiers has already been described in the former section (cf. Sect. 5.2.c). The o-corner (i.e., the Not-all ($\sim(\forall x)\ \Phi x$)) negates two propositions: First, it negates as discordance the existence of $(\forall x)\ \Phi x$. Second, it negates as foreclosure the non-existence of $\sim(\exists x) \sim \Phi x$. However, the Not-all implies existence: $(\exists x) \sim \Phi x$ (cf. Fig. 4). Unfortunately, the question of how we get logically from the o- to i-corner via implication and vice versa (so that we can speak of an equivalent relationship and of existence or non-existence) is not answered yet.

(e) The different meaning of equivalence (implication): Here we turn to the question of how the logical implication on the particular level between the o- and the i-corners and vice versa can be understood in the sense of the square of Le Gaufey (cf. Fig. 4). This also involves the question of existence and non-existence. Of course, the same question arises at the universal level. However, according to our path of questions, we finally ask this question concerning the particular level. We have already seen that Lacan – contrary to the Aristotelian approach – entangles two fundamentally different values, thus shaking the symmetry of the traditional square. Again, where does this asymmetry come from? The different values that we have described so far in terms of existence (left side, the "man-column") and non-existence (right side, the "woman-column") can be explained with the help of set-theory. When we talk about an "all" in terms of the "man-column" ($(\forall x)\ \Phi x$), we assume that there also *exists* the aforementioned set. But if that set exists, then we can pick out an *existing* element or an instance as well. The exception ($(\exists x) \sim \Phi x$) confirms the rule ($(\forall x)\ \Phi x$) via *redoubled negation.* This applies also to the "woman-column": The exception ($\sim(\forall x)$

Φx) confirms the rule (~($\exists x$) ~ Φx) via *redoubled negation*. Here, however, we are dealing with other values. Because "Not-all" negates what does not exist anyway, so we have no existing set with corresponding elements in front of us. That was the reason that led Lacan to say that there is not "the woman." In contrast to the man-side, where the rule is confirmed by exception, thus establishing a symbolism, the Not-all receives the absence of the rule by negation. Not only does Lacan incorporate Bertrand Russell's paradox that there exists no set that contains itself, but Lacan points out that this fact is primarily to be considered. Urban emphasizes this assumption by interchanging the "man-column" (Urban and Le Gaufey call both the formulas of the "man-column" *right deixis*) with the "woman-column" (the *left deixis* in Urban and Le Gaufey) as described above (cf. Sect. 5.1) and also refers to the logic of Gottlob Frege:

> Given that Frege's system leads to such a paradox, the conclusion must be that it cannot be logically sound. So the extensional approach must be seen as taking logical priority to the intensional approach and what makes the Lacanian logical square that much more complete is that it inscribes *both* these approaches. In contrast to the right deixis [the left side of Le Gaufey's *Lacanian logical square* (cf. Fig. 4)] where sets exist, the left deixis [the right side of Le Gaufey's *Lacanian logical square* (cf. Fig. 4)] has priority through its inscription of the deeper truth that some sets do not exist. ([16], p. 175)

So, how should the values of the "woman-column" be characterized? So far, we have been talking in a quite simplistic way about mere non-existence. However, this characterization is not accurate. Because if we are not dealing with a set or a universal, then there exists the non-existence of essence. Nothing else is expressed by the formulas on the right. However, according to Urban, there are consequences for both columns of *the Lacanian logical square:*

> In contrast to the classical logical square, running across the top of the Lacanian square are propositions which deeply damage the universal, for in *both* partitions the universal simply cannot collectivize all the elements which would give rise to a homogeneous unity without exception. ([16], p. 181)

It seems, then, that apart from the mentioned asymmetry in the values of the left and the right, one has to define the gap between the universal and its exception. To find an answer, Le Gaufey addresses Lacan's *L'étourdit.* In this chapter, the latter introduces the term of *limitation.* Lacan turns to the relationship of ($\forall x$) Φx and ($\exists x$) ~ Φx, or, in other words, how the particular affirms the universal. According to Lacan, ($\forall x$) Φx means, "for all x, Φx is satisfied" ([10], p. 56), and ($\exists x$) ~ Φx means, "there is by exception the case, familiar in mathematics (the argument x = 0 in the exponential [*sic*] function 1/x), the case where there exists an x for which Φx, the function, is not satisfied, namely, by not functioning, is in effect excluded" ([10], p. 56). Urban adds a *value table* as an example to the drawing of a *hyperbolic function* by Le Gaufey:

Le Gaufey and Urban basically agree in their explanations regarding *the hyperbolic function,* i.e., the graph in Fig. 5. As we can see, the line of the graph never crosses the y- or the x-axes, even if we would always extend the values of the adjacent *value table.* The value $x = 0$ is, like the exception (($\exists x$) ~ Φx), never reached. "In a word,

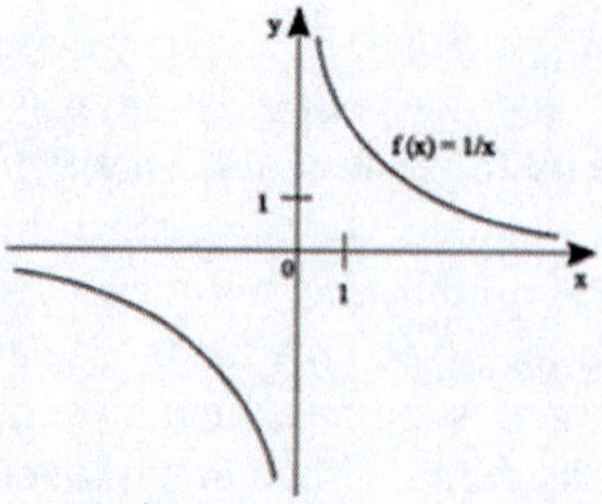

x	-2	-1	-1/2	-1/4	1/4	1/2	1	2
f(x)	-1/2	-1	-2	-4	4	2	1	1/2

Fig. 5 Hyperbolic function and value table. (Redrawn from Hubert Martin Schüler, cf. [16], p. 182)

division by 0 is undefined" ([16], p. 182). However, if the axes never intersect with the line of the graph, how should one know if this undefined belongs to a set or not? This question also drives Urban:

> There is an undecidability: if the submission of All x under the f(x) is conditioned by the fact that at least one escapes it, should this escaped $x = 0$ be counted amongst the All or not? ([16], p. 182)

So again we come to the question of equivalence. If we know that not all x are Φ, then how can we say that there *exists* an x which is not Φ, or formalized: $\sim(\forall x)\ \Phi x \rightarrow (\exists x) \sim \Phi x$ (as well as $(\exists x) \sim \Phi x? \rightarrow \sim(\forall x)\ \Phi x)$? Urban suggests a simple but plausible solution because "[a] choice must be made" ([16], p. 186). He simply asks the question of what condition must be fulfilled that a logical inference can be done. If we only have the aforementioned implications $(\exists x) \sim \Phi x \rightarrow \sim(\forall x)\ \Phi x$ as well as $\sim(\forall x)\ \Phi x \rightarrow (\exists x) \sim \Phi x$, how can we get to $\sim(\forall x)\ \Phi x$ or to $(\exists x) \sim \Phi x$? – only when we as subjects discover one ($(\exists x) \sim \Phi x$) or the other ($\sim(\forall x)\ \Phi x$) antecedent of one of the implications as a second premise and apply the well-known elemental inference *modus ponens*. Because, as Urban argues, "without the actualization of the antecedent this knowledge remains unrealized" ([16], p. 191). The existence of a rule is thus not only borne by its exception, but also the existence of this exception must be recognized. Without the act of knowing of the child in Hans Christian Andersen's fairy tale *The Emperor's New Clothes* that the emperor is naked, there is no concept of swindle.

(f) The difference between the i-corner (cf. Fig. 4) and the I-corner (cf. Fig. 2): How can the difference between the i-corner and the traditional I-corner be summarized from what has been said so far? Le Gaufey replaces – of course in the sense of Lacan – the traditional formula of $(\exists x)\ \Phi x$ with the formula of $(\exists x) \sim \Phi x$. So we are no longer talking about an existing thing with a certain essence. Such an existence is impossible for Lacan. In other words, the green of the apple we hold in our hand is not the green we characterize it with. But only because of what we see is not *the* green, we can call it green. So, if we could not relate to anything outside, we could not refer to a set. The exception confirms the rule via *redoubled negation* (cf. Sect. 5.2.b), but without crossing the line of zero (cf. Sect. 5.2.e). $(\exists x) \sim \Phi x$ confirms that there *exists* a set $((\forall x)\ \Phi x)$, but at the same time, it also confirms that if this set exists for an appreciative subject, there is an open outside $({\sim}(\exists x) \sim \Phi x)$. This opening is the prerequisite for assuming that point of view at all.

(g) The difference between the e-corner (cf. Fig. 4) and the E-corner (cf. Fig. 2): Finally, we take a quick look at what says of itself that there is nothing that is not somehow: $({\sim}(\exists x) \sim \Phi x)$. First of all, it must be said that similar to the i-corner, the Aristotelian negation of a certain set $((\forall x) \sim \Phi x)$ has to be replaced with the negation of an existence that is not somehow. The existence of this "Non-Aristotelian-rule," which has a different meaning than the A-corner of the latter (cf. Figure 2), is confirmed by its exception: ${\sim}(\forall x)\ \Phi x$, and – again: if a subject recognizes its existence and concludes $(\exists x) \sim \Phi x$. The difference between the Aristotelian and the Lacanian meaning of ${\sim}(\exists x) \sim \Phi x$ is that unlike the former there is no set that would imply the existence of an essence. The Lacanian meaning overcomes the limit of existence, but not just on universal level. The *pas-tout* confirms as particular symbolic proposition the open background of the concept of existence.

6 The Naturalness of Lacan's Logic

As we have seen with Blanché, Béziau, and Brunschwig, various problems arise in *the traditional square of opposition* (cf. Sect. 4). These problems basically depend on the topic of naturalness of logical reasoning and language itself. In this section, I will address the issue of naturalness in the context mentioned because in Lacan's logic, we can find a new contribution to this. I would like to focus on two main aspects: First, that Lacan contributes to the debate devoted to the question of what is natural logic. Second, one can see in Lacan's logic of incompleteness an argument for the fact that all logic in the Aristotelian tradition is fragmented insofar as it builds on incompleteness.

Many authors and commentators of the Aristotelian logic and its following tradition discuss – of course not always in an explicit sense – the term of naturalness. For example, Arthur Schopenhauer argues that only with the fourth figure of the Aristotelian logic, unnaturalness is introduced (cf. [19, 20]). Thus, Schopenhauer made a contribution to natural deduction already in the nineteenth century. Today, we find contributions to this topic reaching, e.g., from George Lakoff (cf. [14]) to Johan van Benthem (cf. [3]). Another debate also discusses how to understand the Aristotelian syllogistic. Are we dealing here with some kind of natural deduction? Jan Łukasiewicz and Günter Patzig contradict this thesis and argue for an axiomatic interpretation (cf. [1, 13]). For John Corcoran – for example – "Aristotle's syllogistic is an underlying logic which includes a natural deductive system [...]" ([6], p. 85). Of course, the question how Aristotle should be interpreted is not the question of this chapter. It is all about showing that one has always tried to connect logic with naturalness and Brunschwig's concept of it starts, as shown in Sect. 4, much earlier: With the introduction of the *minimal particular*, Aristotle already excluded the *maximum particular* as an important aspect of *natural* language.

At least, one might have to admit that incompleteness is closer connected to naturalness, not only because Lacan confronted Aristotle with Russell's Paradox. Isn't it appropriate to speak of naturalness in logic only if one necessarily presupposes a subject who also recognizes existence?

> Psychoanalytic ethics goes further. In deciding never to organize individuals or observations under a concept, namely, by deciding that everything we say remains particular, is a maximal particular, and implies also its counterpart in the expression (...), we will confer on existence the power of escaping any concept by which we might have believed we could corner it. ([7], p. 4)

References

1. Patzig, G.: Aristotle's Theory of the Syllogism. Dordrecht, Holland (1968)
2. Badiou, A./ Cassin, B.: Es gibt keinen Geschlechtsverkehr, Ed. by M. Coelen and F. Ensslin, Diaphanes, Zürich (2012)
3. Benthem, J.: A Brief History of Natural Logic. In M. Chakraborty, B. Löwe, M. N. Mitra, S. Sarukkai (ed.), Logic, Navya-Ny¯aya & Applications: Homage to Bimal Krishna Matilal. London (2008) 21–42
4. Brunschwig, J.: La proposition particulière et les preuves de nonconcluance chez Aristote, in: Cahier pour l'Analyse 10, La formalisation, Éditions du Seuil, Paris (1969)
5. Béziau, J.-Y.: The Power of the Hexagon, in: Logica Universalis (vol. 6, issue 1-2, pp. 1-43), Springer, Basel (2012) https://link.springer.com/article/10.1007/s11787-012-0046-9
6. Corcoran, J.: Aristotle's natural deduction system, in J. Corcoran (ed.), Ancient Logic and Its Modern Interpretations. Dordrecht-Holland (1974) 85-131
7. Fierens, C.: The fact of saying notall with reference to Le Gaufey's work: Lacan's notall, logical consistency, clinical consequences, trans. by Cormac Gallagher, no pagination, (2019) http://www.lacaninireland.com/web/wp-content/uploads/2010/06/Autumn-2008-103-121-1.pdf
8. Grigg, R.: Lacan and Badiou: Logic of the pas-tout, in: Filozofski vestnik (vol. 26, no. 2, pp. 53-65), Deakin Research Online (2005) http://hdl.handle.net/10536/DRO/DU:30003276
9. Hegel, G. W. F.: Wissenschaft der Logik I, Gesammelte Werke 5, Ed. by. E. Moldenauer and K. M. Michel, Suhrkamp, Frankfurt (1986)
10. Lacan, J.: L'étourdit, in: The letter 41 (2009) 31-80, trans. by Cormac Gallagher, (2019) https://www.valas.fr/IMG/pdf/EtourditJL-etourdit-CG-Trans-Letter-41.pdf
11. Lacan, J.: The seminar of Jacques Lacan, Book XIX 1971- 1972, trans. by Cormac Gallagher, (2019) http://www.lacaninireland.com/web/wp-content/uploads/2010/06/Book-19-Ou-pire-Or-worse.pdf
12. Lacan, J.: Encore, the seminar of Jacques Lacan, Book XX 1972- 1973, Ed. by Jacques-Alain Miller, W. W. Norton & Company Inc. (1975)
13. Łukasiewicz, J.: Aristotle's Syllogistic: From the Standpoint of Modern Formal Logic, 2nd ed., Oxford (1957)
14. Lakoff, G.: Linguistics and Natural Logic, Synthese 22 (1970-71), 151–271.
15. Le Gaufey, G.: Towards a critical reading ot the formulae of sexuation, trans. by Cormac Gallagher, (2019) http://www.lacaninireland.com/web/wp-content/uploads/2010/06/TOWARDS-A-CRITICAL-READING-2506.pdf
16. Urban, W. J.: Sexuated Topology and the Suspension of Meaning: A Non-Hermeneutical Phenomenological Approach to Textual Analysis, Toronto (2014) https://yorkspace.library.yorku.ca/xmlui/bitstream/handle/10315/27706/

Urban_William_J_2014_Phd.pdf?sequence=2&isAllowed=y
17. Žižek, S.: Der erhabenste aller Hysteriker – Psychoanalyse und die Philosophie des deutschen Idealismus, Turia & Kant, Wien-Berlin (1992)
18. Ragland, E.: The Logic of Sexuation – From Aristotle to Lacan, State University of New York Press (2004)
19. Schopenhauer, A.: Philosophische Vorlesungen, Vol. I. Ed. by F. Mockrauer. (= Sämtliche Werke. Ed. by Paul Deussen, Vol. 9). München (1913)
20. Schüler, H. M./ Lemanski, J.: Arthur Schopenhauer on Naturalness in Logic. In: Language, Logic and Mathematics in Schopenhauer. Birkhäuser. Cham (2019)
21. Forbes, J./ da Costa, N.: sobre psicanálise e lógica. FALO, Revista Brasileira do Campo Freudiano, 1, (1987), 103-111. www.jorgeforbes.com.br/assets/files/Artigos/Sobre-Psicanalise-e-Logica-JF-e-NC_002.pdf

On Modal Opposition Within Some Modal Discussive Logics

Krystyna Mruczek-Nasieniewska and Marek Nasieniewski

Abstract We consider the issue of the square of opposition for two modal extensions of the discussive logic $\mathbf{D_2}$. To this aim we recall some basic information on discussive logic, but also mention some facts concerning the mentioned extensions of $\mathbf{D_2}$. Our idea is to extend the discussive language with modalities, which although are considered in the context of the discussive logic, but are used only auxiliarily and are absent from its object language.

Keywords Discussive logic · Modal logic · Discussive modal logic · Jaśkowski · Extensions of the discussive logic · The square of opposition

Mathematics Subject Classification (2000) Primary 03B53; Secondary 03A05 and 03B45

1 Introduction

Jaśkowski proposed a logical calculus that could be applied to inconsistent systems but would not result in their overfilling. Jaśkowski expressed his calculus with the use of the modal logic **S5**, so, via standard facts, in classical (quantifier) logic (see [2, p. 55]). The aim was to obtain a system that would not be overfilled, that is, would not lead in general to the set of all expressions when applied to inconsistent set of premisses. Additionally, two requirements were stipulated as regards the resulting calculus, that it '(2) would be rich enough to enable practical inference, (3) would

The authors of this work benefited from support provided by the Polish National Science Centre (NCN), grant number 2016/23/B/HS1/00344

K. Mruczek-Nasieniewska (✉) · M. Nasieniewski
Department of Logic, Institute of Philosophy, Faculty of Philosophy and Social Sciences, Nicolaus Copernicus University, Toruń, Poland
e-mail: mruczek@umk.pl; Krystyna.Mruczek-Nasieniewska@umk.pl; mnasien@umk.pl

J.-Y. Beziau, I. Vandoulakis (eds.), *The Exoteric Square of Opposition*, Studies in Universal Logic, https://doi.org/10.1007/978-3-030-90823-2_12

have an intuitive justification' [2, p. 38]. When proposing his solution, he used a model of discussion. During a discussion inconsistent opinions can be formulated, but participants of this discussion as well as external observers are not inclined to deduce every sentence from the set of statements presented in the discussion. Hence, in the model, two types of point of view are considered: internal ones of particular participants of the discussion and an external one of some observers.

Jaśkowski proposed to express some interactions that are taking place between participants, in particular, he proposed connectives of the so-called discussive conjunction and discussive implication (as well as definable discussive equivalence) that were used to represent some interactions that can take place during a discussion. From the point of view of a given debater, the statements of other participants of the discussion have to be differentiated from the debater's own statements. The former is marked by the possibility operator—'$\Diamond$'. The justification for the use of the modal operator is that from debater's point of view, the statements of others do not have to be true. The possibility operator can be treated according to Jaśkowski as saying [2, p. 43]:

> 'in accordance with the opinion of one of the participants in the discourse'.

Next to modalities corresponding to debater's evaluations of statements made by other debaters, in his definition, Jaśkowski also includes a point of view of an external observer. This way of considering other statements is also expressed by possibility operator. It can be seen as expressing the fact that from the point of view of the 'impartial arbiter' (see [1, p. 149], English version) all opinions presented during the discussion are only possible. So ([1, p. 149], English version),

> "if a thesis is recorded in a discursive system, its intuitive sense ought to be interpreted so as if it were preceded by the symbol *Pos*".

Hence, the modal possibility operator is applied in the definition of $\mathbf{D_2}$ on two levels, but none of these two uses is explicitly saved in the resulting language of the logic $\mathbf{D_2}$. So, although $\mathbf{D_2}$ is connected with a modal logic—the logic **S5**, $\mathbf{D_2}$ has got neither $\Diamond$ nor $\Box$ in its language. One can consider an extension of $\mathbf{D_2}$ with the help of modal operators of $\Diamond_{\mathrm{d}}$—possibility and $\Box_{\mathrm{d}}$—necessity (see [6]). The obtained logics is denoted as $\mathbf{mD_2}$. Similarly as the logic $\mathbf{D_2}$, $\mathbf{mD_2}$ is defined by translations referring to the modal logic **S5**. Semantic conditions for $\Diamond_{\mathrm{d}}$ and $\Box_{\mathrm{d}}$ are standard. However, the resulting modal logic as a whole behaves rather in a non-standard way. In the present paper we consider the issue of the square of opposition for this logic. As a result of the given analysis, we consider also a modified version of this extension, where in the discussive model we allow an influence of some general community. For this aim we consider additional modalities.

2 Basic Notions and Facts

2.1 Standard Modal Formulas

Modal formulas are formed standardly from propositional variables: 'p', 'q', 'p_0', 'p_1', 'p_2', ...; truth-value operators: '$\neg$', '$\vee$', '$\wedge$', '$\rightarrow$', and '$\leftrightarrow$' (connectives of negation, disjunction, conjunction, material implication, and material equivalence, respectively); modal operators: the necessity symbol '$\Box$' and the possibility symbol '$\Diamond$'; and the brackets. By $\mathrm{For_m}$ we denote the set of all modal formulas.

The set $\mathrm{For_m}$ includes the set of all classical formulas. Let **Taut** be the set of all classical tautologies. Besides, for any $\varphi, \psi, \chi \in \mathrm{For_m}$, let $\chi[{}^{\varphi}/_{\psi}]$ be any formula that results from χ by replacing one, none, or more than one occurrence of φ, in χ, by ψ.

As usually, modal logics are sets of formulas. By a *modal logic* we mean a set $\boldsymbol{L}$ of modal formulas satisfying following conditions:

- **Taut** $\subseteq \boldsymbol{L}$,
- $\boldsymbol{L}$ includes the following set of formulas

$$\left\{ \ulcorner \chi[{}^{\neg\Box\neg\varphi}/_{\Diamond\varphi}] \leftrightarrow \chi \urcorner \ : \ \varphi, \chi \in \mathrm{For_m} \right\}. \qquad (rep^{\Box})$$

- $\boldsymbol{L}$ is closed under the following two rules: *modus ponens* for '$\rightarrow$':

$$\frac{\varphi \qquad \varphi \rightarrow \psi}{\psi} \qquad \text{(mp)}$$

- and *uniform substitution*

$$\frac{\varphi}{s\,\varphi}, \qquad \text{(sb)}$$

  where $s\,\varphi$ is a result of the uniform substitution of formulas for propositional variables in φ.

If we skip the first condition we can say about modal logics in a broader sense or non-classical modal logics.

By the uniform substitution, every modal logic includes the set **PL** of modal formulas being substitution instances of elements of **Taut**.

By ($rep^{\Box}$), every modal logic has the following thesis:

$$\Diamond p \leftrightarrow \neg\Box\neg\, p. \qquad (\mathrm{df}\,\Diamond)$$

In this paper the term 'modal logic' is always understood as a set of modal formulas. All members of a logic are called its *theses*.

2.2 *Discussive Logic*

Discussive formulas are again formed in the standard way using propositional variables, but this time from truth-value operators: '$\neg$' and '$\vee$' (negation and disjunction); discussive connectives: '$\wedge_d$', '$\rightarrow_d$', '$\leftrightarrow_d$' (conjunction, implication and equivalence); and the brackets. In the case of a discussive modal logic, we use also discussive modal operators $\Diamond_d$ and $\Box_d$—'discussive', since these operators have also an explication in the model of discussion.

Let $\mathrm{For^d}$ ($\mathrm{For^d_m}$) be the set of all discussive formulas (respectively, discussive modal formulas) in this language.

2.2.1 Translation from $\mathrm{For^d}$ into $\mathrm{For_m}$

Let $\mathrm{i_0}$ be the translation from $\mathrm{For^d}$ into $\mathrm{For_m}$ such that:

1. $\mathrm{i_0}(a) = a$, for any propositional variable a,
2. for any $A, B \in \mathrm{For^d}$:
   - $\mathrm{i_0}(\neg A) = \ulcorner \neg\, \mathrm{i_0}(A) \urcorner$,
   - $\mathrm{i_0}(A \vee B) = \ulcorner \mathrm{i_0}(A) \vee \mathrm{i_0}(B) \urcorner$,
   - $\mathrm{i_0}(A \wedge_d B) = \ulcorner \mathrm{i_0}(A) \wedge \Diamond \mathrm{i_0}(B) \urcorner$,
   - $\mathrm{i_0}(A \rightarrow_d B) = \ulcorner \Diamond \mathrm{i_0}(A) \rightarrow \mathrm{i_0}(B) \urcorner$,
   - $\mathrm{i_0}(A \leftrightarrow_d B) = \ulcorner (\Diamond \mathrm{i_0}(A) \rightarrow \mathrm{i_0}(B)) \wedge \Diamond(\Diamond \mathrm{i_0}(B) \rightarrow \mathrm{i_0}(A)) \urcorner$.

2.2.2 Historical Reminder

As it was mentioned, Jaśkowski used discussive operators to express some basic interactions that can hold between debaters. The first interaction has been expressed by Jaśkowski in the following way 'if anyone states that p, then q' (see [1, p. 67]). This phrase is treated as an intuitive understanding of Jaśkowski's discussive implication. As one might see, we apply the custom to denote discussive implication by: '$\rightarrow_d$'. Taking into account the intuitive meaning of discussive implication, Jaśkowski proposes the formula

$$\Diamond p \rightarrow q$$

as the intended understanding of the formula

$$p \rightarrow_d q\,.$$

The technical reason for such interpretation of discussive implication is its ability to ensure the closure of the set of theses on modus ponens. In particular, Jaśkowski observes [2, p. 44], [1, p. 67]:

In every discussive system two theses, one of the form:

$$\mathfrak{P} \rightarrow_{\mathrm{d}} \mathfrak{Q},$$

and the other of the form:

$$\mathfrak{P},$$

entail the thesis

$$\mathfrak{Q},$$

and that on the strength of the theorem

$$\Diamond(\Diamond p \rightarrow q) \rightarrow (\Diamond p \rightarrow \Diamond q). \tag{M$_2$1}$$

In [3][1] a discussive conjunction (notation: $p \wedge_{\mathrm{d}} q$) has been introduced:

$$p \wedge \Diamond q.$$

It is usually understood as a summary made by a debater who expressed p. In the very same paper, discussive equivalence $p \leftrightarrow_{\mathrm{d}} q$ is expressed by the formula:

$$(p \rightarrow_{\mathrm{d}} q) \wedge_{\mathrm{d}} (q \rightarrow_{\mathrm{d}} p).$$

2.3 *The Discussive Logic* $\mathrm{D_2}$ *as a Set of Discussive Formulas*

Jaśkowski's discussive logic $\mathbf{D_2}$ can be treated either as a set of discussive formulas or as some consequence relation on the set of all discussive formulas. Nowadays, the discussive logic $\mathbf{D_2}$ is usually understood in the first way and formulated with the help of the modal logic **S5** as follows:

$$\mathbf{D_2} := \{ A \in \mathrm{For^d} : \ulcorner \Diamond \mathrm{i_0}(A) \urcorner \in \mathbf{S5} \}.$$

As one can see such a formulation corresponds to both levels of the modal interpretation recalled and sketched above. As one can also easily see, the set $\mathbf{D_2}$ is closed under substitution. Besides, as it was planned by Jaśkowski, $\mathbf{D_2}$ is closed on *modus ponens* for '$\rightarrow_{\mathrm{d}}$'. It is achieved by the use of the formula (M$_2$1), which belongs to **S5**. Using this formula we see that for $A, B \in \mathrm{For^d}$, if $A \in \mathbf{D_2}$, $\ulcorner A \rightarrow_{\mathrm{d}} B \urcorner \in \mathbf{D_2}$, i.e., $\Diamond \mathrm{i_0}(A), \Diamond \mathrm{i_0}(A \rightarrow_{\mathrm{d}} B) \in \mathbf{S5}$, then $\Diamond \mathrm{i_0}(B) \in \mathbf{S5}$, so $B \in \mathbf{D_2}$.

Besides, by (2.1) and (2.2),

$$\Diamond(\Diamond p \rightarrow p) \tag{2.1}$$

$$\Diamond(\Diamond p \rightarrow (\Diamond q \rightarrow (p \wedge \Diamond q))) \tag{2.2}$$

---

[1] See also [4].

the following formulas

$$p \rightarrow_{\mathrm{d}} p \qquad ((1)^{\mathrm{d}})$$

$$p \rightarrow_{\mathrm{d}} (q \rightarrow_{\mathrm{d}} (p \wedge_{\mathrm{d}} q)) \qquad ((2)^{\mathrm{d}})$$

belong to $\mathbf{D_2}$.

As a standard counterexample, consider $\ulcorner p \rightarrow (\neg p \rightarrow q)\urcorner$ that is not a thesis of $\mathbf{D_2}$.

3 Discussive Modal Logic

3.1 *Extension of Jaśkowski's Translation*

The first step while working on a modal extension of $\mathbf{D_2}$ is to save modalities in the object language of discussive logic. So, we keep Jaśkowski's intuitive model of discussion but add to the language modalities that are interpreted standardly. By considering this modal extension of $\mathbf{D_2}$ 'we allow' participants of a discussion to explicitly use possibility and necessity operators. Such extension was investigated in [6].

While formulating a modal logic over $\mathbf{D_2}$ we consider an extension i_1 of the translation i_0 onto the set of modal formulas by adding to the previously given conditions, two clauses:

- $\mathrm{i}_1(\Diamond_{\mathrm{d}} A) = \ulcorner \Diamond \mathrm{i}_1(A)\urcorner$,
- $\mathrm{i}_1(\Box_{\mathrm{d}} A) = \ulcorner \Box \mathrm{i}_1(A)\urcorner$.

The obtained logic is denoted as $\mathbf{mD_2}$, where:

$$\mathbf{mD_2} := \{ A \in \mathrm{For}^{\mathrm{d}}_{\mathrm{m}} : \ulcorner \Diamond \mathrm{i}_1(A)\urcorner \in \mathbf{S5} \}.$$

As an outcome, we can consider, for example, formulas of the form:

$$\Diamond_{\mathrm{d}}(p \rightarrow_{\mathrm{d}} q) \rightarrow_{\mathrm{d}} (\Box_{\mathrm{d}} p \rightarrow_{\mathrm{d}} q) \qquad (*)$$

$$\Box_{\mathrm{d}}(p \rightarrow_{\mathrm{d}} q) \rightarrow_{\mathrm{d}} (\Box_{\mathrm{d}} p \rightarrow_{\mathrm{d}} \Box_{\mathrm{d}} q) \qquad (\mathrm{K_d})$$

$$(\Box_{\mathrm{d}} p \rightarrow_{\mathrm{d}} \Diamond_{\mathrm{d}} q) \rightarrow_{\mathrm{d}} \Diamond_{\mathrm{d}}(p \rightarrow_{\mathrm{d}} q). \qquad (\mathrm{K1_d})$$

Using both translations we obtain:

$$\Diamond(\Diamond\Diamond(\Diamond p \rightarrow q) \rightarrow (\Diamond\Box p \rightarrow q)) \qquad (*^{\mathrm{i}})$$

$$\Diamond(\Diamond\Box(\Diamond p \rightarrow q) \rightarrow \Diamond(\Diamond\Box p \rightarrow \Box q)) \qquad (\mathrm{K^i_d})$$

$$\Diamond(\Diamond(\Diamond\Box p \rightarrow \Diamond q) \rightarrow \Diamond(\Diamond p \rightarrow q)) \qquad (\mathrm{K1_d}^{\mathrm{i}})$$

We see that (∗), (K_d) ∈ $\mathbf{mD_2}$ and ($K1_d$) ∉ $\mathbf{mD_2}$, while

$$\Diamond(p \to q) \to (\Box p \to q) \notin \mathbf{S5}$$

$$\Box(p \to q) \to (\Box p \to \Box q) \in \mathbf{S5}$$

and of course

$$(\Box p \to \Diamond q) \to \Diamond(p \to q) \in \mathbf{S5}.$$

To make a readable comparison of $\mathbf{mD_2}$ with standard modal logics, we apply a function e from $\mathrm{For}^{\mathrm{d}}_{\mathrm{m}}$ into $\mathrm{For}_{\mathrm{m}}$ which removes the subscript '$_{\mathrm{d}}$'.

Using the above mentioned examples and definitions we give the following resume of some basic facts on $\mathbf{mD_2}$:

Fact 3.1 ([6])

1. *$\mathbf{S5}$ and e[$\mathbf{mD_2}$] cross each other.*
2. *The set $\mathbf{mD_2}$ is not closed under necessitation rule*

$$\frac{A}{\Box_{\mathrm{d}} A}. \tag{3.1}$$

3. *The set $\mathbf{mD_2}$ is closed under modus ponens (mp) and is closed under uniform substitution (sb), hence $\mathbf{mD_2}$ is a logic. In particular, it is a non-classical modal logic.*
4. *(a)* $\mathbf{CL}^+ \subsetneq$ e[$\mathbf{mD_2}$]
   (b) $\mathbf{CL} \not\subseteq \mathbf{mD_2}$
   (c) $\mathbf{D_2} \subsetneq \mathbf{mD_2}$ *and* $\mathbf{mD_2}$ *is a conservative extension of the logic* $\mathbf{D_2}$.
5. *The following formulas are theses of* $\mathbf{mD_2}$

$$\neg \Diamond_{\mathrm{d}}\, p \to_{\mathrm{d}} \Box_{\mathrm{d}} \neg\, p$$

$$\Box_{\mathrm{d}} \neg\, p \to_{\mathrm{d}} \neg \Diamond_{\mathrm{d}}\, p$$

$$\neg \Box_{\mathrm{d}} \neg\, p \to_{\mathrm{d}} \Diamond_{\mathrm{d}}\, p$$

$$\neg \Diamond_{\mathrm{d}} \neg\, p \to_{\mathrm{d}} \Box_{\mathrm{d}}\, p$$

$$\Box_{\mathrm{d}}\, p \to_{\mathrm{d}} p$$

$$p \to_{\mathrm{d}} \Diamond_{\mathrm{d}}\, p$$

$$\Diamond_{\mathrm{d}}\, p \to_{\mathrm{d}} p$$

$$\neg \Box_{\mathrm{d}} \neg\, p \to_{\mathrm{d}} p.$$

6. *The following formulas are not theses of* $\mathbf{mD_2}$

$$(p \to_{\mathrm{d}} q) \to_{\mathrm{d}} (\neg q \to_{\mathrm{d}} \neg p)$$
$$(\neg q \to_{\mathrm{d}} \neg p) \to_{\mathrm{d}} (p \to_{\mathrm{d}} q)$$
$$p \to_{\mathrm{d}} \Box_{\mathrm{d}} p$$
$$\neg p \to_{\mathrm{d}} \Box_{\mathrm{d}} \neg p$$
$$\Diamond_{\mathrm{d}}(p \to_{\mathrm{d}} \Box_{\mathrm{d}} p)$$
$$\Box_{\mathrm{d}}(\neg p \to_{\mathrm{d}} \Box_{\mathrm{d}} \neg p).$$

7. $\mathbf{mD_2}$ *(as* $\mathbf{D_2}$*) is not closed on rules of contraposition*

$$\frac{A \to_{\mathrm{d}} B}{\neg B \to_{\mathrm{d}} \neg A}$$

$$\frac{\neg B \to_{\mathrm{d}} \neg A}{A \to_{\mathrm{d}} B}$$

(in fact, $(p \vee \neg p) \wedge_{\mathrm{d}} p \to_{\mathrm{d}} p$, $\neg\neg((p \vee \neg p) \wedge_{\mathrm{d}} p) \to_{\mathrm{d}} \neg\neg p \in \mathbf{D_2} \subseteq \mathbf{mD_2}$, *while* $p \to_{\mathrm{d}} (p \vee \neg p) \wedge_{\mathrm{d}} p \notin \mathbf{mD_2}$*).*

8. $\mathbf{mD_2}$ *(as* $\mathbf{D_2}$*) is not extensional. Indeed*

$$p \leftrightarrow_{\mathrm{d}} (\neg p \vee p) \wedge_{\mathrm{d}} p \in \mathbf{D_2} \subseteq \mathbf{mD_2}$$
$$\ulcorner (q \wedge_{\mathrm{d}} \neg p) \leftrightarrow_{\mathrm{d}} (q \wedge_{\mathrm{d}} \neg((\neg p \vee p) \wedge_{\mathrm{d}} p))\urcorner \notin \mathbf{mD_2}$$

since

$$\ulcorner \Diamond(\Diamond(q \wedge \Diamond\neg p) \to (q \wedge \Diamond\neg((\neg p \vee p) \wedge \Diamond p)))\urcorner \notin \mathbf{S5}.$$

9. *For any* $A \in \mathrm{For}^{\mathrm{d}}_{\mathrm{m}}$, *such that* $\mathrm{i}_1(A)$ *is a thesis of* $\mathbf{S5}$, *then* A *is a thesis of* $\mathbf{mD_2}$.

We can see that $\Diamond_{\mathrm{d}}$ and $\Box_{\mathrm{d}}$ are not dual on the basis of $\mathbf{mD_2}$. Points 7 and 8, of the above fact 3.1, lead us to the following observations:

Fact 3.2 ([6]) *The set* $\mathbf{mD_2}$ *is not closed under the congruence and extensionality rules:*

$$\frac{A \leftrightarrow B}{\Box_{\mathrm{d}} A \leftrightarrow \Box_{\mathrm{d}} B} \qquad \frac{A \leftrightarrow B}{C \leftrightarrow C(A /\!\!/ B)}$$

As a result, $\mathbf{mD_2}$ *is also not closed on the monotonicity nor regularity rule:*

$$\frac{A \to B}{\Box_{\mathrm{d}} A \to \Box_{\mathrm{d}} B} \qquad \frac{(A \wedge_{\mathrm{d}} B) \to_{\mathrm{d}} C}{(\Box_{\mathrm{d}} A \wedge_{\mathrm{d}} \Box_{\mathrm{d}} B) \to_{\mathrm{d}} \Box_{\mathrm{d}} C}.$$

3.2 Semantics for the Modal Discussive Logic

We focus here on semantical analysis, however, the logic under consideration can be also expressed purely syntactically—for axiomatisation of the logic $\mathbf{mD_2}$, please see [6]. The given semantics is natural, it uses respective modal meanings of the considered connectives and when applied to the language without modalities can be used to determine of $\mathbf{D_2}$.

A *relational frame* for a discussive modal logic (a frame) is a pair $\langle W, R\rangle$ consisting of a nonempty set W and a binary relation R on W. As usually, elements of sets W are called (accessible) *worlds*, while R is the *accessibility relation.*

A *model* for the logic $\mathbf{mD_2}$ is a triple $\langle W, R, V\rangle$, where $\langle W, R\rangle$ is a frame and the function $V\colon \mathrm{For}^{\mathrm{d}}_{\mathrm{m}} \times W \longrightarrow \{0, 1\}$ preserves classical truth conditions for negation and disjunction

$$V(\neg A, w) = 1 \quad \text{iff} \quad V(A, w) = 0, \tag{3.2}$$

$$V(A \vee B, w) = 1 \quad \text{iff} \quad V(A, x) = 1 \text{ or } V(A, x) = 1, \tag{3.3}$$

standardly understood conditions for discussive connectives:

$$V(A \wedge_{\mathrm{d}} B, w) = 1 \quad \text{iff} \quad V(A, w) = 1 \text{ and } \exists_{x\in R(w)} V(B, x) = 1,$$

$$V(A \rightarrow_{\mathrm{d}} B, w) = 1 \quad \text{iff} \quad \forall_{x\in R(w)} (V(A, x) = 0 \text{ or } V(B, w) = 1),$$

$$V(A \leftrightarrow_{\mathrm{d}} B, w) = 1 \quad \text{iff} \quad V(A \rightarrow_{\mathrm{d}} B, w) = 1 \text{ and } \exists_{y\in R(w)} V((B \rightarrow_{\mathrm{d}} A), y) = 1$$

and usual conditions for modalities:

$$V(\Box_{\mathrm{d}} A, w) = 1 \quad \text{iff} \quad \forall_{x\in R(w)}\ V(A, x) = 1\,,$$

$$V(\Diamond_{\mathrm{d}} A, w) = 1 \quad \text{iff} \quad \exists_{x\in R(w)}\ V(A, x) = 1\,,$$

where $R(w) = \{x \in W : w\, R\, x\}$. As usually, V is determined by its restriction to the set of all propositional variables. We say that the model $\langle W, R, V\rangle$ is based on the frame $\langle W, R\rangle$.

Definition 3.3 A formula A is *discussively true* in a model $M = \langle W, R, V\rangle$ iff for each $w \in W$, there is $x \in R(w)$ such that $V(A, x) = 1$.

We say that a formula $A \in \mathrm{For}^{\mathrm{d}}_{\mathrm{m}}$ is *discussively valid* in a frame iff A is discussively true in all models based on this frame.

Fact 3.4 ([6]) *For any $A \in \mathrm{For}^{\mathrm{d}}_{\mathrm{m}}$ and a model M:*
A is discussively true in M iff $\Diamond \mathrm{i}_1(A)$ is standardly true in M.

We can vary the conditions that are imposed on the relation R.

Definition 3.5 Let $F = \langle W, R\rangle$. The frame F (or the accessibility relation R) is

(i) *trivial* iff $R = \emptyset$,
(ii) *serial* iff $\forall_{x\in W} \exists_{y\in W}\ xRy$,

(iii) *reflexive* iff $\forall_{x \in W}\ xRx$,
(iv) *symmetric* iff $\forall_{x,y \in W}(xRy \implies yRx)$,
(v) *transitive* iff $\forall_{x,y,z \in W}(xRy\ \&\ yRz \implies xRz)$,
(vi) *Euclidean* iff $\forall_{x,y,z \in W}(xRy\ \&\ xRz \implies yRz)$.

Fact 3.6 *The set of all formulas from* $\mathrm{For}^{\mathrm{d}}_{\mathrm{m}}$ *that are discussively valid in every frame is empty. In particular, the set of all formulas that discussively valid in a frame with a world w with no alternatives (that is in a frame which is not serial) is empty.*

As an introductory step to the issue of the square of opposition for $\mathbf{mD_2}$, we recall some facts concerning positive and negative examples of discussive validity for specific classes of frames, extending slightly results given in Fact 3.1. Below, we also refer to the issue of completeness for $\mathbf{mD_2}$.

Fact 3.7 ([6])

1. *The formula*

$$p \rightarrow_{\mathrm{d}} \Diamond_{\mathrm{d}} p$$

is discussively valid in every frame with a serial accessibility relation.

2. *It is not the case that*

$$\Box_{\mathrm{d}} p \rightarrow_{\mathrm{d}} \Diamond_{\mathrm{d}} p$$

is discussively valid in every serial frame.

3. *The formulas*

$$\neg \Box_{\mathrm{d}} p \vee p$$

$$\neg p \vee \Diamond_{\mathrm{d}} p$$

are discussively valid in every reflexive frame but also in those fulfilling the condition:

$$\forall_w \exists_u \Big(wRu \wedge \exists_v (uRv \wedge wRv) \Big).$$

While

$$\Box_{\mathrm{d}} p \rightarrow_{\mathrm{d}} p$$

$$\neg \Diamond_{\mathrm{d}} \Box_{\mathrm{d}} p \vee p$$

is valid in every serial and symmetric frame but also fulfilling the condition

$$\underset{z}{\forall}\underset{u}{\exists}(zRu \wedge \underset{x}{\forall}(uRx \rightarrow \underset{w}{\exists}(xRw \wedge zRw))).$$

4. *For the formula*

$$\Diamond_{\mathrm{d}} \Box_{\mathrm{d}} p \rightarrow_{\mathrm{d}} p$$

it is not the case that it is discussively valid in every symmetric frame or even frames that are both symmetric and serial.

We know that (see [6] and [5]):

Theorem 3.8 ([6])

1. *A formula belongs to* $\mathbf{mD_2}$ *iff it is discussively valid in every reflexive and Euclidean frame.*
2. *The logic* $\mathbf{mD_2}$ *is determined by the class of reflexive and transitive frames.*
3. *The logic* $\mathbf{mD_2}$ *is determined by the class of serial and transitive frames.*
4. *The logic* $\mathbf{mD_2}$ *is determined by the class frames fulfilling*

$$\forall_w \exists_u \Big(wRu \wedge \forall_x (uRx \rightarrow wRx) \Big)$$

$$\forall_w \exists_u \Big(wRu \wedge \forall_x \forall_y (uRx \wedge xRy \rightarrow wRy) \Big).$$

4 Another Modal Extension of $\mathbf{D_2}$

Fact 4.1 ([6]) *The following two formulas belong to* $\mathbf{mD_2}$:

$$\Diamond_{\mathrm{d}}\, p \leftrightarrow_{\mathrm{d}} ((p \vee \neg p) \wedge_{\mathrm{d}} p)$$

$$\Box_{\mathrm{d}}\, p \leftrightarrow_{\mathrm{d}} \neg((p \vee \neg p) \wedge_{\mathrm{d}} \neg p).$$

However, although these equivalences hold, it is still intriguing to consider calculuses with $\Diamond_{\mathrm{d}}$ and $\Box_{\mathrm{d}}$, since $\mathbf{D_2}$ and $\mathbf{mD_2}$ are not closed under the rule of extensionality (see Fact 3.2).

So, the added modalities give a variant of the model of discussion in which a given participant expresses his own modal views. In what follows, we refer to the next extension of $\mathbf{D_2}$, where we extend also the model of discussion.

4.1 Semantics for General/Public Discussive Modalities

We consider a model of discussion in which next to the discussive group, there is possibly a broader community of people whose statements will be used as an intuitive explication of new modalities $\Box^{\mathrm{g}}$ and $\Diamond^{\mathrm{g}}$. In this way we obtain the set $\mathrm{For}^{\mathrm{dg}}_{\mathrm{m}}$ being another extension of $\mathrm{For}^{\mathrm{d}}$ but also of the set $\mathrm{For}^{\mathrm{d}}_{\mathrm{m}}$. The semantics considered here is a natural extension of the previously given semantical conditions for discussive modalities: it uses respective modal meaning of the considered connectives with respect to another—not smaller then W in the sense of inclusion—domain W_g, and when applied to the language without modalities, i.e., when W_g is dropped, it comes down to the semantics that can be used for determination of $\mathbf{D_2}$ or $\mathbf{mD_2}$. Intuitively, the set W_g corresponds to the voices that express the external points of view that can be taken into account by members of the given discussive group.

An *relational frame* for a discussive modal logic with public modalities is a quadruple $\langle W, W_g, W \times W, W_g \times W_g \rangle$ consisting of a nonempty set W, and a set $W_g \supseteq W$. Elements of the set W are still called (accessible) *worlds*, while elements of the set W_g—general worlds.

A *model* for an extended discussive modal logic is any 5-tuple $\langle W, W_g, W \times W, W_g \times W_g, V \rangle$, where $\langle W, W_g, W \times W, W_g \times W_g \rangle$ is a frame and $V\colon \mathrm{For}^{\mathrm{dg}}_{\mathrm{m}} \times W_g \longrightarrow \{0, 1\}$ is a function that preserves classical truth conditions (3.2) and (3.3) for negation and disjunction; discussive conditions for modalities adapted for the current context:

$$V(A \wedge_{\mathrm{d}} B, w) = 1 \quad \text{iff} \quad V(A, w) = 1 \text{ and } \exists_{x \in W} V(B, x) = 1,$$

$$V(A \rightarrow_{\mathrm{d}} B, w) = 1 \quad \text{iff} \quad \forall_{x \in W}(V(A, x) = 0 \text{ or } V(B, w) = 1),$$

and the following conditions for new 'general'/'public' modalities:

$$V(\Box^{\mathrm{g}} A, w) = 1 \quad \text{iff} \quad \forall_{x \in W_g}\ V(A, x) = 1\,,$$

$$V(\Diamond^{\mathrm{g}} A, w) = 1 \quad \text{iff} \quad \exists_{x \in W_g}\ V(A, x) = 1\,.$$

As usually, V is determined by its restriction to the set of all propositional variables. As we observed, we can consider discussive modalities expressed by discussive connectives:

$$\Diamond_{\mathrm{d}} A = \quad (p \vee \neg p) \wedge_{\mathrm{d}} A \qquad \text{(df } \Diamond_{\mathrm{d}}\text{)}$$

$$\Box_{\mathrm{d}} A = \quad \neg A \rightarrow_{\mathrm{d}} \neg(p \vee \neg p). \qquad \text{(df } \Box_{\mathrm{d}}\text{)}$$

We say that the model $\langle W, W_g, W \times W, W_g \times W_g, V \rangle$ is based on the frame $\langle W, W_g, W \times W, W_g \times W_g \rangle$.

Since we consider full accessibility relations, in fact as frames we could treat just pairs $\langle W, W_g \rangle$, while $\langle W, W_g, V \rangle$—as models, stipulating only that $W \subseteq W_g$. We repeat the notation of validity for the introduced notions of model and frame referred to definitions given in this subsection:

Definition 4.2 A formula A is *discussively true* in a model $M = \langle W, W_g, W \times W, W_g \times W_g, V \rangle$ iff for each $w \in W$, there is $x \in W$ such that $V(A, x) = 1$.

We say that a formula is *discussively valid* in a frame iff it is discussively true in all models based on this frame.

The set of all formulas discussively valid in the class of frames of the form $\langle W, W_g \rangle$ is noted by $\mathbf{mgD_2}$. We easily see:

Fact 4.3 $\mathbf{D_2} \subseteq \mathbf{mgD_2}$.[2]

[2] The logic $\mathbf{mgD_2}$ can be of course characterised syntactically but this exceeds the aim of the present paper. For details see [7].

5 Square of Opposition

Now we consider theses of $\mathbf{mD_2}$ that are connected to the square:[3]

Fact 5.1 *Suppose that* $A \in \mathrm{For}^{\mathrm{d}}_{\mathrm{m}}$, $\mathrm{e}(A)$ *is a thesis of* **S5** *and each atom* a *of* A *occurs in subformulas of the form* $\ulcorner \Diamond_{\mathrm{d}}\, a \urcorner$, $\ulcorner \Diamond_{\mathrm{d}} \neg a \urcorner$, $\ulcorner \Box_{\mathrm{d}}\, a \urcorner$ *or* $\ulcorner \Box_{\mathrm{d}} \neg a \urcorner$. *Then* $A \in \mathbf{mD_2}$.

Below, a formula of the form $\neg A \vee B$ is denoted as $A \rightarrow_{\mathrm{c}} B$.

Fact 5.2 *The following formulas are theses of* $\mathbf{mD_2}$*:*

$$\neg \Box_{\mathrm{d}}\, p \vee \neg \Box_{\mathrm{d}} \neg p$$

$$\Diamond_{\mathrm{d}}\, p \vee \Diamond_{\mathrm{d}} \neg p$$

$$\neg \Box_{\mathrm{d}}\, p \vee \Diamond_{\mathrm{d}}\, p$$

$$\neg \Box_{\mathrm{d}} \neg p \vee \Diamond_{\mathrm{d}} \neg p$$

$$\neg \Box_{\mathrm{d}}\, p \vee \neg \Diamond_{\mathrm{d}} \neg p$$

$$\Box_{\mathrm{d}}\, p \vee \Diamond_{\mathrm{d}} \neg p$$

$$\neg \Box_{\mathrm{d}} \neg p \vee \neg \Diamond_{\mathrm{d}}\, p$$

$$\Box_{\mathrm{d}} \neg p \vee \Diamond_{\mathrm{d}}\, p$$

or using $\rightarrow_{\mathrm{c}}$ *for the case of contraries, subalternation, and contradictories*

$$\Box_{\mathrm{d}}\, p \rightarrow_{\mathrm{c}} \neg \Box_{\mathrm{d}} \neg p$$

$$\neg \Diamond_{\mathrm{d}}\, p \rightarrow_{\mathrm{c}} \Diamond_{\mathrm{d}} \neg p$$

$$\Box_{\mathrm{d}}\, p \rightarrow_{\mathrm{c}} \Diamond_{\mathrm{d}}\, p$$

$$\Box_{\mathrm{d}} \neg p \rightarrow_{\mathrm{c}} \Diamond_{\mathrm{d}} \neg p$$

$$\Box_{\mathrm{d}}\, p \rightarrow_{\mathrm{c}} \neg \Diamond_{\mathrm{d}} \neg p$$

$$\neg \Box_{\mathrm{d}}\, p \rightarrow_{\mathrm{c}} \Diamond_{\mathrm{d}} \neg p$$

$$\Box_{\mathrm{d}} \neg p \rightarrow_{\mathrm{c}} \neg \Diamond_{\mathrm{d}}\, p$$

$$\neg \Box_{\mathrm{d}} \neg p \rightarrow_{\mathrm{c}} \Diamond_{\mathrm{d}}\, p.$$

Moreover the respective connections also hold if the discussive connectives are used:

$$\neg(\Box_{\mathrm{d}}\, p \wedge_{\mathrm{d}} \Box_{\mathrm{d}} \neg p) \text{ or } \Box_{\mathrm{d}}\, p \rightarrow_{\mathrm{d}} \neg \Box_{\mathrm{d}} \neg p$$

[3] Some of these formulas are recalled from [6].

$$\neg \Diamond_{\mathrm{d}}\, p \rightarrow_{\mathrm{d}} \Diamond_{\mathrm{d}} \neg\, p$$

$$\Box_{\mathrm{d}}\, p \rightarrow_{\mathrm{d}} \Diamond_{\mathrm{d}}\, p$$

$$\Box_{\mathrm{d}} \neg\, p \rightarrow_{\mathrm{d}} \Diamond_{\mathrm{d}} \neg\, p$$

$$\Box_{\mathrm{d}}\, p \rightarrow_{\mathrm{d}} \neg \Diamond_{\mathrm{d}} \neg\, p$$

$$\neg \Box_{\mathrm{d}}\, p \rightarrow_{\mathrm{d}} \Diamond_{\mathrm{d}} \neg\, p$$

$$\Box_{\mathrm{d}} \neg\, p \rightarrow_{\mathrm{d}} \neg \Diamond_{\mathrm{d}}\, p$$

$$\neg \Box_{\mathrm{d}} \neg\, p \rightarrow_{\mathrm{d}} \Diamond_{\mathrm{d}}\, p.$$

- Taking into account the above theses and the fact that $\mathbf{mD_2}$ as a logic is closed on substitution, we have the square for discussive modalities within $\mathbf{mD_2}$, where respective connections can be understood both: classically and discussively.

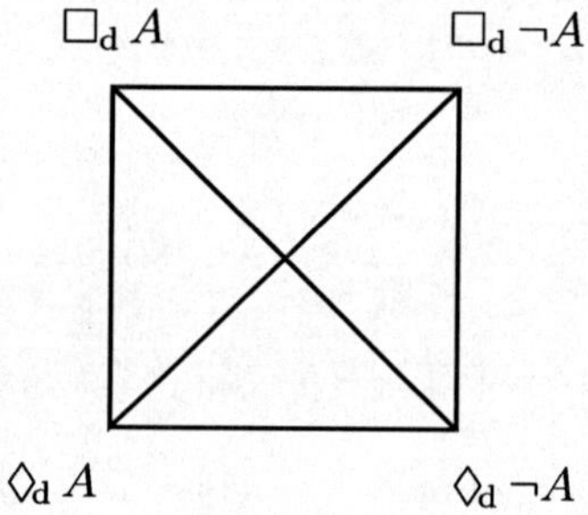

One can observe that similar result holds for the logic $\mathbf{m_gD_2}$. Again e is the function that removes subscripts refereeing to discussive connectives but also indexes referring to the 'global' modalities:

Fact 5.3 *Given* $A \in \mathrm{For}^{\mathrm{dg}}_{\mathrm{m}}$ *formulated in the sublanguage solely with* $\vee$, $\neg$, $\Box^{\mathrm{g}}$, $\Diamond^{\mathrm{g}}$, *and such that* $\mathrm{e}(A)$ *is a thesis of* **S5**, *then* $A \in \mathbf{m_gD_2}$.

So, similarly as for $\mathbf{mD_2}$, we have:[4]

Fact 5.4 *The following formulas are theses of* $\mathbf{m_gD_2}$

$$\neg \Box^{\mathrm{g}} p \vee \neg \Box^{\mathrm{g}} \neg\, p$$

$$\Diamond^{\mathrm{g}}\, p \vee \Diamond^{\mathrm{g}} \neg\, p$$

$$\neg \Box^{\mathrm{g}} p \vee \Diamond^{\mathrm{g}}\, p$$

$$\neg \Box^{\mathrm{g}} \neg\, p \vee \Diamond^{\mathrm{g}} \neg\, p$$

[4] For details see [7].

$$\neg \Box^{g} p \vee \neg \Diamond^{g} \neg p$$

$$\Box^{g} p \vee \Diamond^{g} \neg p$$

$$\neg \Box^{g} \neg p \vee \neg \Diamond^{g} p$$

$$\Box^{g} \neg p \vee \Diamond^{g} p$$

or using $\rightarrow_{c}$ *for the case of contraries, subalternation, and contradictories*

$$\Box^{g} p \rightarrow_{c} \neg \Box^{g} \neg p$$

$$\neg \Diamond^{g} p \rightarrow_{c} \Diamond^{g} \neg p$$

$$\Box^{g} p \rightarrow_{c} \Diamond^{g} p$$

$$\Box^{g} \neg p \rightarrow_{c} \Diamond^{g} \neg p$$

$$\Box^{g} p \rightarrow_{c} \neg \Diamond^{g} \neg p$$

$$\neg \Box^{g} p \rightarrow_{c} \Diamond^{g} \neg p$$

$$\Box^{g} \neg p \rightarrow_{c} \neg \Diamond^{g} p$$

$$\neg \Box^{g} \neg p \rightarrow_{c} \Diamond^{g} p$$

Again, the respective connections also hold, if instead of '$\rightarrow_{c}$', discussive implication is used. To see this, using the semantical characterisation of $\mathbf{m_{g}D_{2}}$, first we easily can observe that:

Fact 5.5 *The following formulas are theses of* $\mathbf{m_{g}D_{2}}$*:*

$$\Box^{g} p \rightarrow_{c} \Box_{d} p$$

$$\Box^{g} p \rightarrow_{d} \Box_{d} p$$

$$\Diamond_{d} p \rightarrow_{c} \Diamond^{g} p$$

$$\Diamond_{d} p \rightarrow_{d} \Diamond^{g} p$$

Hence, we have:

Lemma 5.6 *The following formulas are theses of* $\mathbf{m_{g}D_{2}}$*:*

$$\Box^{g} p \rightarrow_{d} \neg \Box^{g} \neg p$$

$$\neg \Diamond^{g} p \rightarrow_{d} \Diamond^{g} \neg p$$

$$\Box^{g} p \rightarrow_{d} \Diamond^{g} p$$

$$\Box^{g} \neg p \rightarrow_{d} \Diamond^{g} \neg p$$

$$\Box^{g} p \rightarrow_{d} \neg \Diamond^{g} \neg p$$

$$\neg \Box^{g} p \rightarrow_{d} \Diamond^{g} \neg p$$
$$\Box^{g} \neg p \rightarrow_{d} \neg \Diamond^{g} p$$
$$\neg \Box^{g} \neg p \rightarrow_{d} \Diamond^{g} p$$

Hence, we have the similar square for the pair of 'public' modalities in $\mathbf{m_g D_2}$.

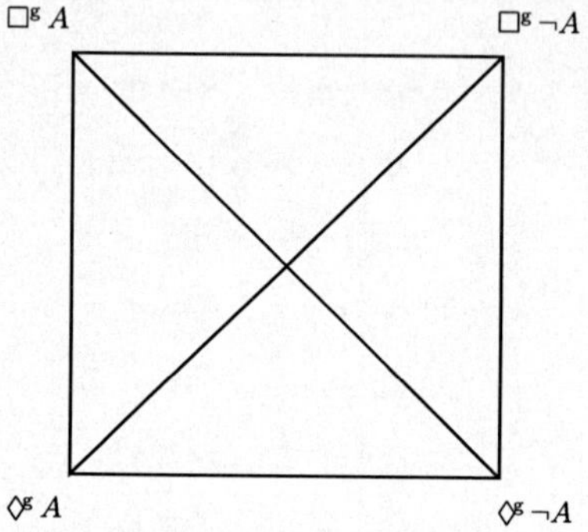

Taking into account Fact 5.5 we obtain the extended square, where again, the respective relations can be expressed either in terms of classical implication, but—what is more interesting—also with the help of discussive implication.

We can connect both these squares by putting one onto another:

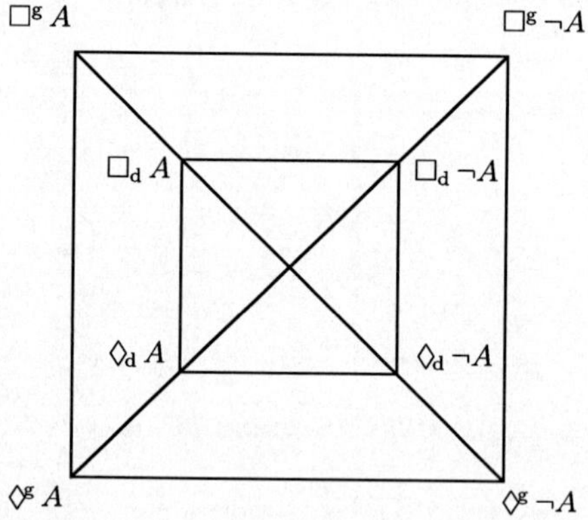

Acknowledgments The authors would like to thank the referee of this paper for his/her careful reading of the manuscript.

References

1. S. Jaśkowski, *Rachunek zdań dla systemów dedukcyjnych sprzecznych.* Studia Societatis Scientiarum Torunensis, Sect. A, **I** no. 5 (1948), 57–77. The first English version: *Propositional calculus for contradictory deductive systems.* Studia Logica **24** (1969), 143–157. https://doi.org/10.1007/BF02134311
2. S. Jaśkowski, *A propositional calculus for inconsistent deductive systems.* Logic and Logical Philosophy **7** (1999), 35–56; the second English version of [1]. https://doi.org/10.12775/LLP.1999.003
3. S. Jaśkowski, *O koniunkcji dyskusyjnej w rachunku zdań dla systemów dedukcyjnych sprzecznych.* Studia Societatis Scientiarum Torunensis, Sect. A, **I** no. 8 (1949), 171–172.
4. S. Jaśkowski, *On the discussive conjunction in the propositional calculus for inconsistent deductive systems.* Logic and Logical Philosophy **7** (1999), 57–59; the English version of [3]. https://doi.org/10.12775/LLP.1999.004
5. M. Nasieniewski, *A comparison of two approaches to parainconsistency: Flemish and Polish.* Logic and Logical Philosophy **9** (2001): 47–74. https://doi.org/10.12775/LLP.2001.004
6. K. Mruczek-Nasieniewska, M. Nasieniewski and A. Pietruszczak, *A modal extension of Jaśkowski's discussive logic* $\mathbf{D_2}$. Logic Journal of the IGPL **27** No. 4 (2019), 451–477. https://doi.org/10.1093/jigpal/jzz014
7. K. Mruczek-Nasieniewska, M. Nasieniewski and A. Pietruszczak, *On a Modal Discussive Logic corresponding to the Extended Model of Discussion.* In preparation.

On the Transformations of the Square of Opposition from the Point of View of Institution Model Theory

Yiannis Kiouvrekis, Petros Stefaneas, and Ioannis Vandoulakis

Abstract In recent decades, research in the square of opposition has been increased. New interpretations, extensions, and generalizations have been suggested, both Aristotelian and non-Aristotelian ones. The paper attempts at comparing different versions of the square of opposition. For this reason, we appeal to the wider categorical model-theoretic framework of the theory of institutions.

Keywords Square of opposition · Classification of judgements · Abstract model theory · Category theory · Institutions theory · Galois connection

Mathematics Subject Classification (2010) Primary 03B22, 03C95, 18A15

1 Introduction

During the second half of the twentieth century, the research in the square of opposition was revived. First, Augustin Sesmat [7] and Robert Blanché [3] extended independently the square of opposition to a logical hexagon which includes the relationships of six statements. This was followed by an extension to a "logical cube" that paved the way to the development of a series of n-dimensional objects called logical bi-simplexes of dimension n [5, 6].

The second line of research was developed during the last 20 years by Jean-Yves Beziau's attempts to find an intuitive basis for paraconsistent negation, which is

Y. Kiouvrekis (✉)
Department of Public and One Health, University of Thessaly, Volos, Greece

University of Nicosia, Nicosia, Cyprus
e-mail: kiouvrekis.y@uth.gr

P. Stefaneas
Department of Mathematics, National Technical University of Athens, Athens, Greece

I. Vandoulakis
The Hellenic Open University, Patras, Greece

J.-Y. Beziau, I. Vandoulakis (eds.), *The Exoteric Square of Opposition*, Studies in Universal Logic, https://doi.org/10.1007/978-3-030-90823-2_13

the O-corner of the square of opposition [1, 2]. In this connection, the third of the authors posed the question how to compare all versions of the square of opposition and, if possible, their different illustrations by various configurations. This required the passage to a wider framework, where the different versions of the theory within different logics could be compared. To this effect, the authors appealed to abstract categorical model theory and specifically to the theory of institutions [13].

Within this wider framework, we can compare the various variants of the square of opposition and their different configurations within the various underlying logics and provide a formal description and characterization of their relations. Moreover, this general framework can serve as a methodology for the formal analysis of eventual variations that can be constructed in the future.

The concept of the institution was introduced by Joseph Goguen and Rod Burstall in the late 1970s, to deal with the vast variety of logical systems developed and used in computer science. The concept tries to capture the essence of the concept of "logical system" [4]. Informally speaking, an institution is a mathematical structure for "logical systems," based on the concept of satisfaction between sentences and models.

In the first section of the paper, we introduce the concept of the square of opposition. In the second section, we expose fundamental concepts from category theory and institution theory that we will use to treat the square of opposition within a wider institution-theoretic framework. The third section introduces the concept of the rhombus of opposition and examines certain aspects of the configurational change of the squares of opposition inside and between logical systems.

In the fourth section, we use the concept of the Galois connection, which is a useful generalization of correspondence between subgroups and subfields that are studied in Galois theory, to show the equilibrium that one can establish between the standard square of opposition (of sentences) and the internal semantics of Boolean connectives at a meta-level. Finally, we introduce the concept of the dual square that can give us not only squares for sentences but also squares for sets of sentences.

2 Squares of Opposition

The theory of the opposition is exposed by Aristotle in *De Interpretatione* 6–7, 17 b 17–26 and *Prior Analytics* I.2, 25 a 1–25 to describe the logical relations between the four basic categorical judgements. During the Middle Ages, Aristotle's theory was represented by a square diagram. This was done by altering the semantics of the O form. During the nineteenth to twentieth centuries, it assumed two major reinterpretations:

(a) within the context of the algebra of logic (see Fig. 1) (Boole, Venn, and others),
(b) within the second-order predicate logic, by using the newly introduced concept of quantified variables by Frege (see Figs. 2 and 3). Within these interpretations, the shape of the "square" remains unaffected.

Beginning with Nicolai A. Vasiliev (1880–1940), the traditional "square" loses for the first time its original square shape; it is transformed into "triangle." This was by a new alteration of semantics of the O form, based on Aristotelian concepts that were neglected in the Aristotelian tradition of logic, notably the concept of indefinite judgment [8].

During the twentieth century, new transformations of the "square" into various shapes appear, i.e., into "hexagon" [3], or "cube" (Fig. 3), by altering the semantics and establishing relationships between truth-values. The new objects admit various interpretations in terms of traditional logic, quantification theory, modal logic, order theory, or paraconsistent logic.

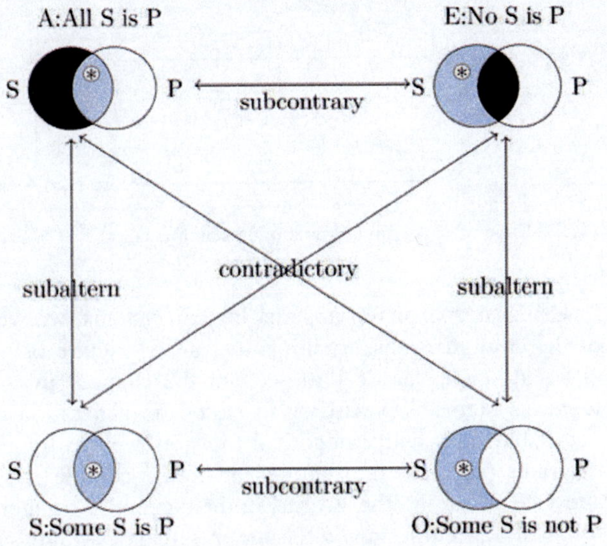

Fig. 1 Representation of the Square of Opposition in algebra of logic (Boole, Venn, and others)

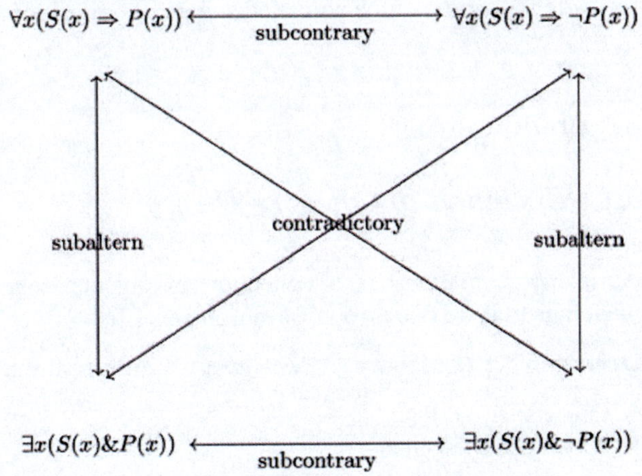

Fig. 2 Representation of the square of opposition in the functional tradition of logic (Frege and others)

However, a natural question arises: how do all these configurations often realized within different logics, are related? Can we describe these transformations in logical terms? What changes and what remains invariant in these transformations?

To examine these questions, we appeal to the concepts of the theory of institutions, introduced by Goguen and Burstall [4]. The theory of institution has the advantage that is not committed to any specific logical system. Moreover, its high level of abstraction allows the accommodation of not only classical, but also non-classical logical systems. A structure-preserving mapping, called *morphism of institutions*, is defined by Goguen and Burstall [4], that operates as projection from a more complex institution into a simpler one.

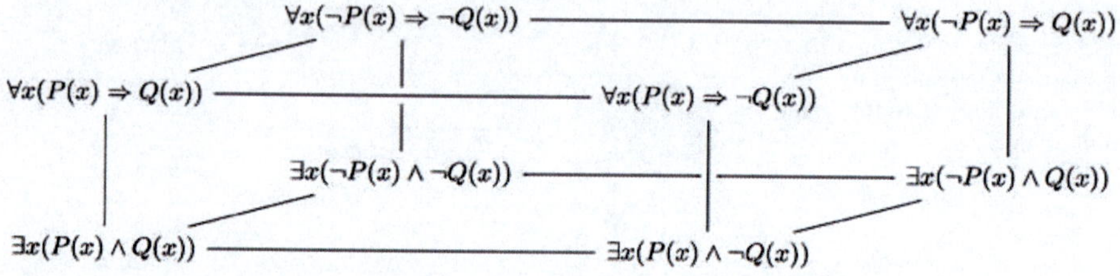

Fig. 3 Cube of opposition of quantified statements

Using morphisms of abstract logical systems, we will study the transformation of the configurations, traditionally called "square of opposition" into entities of different shape, taking into account the changes in semantics of the underlying logical systems. We will try to study the generation of new entities (diagrams) out of old ones with categorical tools, as well as by encoding/embedding simple diagrams (squares) into entities of higher complexity (polygons or 3D objects), and vice versa. In other words, in the context of category theory we will study the following question: how a change in semantics might generate different outcomes (of various shapes) of the so-called square of opposition.

3 Preliminaries

3.1 Essentials of Category Theory

An *abstract category* is a collection of objects, together with a collection of mappings that we call arrows or morphisms [9].

Definition 3.1 (Category) A *category* $\mathbb{C}$ consists of the following:

- Objects: $A, B, C, \ldots$
- Arrows or Morphisms: $f, g, h, \ldots$

that satisfy the following axioms:

1. for every arrow f, there are objects

$$A = \text{dom}(f), \quad B = \text{codom}(f)$$

   called the *domain* and the *codomain* of f, and we write

$$f : A \to B$$

2. Let $f : A \to B$ and $g : B \to C$ such that codom(f) = dom(g), then there exists an arrow/morphism h such that

$$h = g \circ f : A \to C$$

   called the *composite* of f and g.

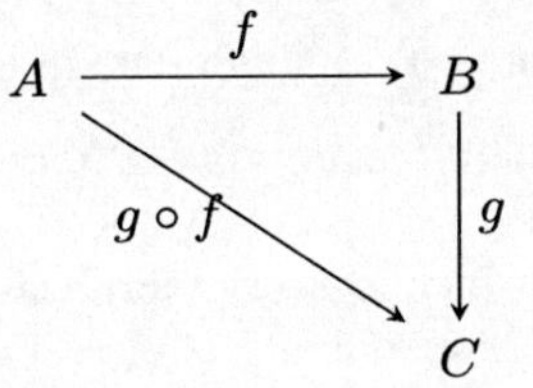

3. For every object A, there exists an arrow

$$1_A : A \to A$$

   called the *identity arrow* of A
4. For all $f : A \to B, \; g : B \to C, \; h : C \to D$

$$h \circ (g \circ f) = (h \circ g) \circ f$$

   with codom(f) = dom(g) and codom(g) = dom(h)
5. and for all $f : A \to B$

$$f \circ 1_A = f = 1_B \circ f$$

A category is anything that satisfies the above definition. Let us illustrate this concept by two elementary, but fundamental examples.

Example 1 (Sets) The first example concerns the category of sets. The objects of this category are sets and the morphisms are the functions between the sets. It is obvious that the identity arrow or identity morphism is the identity function of the set onto itself. The triangle below represents the usual composition of functions.

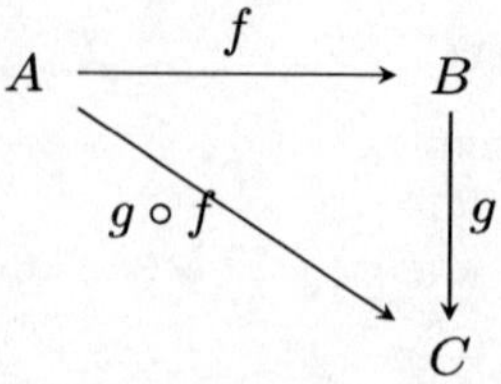

Example 2 (Partial Ordered Sets) A slightly more complex category than the previous one, is that of partially order sets, (abbreviated as posets). A space is called partially ordered, whenever we can define a partial order in this space. What special feature does this category have? It is one of the most typical examples of categories and help us to understand why we will use the language of the category theory in the rest of our paper. Morphisms, the arrows, are functions that preserve the structure.

Another concept in category theory that plays a fundamental role is that of the functor.

Definition 3.2 (Functor) Let $\mathbb{C}$, $\mathbb{D}$ be two categories, then a functor F with domain $\mathbb{C}$ and codomain $\mathbb{D}$ consists of two suitably related functions:

- the object function F, which to every object X of $\mathbb{C}$ assigns an object $F(X)$ of $\mathbb{D}$;
- and the arrow function F which to every morphism $f : X \to Y$ of $\mathbb{C}$ assigns the $F(f) : F(X) \to F(Y)$ of $\mathbb{D}$ such that $F(1_X) = 1_{F(X)}$ and:

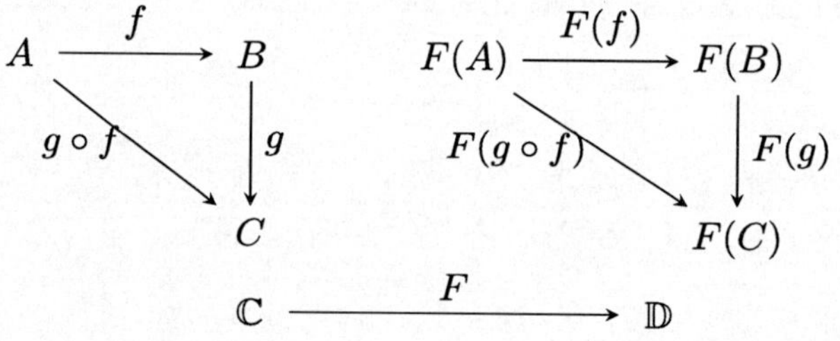

Example 3 (Power Set Functor) One of the most simple examples is the power set functor $\mathcal{P}$: **Set** $\to$ **Set**. The object function for every set X assigns its powerset $\mathcal{P}(X)$ and the morphism function to every $f : X \to Y$ assigns the $\mathcal{P}(f) : \mathcal{P}(X) \to \mathcal{P}(Y)$ which sends every subset $Z \subseteq X$ to its image $F(Z) \subseteq Y$.

Definition 3.3 (Natural Transformation) Let $\mathbb{C}$, $\mathbb{B}$ be two categories and S, T be two functors $S : \mathbb{C} \mapsto \mathbb{D}$, $T : \mathbb{C} \mapsto \mathbb{D}$. A *natural transformation* $\tau : S \Rightarrow T$ is an action which to every object $c \in \mathbb{C}$ assigns an arrow $\tau_c : S(c) \to T(c)$ of the category $\mathbb{B}$ such that for every morphism $f : c \to c'$ in $\mathbb{C}$ the diagram in Fig. 4 is commutative.

Thus a natural transformation is a set of morphisms translating (mapping) the picture S to the picture T with all squares and parallelograms be commutative [9].

Fig. 4 Fundamental Square

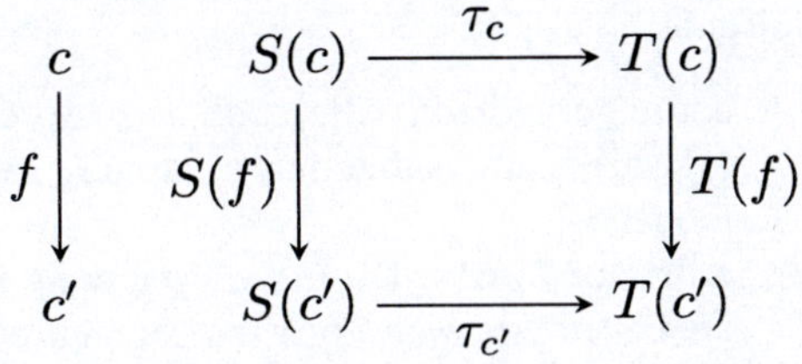

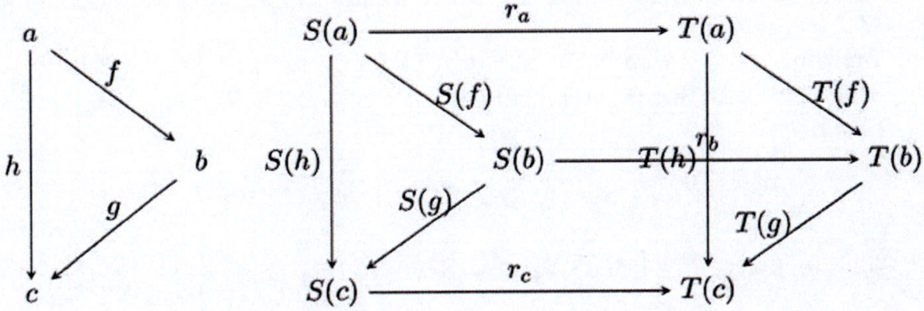

Fig. 5 Natural transformation scheme

A natural transformation is often called *morphism of functors* (Fig. 5).

Example 4 A determinant is a natural transformation. Let $\det_K M$ be the determinant of the $n \times n$ matrix M with entries in the commutative ring K. If K^* denotes the group of units of K, then M is non-singular when $\det_K M$ is a unit, and $\det_K$ is a morphism $GL_n K \mapsto K^*$ of groups. Since the determinant is defined by the same formula for all rings K, each morphism $f : K \mapsto K'$ of commutative rings gives a commutative diagram.

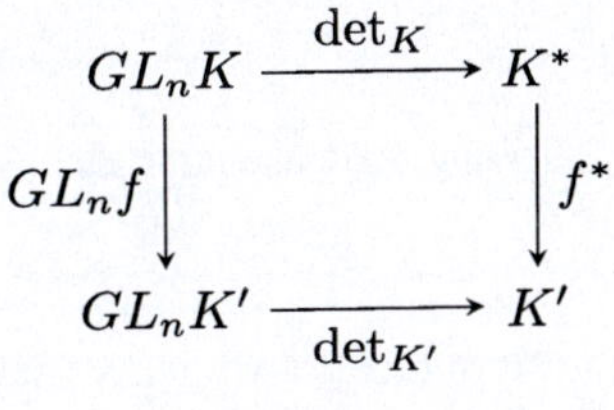

This means that the map $\det : GL_n \mapsto (\ )^*$ is natural between the functors $\boldsymbol{CRng} \to \boldsymbol{Grp}$.

3.2 Essentials of Institution Theory

Definition 3.4 (Institutions) An *Institution* $\mathcal{I} = \left(\mathbb{S}ig^{\mathcal{I}}, \boldsymbol{Sen}^{\mathcal{I}}, \boldsymbol{Mod}^{\mathcal{I}}, \models^{\mathcal{I}}\right)$ consists of:

1. a category $\mathbb{S}ig^{\mathcal{I}}$, whose objects are called *signatures*, that defines the (non-logical) symbols that may be used in sentences and that need to be interpreted in models;
2. a functor $\boldsymbol{Sen}^{\mathcal{I}} : \mathbb{S}ig^{\mathcal{I}} \to \mathbb{S}et$ giving for every signature a set whose elements are called *sentences* over that signature;
3. a functor $\mathbf{Mod}^{\mathcal{I}} : \left(\mathbb{S}ig^{\mathcal{I}}\right)^{op} \to \mathbb{CAT}$ such that for every signature Σ assigns a category which objects are called Σ*-models* and which arrows are called Σ-*morphisms*, and for every signature morphism $\sigma : \Sigma \mapsto \Sigma'$, the reduct functor $\mathbf{Mod}(\sigma) : \mathbf{Mod}(\Sigma') \mapsto \mathbf{Mod}(\Sigma)$, where $\mathbf{Mod}(\sigma)(M')$ is often designated by $M' \upharpoonright \sigma$ (Fig. 6a).
4. a relation $\models^{\mathcal{I}}_{\Sigma} \subseteq \left|\boldsymbol{Mod}^{\mathcal{I}}(\Sigma)\right| \times \boldsymbol{Sen}^{\mathcal{I}}(\Sigma)$ for every $\Sigma \in \left|\mathbb{S}ig^{\mathcal{I}}\right|$, called Σ-*satisfaction* such that for every morphism $\phi : \Sigma \to \Sigma'$ in $\mathbb{S}ig^{\mathcal{I}}$, the *satisfaction condition*

$$M' \models^{\mathcal{I}}_{\Sigma'} \boldsymbol{Sen}^{\mathcal{I}}(\phi)(\rho) \text{ iff } \mathbf{Mod}^{\mathcal{I}}(\phi)(M') \models^{\mathcal{I}}_{\Sigma} \rho$$

holds for every $M' \in \left|\mathbf{Mod}^{\mathcal{I}}\right|$ and $\rho \in \boldsymbol{Sen}^{\mathcal{I}}(\Sigma)$ (Fig. 6b).

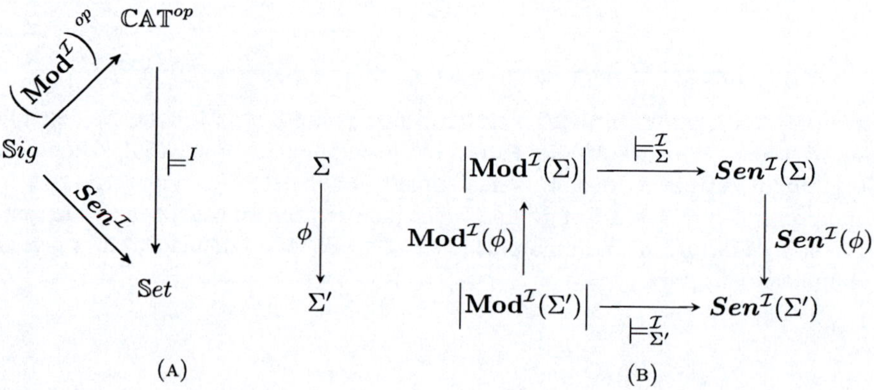

Fig. 6 Institution fundamental diagrams. **(a)** Institution fundamental triangle. **(b)** Institution fundamental square

In words, we assume that signature morphisms map sentences and models in such a way that satisfaction is preserved. Sentences are mapped along with signature morphisms and models are reduced against signature morphisms. The satisfaction condition expresses that truth is invariant under change of notation. One of the most useful examples is that of standard Propositional Logic (**PL**).

3.3 Examples of Institutions

Example 5 (Propositional Logic) The institution PL of Propositional Logic is defined in the following way:

- The category $\mathbb{S}ig^{PL}$ has as objects sets of propositional variables and as arrows the functions between them.
- A signature morphism σ is a mapping between the propositional variables.
- The functor $\boldsymbol{Sen}^{PL}$ maps every signature Σ to $\boldsymbol{Sen}^{PL}(\Sigma)$ which consists of propositional formulas with propositional variables from Σ and connectives for conjunction, disjunction, implication, and negation.
- The $\boldsymbol{Sen}^{PL}(\sigma)$ is the extension of σ to all formulas.
- Models of Σ are truth valuations, i.e., mappings from Σ into the standard Boolean algebra $Bool = \{0, 1\}$.
- A model morphism between Σ-models M and M' exists iff for all $p \in \Sigma,\ M(p) \leq M'(p)$.
- Given $\sigma : \Sigma_1 \to \Sigma_2$ and a Σ_2-model $M_2 : \Sigma_2 \to Bool$, then the reduct $M_2 \restriction_\sigma$ is the composition $M_2 \circ \sigma$.
- $M \models^{PL}_{\Sigma} \phi$ if and only if ϕ evaluates 1 under the standard extension of M to all formulas.

Example 6 (Temporal Logic) The following example is a simplified version of Temporal Logic (**TL**) and its formalization in the theory follows [10].

- The signatures $\mathbb{S}ig^{TL}$ consists of sets of actions;
- The models $\mathbf{Mod}^{TL}(\Sigma)$ consist of sets of runs, which are finite or infinite sequences of (sets of) actions;
- The sentences $\boldsymbol{Sen}(\Sigma)$ are sets of sentences which are build up from atomic sentences p using the standard propositional and temporal connectives;
- A satisfaction relation $M \models^{TL}_{\Sigma} \phi$ holds if and only if ϕ holds at the beginning of every run in M;

This section is devoted to examples of institutions.

Example 7 (First Order Logic) Possibly, the most important example of institution is First Order Logic (**FOL**). A *signature* in **FOL** is a triple (S, F, P) that consists of:

- S is the set of *sort* symbols, for example, $S = \{N, Z\}$ where N denotes the natural numbers ($\mathbb{N}$) and Z the integers ($\mathbb{Z}$);
- $F = \{F_{w \to s} \mid w \in S^*, s \in S\}$ is a family of sets of *operation* symbols such that $F_{w \to s}$ denotes the set of operations with arity w and sort s, for example, $F_{NN \to N} = \{+\}, F_{ZZ \to Z} = \{+, -\}$;

- $P = \{P_w \mid w \in S^*\}$ a family of sets relation symbols where P_w denotes the set of relations with arity w, for example, $P_{NN} = P_{ZZ} = \{\leq\}$ or $P_w = \emptyset$;

The models of **FOL** are related with the sort symbols in a natural way and the sentences are the standard expansion of **PL**. Furthermore, the institution of **PL** can be seen as a sub-institution of **FOL** obtained by restricting the signatures to those with an empty set of sort symbols [10].

Example 8 (Weak Propositional Logic) The weak propositional logic (denoted by **WPL**) designates a variation of Propositional Logic. The sentences are the same as in **PL**, but the models are valuations $M : \boldsymbol{Sen}(P) \mapsto \{0, 1\}$ the standard truth table semantics of all Boolean connectives except negation, i.e.:

- $M(\phi \wedge \psi) = 1$ if and only if $M(\phi) = M(\psi) = 1$;
- $M(\phi \vee \psi) = 0$ if and only if $M(\phi) = M(\psi) = 0$;
- if $M(\phi) = 1$, then $M(\neg\phi) = 0$;

Example 9 (Modal First Order Logic) The last example is the standard Modal First Order Logic (**MFOL**), with modalities $\Box, \Diamond$ and Kripke semantics. The **MFOL** *signatures* are sixples (S, S_0, F, F_0, P, P_0), where (S, F, P) is the signature of **FOL** and (S_0, F_0, P_0) is a sub-signature of (S, F, P) of rigid symbols [10]. A **MFOL** model (W, R), called Kripke model, consists of

- a family $W = \{W_i\}_{i \in I_W}$ of possible worlds, which are models in **FOL**;
- a binary relation $R \subseteq I_W \times I_W$ between the possible worlds such that the following sharing condition holds:
  $\forall i, j \in I_W$ we have that $W_i^x = W_j^x$ for each x.

The satisfaction of **MFOL** sentences by the Kripke models is defined in the following way:

$$(W, R) \models \phi \Leftrightarrow (W, R) \models^i \phi \; \forall i \in I_W \tag{3.1}$$

where $(W, R) \models^i \phi$ is defined by induction:

- $(W, R) \models^i \phi$ if and only if $W_i \models^{FOL} \phi$ for all atom ϕ and each $i \in I_W$;
- $(W, R) \models^i \phi \wedge \psi$ if and only if $W_i \models^{FOL} \phi$ and $W_i \models^{FOL} \psi$;
- $(W, R) \models^i \Box\phi$ if and only if $W_i \models^k \phi$ for $< i, k > \in R$;
- $(W, R) \models^i \forall X\phi$ if and only if $(W', R) \models^i \phi$ for all expansions (W', R) of (W, R) to a Kripke model;
- $\Diamond\phi$ abbreviates $\neg\Box\neg\phi$;

Like **PL** we get the institution of Modal Propositional Logic (**MPL**) as a sub-institution of **MFOL** defined by the signatures with an empty set of sort symbols and empty set of rigid relation symbols.

3.4 Morphisms

The concept of institution morphism formalizes the mapping from a more complex to a simpler institution.

Definition 3.5 (Institution Morphism) Let $\mathcal{I}$ and $\mathcal{I}'$ be two institutions, then an *institution morphism* $\Phi : \mathcal{I} \to \mathcal{I}'$ consists of:

1. a functor $\Phi : \mathbb{S}ig \to \mathbb{S}ig'$
2. a natural transformation $\alpha : \Phi; \boldsymbol{Sen}' \Rightarrow \boldsymbol{Sen}$ and
3. a natural transformation $\beta : \mathbf{Mod} \Rightarrow \Phi^{op}; \mathbf{Mod}'$

such that the following Satisfaction Condition holds

$$m \models_{\Sigma} \alpha_{\Sigma}(e') \quad \text{iff} \quad \beta_{\Sigma}(m) \models_{\Phi(\Sigma)} e' \tag{3.2}$$

for any Σ-model m from $\mathcal{I}$ and any $\Phi(\sigma)$-sentence e' from $\mathcal{I}'$

Remark 3.6 Φ; $\boldsymbol{Sen}'$ is the composition of the functors Φ and $\boldsymbol{Sen}'$, for more information see [10].

Figures 7 and 8 show a representation of the natural transformations α_{Σ} and β_{Σ}. The institution morphisms are suitable to formalize "forgetful" maps between more complex institutions to simpler ones.

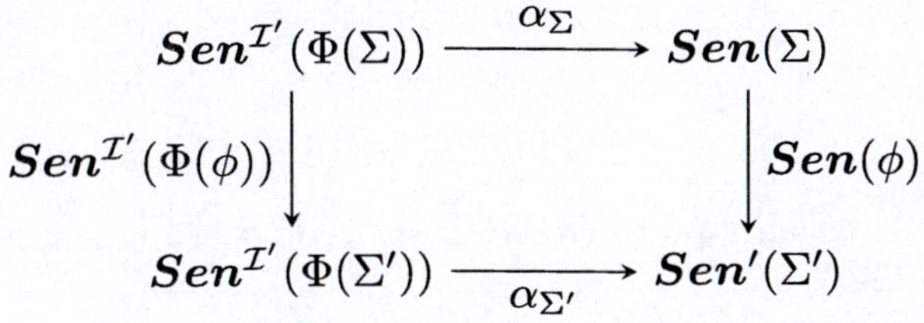

Fig. 7 First Morphism square

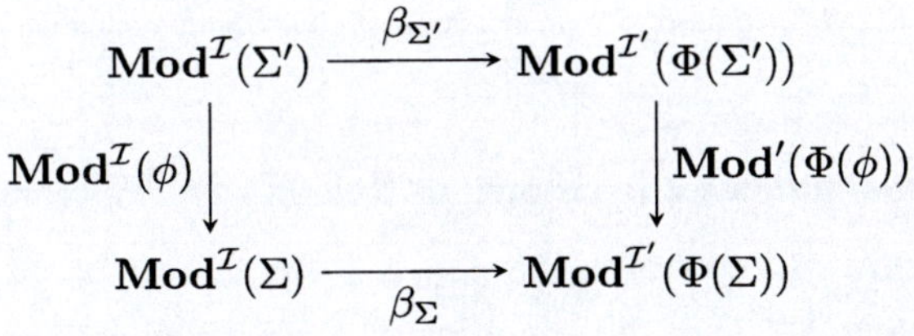

Fig. 8 Second Morphism square

*Example 10 (The Morphism Between **FOL** and **MFOL**)* Regarding these two institutions we can define the morphism $\Phi : \boldsymbol{FOL} \mapsto \boldsymbol{MFOL}$ which maps the $\boldsymbol{FOL} - (S, F, P)$ signature to $\boldsymbol{MFOL} - (S, S, F, F, P, P)$ signature, such that the natural transformation α erases the modalities from the sentences.

4 Institution-Theoretic Square of Opposition

4.1 *The Aristotelian Relations of Sentences in Institution-Theoretic Setting*

The square of opposition is commonly known as a diagram for which many extensions have been proposed in the second half of the twentieth century. However, most of them are discussed at informal level.

Using the formalism introduced by Hans Smessaert and Lorenz Demey [7], we generalize these concepts to the level of theory of institutions. It should be emphasized that the theory of institutions guarantees that the subsequent definitions apply to all logical systems under consideration. For reasons of convenience, we assume that the logical systems contain the classical connectives as a syntactic method of constructing sentences. The authors cited above define Aristotelian Geometry as a logical system, which has the links of denial, conjunction, and implication. Hence, the following definitions are naturally given:

Definition 4.1 (Aristotelian Relation of Contradictoriness) Let $\mathcal{I} = \left(\mathbb{S}ig^{\mathcal{I}}, \boldsymbol{Sen}^{\mathcal{I}}, \boldsymbol{Mod}^{\mathcal{I}}, \models^{\mathcal{I}}\right)$ be an arbitrary institution, $\Sigma \in \mathbb{S}ig^{\mathcal{I}}$ and ϕ, ψ be sentences in $\boldsymbol{Sen}(\Sigma)$. Then the propositions ϕ, ψ are called *contradictory*, if the truth of one implies the falsity of the other, and conversely.

$$\models^{\mathcal{I}}_{\Sigma} (\phi \Rightarrow \neg\psi) \wedge (\neg\psi \Rightarrow \phi) \tag{4.1}$$

$$\models^{\mathcal{I}}_{\Sigma} (\neg\phi \vee \neg\psi) \wedge (\psi \vee \phi) \tag{4.2}$$

$$\models^{\mathcal{I}}_{\Sigma} \neg(\phi \wedge \psi) \text{ and } \models^{\mathcal{I}}_{\Sigma} \neg(\neg\phi \wedge \neg\psi) \tag{4.3}$$

We denote the relation between two contradictory sentences $R_C(\phi, \psi)$ or by a graph (Fig. 9)

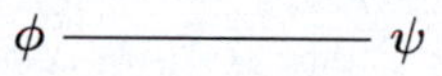

Fig. 9 Geometrical representation of contradictory sentences

Definition 4.2 (Aristotelian Relation of Contrariety) Let $\mathcal{I} = \left(\mathbb{S}ig^{\mathcal{I}}, \boldsymbol{Sen}^{\mathcal{I}}, \boldsymbol{Mod}^{\mathcal{I}}, \models^{\mathcal{I}}\right)$ be an arbitrary institution, $\Sigma \in \mathbb{S}ig^{\mathcal{I}}$ and ϕ, ψ be sentences in $\boldsymbol{Sen}(\Sigma)$. Then the sentences ϕ, ψ are called *contrary*, if they cannot both be true.

$$\models^{\mathcal{I}}_{\Sigma} \neg(\phi \wedge \psi) \text{ and } \not\models^{\mathcal{I}}_{\Sigma} \neg(\neg\phi \wedge \neg\psi) \tag{4.4}$$

We denote the relation between two contrary sentences $R_c(\phi, \psi)$ or by a graph (Fig. 10)

ϕ - - - - - - - ψ

Fig. 10 Geometrical representation of contrary sentences

Definition 4.3 (Aristotelian Relation of Subcontrariety) Let $\mathcal{I} = \left(\mathbb{S}ig^{\mathcal{I}}, \boldsymbol{Sen}^{\mathcal{I}}, \boldsymbol{Mod}^{\mathcal{I}}, \models^{\mathcal{I}}\right)$ be an arbitrary institution, $\Sigma \in \mathbb{S}ig^{\mathcal{I}}$ and ϕ, ψ be sentences in $\boldsymbol{Sen}(\Sigma)$. Then the sentences ϕ, ψ are called *subcontrary*, if it is impossible for both to be false.

$$\not\models_{\Sigma}^{\mathcal{I}} \neg(\phi \wedge \psi) \text{ and } \models_{\Sigma}^{\mathcal{I}} \neg(\neg\phi \wedge \neg\psi) \tag{4.5}$$

We denote the relation between two subcontrary sentences $R_s(\phi, \psi)$ or by a graph (Fig. 11)

ϕ .... ψ

Fig. 11 Geometrical representation of subcontrary sentences

Definition 4.4 (Aristotelian Relation of Subalternation) Let $\mathcal{I} = \left(\mathbb{S}ig^{\mathcal{I}}, \boldsymbol{Sen}^{\mathcal{I}}, \boldsymbol{Mod}^{\mathcal{I}}, \models^{\mathcal{I}}\right)$ be an arbitrary institution, $\Sigma \in \mathbb{S}ig^{\mathcal{I}}$ and ϕ, ψ be sentences in $\boldsymbol{Sen}(\Sigma)$. Then the sentences ϕ, ψ are called *subalternate*, if the truth of the first (the "superaltern") implies the truth of the second (the "subaltern"), but not conversely.

$$\models_{\Sigma}^{\mathcal{I}} \phi \rightarrow \psi \text{ and } \not\models_{\Sigma}^{\mathcal{I}} \psi \rightarrow \phi \tag{4.6}$$

We denote the relation between two subalternate sentences $R_S(\phi, \psi)$ or by a graph (Fig. 12)

$\phi \longrightarrow \psi$

Fig. 12 Geometrical representation of subalternate sentences

Definition 4.5 (Boethian Diagram) Let an arbitrary institution $\mathcal{I} = \left(\mathbb{S}ig^{\mathcal{I}}, \boldsymbol{Sen}^{\mathcal{I}}, \boldsymbol{Mod}^{\mathcal{I}}, \models^{\mathcal{I}}\right)$ and $\Sigma \in \mathbb{S}ig$. Then a *Boethian diagram* is an edge-labeled graph. The vertices of the graph are pairwise non-equivalent sentences $e_1, e_2, \ldots, e_n \in \boldsymbol{Sen}(\Sigma)$ and the edges of the graph are the Aristotelian relations (see Fig. 13).

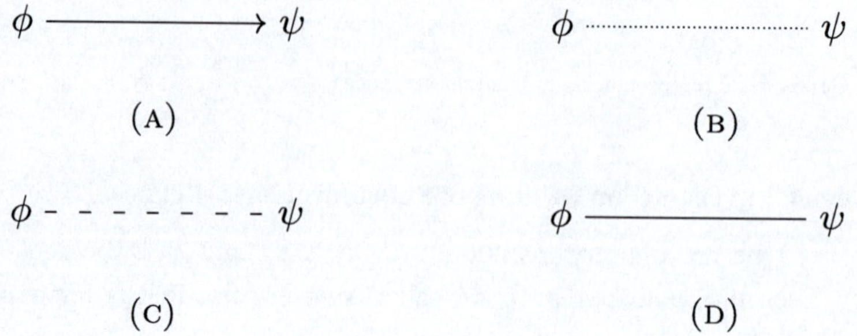

Fig. 13 The fundamental Aristotelian relations. (**a**) Relation of subalternation. (**b**) Relation of subcontrariety. (**c**) Relation of contrariety. (**d**) Relation of contradictoriness

Concerning the Definitions 4.6 and 4.7 below, we should note that Definition 4.6 is the common traditional square of opposition. However, the shape in the second Definition 4.7 is introduced, as we will see in a natural way so that we could see how the square of opposition changes from one logical system to another logical system. For this reason, we call it *rhombus of opposition*.

Definition 4.6 (Aristotelian Square) Let an arbitrary institution $\mathcal{I} = \left(\mathbb{S}ig^{\mathcal{I}}, \boldsymbol{Sen}^{\mathcal{I}}, \boldsymbol{Mod}^{\mathcal{I}}, \models^{\mathcal{I}}\right)$, $\Sigma \in \mathbb{S}ig$ and $p, q \in \boldsymbol{Sen}(\Sigma)$. Then an *Aristotelian Square* is a graph of the following form (Fig. 14):

Fig. 14 Aristotelian square of opposition

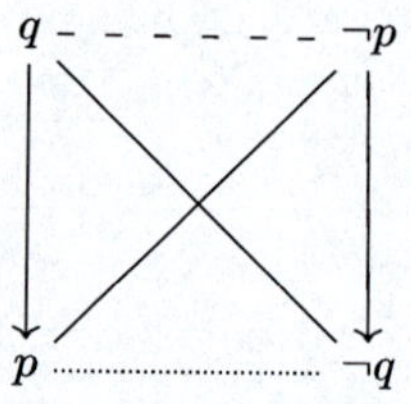

Definition 4.7 (Rhombus of Opposition) Let an arbitrary institution $\mathcal{I} = \left(\mathbb{S}ig^{\mathcal{I}}, \boldsymbol{Sen}^{\mathcal{I}}, \boldsymbol{Mod}^{\mathcal{I}}, \models^{\mathcal{I}}\right)$, $\Sigma \in \mathbb{S}ig$ and $p, q \in \boldsymbol{Sen}(\Sigma)$. Then a *rhombus of opposition* is (Fig. 15)

Fig. 15 Aristotelian Rhombus of opposition

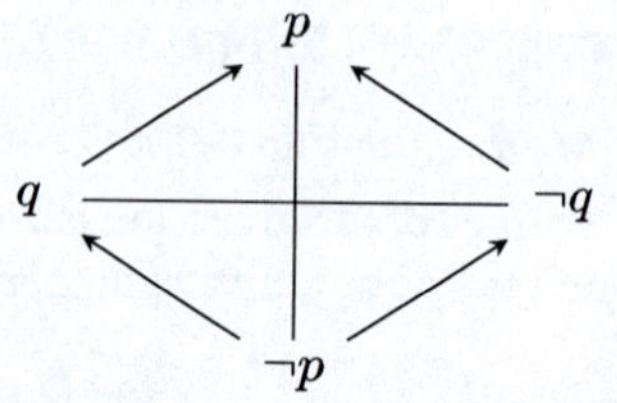

4.2 *The Example of PL: Square of Opposition*

In this subsection, we study the action of signature morphisms, i.e., how they affect the corresponding configurations.

The first example that we examine is that of propositional calculus. As shown in Figs. 16 and 17, after a signature morphism, the square structure is retained, if and only if the relationship remains unaltered; otherwise the square turns into a straight line segment. For the case of the straight line segment it is sufficient to imagine the possibility where $\sigma(p) = \sigma(q) = \chi$.

Fact Let $\mathcal{I} = \left(\mathbb{S}ig^{\mathcal{I}}, \boldsymbol{Sen}^{\mathcal{I}}, \boldsymbol{Mod}^{\mathcal{I}}, \models^{\mathcal{I}}\right)$ be an institution with the traditional square, then for every $\sigma : \Sigma \to \Sigma'$ the square either remains invariant (Fig. 16 or it terms of a line Fig. 17).

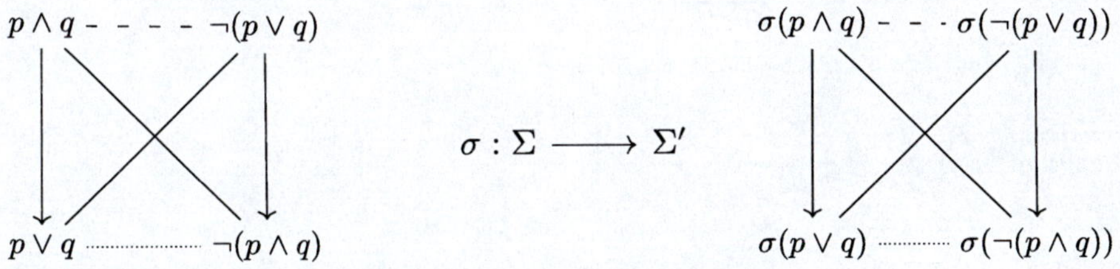

Fig. 16 Square of opposition in institution with Boolean connectives

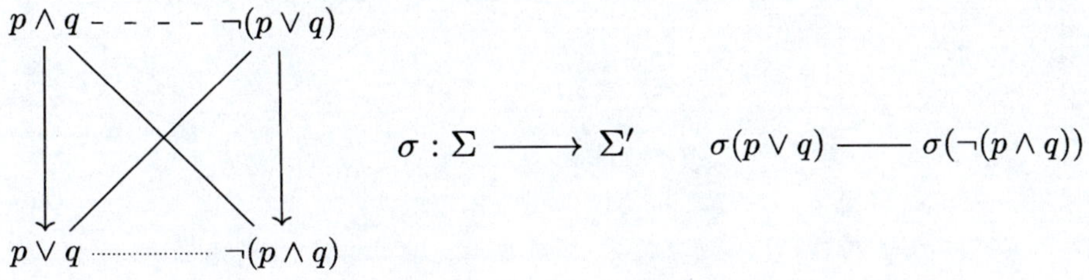

Fig. 17 Line of opposition in institution with Boolean connectives

4.3 *After the Action of Morphisms*

In this section, we present several examples (shown in Figs. 18, 19, 20, and 21) in which the square of opposition changes under the action of certain functors, i.e., we will illustrate how the square of opposition changes when we pass from one logical system over another logical system.

The square of opposition of the modal logic S5 changes when the "forgetful" functor acts and assigns the shape to the primary logic system. In the case of the square of opposition, the configuration changes and becomes a straight line segment, as the Sherwood-Czezowski hexagon does. On the other hand, the Sesmat-Blanche and the Beziau hexagons become a rhombus. As we mentioned earlier, we can have a morphism $\Phi : \boldsymbol{FOL} \mapsto \boldsymbol{MFOL}$ which represents the projection of Modal Logic into First Order Logic.

Fig. 18 The modal system S5

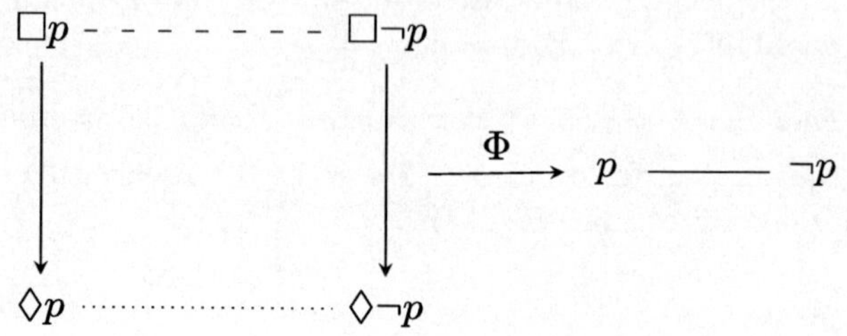

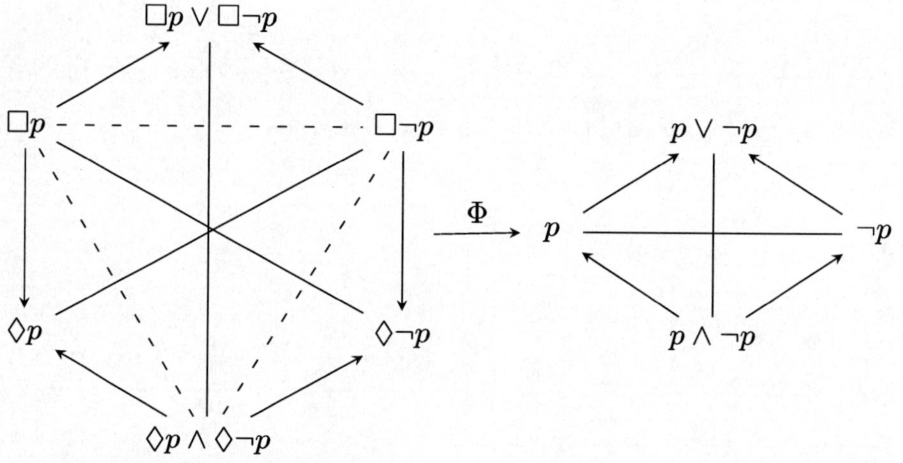

Fig. 19 Sesmat-Blanche hexagon

5 Institution-Theoretic Treatment of the Square of Opposition

In general, in the square of opposition we have a relation between two sentences. We have defined the relations of sentences $R_i(\phi, \psi)$ where i belongs to $\{C, c, S, s\}$.

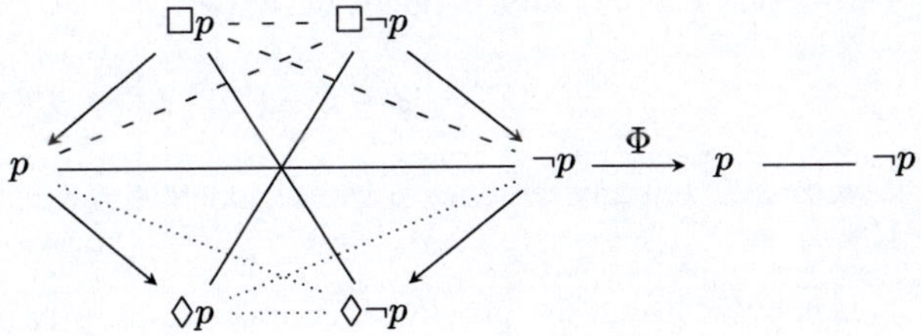

Fig. 20 Sherwood-Czezowski hexagon

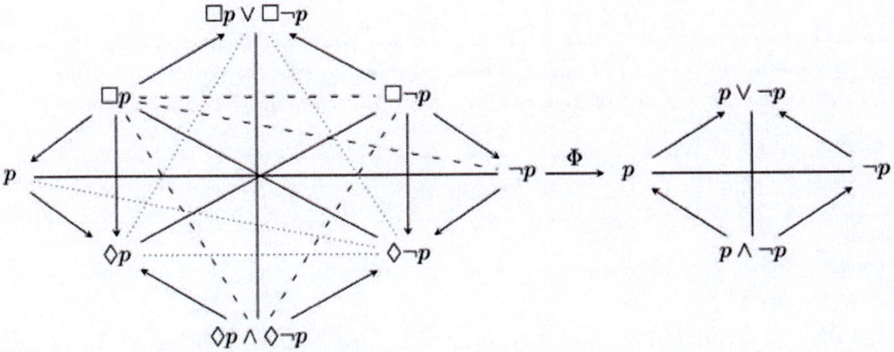

Fig. 21 Beziau octagon for S5

In order to pass to dual relations $R_i^*(\phi^*, \psi^*)$, we appeal to the concept of Galois connection, which is defined as follows:

$$* : R(\phi, \psi) \mapsto R^*(\phi^*, \psi^*) \tag{5.1}$$

Thus, the Galois connection forms way the dual relation, as well as the dual square of opposition in a natural way [9, 10].

5.1 *Galois Connection*

Let Σ be a signature in an arbitrary institution $\mathcal{I} = \left(\mathbb{S}ig^{\mathcal{I}}, \boldsymbol{Sen}^{\mathcal{I}}, \boldsymbol{Mod}^{\mathcal{I}}, \models^{\mathcal{I}}\right)$. Then we know that if E is set of sentences we have that the models of E is the class of models such that $M \models \phi$ for every sentence in E. Moreover, the **theory** of a class of models $\mathbb{M}$ is the set of sentences ϕ such that $M \models \phi$ for every model in this class. In formal language, this is expressed as follows:

- for every set of Σ-sentences E, we have

$$E^* = \{M \in \mathbf{Mod}(\Sigma) \mid M \models_\Sigma \phi \; \forall \phi \in E\}$$

- for every class $\mathbb{M}$ of Σ-models, we have

$$\mathbb{M}^* = \{\phi \in \boldsymbol{Sen}(\Sigma) \mid M \models \phi \ \forall M \in \mathbb{M}\}$$

Remark 5.1 For any sentence ϕ and a model M, we denote $\{\phi\}^* = \phi^*$ and $\{M\}^* = M^*$

It is evident that the previous definition implies two functions $(-)^*$, where "-" denotes an empty place, which are known as *Galois Connection.*

Lemma 1 *The two functions denoted $*$ in the previous paragraph determine a Galois connection (see [11, 12]), whenever they satisfy the following properties, for any collections E, E' of Σ-sentences and collections $\mathbb{M}$, $\mathbb{M}'$ of Σ-models:*

1. $E \subseteq E' \Rightarrow E'^* \subseteq E^*$.
2. $\mathbb{M} \subseteq \mathbb{M}' \Rightarrow \mathbb{M}'^* \subseteq \mathbb{M}^*$.
3. $E \subseteq E^{**}$.
4. $\mathbb{M} \subseteq \mathbb{M}^{**}$.
5. $E^* = E^{***}$.
6. $\mathbb{M}^* = \mathbb{M}^{***}$.
7. *There is a dual (i.e., inclusion reversing) isomorphism between the closed collections of sentences and the closed collections of models. This isomorphism maps unions to intersections and intersections to unions.*

   (a) $\bigcap_n E_n^* = \left(\bigcup_n E_n\right)^*$

   (b) $\left(\bigcap_n E_n^*\right)^{**} = \left(\bigcup_n E_n\right)^*$

   (c) $\left(\bigcup_n E_n^{**}\right)^* = \bigcap_n E_n^*$

   (d) $\left(\bigcup_n E_n^{**}\right)^* = \left(\bigcup_n E_n\right)^*$

   (e) $\left(\bigcap_n E_n^{**}\right)^* = \left(\bigcup_n E_n^*\right)^{**}$

   There are also dual identities to (a)–(e) for collections of models.

Proof We prove the first, the second and the 7(a). Let E be a set of Σ-sentences and

$$E^* = \{M \in \mathbf{Mod}(\Sigma) \mid \forall \phi \in E \ M \models \phi\}$$

a collection of models. Let $E_1 \subseteq E_2 \subseteq |Sen(\Sigma)|$ then from Galois Connection we take two collections of set of models, E_1^* and E_2^*.

$$
\begin{aligned}
M \in E_2^* &\Rightarrow \\
(\forall \phi \in E_2)\ M \models \phi &\Rightarrow \\
(\forall \phi \in E_1)\ M \models \phi &\Rightarrow \\
M \in E_1^* &\Rightarrow \\
E_2^* &\subseteq E_1^*
\end{aligned}
\tag{5.2}
$$

For the second, if $M \subseteq M' \subseteq |\mathbf{Mod}(\Sigma)|$ then

$$
\begin{aligned}
\phi \in M'^* &\Rightarrow \\
(\forall m \in M')\ m \models_\Sigma^{\mathcal{I}} \phi &\Rightarrow \\
(\forall m \in M)\ m \models_\Sigma^{\mathcal{I}} \phi &\Rightarrow \\
\phi \in M &\Rightarrow \\
M'^* &\subseteq M^*
\end{aligned}
\tag{5.3}
$$

And for the conjunction, if $M \in E_1^* \cap E_2^*$ then

$$
\begin{aligned}
M \in E_1^* \ \&\ M \in E_2^* &\Leftrightarrow \\
(\forall \phi \in E_1)\ M \models_\Sigma^{\mathcal{I}} \phi \ \&\ (\forall \phi \in E_2)\ M \models_\Sigma^{\mathcal{I}} \phi &\Leftrightarrow \\
(\forall \phi \in (E_1 \cup E_2))\ M \models_\Sigma^{\mathcal{I}} \phi &\Leftrightarrow \\
M \in (E_1 \cup E_2)^*
\end{aligned}
\tag{5.4}
$$

Thus we conclude that $\phi^* \cap \psi^* = (\phi \cup \psi)^*$ □

5.2 Aristotelian Relations and the Galois Connection

According to the previous section, we have four fundamental relations $R_i(\_,\_)$, where i belongs to $\{C, c, S, s\}$. Now we give an institution-independent form of these definitions using the Galois connection. First, we translate these relations in terms of the Galois connection.

(1) $\models_\Sigma^{\mathcal{I}} \neg(\phi \wedge \psi)$ In terms of Galois connection this means that

$$
\begin{aligned}
\forall M \in \mathbf{Mod}(\Sigma)\left(M \models_\Sigma^{\mathcal{I}} \neg\phi \text{ or } M \models_\Sigma^{\mathcal{I}} \neg\psi\right) &\Leftrightarrow \\
\forall M \in \mathbf{Mod}(\Sigma)\left(M \in \overline{\phi^*} \text{ or } M \in \overline{\psi^*}\right) &\Leftrightarrow \\
\overline{\phi^* \cap \psi^*} = \overline{\phi^*} \cup \overline{\psi^*} &= \mathbf{Mod}(\Sigma)
\end{aligned}
\tag{5.5}
$$

(2) $\models^{\mathcal{I}}_{\Sigma} \neg(\neg\phi \wedge \neg\psi)$ In terms of Galois connection this means that

$$\begin{aligned}\forall M \in \mathbf{Mod}(\Sigma)\left(M \models^{\mathcal{I}}_{\Sigma} \phi \text{ or } M \models^{\mathcal{I}}_{\Sigma} \psi\right) &\Leftrightarrow \\ \forall M \in \mathbf{Mod}(\Sigma)\left(M \in \phi^* \text{ or } M \in \psi^*\right) &\Leftrightarrow \\ \phi^* \cup \psi^* = \mathbf{Mod}(\Sigma)\end{aligned} \tag{5.6}$$

(3) $\not\models^{\mathcal{I}}_{\Sigma} \neg(\phi \wedge \psi)$ In terms of Galois connection this means that

$$\begin{aligned}\exists M \in \mathbf{Mod}(\Sigma):\ M \models^{\mathcal{I}}_{\Sigma} \phi \wedge \psi &\Leftrightarrow \\ \exists M \in \mathbf{Mod}(\Sigma):\ M \models^{\mathcal{I}}_{\Sigma} \phi \ \&\ M \models^{\mathcal{I}}_{\Sigma} \psi &\Leftrightarrow \\ \exists M \in \mathbf{Mod}(\Sigma):\ M \in \phi^* \ \&\ M \in \psi^* &\Leftrightarrow \\ \exists M \in \mathbf{Mod}(\Sigma):\ M \in \phi^* \cap \psi^* &\Leftrightarrow \\ \overline{\phi^*} \cup \overline{\psi^*} \subset \mathbf{Mod}(\Sigma) &\Leftrightarrow \\ \overline{\phi^* \cap \psi^*} = \overline{\phi^*} \cup \overline{\psi^*} \neq \mathbf{Mod}(\Sigma)\end{aligned} \tag{5.7}$$

(4) $\not\models^{\mathcal{I}}_{\Sigma} \neg(\neg\phi \wedge \neg\psi)$ In terms of Galois connection this means that

$$\begin{aligned}\exists M \in \mathbf{Mod}(\Sigma):\ M \models^{\mathcal{I}}_{\Sigma} \neg\phi \wedge \neg\psi &\Leftrightarrow \\ \exists M \in \mathbf{Mod}(\Sigma):\ M \in \overline{\phi^*} \ \&\ M \in \overline{\psi^*} &\Leftrightarrow \\ \exists M \in \mathbf{Mod}(\Sigma):\ M \in \overline{\phi^*} \cap \overline{\psi^*} &\Leftrightarrow \\ \exists M \in \mathbf{Mod}(\Sigma):\ M \in \overline{\phi^* \cup \psi^*} &\Leftrightarrow \\ \phi^* \cup \psi^* \subset \mathbf{Mod}(\Sigma) &\Leftrightarrow \\ \phi^* \cup \psi^* \neq \mathbf{Mod}(\Sigma)\end{aligned} \tag{5.8}$$

We should note that in the initial definition we talked about relations between sentences. However, by introducing the concept of the Galois Connection, we talk now about relations between collections of models. Then, applying again the concept of Galois Connection, we pass to collections of sentences, i.e., essentially to relations of sentences again (Figs. 22 and 23). Thus, in terms of relations we have the following scheme:

$$R(\phi, \psi) \xrightarrow{\ *\ } R^*(\phi^*, \psi^*) \xrightarrow{\ *\ } R^{**}(\phi^{**}, \psi^{**}) \tag{5.9}$$

This scheme is transferred in a natural way to the square's schemes. According to the following definitions, we have:

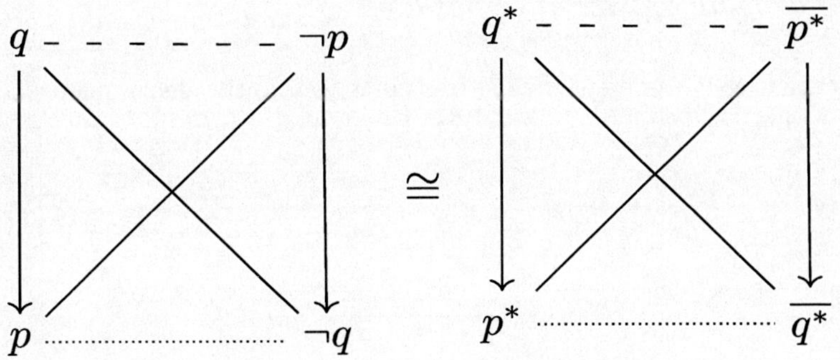

Fig. 22 Transformation of square

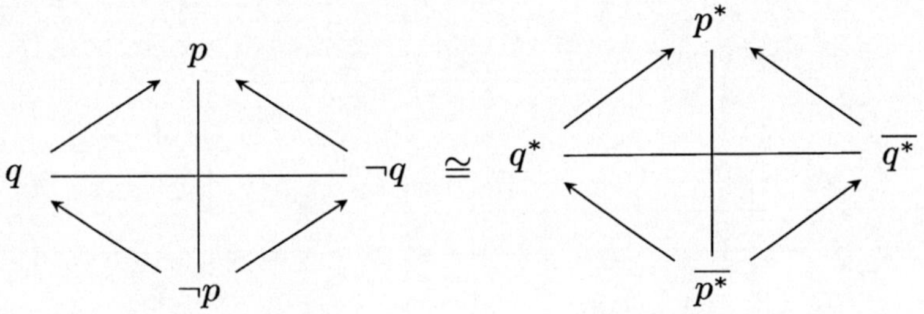

Fig. 23 Transformation of rhombuses

Definition 5.2 Two sets of models ϕ^*, ψ^* are in *dual contradictory* relation $R^*_C(\phi^*, \psi^*)$ if $\overline{\phi^* \cap \psi^*} = \overline{\phi^*} \cup \overline{\psi^*} = \mathbf{Mod}(\Sigma)$ and $\phi^* \cup \psi^* = \mathbf{Mod}(\Sigma)$ which is equivalent to

$$\overline{\phi^*} = \psi^* \tag{5.10}$$

Definition 5.3 Two sets of models ϕ^*, ψ^* are in *dual contrary* relation $R^*_c(\phi^*, \psi^*)$ if

$$\overline{\phi^* \cap \psi^*} = \overline{\phi^*} \cup \overline{\psi^*} = \mathbf{Mod}(\Sigma) \text{ and } \overline{\phi^*} \cap \overline{\psi^*} \neq \emptyset \tag{5.11}$$

Definition 5.4 Two sets of models ϕ^*, ψ^* are in *dual subcontrary* relation $R^*_s(\phi^*, \psi^*)$ if

$$\phi^* \cup \psi^* = \mathbf{Mod}(\Sigma) \text{ and } \phi^* \cap \psi^* \neq \emptyset \tag{5.12}$$

Definition 5.5 Two sets of models ϕ^*, ψ^* are in *dual subalternate* relation $R^*_S(\phi^*, \psi^*)$ if

$$\phi^* \subset \psi^* \tag{5.13}$$

5.3 *The Dual Square of Opposition*

The scheme above is transferred in a natural way to the dual square's schemes (Figs. 24 and 25).

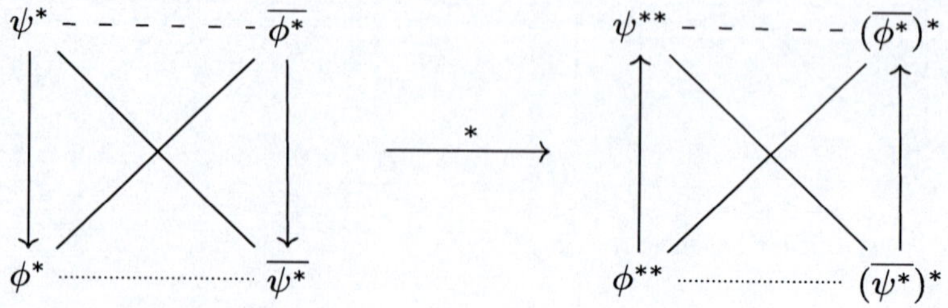

Fig. 24 Dual square of opposition

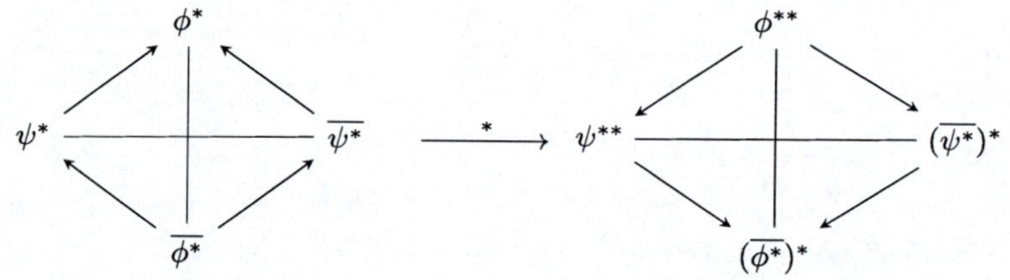

Fig. 25 Dual rhombus of opposition

Definition 5.6 Two sets of sentences ϕ^{**}, ψ^{**} are in a *dual dual contradictory* relation $R_C^{**}(\phi^{**}, \psi^{**})$, if $R_C^{*}(\phi^{***}, \psi^{***})$;
Two sets of sentences ϕ^{**}, ψ^{**} are in a *dual dual contrary* relation $R_c^{**}(\phi^{**}, \psi^{**})$, if $R_c^{*}(\phi^{***}, \psi^{***})$;
Two sets of sentences ϕ^{**}, ψ^{**} are in a *dual dual subcontrary* relation $R_s^{**}(\phi^{**}, \psi^{**})$, if $R_s^{*}(\phi^{***}, \psi^{***})$;
Two sets of sentences ϕ^{**}, ψ^{**} are in a *dual dual subalternate* relation $R_S^{**}(\phi^{**}, \psi^{**})$, if $R_S^{*}(\phi^{***}, \psi^{***})$

We know that $E^{*} = E^{***}$ and $\mathbb{M}^{*} = \mathbb{M}^{***}$. So we can obtain the following generalization for abstract set of models and sentences (Figs. 26 and 27).

Definition 5.7 Two sets of models $\mathbb{D}, \mathbb{E}$ are in a *dual contradictory* relation $R_C^{*}(\mathbb{D}, \mathbb{E})$, if

$$\overline{\mathbb{D} \cap \mathbb{E}} = \overline{\mathbb{D}} \cup \overline{\mathbb{E}} = \mathbf{Mod}(\Sigma) \text{ and } \mathbb{D} \cup \mathbb{E} = \mathbf{Mod}(\Sigma) \tag{5.14}$$

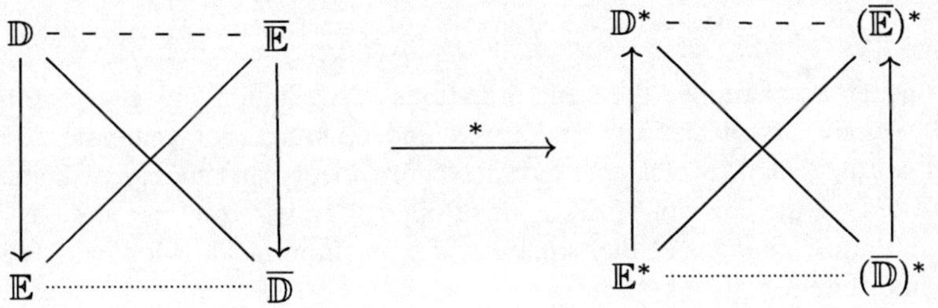

Fig. 26 Generalized dual square of opposition

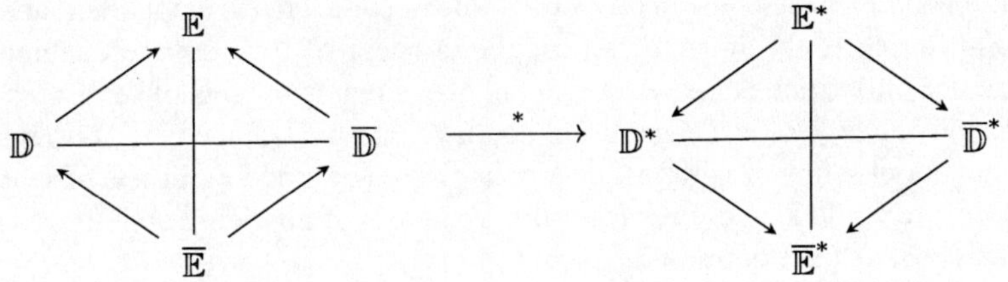

Fig. 27 Generalized dual rhombus of opposition

Two sets of models $\mathbb{D}$, $\mathbb{E}$ are in a *dual contrary* relation $R_c^*(\mathbb{D}, \mathbb{E})$, if

$$\overline{\mathbb{D} \cap \mathbb{E}} = \overline{\mathbb{D}} \cup \overline{\mathbb{E}} = \mathbf{Mod}(\Sigma) \text{ and } \overline{\mathbb{D}} \cap \overline{\mathbb{E}} \neq \emptyset \tag{5.15}$$

Two sets of models $\mathbb{D}$, $\mathbb{E}$ are in a *dual subcontrary* relation $R_s^*(\mathbb{D}, \mathbb{E})$, if

$$\mathbb{D} \cup \mathbb{E} = \mathbf{Mod}(\Sigma) \text{ and } \mathbb{D} \cap \mathbb{E} \neq \emptyset \tag{5.16}$$

Two sets of models $\mathbb{D}$, $\mathbb{E}$ are in a *dual subalternate* relation $R_S^*(\mathbb{D}, \mathbb{E})$, if

$$\mathbb{D} \subset \mathbb{E} \tag{5.17}$$

Two sets of sentences D, E are in a *dual dual contradictory* relation $R_C^{**}(D, E)$, if their duals D^*, E^* are in a dual contradictory relation $R_C^*(D^*, E^*)$;
Two sets of sentences D, E are in a *dual dual contrary* relation $R_c^{**}(D, E)$, if their duals D^*, E^* are in a dual contrary relation $R_c^*(D^*, E^*)$;
Two sets of sentences D, E are in a *dual dual subcontrary* relation $R_s^{**}(D, E)$, if their duals D^*, E^* are in a dual subcontrary relation $R_s^*(D^*, E^*)$;
Two sets of sentences D, E are in a *dual dual subalternate* relation $R_S^{**}(D, E)$, if their duals D^*, E^* are in a dual subalternate relation $R_S^*(D^*, E^*)$;

6 Conclusions

In this paper we examined the transformations of the logical object conventionally called "square of opposition" undergoes under changes of semantics. For this reason, we appealed to concepts from category theory and theory of institutions. By introducing the concept of rhombus of opposition we examined the basic cases of configuration changes of the 'squares' of opposition inside a logical system and between different logical systems.

Further, by introducing the concept of Galois connection we showed the equilibrium that can be established between the sentences of the traditional square of opposition and the internal semantics of Boolean connectives, using them at a meta-level. Furthermore, the introduction of the concept of dual square enabled us to examine not only squares for sentences but also squares for sets of sentences.

Since quite a few logical systems do not have internal connectives, it is legitimate to talk not about sentences graphs, but about classes of models and sets of sentences. Therefore, we could also escape from the weakness of not being able to write basic relationships, such as contradiction.

This is the first time that different 'squares' of opposition were compared within a unified framework by using abstract model theory. We aim at integrating the different versions of the 'square' of opposition into this universal framework.

Acknowledgments This research is funded in the context of the project "MIS5005844" under the call for proposals supporting young researchers with emphasis on new researchers" (EDULL 34). The project is co-financed by Greece and the European Union (European Social Fund) by the Operational Programme Human Resources Development, Education and Lifelong Learning 2014–2020.

References

1. Béziau, J.-Y. New Light on the Square of Oppositions and its Nameless Corner, *Logical Investigations*, 10, (2003), pp. 218–233.
2. Béziau, Jean-Yves; Basti, Gianfranco (Eds.). The Square of Opposition. A Cornerstone of Thought. Springer (2017).
3. Blanché, R., Sur l'opposition des concepts, *Theoria*, 19 (1953).
4. Goguen, J, Burstall, R.: Institutions: Abstract model theory for specification and programming. *Journal of the Association for Computing Machinery*, 39, 95–146, (1992)
5. Moretti, Alessio: The Crowdy Logical Zoo Inhabited by the Old Square of Oppositions and the Many Strange Visitors of it. Available online at http://www.square-of-opposition.org/alessio%20moretti.html (Accessed 16-9-2018).
6. Moretti, Alessio: Non-linear Modal Graphs: the Simplest Bifurcation inside n-Opposition Theory, *UNILOG* 2007, Xi'an, China (August 2007), p. 87.
7. Smessaert, Hans & Lorenz Demey Logical Geometries and Information in the Square of Oppositions, *Journal of Logic, Language and Information* 23 (4) 2014:527–565.
8. Vandoulakis Ioannis and Tatiana Denisova, "On the Historical Transformations of the Square of Opposition as Semiotic Object", *Logica Universalis* 14 (2020), 7–26.
9. S. Mac Lane, Categories for the Working Mathematician (Graduate Texts in Mathematics 5)

10. Razvan Diaconescu, Institution-independent Model Theory, *Studies in Universal Logic*, Birkhaüser Basel, 2008, pp. 370. ISBN 978-3-7643-8707-5
11. Birkhoff, G. and Mac Lane, S. "The Galois Group." §15.2 in *A Survey of Modern Algebra*, 5th ed. New York: Macmillan, pp. 397–401, 1996.
12. Jacobson, N. Basic Algebra I, 2nd ed. New York: W. H. Freeman, p. 234, 1985.
13. Yiannis Kiouvrekis, Petros Stefaneas and Ioannis Vandoulakis "On the transformations of the Square of Opposition", Jean-Yves Beziau, Arthur Buchsbaum, Ioannis Vandoulakis (eds) Sixth World Congress on the Square of Opposition Crete, November 1–5, 2018, Handbook of Abstracts, 50–51.

Color-Coded Epistemic Modes in a Jungian Hexagon of Opposition

Julio Michael Stern

Dedicated to Eva Leonore Fanny Stern, my mother and translator of C.G. Jung's works to Portuguese.
Colors symbolize qualities, which can be interpreted in various ways.
Psychologically this points to orienting functions of consciousness, of which at least one is unconscious and therefore not available for conscious use.

C.G. Jung (CW, IX, pr.582, abridged) [77].

Abstract This article considers distinct ways of understanding the world, referred in psychology as *functions of consciousness* or as *cognitive modes*, having as scope of interest epistemology and natural sciences. Inspired by C.G. Jung's *simile of the spectrum*, we consider three basic cognitive modes, associated with: (R) embodied instinct, experience, and action; (G) reality perception and learning; and (B) concept abstraction, rational thinking, and language. RGB stands for the primary colors: red, green, and blue. Accordingly, a conceptual map between cognitive modes and primary and secondary colors is build based on physics and physiology of color perception and epistemological characteristics of the aforementioned cognitive modes, leading to logical relations structured as an *hexagon of opposition*.

Keywords Bit-strings · Cognitive modes · Color-coded logic · Epistemology · Hexagon of opposition · C.G. Jung · Simile of the spectrum

---

Thanks to Jean-Yves Beziau and Ioannis Vandoulakis, organizers of the 6th World Congress on the Square of Opposition, in November 01-05, 2018, at Kolymbari, Crete.

J. M. Stern (✉)
IME-USP Institute of Mathematicsáticas and Statisticsísticas, University of São Paulo, São Paulo, Brazil
e-mail: jstern@ime.usp.br

J.-Y. Beziau, I. Vandoulakis (eds.), *The Exoteric Square of Opposition*, Studies in Universal Logic, https://doi.org/10.1007/978-3-030-90823-2_14

Mathematics Subject Classification (2000) Primary 03A10; Secondary 62A01

1 Introduction

Epistemology or knowledge theory is the branch of philosophy concerned with studying how we learn about our environment and how we verify and justify the acquired knowledge. In this article, I restrict my interest in epistemology to the scope of natural sciences. Nevertheless, my interests also take in consideration the human subject, observer, or agent of learning, and how he or she uses and integrates distinct ways of understanding the world—ways often refereed in psychology as *functions of consciousness* or as *cognitive modes*, see [176]. With this goal in mind, I follow in the footsteps of Swiss psychologist Carl Gustav Jung (1875–1961), who used conceptual models where *colors symbolize qualities* constituting a color-coded system that *points to orienting functions of consciousness*, as stated in the opening quotation. The best known of these systems concerns Jung's categorization of *psychological types*— that is *not* a system used in this article. Instead, I develop in the sequel an alternative system of color-coded cognitive modes based on Jung's celebrated *simile of the spectrum*.

The systems of color-coded cognitive modes used by Jung are in no way arbitrary: First, these colors and modes relate to associations Jung frequently found in patient's dreams or in historically recorded imagery that also relate to the etymology of color terms and the evolution and organizational patterns of these terms found in human languages. Second, these color-codings have significant connections to the physics of color formation and; Third, these color-codings have significant connections to the physiology of color perception. These physical and physiological connections are frequently overlooked in the psychology literature. Nevertheless, the aforementioned connections are specially interesting for the epistemological applications I have in mind, for they correspond to, respectively, external vs. internal or objective vs. subjective aspects of color processing, in particular, or knowledge representation in general.

Section 2 reviews basic notions of modern color theory. Section 3 relates color theory and logical structures. Section 4 develops a model inspired by Jung's simile of the spectrum in which color-coded cognitive modes and the logical structure of their interrelations are interpreted in the context of epistemology and philosophy of science. Sections 5, 6, and 7 examine some examples of how these cognitive modes are interpreted in the scope of scientific disciplines. Section 8 presents some directions for further research and final remarks.

Fig. 1 Newton's [106] *Opticks* (updated) color wheel (circular perception structure) vs. linear structure of light spectrum; Helmholtz's [61] cone receptors response curves

2 Modern Color Theory

This section presents an abridged and selective chronology of modern color theory, focusing on relevant concepts needed for this paper. Modern color theory starts with the publication of Isaac Newton's *Opticks* (1704), where he showed how (a) a ray of white sun light can be decomposed by a prism into a *spectrum* of color hues, forming a linear continuum ranging from red to violet, as commonly seen in a rainbow, see Fig. 1r.[1] Moreover, Newton showed that (b) different color sensations can be generated by mixing light of specific spectral hues. For example, a sensation of violet can be generated by mixing red and blue. Furthermore, (c) color sensations like magenta or purple are not produced by light from any single locus in the linear spectrum, but they can only be produced by various mixtures of red and blue. Hence, Newton suggested that (d) human perception of colors is better represented by a *color wheel*, where the red and violet ends of the linear spectrum are joined to form a circle. Figure 1l depicts an updated version of Newton's color wheel, see [93]. In this article, the violet-magenta-purple region joining the extremities of the linear spectrum is called as the *paradoxical region* of the color wheel, while magenta-purple hues span the more restricted *non-spectral* region. In his famous *simile of the spectrum*, C.G.Jung (CW, VIII, pr.414–416, pp.3167–3169) compared this representation to an Ouroboros—a serpent biting its own tail at the paradoxical region of the color wheel, see Fig. 1c and [2].

Using a simple thermometer, in 1801, William Herschel was able to detect infra-red radiation, that is, radiation located beyond the red end of the spectrum that is invisible to the human eye, see [151]. In a similar way, using photo-chemical reactions, in 1801, Johann Wilhelm Ritter detected ultra-violet radiation beyond the violet end of the visible spectrum. Hence, using the color wheel representation, it is an understandable *façon de parler* to speak of hues at the paradoxical region as *neither ultra-violet nor infra-red but an undivided blend of both*, see [132, p.23].

[1] Positional figure locators: c=center, t=top, b=bottom, l=left, r=right.

Meanwhile, Thomas Young [181] postulated that the human perception of color is based on three types of light receptors at the eye's retina. Hermann von Helmholtz [61] and James Clerk Maxwell (1857) were able to verify Young's intuition in a series of experiments designed to elucidate peculiarities of human perception of color. These three receptors are nowadays denominated L, M, S *cones* that are sensible to radiation roughly located, respectively, at red, green, blue regions of the spectrum, see Fig. 1r. Maxwell's [95] triangle uses a convenient system of coordinates to specify color hues by their red, green, and blue (RGB) components. This system of coordinates is known in mathematics as (de Finetti's) compositional diagram, where each coordinate is in the [0, 1] interval and all coordinates add up to 1, see Fig. 2tl, [46, S.77], [91, 160].

The sensitivity curves of LMS/RGB receptors depicted at Fig. 1r are normalized, i.e., these curves are plotted with maxima of same height. In fact, their absolute sensitivities are quite different: S/B receptors have a much smaller (neural output density) response than M/G receptors that, in turn, have a smaller response than L/R receptors. Furthermore, these receptors have distinct and non-linear response curves to color hue, resulting in highly non-linear combined response curves for brightness and color saturation, or for other qualitative aspects of color perception around the

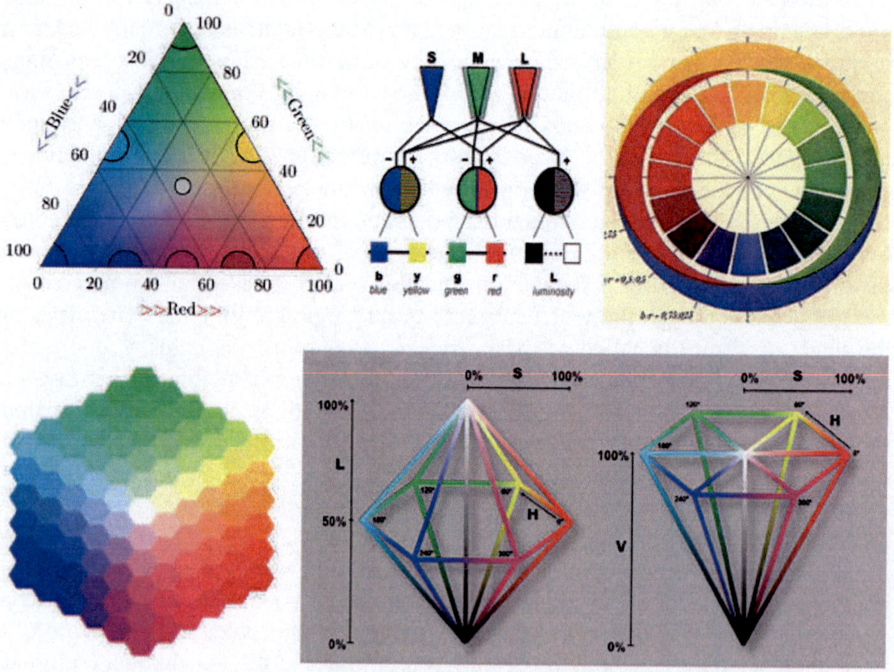

Fig. 2 Top: Maxwell [95] RGB compositional diagram including approximate position of violet, purple, primary and secondary colors; Neural network re-encoding trichromatic (RGB) inputs into oppositional (RGBY) outputs; Hering [65] circle of four antagonistic archetypal colors (RGBY). Bottom: Hexagonal tiling and color schemata, from Runge [129] and Wundt [179] to HSL/HSV encoding by Smith, Joblove and Greenberg [73].

hue circle. Hence, additional calibration points around the hue circle are needed for good color encoding systems. For this purpose, the hexcone (hexagonal cone) color encoding system includes calibration points located midway between the primary colors (RGB), corresponding to the secondary colors cyan, magenta, and yellow (CMY), see Fig. 2b.

The primary colors constitute an additive basis, that is, different colors hues can be generated by mixing RGB light sources of different intensities. In particular, each one of the secondary colors is generated by mixing two primary colors, namely C=G+B, M=R+B, and Y=R+G. In contrast, the secondary colors constitute a subtractive basis, that is, different color hues can be generated by sending withe light through CMY filters of different intensities (like artists do by mixing paints).

The hexcone and similar color encoding systems were first envisioned by Philipp Otto Runge [129], further explored by Wilhelm Wundt [179, 180], and greatly developed for TV broadcasting and computer graphics in order to achieve good quality renderization of color images at high processing speed, see [58, 148, 150]. Hexcone encoding and similar systems are now ubiquitous, underlying color information structure in the modern world. The logical structure of such hexagonal color models is further examined in the next section.

The tripolar color model, developed by Maxwell and Helmholtz, was able to explain how distinct physical light sources and filters can be combined to obtain different colors. Meanwhile, Ewald Hering [65] developed an alternative quadripolar color model based on four archetypal colors or *Urfarben*, organized as antagonistic processes opposing red vs. green and yellow vs. blue, see Fig. 2tr. Hering's model was able to explain some color phenomena related to perception latency, see [65, 172]. Hering's model could also explain recurring organization patterns for color words found in human languages. Interestingly, exactly the same colors and structure are used by Jung to color-code oppositional cognitive modes in his theory of personality types, see [77], [76, p.48] [88] and [176].

At the beginning of XX century, Erwin Schrödinger [134–137] showed how to combine the aforementioned tripolar and quadripolar models into an integrated color theory, but the functional transforms underlying this integration are still a matter of current research. For example, Chittka and coauthors [20–22] show how neural networks responsible for post-processing of signals generated by cone receptors conform to the oppositional structure anticipated by Hering, see Fig. 2tc. From a logical point of view, the simplest structure able to integrate the aforementioned tripolar and quadripolar models is the hexagon of opposition, studied in the next section. For general overviews of color theory and its historical development, see [83, 93].

The aforementioned tripolar models describe color processing at the interface between the human eye and the external environment, while quadripolar models describe processes at a corresponding interface to the internal world of an embodied human mind. My interest in epistemology demands simultaneous attention to both external phenomena and their internal representation. Following Jung's intuition, I use for this purpose the framework provided by color theory, for vision is arguably the most important human sense for perception of phenomena in the external environment, and it should therefore have a comparable influence and importance in human internal representation and psychological processing. Accordingly, Sect. 4 develops a model in which primary and secondary colors equipped with a hexa-polar structure are interpreted as cognitive modes in the context of epistemology and philosophy of science.

3 Logic Structures and Color Theory

The superposition or compositional properties of primary and secondary colors entail a rich and intuitive algebraic structure that has been extensively explored in mathematical and philosophical studies, see [70, 71, 149]. Formally, a bit-string $\langle r, g, b \rangle$ in the 3-dimensional Boolean space $\{0, 1\}^3$ is used to represent the colors Red (R), Green (G), Blue (B), Cyan (C), Yellow (Y), Magenta (M), Black (K), and White (W), as shown in the cubic diagram at Fig. 3tl. Analogously, a vector $\langle r, g, b \rangle$ in the 3-dimensional Euclidean unit cube $[0, 1]^3$ is used to represent a continuum of color hues, as (partially) depicted in Fig. 3tl. Arrows in these diagrams represent color intensity gradients for the Euclidean color cube, and entailment or inferiority relations for the Boolean color cube.

The entailment relations in the Boolean color cube impose a (transitive) order structure captured by the algebraic lattice depicted in Fig. 3tr; for further details, see [8, 32–35, 70, 71]. The geometric orthogonal projection of the color cube along the K-W axis generates the color hexagon, as depicted in Fig. 3l. In addition to the entailment relations directly inherited from the color cube, the color hexagon includes other important logical relations corresponding to color theoretic properties: *Contrariety* relations, represented in the hexagon by dashed lines (– –), interconnect elements of the additive color basis; Meanwhile, *sub-contrariety relations*, represented in the hexagon by dotted lines ($\cdots$), interconnect elements of the subtractive color basis, see Fig. 3bl. Accordingly, bit-string codes of any two contrary colors have null or K=$\langle 0, 0, 0 \rangle$ intersection or minimum, while bit-string codes of any two sub-contrary colors have full or W=$\langle 1, 1, 1 \rangle$ union or maximum. Finally, *complementarity* relations, represented in the color hexagon by parallel lines (═), interconnect colors with complementary bit-string codes.

Curiously (or insightfully), one can observe a synchronic evolution of the human understanding and the historical development, on the one hand, of color theories and their logical structures and, on the other hand, of inference systems formalizing human reasoning and their logical structures. Classical and medieval logic orbits around tripolar and quadripolar structures known as triangles and squares of opposition, see Fig. 4tl,tc. Only in modern times, since [10], were these structures

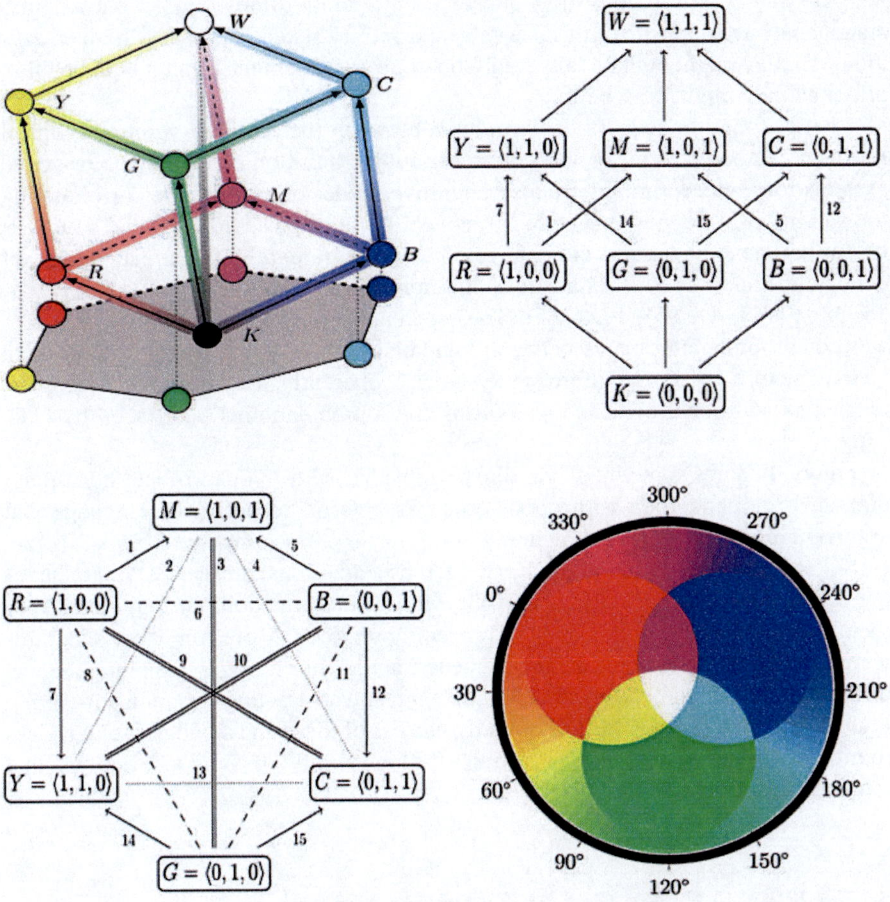

Fig. 3 Top left: Cubic diagram of color entailment relations projected into the hexagon; Top right: Hasse diagram for (transitive) mereological relations of entailment or inferiority ($\longrightarrow$); Bottom-left: Hexagon of opposition for additive (RGB) and subtractive (CMY) colors with corresponding mereological or bit-string relations of complementarity ($=$), contrariety ($--$), and sub-contrariety ($\cdots$); Bottom-right: Color wheel showing hue continuum in standard angular coordinate

generalized so to integrate tripolar and quadripolar oppositional relations, see also [52, 53, 72, 138]. The simplest structure of this kind is the logical hexagon of opposition, depicted in Fig. 4br.

Figure 4br illustrates oppositional relations in the logical hexagon either by arithmetic equality and inequality operators, ($<$, $>$, $=$, $\neq$), or by modal logic operators of necessity, possibility, and negation, ($\Box$, $\Diamond$, $\neg$). Applying to the logical hexagon the same convention used in the color hexagon: Implication or subalternation relations are represented by arrows ($\longrightarrow$); Contrariety relations are represented by dashed lines ($--$); Sub-contrariety relations are represented by dotted lines ($\cdots$);

and Contradiction relations are represented by parallel lines (=). Contradictory statements have opposite truth-false values; Contrary statements cannot both be true, although they might both be false; and Sub-contrary statements cannot both be false, although they might both be true.

I believe that the existing isomorphism between the color hexagon and logical hexagon can be taken as a sign reinforcing Jung's intuition of seeking color-coded systems for representing interrelated cognitive modes or as evidence corroborating the validity of following this path. Moreover, the same basic oppositional structure, or further generalizations thereof, can be used to represent a great variety of deductive and inductive (statistical) inference systems, see [6, 7, 16, 17, 32–35, 38, 39, 43, 44, 101, 102, 165]. These extensions and generalizations engender additional homeomorphisms between logical structures found in color theory and (sub-)structures of those inference systems. Coherently, I take these homeomorphisms as additional evidence supporting the aforementioned path taken by C.G. Jung.

Figure 4 shows some medieval illustrations: These tri-, quadri-, and hexa-polar diagrams are concerned with oppositional aspects of, respectively, language and argumentation, see [32], and alchemy and gnostic philosophy, see [36, p.45], [56, p.184], [109, pl.1], [114, pl.6]. Each of these diagrams presents a fragment of the full hexagon of opposition, Fig. 4br, whose interpretations in logic and color theory were analyzed in this and the preceding sections. Moreover, these diagrams were conceived as *conceptual maps*, hence acting like bridges that interconnect different fields of study by seeking, identifying, and abstracting common underlying logical structures, see [170]. The obvious success of these and analogous enterprises reinforces, once again, my conviction of the validity of Jung's intuitions that motivate this article.

4 Epistemic Color-Coded Cognitive Modes

Eugen Bleuler (1857–1939) was the director of Burghölzli psychiatric hospital from 1898 to 1927. Jung worked at Burghölzli from 1900 to 1909 where he developed several key ideas of analytical psychology. Bleuler [11, 12] had a special interest in chromesthesia and other paradoxical phenomena related to color perception. Jung was also aware of Wilhelm Wundt's [179] psychometric studies, including color theory and perception. Hence, we can safely assume Jung had a good understanding of the complex structure and rich interconnections implied by his simile of the spectrum. Surprisingly, some interpretations found in psychology textbooks present Jung's simile in over-simplified fashion, sometimes even reducing it to a linear structured allegory presented a few years earlier by Frederic Myers [103, 1891, pp.298–306; 1892, 333–336] and, in so doing, fail to capture essential aspects of Jung's simile. The next abridged quotation presents, in a condensed form, Jung's own formulation of the simile of the spectrum:

Fig. 4 Top and bottom-left: Medieval diagrams of tri-, quadri- and hexa-polar oppositional structures; Bottom-right: Blanche [9] hexagon of opposition for ($\Box$, $\Diamond$, $\neg$) modal logic operators of necessity, possibility, and negation, or ($<$, $>$, $=$, $\neq$)(in)equality relations, including oppositional relations of contradiction ($=\!=$), contrariety ($- -$), sub-contrariety ($\cdots$) and subalternation ($\longrightarrow$)

> *[We] employ once more the simile of the spectrum... The dynamism of instinct is lodged as it were in the infra-red part of the spectrum, whereas the instinctual image lies in the ultra-violet part. If we remember our color symbolism, then, as I have said, red is not such a bad match for instinct. But for spirit, as might be expected, blue would be a better match than violet. Violet is the 'mystic' color, and it certainly reflects the indubitably 'mystic' or paradoxical quality of the archetype in a most satisfactory way. Violet is a compound of blue and red, although in the spectrum it is a color in its own right. ... Because the archetype is a formative principle of instinctual power, its blue is contaminated with red: it appears to be violet... The creative fantasy of the alchemists sought to express this abstruse secret of nature by means of another, no less concrete, symbol: the Ouroboros, or tail-eating serpent.*
>
> Jung (CW, VIII, pr.414–416, pp.3167–3169, abridged).

Jung's simile of the spectrum is a metaphor used to explain essential aspects of *archetypes*, a concept we further discuss at Sect. 7. At this point we focus on specifics of the color symbolism used in the simile, involving the colors red, blue, and violet. Red and blue correspond to the color receptors of the human retina closer to the extremes of the visible linear spectrum, while violet lies in the paradoxical region of the color wheel where the Ouroboros bites its tail, see Fig. 11,c. In the topology of the color wheel, opposite to violet and midway in the linear spectrum between red and blue is the locus of color green, a color that, like the colors used in the simile, finds a consistent symbolic meaning in Jung's work, as expressed in the following abridged quotations:

> *Red, the blood color, has always signified emotion and instinct.*
> Jung (CW, VIII, pr.384, p.3143) [77].

> *Blue, the color of air and sky, is most readily used for depicting spiritual contents.*
> Jung (CW, VIII, fn.122, p.3167) [77].

> *Statistically, at least, green is correlated with the sensation function [... in...] relation to the real world.*
> Jung (CW, IX, fn.130, pr.582, p.3840).

> *Of the essence of things, of absolute being, we know nothing. But we experience various effects: from 'outside' by way of the senses, from 'inside' by way of fantasy. ... the color 'green'... is an expression, an appearance standing for something unknown but real.*
> Jung (CW, VII, pr.355, p.2862).

Table 1 presents the symbolic meanings of primary and secondary colors as they are used in the hexa-polar epistemological model under construction in this article. The three primary colors plus violet are reinterpreted in the context of epistemology, our targeted application field, but still preserving (I hope) much of Jung's original interpretations in the context of psychology.

As far as I know, cyan never found in Jung's work a distinct symbolic meaning. This is not surprising for, outside the terminology of modern color theory, few human languages (like Russian, Mongolian, Italian and Hebrew, but neither German nor English) have a distinct traditional word for this color, using instead compound expressions like light-blue or greenish-blue; for pertinent references in etymology, evolutionary linguistics and grammar of color terms, see [5, 41, 80, 96, 153], and also Jung's grandfather, Samuel Preiswerk [118].

Yellow (citrinus or $\xi\alpha\nu\theta o\varsigma$) was Jung's "missing" color, used to reestablish oppositional symmetry and complete his quadripolar basis for psychological types (that can then be unfolded in 2^k-polar models, for $k = 3, 4, 5$), see Jung (CW, XII, pr.333) [77]; [76, p.48] [176]. In the same way, Yellow is the color still missing in our hexa-polar model, where it takes a symbolic meaning specific to the model at hand, see Table 1.

In real life experience, it is difficult to spot pure spectral colors, for processes that naturally generate light produce either a mixture of isolated frequencies (like chemical spectra) or, after some interaction in the environment, (like reflection and scattering) complex mixtures in the color space. Likewise, in our epistemological analog, it is difficult to spot examples of scientific models or theories that would

Table 1 Color-coded epistemological cognitive modes

Red: Color of blood, symbol of (e)motion and instinct. Capacity to maintain embodied life (grounded existence and autopoiesis), of well-adapted reactions or purposive interactions with objects in a scope of interest.

Yellow: Color of metallic gold: symbol of craft work, fine artisanry, precise manufacture, industry, and technology.

Green: Color of vegetation, symbol of sensory perception and sense of reality: Ability to perceive and learn existing qualitative relations in the scope of interest; Capacity to discern, detect, and evaluate independence, correlation or other forms of statistical association between quantities of interest.

Cyan: Light-blue, symbol of reliable empirical statements: Ability to build, use, and communicate good descriptive or predictive models of reality.

Blue: Color of the sky, symbol of thinking and the rectified spirit: Capacity to distill conceptual notions or sublimate abstract ideas; Ability to relate and interconnect such concepts and retrieve or communicate pertinent relational chains in organized conceptual networks. A lexicon used to express and communicate such concepts is called (in computer science) an *ontology*.

Violet: Spectral hue in the - - purple-magenta-violet paradoxical region of the color wheel. Symbol of the cryptic (or psychoid) nature of archetypal forms, half-way between adaptive instincts and their teleological representation as conscious images or ideas: Ability to find, seek, or suggest meaningful associations, symbolic connections or causal relations.

be well described by an isolated primary color. Far easier is to give good examples related to secondary colors (CYM), corresponding to coordinated operations in the space spanned by (at least) two primary colors.

As should be expected, well-developed scientific theories integrate all primary and secondary colors (*cauda pavonis*), hence providing the clearest views in their areas of application. Nevertheless, those theories never drop from the sky *fix-und-fertig* (already fully assembled and ready to go). Usually, they are first noticed while in a dark shade of a secondary color and, from there, progressively evolve so to better illuminate their fields of study. In the following sections we discuss some examples of this kind, discerning positive aspects of scientific models or theories in an evolutionary stage appropriately described by a secondary color, as well as corresponding negative effects due to the missing primary color.

5 Yellow: Invisible Carriers in Charge

This section presents case studies of technological development that, according to our epistemological model for color-coding cognitive modes, could be characterized as yellow—the secondary color made by adding red and green. The technological devices under study had to be manufactured and employed for specific purposes where they had to perform according to strict objective criteria. However, these case studies also illustrate partial successes made by trial-and-error, as well as the overcoming of deficiencies in cognitive mode blue, namely how overcoming a paralyzing deadlock required a breakthrough that, in turn, could only be achieved when key concepts could be abstracted and ensuant metaphors were developed and used to illuminate blind-spots previously dark to consciousness.

The twentieth century spans the development of electronics—the technology of generating, amplifying, and precisely controlling electrical currents. The evolution of electronics came in two great waves, characterized by the key device used to exert this control, namely vacuum tube triodes and semiconductor transistors. Studying electronics' history is facilitated by abundant documentation, including laboratory notebooks of experimental pioneers, scientific articles reporting important breakthroughs, textbooks on the subject written by main protagonists, and even audio and video recordings of interviews with those personalities. Finally, there are good collections of early prototypes and production samples of these artifacts, and a good literature dedicated to the history of these technologies. For general references see [67, 108, 122, 123, 140, 144, 173]. For additional details relevant for this section, see [4, 14, 15, 29, 31, 66, 86, 87, 124–126, 141, 144–147], and also references [184–190] listed as videos and simulations.

Triodes and transistors, also called valves or amplifiers, use a small input, the emitter (or cathode) to base (or grid) electric signal, to regulate a much larger output, the emitter to collector (or anode) electric current. Figure 5t,cr depicts modern diagrammatic representations and shows photographs of the earliest prototypes of these devices. In both cases (triodes and transistors), pioneering inventors had a poor understanding of the fundamental science involved: They were severely misguided by inappropriate concepts and metaphors that generated intellectual blind-spots that, in turn, temporarily halted further development. In both cases, electrically charged particles flow through these devices, but the nature and behavior of these particles was a source of confusion and misunderstanding.

In the case of vacuum tubes, early researchers thought that charged particles flowing through the vacuum tube resulted from the chemical decomposition of gas molecules into positively charged cations and negatively charged anions. Figure 5crt shows an Audion, whose patent explicitly required some residual gas left in the tube for ionization, resulting in a working but very noisy and unreliable amplifier. Later on, the development of theoretical and experimental means and methods of physics and chemistry demonstrated that electrons traveling through vacuum were the carriers in charge of the relevant transport processes, a hypotheses formerly perceived as incoherent, for there is an apparent contradiction in having a current of something in empty space. The apparent paradox was solved by realizing that the electrons in question were sub-atomic particles orders of magnitude smaller than chemical molecules of ordinary matter, see [1]. Triodes and their variants build using high-vacuum tubes were reliable, had good signal to noise characteristics, and became the backbone of subsequent developments in electronic technology.

In the case of semiconductors, researchers had to follow a path in the opposite direction, that is, they had to realize that not only electrons, but also (at least initially) mysterious positively charged (quasi-)particles called *holes* had to be invoked in order to understand and control the relevant electrical flows. The concept of electron-holes, or just holes, was made explicit for the first time by Werner Heisenberg [60]. Emerging from quantum mechanics mathematical formalisms for solid-state physics, this is easy to visualize metaphor often offers the best way to answer Heisenberg's signature question, see [67, p.113,120]: *How can we make that physically insightful (anschaulisch) or intuitive?*

Figure 5cl, resembling Shockley (1950, p.57), depicts free electrons (−) and positive charged holes (+) flowing as missing electrons in covalent bonds in a (doped) Silicon crystal lattice, see also [187, 188]. Figure 5bl, resembling Shockley (1950, p.8,9), depicts his famous *two-story parking garage* for the flow of holes (+) and electrons (−): Electric flow is impossible in the perfect crystal lattice of pure 4-valent Silicon or Germanium, but possible if the crystal is "doped" with scattered impurities of either a 5-valent element, like Phosphorus or Arsenic, introducing a free electron in the crystal, or a 3-valent element, like Boron, introducing a missing electron or hole in the lattice of covalent bonds.

Pioneering researchers trying to build a semiconductor triode were fully aware of the existence of excess electrons and holes in crystalline structures that could be conceived as negatively and positively charged particles. Moreover, they knew that, depending on the type of semiconductor, the number of particles of one kind far exceeded the other, whence called majority and minority carriers. Furthermore, they implicitly hold a majority only premise, namely they tried to build semiconductor devices relying only on majority carriers, minority carriers having a superficial or no role to play, see [64] for such a device. Appreciating the importance of minority carriers was the conceptual blind-spot to overcome in order to achieve a viable solid-state triode.

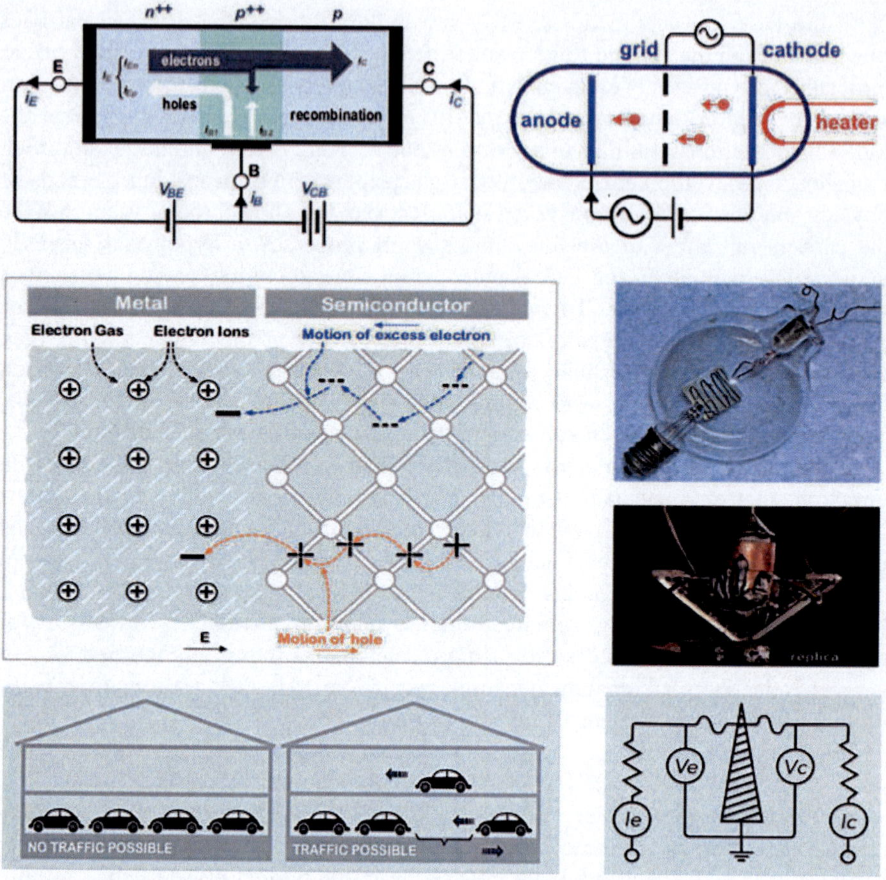

Fig. 5 Top: Diagrammatic representation of Transistor and Triode vacuum tube. Right: Early devices relying on misleading metaphors: Audion, on gas ionization; Contact point transistor, on surface effects. Left: Shockley (1950) parking garage metaphor for the flow of holes (+) and free electrons (−) in a crystal lattice

Figure 5crb depicts a contact point transistor, invented by John Bardeen and Walter Brattain at Bell-Labs in 1949. Like the Audion, this pioneering device worked, but just barely. Its invention was a fruit of much trial-and-error experimentation guided by fuzzy ideas about the role played by minority carriers—supposed to be trapped at a semiconductor's surface or confined to its interfaces. Retrospectively, Brattain stated he had *an intuitive feel for what you could do in semiconductors, not a theoretical understanding*, see [15, p.40] and [142]. Figure 5br, depicts John Shive's [141] double-surface triode, used to demonstrate the importance of in-depth (non-superficial) currents of minority carriers, the conceptual breakthrough needed for William Shockley [143] to invent the Junction Transistor. Figure 5tl gives a diagrammatic representation of a junction transistor, where majority carriers (electrons) are responsible for the main current through the device. Nevertheless this flow of majority carriers is controlled by a secondary current of minority carriers (holes) injected at the base (or grid). The interaction of holes from this much smaller secondary current with electrons flowing in the semiconductor constitutes the key mechanism used to efficiently and reliably control the main flow, see [187, 188].

Vacuum as a transport medium is a difficult thing to "see", and so is a flow of empty holes! Nevertheless, in the aforementioned case studies, overcoming associated blind-spots was the pivotal step to progress, see next quotations. Not surprisingly, in 1906, 1928 and 1932, Joseph Thomson, Owen Richardson and Irving Langmuir were all awarded a Nobel Prize for elucidating the nature and laws of thermionic emission, the theoretical foundation of vacuum tube technology. Contemporary textbooks in solid-state physics are fully immersed in the quantum mechanics theoretical framework, see and compare [79]. In contrast, John Bardeen, Walter Brattain and William Shockley shared a Nobel Prize (1956) for inventing the transistor using simplified (semi-classical) models for the dynamics of flow and interaction (drift, diffusion and recombination) of majority and minority carriers in semiconductors. Essentially, they "only had to see" interacting flows of electrons and holes; see [107, 128, 130, 140, 141, 143–147], and references [184–190] listed as videos and simulations. The next quotations reveal this mindset:

> *The explanation of these effects involved both the majority and the minority carriers. The fact that minority carriers might play an important role in the understanding of semiconductor phenomena was more or less overlooked by other investigators. As we shall see later, this was another blind-spot. ... In the course of these experiments it became evident that the minority carrier, even in small concentrations, played a very important role. ... It is of course not surprising that this blind-spot persisted for so long. The minority carriers were, after all, present in too small concentrations in most semiconductors to matter very much.* Pearson and Brattain (1955, p.1797,1801,1802)[110].

> *The hole, or deficit produced by removing an electron from the valence-bond structure of a crystal, is the chief reason for existence of this book.*
> Shockley (1950, Preface, 1st line).

At this point, it is worth to remember Heinz von Foerster's [49] Principle of *The blind spot: One does not see what one does not see*. As explained in [159], if we lack an appropriate conceptual framework to represent a specific "pattern of reality", our "mind's eye" will not be able discern this pattern, even when the conditions for its occurrence are directly available in our environment. This notion is also in tune with the etymological origin of the word theory, from Ancient Greek: $\Theta\epsilon\omega\rho\iota\alpha = \Theta\epsilon\alpha\nu + o\rho\alpha\omega$, *theoria* = *thean* (a view) +*horao* (I see). Retrospectively, once we are able to see what was hidden in a former blind-spot, it may be hard to believe that someone (possibly ourselves) could not see "that" what had always been there! Even so, incorporating and integrating new theories, adopting new ways of seeing the world, and accepting its consequences, may not be easy. We often cling to old blind-spots, resist change, hold on to old ideas and/or to the old habits, modus operandi or ways of being that grew with them. Furthermore, these inertial effects can be various, complex, multi-layered, mutually reinforcing and, therefore, can be easily misunderstood—sometimes even misinterpreted as intentional efforts aiming to suppress innovation and progress, see [47, 159, 177, 178].

6 Cyan Science: As in Heaven Not on Earth

In Ptolemy's astronomy, a planet moves around its epicycle, a small circle whose center moves around a larger one, the planet's deferent, see [183]. All motions in heaven are explained by a composition of circular motions of this sort. Ptolemy model can be displayed by planetaria—gear driven mechanical simulators, see Fig. 6l, [51, 117]. Ptolemy astronomy provides a *Kinematic* description of planetary motions, namely it presents a model of orbital trajectories without regard to their *causes*, that is, without answering the question of *why* these trajectories are the way they do. Moreover, the heavenly world is conceived as an ideal reality inaccessible and alien to human beings—confined to the imperfect sub-lunar world. Hence, the astronomer is an observer completely detached from the reality he or she observes.

Newtonian Mechanics presents a *Dynamic* model that derives the trajectory followed by a material body from the physical forces acting upon it. Hence, these forces are conceived as the *causes* producing and determining a given trajectory exactly the way it is. Moreover, under appropriate circumstances, these forces can be precisely measured and manipulated, so that the trajectories of the bodies they impel can be controlled according to our will and power. Figure 6r shows a diagram from Newton's magnum opus, *Philosophiae Naturalis Principia Mathematica*, illustrating the smooth transition from sub-orbital to orbital trajectories of a cannonball. This diagram is reproduced in a Hungarian postage label (Michel HU 3199AZf, 1977, with highlighted sub-orbital trajectories), near the lift-off of a Soyuz rocket impelling an artificial satellite to orbit. It is perfectly feasible to build mechanical simulators of such forces and consequent orbits. However, these models are useful to illustrate the dynamics of Newtonian systems, not as analog computers used for orbit calculations, a task better suited to the mathematics of differential equations, see [19, 98, 99, 171].

Ptolemy's astronomy is cyan science: Blue because it is based on well-established concepts and metaphors and it is expressed in the formal language of Greek geometry; and Green because its descriptions and predictions are in excellent agreement with empirical data—up to the observational precision attainable at that time (and for many centuries later). However, it lacks the color red, for it does not admit any possible interaction between the observer and the (kind of) objects he or she observes.

In contrast, Newtonian physics provides a much clearer light: Blue because it is based on new but well-established concepts like positional coordinates, velocity, acceleration, and force, and it is expressed in the formal language of differential and integral calculus, see [106]; Green because it agrees with the most accurate empirical data available, surpassing in this respect Ptolemaic astronomy; and Red because the same universal laws govern heaven and earth, where humans are no longer dis-empowered voyeurs of the sky, but partakers in a universe in which they eventually become spacecraft builders, astronauts, or cosmonauts.

7 Purple-Violet: Suggestive Instincts-Insights

> *[There] are essential phenomena of life which express themselves psychically, just as there are other inherited characteristics which express themselves physiologically. ... Among these inherited psychic factors there [are] universal dispositions of the mind, and they are to be understood as analogous to Plato's forms (eidola), in accordance with which the mind organizes its contents. One could also describe these forms as categories analogous to the logical categories which are always and everywhere present as the basic postulates of reason. Only, in the case of our "forms", we are not dealing with categories of reason but with categories of the imagination. ... following St. Augustine, I call them "archetypes".*
>
> Jung (CW, IX, pr.845, pp.5401–5402, abridged).

> *The archetypal representations (images and ideas) mediated to us by the unconscious should not be confused with the archetype as such... It seems to me probable that the real nature of the archetype is not capable of being made conscious, that it is transcendent, on which account I call it psychoid.*
>
> Jung (CW, VIII, pr.417, pp.3169, abridged).

Pythagoras' theorem, one of the best known results of Euclidean geometry, establishes an invariant relation between the lengths of the edges in a right triangle, namely the sum of the squares of the lengths of the catheti is equal to the square of the length of the hypotenuse; for illustrative images, see Fig. 7. For an intuitive understanding and beautiful visual proofs of Pythagoras theorem, see [104]; for its history, see [121]. Felix Klein [81, 82] *Erlangen program* to the study of geometry is based on the following question: What kind of *transformations* can be applied

Fig. 6 Blasius [183] mechanism based on Hipparchus of Nicaea (190–120 BC) or Claudius Ptolemaeus (100–170 AD) deferent plus epicycles astronomical models; Newton's [105] diagram of cannonball sub-orbital and orbital trajectories

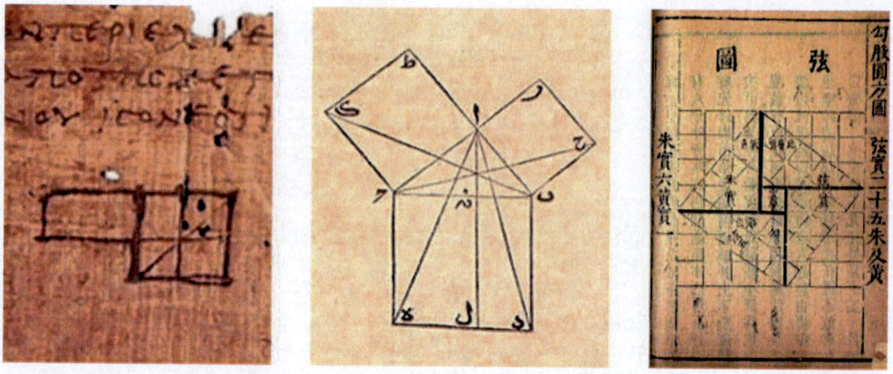

Fig. 7 Euclid of Alexandria (300 BC) and Zhoubi Suanjing (100 BC) diagrammatic demonstrations of Pythagorean theorem

to geometric figures that preserve their essential characteristics? For example: How can the position of each vertex of a triangle be moved so that the size (Pythagoras theorem suggests the quadratic norm) of its edges and its angles remain invariant? In the case of Euclidean geometry, the answer to the last question is: By composition of a translation (linear displacement along a given direction) and a rotation (angular displacement around a given direction). A standard mathematical representation of this class of movements is given by the algebra of Complex numbers in the Euclidean (two dimensional) plane, and by the algebra of Quaternion numbers in the Euclidean (three dimensional) space. For a readable introduction to Klein's approach to geometry, see [58], for extensions of this program to physics, see [174] and [156, 160, 163]. Complex numbers are covered by high-school or college books, [46] is my favorite. For an intuitive introduction to Quaternions, see [59], or [25] for more abstract views.

Complex and Quaternion arithmetic are standard tools of Computer Graphics and Robotics, exactly because they efficiently encode the possibilities and constraints that govern the movement of physical objects in two and three dimensional space. Nevertheless, mechanical robots (robota= slave worker) are machines conceived to emulate the movements human workers are capable of, and computer graphics emulates human visual perception of physical objects as they are moved or illuminated under changing conditions. Hence, we humans must have internal means and methods, like neural networks, that biologically encode equivalent algebraic structures. Every time we do manual labor, be it a plumber or a surgeon, we coordinate our visual perceptions and fine motor skills by using phylogenetically inherited capabilities that are ontogenetically trained and developed during our lives. Using Jung's terminology, in this context far removed from his original field of psychology, we could say that Complex and Quaternion algebras are good descriptions of archetypal forms of movement the human body if capable of.

Abraham Kaplan's *Law of the Instrument* states: *If your only tool is a hammer, then every problem looks like a nail*. Humans are finite beings that have quite limited resources. If we have a tool that works well in a context, it is only natural to try and test it everywhere we can. In the case at hand, Complex and Quaternion algebras are archetypal forms of movement that seem to be well-adapted to our environment, that is, it seems they efficiently encode essential geometric properties of the space we live in. Moreover, these archetypes are great contributors to human intuition, for we use them all the time in our daily activities. As predicted by Kaplan's law of the instrument, we naturally try to use the same archetypal forms to study different phenomena and, behold, sometimes it works miraculously well! James Clerk Maxwell (1831–1879) equations of electromagnetism can be written as a quaternion differential equation—although vector calculus is an equivalent and nowadays more popular formalism, see [27, 40, 119]. As an applied tool, the same equations are at the core of electronic engineering. As basic physics, Maxwell equations can be verified by extremely precise empirical experiments. The extraordinary precise agreement between simple and compact mathematical formulation of physical theories and empirical tests motivated Wigner's [175] famous comments on the *Unreasonable effectiveness of mathematics in the natural sciences*, see [156, 158] and references therein.

The suggestive power of archetypal insights has, however, a double-edged nature: It may either inspire and drive a work of genius, or else engender persistent and misleading mirages. Abraham Kaplan's aforementioned aphorism—also known as *Law of the Hammer* when applied with a pejorative meaning, can explain the conceptual opposite of a blind-spot, namely some persistent forms of wishful thinking and self-illusion. The term *apophania* (from $\alpha\pi o$ = away $+\phi\alpha\iota\nu\omega$ = bring to light, show, reveal) was coined by Klaus Conrad to describe the frequent misidentification of patterns and meanings at the onset of schizophrenia, see [24, 42, 100]. The closely related *Gambler's fallacy* or *pareidolia* (from $\pi\alpha\rho\alpha$ = beside, instead $+\epsilon\iota\delta\omega\lambda o\nu$ = form, shape) refers to perceptions of inexistent patterns in random data. Pareidolia explains some misleading beliefs or pathological behaviors of gamblers, see [154] and references therein. *Statistical retrospective fishing expeditions* and other variations of the gambler's fallacy are the root cause of many misconceived experimental designs or mistaken statistical analyses. Such spurious chains of argumentation are, unfortunately, sometimes used to justify pseudo-scientific theories, academic deception, or professional malpractice. Jung himself warns about the double-edged power of archetypal insights, a source of inspiration for genius and fools alike:

> *The golden apples drop from the same tree, whether they be gathered by an imbecile locksmith's apprentice or by a Schopenhauer.*
>
> Jung (CW, VII, p.2789, pr.229).

Notwithstanding Jung's harsh warning, I must say that even the most brilliant scientists I know—those who have had the grace of their Eureka or Schopenhauer moments, also had plenty more of dumb locksmith's apprentice moments—trying to use the wrong key to open a door, or even struggling to properly use a good working key. In [157, 164] we carefully dissect some paradigmatic cases of pseudo-scientific studies concerning parapsychology, extra-sensory perception, and the medical (ab)use of phosphoethanolamine and hydroxychloroquine. The strong

Fig. 8 Blindfolded Fortuna (lady luck), [116], and Justice, [28, 54]; Francis Galton [116] Quincunx, demonstrating convergence to Normal (Gaussian) distribution

insights and suggestive power offered by intuitive (violet) archetypal ideas—that is, archetypal forms that correspond to firmly embodied (red) instincts that are also represented in well-established (blue) conceptual ontologies—may shed some light on psychological aspects of these bizarre cases.

Double-blind and randomized statistical trials are the gold standard used to test and accept or reject statistical hypotheses. Figure 8l depicts a medieval personification of Luck ($T\upsilon\chi\eta$, Tyche), blindfolded and spinning the wheel of fortune. Figure 8c depicts Justice ($\Delta\iota\kappa\eta$, Dike) holding her classical instruments, sword and scales, and also blindfolded—representing impartial judgment, an iconographic innovation of that time. Figure 8r shows Francis Galton (1822–1911) Quincuncx machine, used to demonstrate the asymptotic convergence of means of random variables to the Normal or Gaussian distribution, a core result of Mathematical Statistics, see [55, 84]. These three images provide some intuition for the key ideas supporting double-blind and randomized statistical trials; for technical details see references in the next paragraphs.

All the case studies analyzed in [157, 164] involve blunt denials of (green) statistical theory and practice, either by contesting the validity of standard mathematical reasoning, or by disputing the ethics of conducting double-blind and randomized experimental trials, or by recourse to unfounded conspiracy theories, etc. Hence, *Caveat emptor*: Any pragmatic or rhetorical attempt to avoid submitting an empirical model to test at this crucible—in which predictive models are validated or falsified—should be taken as a warning flag for pseudo-science; for related discussions, see [26, 85, 115]. Nevertheless, there are many more important aspects of pseudo-science, some of them, I suspect, relating to the suggestive power of archetypal ideas.

Notwithstanding the former caveat, there are valid methodological and ethical concerns regarding clinical (and similar) trials, that should be addressed using state of the art means and methods. For example, contemporary clinical trials should: Dynamically optimize (minimize) sample sizes, see [50, 89, 90]; Cryptographically secure, traceable, and auditable randomization procedures, see [94, 131], Stern et al. (2020); Provide information and conclusions that are, on the one hand, logically coherent and, on the other hand, understandable and consistently interpretable, see [13, 111–113, 154–167]; Protect participants against discernible sub-optimal treatments or practices; etc. Moreover, in my opinion, clinical trials should inform participating patients and agents of the general framework (including goals and ethics) of clinical trials, and how they differ (and so they must) from standard medical practice, in a way that is far more comprehensive than often done.

8 Final Remarks

As it is the case in color theory, both tripolar and quadripolar structures coexist in Jung's work for the analysis of psychic functions, polarities that, we argued, can only be reconciled using a hexagonal logical structure. For example, as already mentioned, Jung categorization of psychological types has a basic quadripolar oppositional structure. Nevertheless, Jung (CW, X, pr.555–557, pp.4591–4592) suggest that *all man's psychic functions have an instinctual foundation* and that, in turn, *the world of unconscious instincts* has a tripolar structure corresponding to:

- Self-assertion – associated with Nietzsche's *Wille zur Macht* or with will power, the Adlerian standpoint in psychology, and Augustinian *Superbia*;
- Imitation impulse – a reality principle associated with *the Learning capacity, a quality almost exclusive to man, based on the instinct for imitation found in animals. It is in the nature of this instinct to disturb other instinctive activities and eventually to modify them*;
- Sex drive – associated with *preservation of the species*, Freudian libido, and Augustinian *Concupiscentia*.

Considering our color-coding of cognitive modes, it seems natural to associate power and learning to the colors red and green. Finally, the association of sex to the color blue can be motivated by the following analogy: From a biological point of view, the most archaic forms of sex are, in essence, exchange of genetic information (horizontal gene transfer mechanisms are much older than genetic recombination of sexual reproduction), see [30, 69, 97, 152]. Moreover, genetic information is organized around basic units of meaningful information that are encoded in DNA as genes. Analogously, conceptual thinking is organized around basic units of meaningful information that are encoded in language as words. Each in its respective domain, genes and words constitute a basic linguistic vocabulary or a basic repertoire of abstractions used for dealing with life and for communication, that is, they constitute basic *ontologies* for their respective domains, see [159, 160, 163].

The possible similarities or parallelisms between psychology, evolution biology and epistemology suggested by the analogies or correspondences considered so far motivates[2] a few lines of future research:

Psychology has a great expertise in developing qualitative and quantitative *instruments* for the purpose of sketching psychological profiles of human subjects, see for example [176]. In future research we would like to explore the possibility of developing similar instruments to sketch epistemological profiles of scientific theories (as they stand in a given instance). We consider some tools of Bayesian statistics, like survey techniques for elicitation, aggregation, and statistical analysis of expert opinion, to be specially promising for the task at hand.

Recent studies in neuroscience suggest that the neural networks in charge of fine motor skills, used for specialized brain processes that are described (approximately) by Complex and Quaternion algebras, are reused for other tasks, a phenomenon related to what is known in computer science as *code reuse*. For example, [63] suggests that the same code developed to motor-visual perception, control, and coordination of human fine motor skills, is reused for simulating and anticipating actions and intentions of other individuals. Furthermore, [127] advance a *linguistic hypothesis*, suggesting that the same code is reused for language processing. As a consequence, a *pre-linguistic grammar* closely related to the aforementioned algebras lies underneath the basic structure of human language. Furthermore, [120] suggests that the same code is reused, once again, to support abstract concepts related to consciousness and self-awareness.

Complex and Quaternion numbers are members of the small but important family of *normed division algebras*, that also include Real numbers and Octonions. These algebras represent translations and rotations in 1, 2, 3, and higher dimensional spaces. In modern science, we "keep finding" those algebras everywhere we look, see for example [18, 37, 62, 74, 92, 154]. Hence, the questions: Do we keep finding these structures in the universe because they really are out there? Or is that what we keep seeing because these structures are a priory encoded in the equipment we have to perceive and interact with the world? Or is it the case that these are the structures (or a priory categories) that we have because those were the ones selected along our phylogenetic evolutionary path as the best fit or the better adapted to the world as it is? For further considerations on the interplay between archetype theory and evolution biology/psychology, see [68, 133, 168, 169]. Finally: To what extent must the explanations we find most intuitive, see for example [139], be related to our inventory of inborn mental structures or archetypal categories? Perhaps these investigations may help us to better understand a celebrated statement by Johanes Kepler,[3] as quoted and translated by Wolfgang Pauli [109, p.163–164]:

---

[2] Perhaps these analogies can also shed some light on the dual character of $E\rho\omega\varsigma$, namely on the one hand, the young Eros (desire)—the playful god of love and, on the other hand, Eros the elder—equated in Orphic tradition to $\Phi\alpha\nu\eta\varsigma$ (Phanes, from $\phi\alpha\iota\nu\omega$= bring to light, show, reveal), a primordial god generator of life and the first to bring light to human consciousness.

[3] *Geometriae vestigia in mundo expressa, sic ut geometria sit quidam quasi mundi archetypus* [78].

The traces of geometry are expressed in the world so that geometry is, so to speak, a kind of archetype of the world.

Acknowledgments I am grateful for comments received from participants of the 6th World Congress on the Square of Opposition—organized by Jean-Yves Beziau and Ioannis Vandoulakis in November 01–05, 2018, at the Orthodox Academy of Crete in Kolymbari, Greece. I am thankful to Maria Cristina Mariante Guarnieri for advice concerning Jungian psychology, and to *Labô* study group on Jungian psychology at PUC-SP—the Pontifical Catholic University of São Paulo. I have benefited from comments by Ana Carolina Bassi Stern, and from discussions at my Scientific Methodology classes for the A.C. Camargo Cancer Center graduate program. Special thanks to Alex Freitas for his talent and skill as graphic artist, patience, generosity, and good will.

Figures[4] Images I generated using LaTeX and TikZ or pb-diagram: Figs. 1rb; 2tl, 3tl,tr,bl,br (thanks Henrik Midtiby); Fig. 4br. Images prepared for this paper by graphic artist Alex Freitas: Figs. 2bl,bc,br; 5tl,tr,cl,bl,br. Images used with the kind permission of the author or responsible for copyright: [93]: fig.1l; fig.2tc,tr; [32]: fig.4tc; [57]: fig.5crt; [186]: fig.5crb; [183]: fig.6l; Imre Marozsán for Magyar Posta: fig.6r. Images from books in public domain, more than 70 years past after dates of publication and author's death: fig.1c, from [23, p.128]; fig.1rt, from [61, 291]; fig.4tl, from [114, pl.6]; fig.4tr, from [48, p.21]; fig.4bl, from [3, p.128]; fig.6r, from [105, p.195]; fig.7lc, from [45]; fig.7r, from [182]; fig.8l, from [116, p.129]; fig.8c, from [28, p.122]; fig.8r, from [54, p.63].

Funding The author received financial support from IME-USP—the Institute of Mathematics and Statistics of the University of São Paulo; FAPESP—the State of São Paulo Research Foundation (grants CEPID Shell-RCGI 2014-50279-4, CEPID CeMEAI 2013-07375-0), and CNPq—the Brazilian National Counsel of Technological and Scientific Development (grant PQ 307648-2018-4).

References

1. D.L.Anderson (1964). *The Discovery of the Electron: The Development of the Atomic Concept of Electricity*. Princeton, NJ: van Nostrand.
2. H.Atmanspacher (1996). The Hidden Side of Wolfgang Pauli: An Eminent Physicist's Extraordinary Encounter w. Depth Psychology. *J.of Consciousness Studies*, 3, 2, 112–126.
3. Audomaro, Lambertus a S. (1121). *Liber Floridus*. Saint-Omer: St-Omer. https://books.google.be/books?vid=GENT900000106992&printsec=frontcover#v=onepage&q&f=false
4. J.Bardeen, W.H.Brattain (1949). Physical Principles Involved in Transistor Action. *Physical Review*, 75, 8, 1208–1225.
5. B.Berlin, P. Kay (1999). *Basic Color Terms: Their Universality and Evolution*. Stanford: CSLI—Center for the Study of Language and Information.
6. J.Y.Béziau (2015). Opposition and order. pp.1–11 in J.Y.Béziau, K.Gan-Krzywoszynska, *New Dimensions of the Square of Opposition*. Munich: Philosophia Verlag.
7. J.Y.Béziau (2012). The power of the hexagon. *Logica Universalis*, 6, 1–43.
8. G.Birkoff, S.MacLane (1953). *A Survey of Modern Algebra*. NY: MacMilan.
9. R.Blanché (1966). *Structures Intellectuelles: Essai sur l'Organisation Systématique des Concepts*. Paris: Vrin.

[4] Positional locators: c=center, t=top, b=bottom, l=left, r=right.

10. R.Blanché (1953). Sur l'Opposition des Concepts. *Theoria*, 19, 89–130.
11. E.Bleuler, K.Lehmann (1881). *Zwangsmässige Lichtempfindungen durch Schall und verwandte Erscheinungen auf dem Gebiete der andern Sinnesempfindungen.*Leipzig: Fues.
12. E.Bleuler (1925). *Die Psychoide als Prinzip der Organischen Entwicklung*. Berlin: Julius Springer.
13. W.S.Borges, J.M.Stern (2007). The Rules of Logic Composition for the Bayesian Epistemic e-Values. *Logic J.of the IGPL*, 15, 5–6, 401–420. https://doi.org/10.1093/jigpal/jzm032
14. E.Braun (1980). The Contribution of the Gottingen School to Solid State Physics: 1920–40. *Proceedings of the Royal Society A: Mathematical, Physical and Engineering Sciences*, 371, 1744, 104–111.
15. E.Braun, S.MacDonald (1982). *Revolution in Miniature: The History and Impact of Semiconductor Electronics*. Cambridge Univ. Press.
16. J.Bueno-Soler, W.Carnielli (2016). Paraconsistent Probabilities: Consistency, Contradictions and Bayes' Theorem. *Entropy*, 18, 9, 325, 1–18;
17. W.Carnielli, M.E.Coniglio (2016). *Paraconsistent Logic: Consistency, Contradiction and Negation*. Vol. 40 of Logic, Epistemology, and the Unity of Science. Springer.
18. G.Casanova (1976). *L'Algèbre Vectorielle*. Paris: Presses Universitaires de France.
19. S.Chapman (1969). Kepler's Laws: Demonstration and Derivation without Calculus. *American J.of Physics*, 37, 11, 1134–1144.
20. L.Chittka (1996). Optimal Sets of Color Receptors and Color Opponent Systems for Coding of Natural Objects in Insect Vision. *J.of Theoretical Biology*, 181, 2, 179–196.
21. L.Chittka (1992). The Colour Hexagon: A Chromaticity Diagram Based on Photoreceptor Excitations as a Generalized Representation of Colour Opponency. *J.of Comparative Physiology A*, 170, 5, 533–543.
22. L.Chittka, W.Beier, H.Hertel, E.Steinmann, R.Menzel (1992). Opponent Colour Coding is a Universal Strategy to Evaluate the Photoreceptor Inputs in Hymenoptera. *J.of Comparative Physiology A*, 170, 545–563.
23. Cleopatra the Alchemist (c.1000AC). *Chrysopoea of Cleopatra.* Codex Marcianus graecus, 299, fol.188v. Reprinted in p.128 of M.Berthelot (1887). Collection des Ancien Alchimistes Grecs. Tome 1. Paris: Steinheil.
24. K.Conrad (1958). *Die beginnende Schizophrenie. Versuch einer Gestaltanalyse des Wahns.* Stuttgart: Georg Thieme Verlag.
25. J.H.Conway, D.A.Smith (2003). *On Quaternions and Octonions: Their Geometry, Arithmetic and Symmetry*. Wellesley, MA: A.K.Peters.
26. H.L.Coulter (1991). *The Controlled Clinical Trial: An Analysis*. Washington, DC: Center for Empirical Medicine.
27. M.J.Crowe (1967). *A History of Vector Analysis: The Evolution of the Idea of a Vectorial System*. NY: Dover.
28. Dammartin (1700). Recueil de prose et de vers concernant Alexandre, Annibal et Scipion, Hector et Achille, Dagobert, Clovis II et Charles VIII, Philippe le Beau, roi d'Espagne, les comtes de Dammartin. Bibliothèque Nationale de France, MS.4962. https://gallica.bnf.fr/ark:/12148/btv1b103186213/f270.item
29. B. Davydov (1938). On the rectification of current at the boundary between two semiconductors; On the theory of solid rectifiers. *Acad. Sci. URSS*, 20, 279–282; 20, 283–285.
30. R.Dawkins (1976, 2016). *The Selfish Gene*, 4th ed. Oxford Univ. Press.
31. L.De Forest (1906). The Audion; A New Receiver for Wireless Telegraphy. *Transactions of the American Institute of Electrical Engineers*, XXV, 735–763.
32. L.Demey (2020). Mathematization and Vergessenmachen in the Historiography of Logic. *History of Humanities*, 5, 1, 51–74. Includes Figure 4 from: Student notebook (1698–1699, Ms.C24, fol.70r.). Archives de l'université Catholique de Louvain, Louvain-la-Neuve.
33. L.Demey, H.Smessaert (2018). Combinatorial Bitstring Semantics for Arbitrary Logical Fragments. *J.of Philosophical Logic*, 47, 325–363.
34. L.Demey, H.Smessaert (2016). Metalogical Decorations of Logical Diagrams. *Logica Universalis*, 10, 233–292.

35. L.Demey, H.Smessaert (2014). The Relationship between Aristotelian and Hasse Diagrams. p.213–227 in T.Dwyer, H.Purchase, A.Delaney, eds. Diagrammatic Representation and Inference. *Lecture Notes in Computer Science*, 8578.
36. D.Diotallevi (2018). The Case of Robert Fludd. *The Wollesen University of Toronto Art Journal*, 6, 33–54.
37. G.Dixon (1994). *Division Algebras: Octonions, Quaternions, Complex Numbers and the Algebraic Design of Physics*. NY: Springer.
38. D.Dubois, H.Prade (1982). On Several Representations of an Uncertain Body of Evidence. p.167–181 in M.Gupta, E.Sanchez, *Fuzzy Information and Decision Processes*. Amsterdam: Elsevier.
39. D.Dubois, H.Prade (2012). From Blanché Hexagonal Organization of Concepts to Formal Concept Analysis and Possibility Theory. *Logica Universalis*, 6, 1, 149–169.
40. J.D.Edmonds (1998). *Relativistic Reality: A Modern View*. Singapore: World Scientific.
41. A.J.Elliot, M.D.Fairchild, A.Franklin (2015). *Handbook of Color Psychology*. Cambridge Univ.
42. M.Escamilla (2016). *Bleuler, Jung, and the Creation of Schizophrenias*. Einsiedeln: Daimon.
43. L.G.Esteves, R.Izbicki, R.B.Stern, J.M.Stern (2019). Pragmatic Hypotheses in the Evolution of Science. *Entropy*, 21, 883, 1–17. https://doi.org/10.3390/e21090883
44. L.G.Esteves, R.Izbicki, J.M.Stern, R.B.Stern (2017). The Logical Consistency of Simultaneous Agnostic Hypothesis Tests. *Entropy*, 18, 256, 1–22. https://doi.org/10.1093/jigpal/jzx024
45. Euclid of Alexandria (c.300 BC). Elements of Geometry, Book II, Proposition 5. (a) Fragment of Oxyrhynchus papirus (c.100 AD) at University of Pennsylvania. (b) Arabic version of Euclid's Elements by Nasir al-Din al-Tusi (1258), reprinted in 1594 by Medici Press, Rome.
46. B.de Finetti (1957). *Matematica Logico-Intuitiva*. Roma: Edizioni Cremonese.
47. D.Fingermann (2014). *Os Paradoxos da Repetição*. São Paulo: Annablume.
48. R.Fludd (1617). *Utriusque Cosmi Maioris Scilicet et Minoris Metaphysica, Physica atque Technica Historia*. Frankfurt: Johann Theodor de Bry. https://books.google.be/books?vid=GENT900000070435
49. H.von Foerster (2003). *Understanding Understanding: Essays on Cybernetics and Cognition*. NY: Springer.
50. V.Fossaluza, M.S.Lauretto, C.A.B.Pereira, J.M.Stern (2015). Combining Optimization and Randomization Approaches for the Design of Clinical Trials. *Springer Proceedings in Mathematics and Statistics*, 118, 173–184. https://doi.org/10.1007/978-3-319-12454-4_14
51. T.Freeth (2009). Decoding and Ancient Computer. *Scientific American*, 301, 6, 76–83.
52. P.Gallais (1982). *Dialectique du Récit Mediéval: Chrétien de Troyes et l'Hexagone Logique*. Amsterdam: Rodopi.
53. P.Gallais, V.Pollina (1974). Hexagonal and Spiral Structure in Medieval Narrative. *Yale French Studies* 51, 115–132.
54. F.Galton(1889). *Natural Inheritance*. London: Macmillan.
55. A.Gelman, J.B.Carlin, H.S.Stern, D.B.Rubin (2003). *Bayesian Data Analysis*, 2nd ed. NY: Chapman and Hall / CRC.
56. S.Gieser (2005). *The Innermost Kernel*. Berlin: Springer.
57. E.Godshalk (2009). The Spherical Audion: The Gateway to the Golden Age of Radio. *IEEE Microwave Magazine*, 10, 5, 142–144.
58. M.J.Greenberg (1993). *Euclidean and Non-Euclidean Geometries: Development and History*. NY: W.H.Freeman.
59. A.J.Hanson (2006). *Visualizing Quaternions*. San Francisco: Morgan Kaufmann.
60. W.Heisenberg (1931). Zum Paulischen Ausschliessungsprinzip. *Annalen der Physik*, 10, 888–904.
61. H.von Helmholtz (1867, 1909, 2013). *Handbuch der Physiologischen Optik*, 1st ed. Leipzig: Leopold Voss. 3rd ed. translated as *Treatise on Physiological Optics*. Courier Corporation. https://books.google.com.br/books?id=E3EZAAAAYAAJ&printsec=frontcover&redir_esc=y#v=onepage&q&f=false

62. N.D.Hemkumar, J.R.Cavallo (1994). Redundant and On-Line CORDIC for Unitary Transformations. *IEEE Transactions on Computers*, 43, 8, 941–954.
63. G.Hesslow (2002). Conscious Thought as Simulation of Behavior and Perception. *Trends in Cognitive Science* 6, 6, 242–247.
64. R.Hilsch, R.W.Pohl (1938). Steuerung von Elektronenströmen mit einem Dreielektrodenkristall und ein Modell einer Sperrschicht. *Zeitschrift für Physik*, 1938, 3, 399–408.
65. E.Hering (1878, 1964). *Zur Lehre vom Lichtsinne*. Translated as *Outlines of a Theory of the Light Sense*. Cambridge, MA: Harvard University Press.
66. H.R.Huff (2001). *John Bardeen and Transistor Physics. American Institute of Physics Conference Proceedings*, 550, 3, 1–29.
67. L.Hoddeson, E.Braun, J.Teichmann, S.Weart (1992). *Out of the Crystal Maze: Chapters in the History of Solid State Physics*. NY: Oxford Univ. Press.
68. G.B.Hogenson (2019). The Controversy Around the Concept of Archetypes. *J.of Analytical Psychology*, 64, 5, 682–700.
69. R.Inhasz, J.M.Stern (2010). Emergent Semiotics in Genetic Programming and the Self-Adaptive Semantic Crossover. *Studies in Computational Intelligence*, 314, 381–392. https://doi.org/10.1007/978-3-642-15223-8_21
70. D.Jaspers (2017). Logic and Colour in Cognition, Logic and Philosophy. p.249–271 in Silva (2017).
71. D.Jaspers (2012). Logic and Colour. *Logica Universalis*, 6, 227–248.
72. D.Jaspers, P.A.M.Seuren (2016). The Square of Opposition in Catholic Hands: A Chapter in the History of 20th-Century Logic. *Logique & Analyse*, 233, 1–35.
73. G.H.Joblove, D.Greenberg (1978). Color Spaces for Computer Graphics. *Computer Graphics*, 12, 3, 20–25.
74. M.Josipovic (2019). *Geometric Multiplication of Vectors: An Introduction to Geometric Algebra in Physics*. Cham: Birkhäuser.
75. C.G.Jung (1991). *C.G.Jung Collected Works*. London: Routledge. vol. VIII - The Structure and Dynamic of the Psyche, vol. XI - Psychology and Religion: West and East. Original titles: *Gesammelte Werke*, Stuttgart: Patmos-Verlag. Bd. 08 - Die Dynamik des Unbewussten (1967) Bd. 11 - Zur Psychologie westlicher und östlicher Religion (1963).
76. C.G.Jung (1940). *The Integration of Personality*. London: Kegan, Trench & Trubner.
77. C.G.Jung (1939). ETH Lectures XII and XIII. Zurich: *Eidgenössische Technische Hochschule Lecture Notes*. Zurich: K.Schippert.
78. J.Kepler (1606). *De Stella Nova in Pede Serpentarii*. Prague: Pauli Sessii.
79. Ch.Kittel (1953, 1st ed; 1976, 5th ed). *Introduction to Solid State Physics*. NY: Wiley.
80. E.Klein (1987). *A Comprehensive Etymological Dictionary of the Hebrew Language*. Jerusalem: CARTA.
81. F.Klein (1872). *Vergleichende Betrachtungen über neuere geometrische Forschungen*. translation: A Comparative Review of Recent Researches in Geometry. arXiv:0807.3161
82. F.Klein (1948). *Elementary Mathematics from an Advanced Standpoint*. NY: Dover.
83. R.G.Kuehni, A.Schwarz (2008). *Color Ordered: A Survey of Color Order Systems from Antiquity to the Present*. Oxford Univ. Press.
84. J.Kunert, A.Montag, S.Pühlmann (2001). The Quincunx: History and Mathematics *Statistical Papers*, 42, 143–169.
85. C.Kurz (2005). *Imagine Homeopathy: A Book of Experiments, Images & Metaphors*. NY:Thieme.
86. I.Langmuir (1913). The Effect of Space Charge and Residual Gases on Thermionic Currents in High Vacuum. *Physical Review*, 2, 6, 450–486.
87. I.Langmuir (1919). The Arrangement of Electrons in Atoms and Molecules. *J.of the American Chemical Society*, 41, 6, 868–934.
88. K.Q.Laughlin (2015). The Spectrum of Consciousness: Color Symbolism in the Typology of C.G. Jung. https://typeindepth.com/2015/10/the-spectrum-of-consciousness/
89. M.S.Lauretto, F.Nakano, C.A.B.Pereira, J.M.Stern (2012). Intentional Sampling by Goal Optimization with Decoupling by Stochastic Perturbation. *American Institute of Physics Conference Proceedings*, 1490, 189–201. https://doi.org/10.1063/1.4759603

90. M.S.Lauretto, R.B.Stern, K.L.Morgam, M.H.Clark, J.M.Stern (2017). Haphazard Intentional Allocation and Rerandomization to Improve Covariate Balance in Experiments. *American Institute of Physics Conference Proceedings*, 1853, 050003, 1–8. https://doi.org/10.1063/1.4985356
91. M.S.Longair (2008). Maxwell and the Science of Colour. *Philosophical Transactions of the Royal Society*, A, 366, 1871, 1685–1696.
92. P.Lounesto (2001). *Clifford Algebras and Spinors*. 2nd ed. Cambridge University Press.
93. B.MacEvoy (2005). *Color Vision*. Version last revised 08.01.2005 downloaded from https://hosting.iar.unicamp.br/lab/luz/ld/Cor/color_vision.pdf. Latest version available at https://www.handprint.com/LS/CVS/color.html
94. D.R.Marcondes, C.M.Peixoto, J.M.Stern (2019). Assessing Randomness in Case Assignment: The Case Study of the Brazilian Supreme Court. *Law, Probability and Risk*, 18, 2/3, 97–114. https://doi.org/10.1093/lpr/mgz006
95. J.C.Maxwell (1860). On the Theory of Compound Colours, and the Relations of the Colours of the Spectrum. *Philos. Trans. of the Royal Society of London*, 150, 57–84.
96. R.E. MacLaury, G.V. Paramei, D.Dedrick (2007). *Antropology of Color*. Amsterdam: John Benjamins.
97. R.E.Michod. B.R.Levin (1988). *The Evolution of Sex*. Sunderland, MA: Sinauer.
98. K.J. Mirenberg (1968). Gravity Analogue. *Spectra*, 4, 2, 29–34.
99. K.J.Mirenberg (2021). *Introduction to Gravity-Well Models of Celestial Mechanics: Visual Insights to Space-Time and Gravitation*. Retrieved online from: https://www.spiralwishingwells.com/guide/Gravity_Wells_Mirenberg.pdf
100. A.Mishara (2010). Klaus Conrad (1905–1961): Delusional Mood, Psychosis and Beginning Schizophrenia. *Schizophrenia Bulletin*, 36, 1, 9–13.
101. A.Moretti (2012). Why the Logical Hexagon? *Logica Universalis* 6, 69–107.
102. A.Moretti (2009). *The Geometry of Logical Opposition*. Ph.D. Thesis, Univ.of Neuchâtel.
103. F.Myers (1891, 1892). The subliminal consciousness. *Proceedings of the Society for Psychical Research*, 7, 298–355 and 8, 333–404.
104. R.B.Nelsen I, II and III (1993, 2020, 2015). *Proofs without Words: Exercises in Visual Thinking*. Mathematical Association of America.
105. I.Newton (1687, 1728). *Philosophiae Naturalis Principia Mathematica*. London: Joseph Streater. Translated as: *A Treatise of the System of the World*. London: F.Fayram. https://books.google.com.br/books?id=rEYUAAAAQAAJ&redir_esc=y
106. I.Newton (1704). *Opticks: or, A Treatise of the Reflexions, Refractions, Inflexions and Colours of Light. Also Two Treatises of the Species and Magnitude of Curvilinear Figures*. London: Smith and Walford. https://books.google.com.br/books?id=mxhfAAAAcAAJ&redir_esc=y
107. T.H.Ning (1997). On Shockley's 1952 IRE paper. *Proc. of the IEEE*, 85, 12, 2052–2054.
108. J.Orton (2004). *The Story of Semiconductors*. NY: Oxford Univ. Press.
109. W.Pauli (1955). The Influence of Archetypal Ideas on the Scientific Theories of Kepler. p.147–212 in C.G.Jung, W.Pauli (1955). *The Interpretation of Nature and the Psyche*. NY: Pantheon.
110. G.L.Pearson, W.H. Brattain (1955). History of Semiconductor Research. *Proceedings of the IRE*, 1955, 12, 1794–1806.
111. C.A.B.Pereira, J.M.Stern, (1999). Evidence and Credibility: Full Bayesian Significance Test for Precise Hypotheses. *Entropy*, 1, 69–80. https://doi.org/1099-4300/1/4/99
112. C.A.B.Pereira, J.M.Stern, S.Wechsler (2008). Can a Significance Test be Genuinely Bayesian? *Bayesian Analysis*, 3, 79–100. https://doi.org/10.1214/08-BA303
113. C.A.B.Pereira, J.M.Stern (2020). The e-value: A Fully Bayesian Significance Measure for Precise Statistical Hypotheses and its Research Program. *São Paulo J.of Mathematical Sciences*, https://doi.org/10.1007/s40863-020-00171-7
114. Petraeus, Cornelius (1550). *Sylva Philosophorum*. Leiden Ms Vossianus Chym Q 61. https://le-miroir-alchimique.blogspot.com/2011/02/anonyme-sylva-philosophorum.html
115. M.Pigliucci, M.Boudry (2013). *Philosophy of Pseudoscience: Reconsidering the Demarcation Problem*. Univ.of Chicago Press.

116. Pizan, Christine de (1414). *The Book of the Queen*. British Library, Harley MS.4431. Harley MS 4431. http://www.pizan.lib.ed.ac.uk/gallery/pages/129r.htm
117. D.S.Price (1974). Gears from the Greeks. The Antikythera Mechanism: A Calendar Computer from 80 BC. *Transactions of the American Philosophical Society*, 64, 7, 1–70.
118. S.Preiswerk (1871). *Grammaire Hébraïque Précédé d'un Précis Historique sur la Langue Hébraïque*, 3rd ed. Genève: H.Georg.
119. E.M.Purcell (1963, 2013). *Electricity and Magnetism* 3rd ed. Cambridge Univ. Press.
120. V.S.Ramachandran (2007). The Neurology of Self-Awareness. www.edge.org/3rd_culture/ramachandran07/ramachandran07_index.html
121. B.Ratner (2009). Pythagoras: Everyone knows his famous theorem, but not who discovered it 1000 years before him. *J.of Targeting, Measurement and Analysis for Marketing*, 17, 3, 229–242.
122. P.A.Redhead (1998). The Birth of Electronics: Thermionic Emission and Vacuum. *J.of Vacuum Science and Technology A*, 16, 3, 1394–1401.
123. P.A.Redhead (2000). History of Vacuum Devices. p.281–290 in S. Turner. *Proc. of the CERN Accelerator School on Vacuum Technology*. Snekersten, Denmark.
124. O.W.Richardson (1916). *Emission of Electricity from Hot Bodies*. London: Longmans.
125. M.Riordan, L.Hoddeson (1997). Minority Carriers and the First Two Transistors. p.1–33 in A.Goldstein, W.Aspray, *Facets: New Perspectives on the History of Semiconductors*. New Brunswick: IEEE Center for the History of Electrical Engineering.
126. M.Riordan, L.Hoddeson, C.Herring (1999). The invention of the transistor. *Reviews of Modern Physics*, 71, 2, S336-S345.
127. G.Rizzolatti, M.A.Arbib (1998). Language Within our Grasp. *TINS*, 21, 5, 188–194.
128. W.van Roosbroeck (1950). Theory of the Flow of Electrons and Holes in Germanium and Other Semiconductors. *Bell Systems Technical J.*, 29, 4, 560–607.
129. P.O.Runge (1810). *Die Farben-Kugel, oder Construction des Verhaeltnisses aller Farben zueinander* Hamburg: Perthes.
130. R.M.Ryder, R.J.Kircher (1949). Some Circuit Aspects of the Transistor. *Bell Systems Technical J.*, 28, 3, 367–400.
131. O.T.Saa, J.M.Stern (2019). Auditable Blockchain Randomization Tool. *Proceedings*, 33, 17, 1–6. https://doi.org/10.3390/proceedings2019033017
132. M.Sabini (2000). The Bones in the Cave: Phylogenetic Foundations of Analytical Psychology. J.of Jungian Theory and Practice, 2, 17–33.
133. R.Samuels (1998). Evolutionary Psychology and the Massive Modularity Hypothesis. *British J.of Philosophy of Science*, 49, 575–602.
134. E.Schroedinger (1920a). Grundlinien einer Theorie der Farbenmetrik im Tagessehen I. *Annalen der Physik*, 63, 397–426. 427–456.
135. E.Schroedinger (1920b). Grundlinien einer Theorie der Farbenmetrik im Tagessehen II. *Annalen der Physik*, 63, 481–520.
136. E.Schrödinger (1925, 1994). Über das Verhältnis der Vierfarben zur Dreifarbentheorie. *Sitzungsberichte der Kaiserlichen Akademie der Wissenschaften in Wien*, IIa, 134, 471–490. Translated as: On The Relationship of Four-Color Theory to Three-Color Theory. *Color Research and Application*, 19, 1, 37–47.
137. E.Schrödinger, K.K. Niall (2017). *Erwin Schrödinger's Color Theory*. Berlin: Springer.
138. A.Sesmat (1951). *Logique II: Les Raisonnements, La Logistique*. Paris: Hermann.
139. J.N.Shive, R.L. Weber (1982). *Similarities in Physics*. NY: Wiley.
140. J.N.Shive (1959). *The Properties, Physics and Design of Semiconductor Devices*. Princeton, NJ: van Nostrand.
141. J.N.Shive (1949). The Double-Surface Transistor. *Physical Review*, 75, 689–690.
142. W.Shockley (1976). The Path to the Conception of the Junction Transistor. *IEEE Transactions on Electron Devices*, 23, 7, 597–620.
143. W.Shockley (1950). *Electrons and Holes in Semiconductors*. Princeton, NJ: van Nostrand.
144. W.Shockley (1950). Holes and Electrons. *Physics Today*, 3, 10, 16–24.

145. W.Shockley (1952). Transistor Electronics: Imperfections, Unipolar and Analog Transistors. *Proceedings of the IRE*, 40, 11, 1289–1313.
146. W.Shockley, G.I.Pearson, J.R.Haynes (1949). Hole Injection in Germanium: Quantitative Studies and Filamentary Transistors. *Bell Systems Technical J.*, 28, 3, 344–366.
147. W.Shockley, M.Sparks, G.Teal (1951). p-n Junction Transistors. *Physical Review*, 83, 1, 151–162.
148. L.Silberstein (1942). A Fundamental Criterion of Uniform Representability of Equiluminous Colors on a Geometrical Surface. *J.of the Optical Society of America*, 32, 552–556.
149. M.Silva (2017). *How Colours Matter to Philosophy*. Cham: Springer.
150. A.R.Smith (1978). Color Gamut Transform Pairs. *ACM SIGGRAPH Computer Graphics*, 12, 3, 12–19.
151. I.Simon (1966). *Infared Radiation*. van Nostrand.
152. S.Spielrein (1912, 2018). Die Destruktion als Ursache des Werdens. *Jahrbuch für psychoanalytische und psychopathologische Forschungen*, 4, 465–503. Translated as: Destruction as the Cause of Becoming; pp.97–134 in *The Essential Writings of Sabina Spielrein*. NY: Routledge.
153. B.Sterman, J.Taubes (2012). *The Rarest Blue: The Remarkable Story of an Ancient Color Lost to History and Rediscovered*. Guilford, CT: Lyons Press.
154. J.M.Stern (2008a). *Cognitive Constructivism and the Epistemic Significance of Sharp Statistical Hypotheses in Natural Sciences*. `arXiv:1006.5471`
155. J.M.Stern (2008b). Decoupling, Sparsity, Randomization, and Objective Bayesian Inference. *Cybernetics and Human Knowing*, 15, 2, 49–68.
156. J.M.Stern (2011a). Symmetry, Invariance and Ontology in Physics and Statistics. Symmetry, 3, 3, 611–635. https://doi.org/10.3390/sym3030611
157. J.M.Stern (2011b). Spencer-Brown vs. Probability and Statistics: Entropy's Testimony on Subjective and Objective Randomness. *Information*, 2, 2, 277–301. https://doi.org/10.3390/info2020277
158. J.M.Stern (2011c). Constructive Verification, Empirical Induction, and Falibilist Deduction: A Threefold Contrast. *Information*, 2, 4, 635–650. https://doi.org/10.3390/info2040635
159. J.M.Stern (2014). Jacob's Ladder and Scientific Ontologies. *Cybernetics and Human Knowing*, 21, 3, 9–43.
160. J.M.Stern (2017). Continuous Versions of Haack's Puzzles: Equilibria, Eigen-States and Ontologies. Logic J.of the IGPL, 25, 4, 604–631. https://doi.org/10.1093/jigpal/jzx017
161. J.M.Stern (2018). Verstehen (causal/ interpretative understanding), Erklären (law-governed description/ prediction), and Empirical Legal Studies. *JITE – Journal of Institutional and Theoretical Economics/Zeitschrift für die Gesamte Staatswissenschaft*, 174, 1, 105–114. https://doi.org/10.1628/093245617X15120238641866
162. J.M.Stern (2020a). Jacob's Ladder: Logics of Magic, Metaphor and Metaphysics. *Sophia*, 59, 365–385. https://doi.org/10.1007/s11841-017-0592-y
163. J.M.Stern (2020b). A Sharper Image: The Quest of Science and Recursive Production of Objective Realities. *Principia: An International J.of Epistemology*, 24, 2, 255–297. https://doi.org/10.5007/1808-1711.2020v24n2p255
164. J.M.Stern (2021). Deus sive Natura, Relegere, Religare, and Leadership in Times of Pandemia. Submitted for publication.
165. J.M.Stern, R.Iznicki, L.G.Esteves, R.B.Stern (2017). Logically Consistent Hypothesis Testing and the Hexagon of Oppositions. *Logic J.of the IGPL*, 25, 5, 741–757. https://doi.org/10.1093/jigpal/jzx024
166. J.M.Stern, C.A.B.Pereira (2014). Bayesian Epistemic Values: Focus on Surprise, Measure Probability! *Logic J.of the IGPL*, 22, 2, 236–254. https://doi.org/10.1093/jigpal/jzt023
167. J.M.Stern, M.A.Simplicio, M.V.M Silva, R.A.C.Pfeiffer (2020). Randomization and Fair Judgment in Law and Science. p.399–418 in J.A.Barros, D.Krause (2020). A True Polymath: A Tribute to Francisco Antonio Doria. Rickmansworth, UK: College Publications. `arXiv:2008.06709`
168. A.Stevens (1998). Response to P.Pietikainen. *J.of Analytical Psychology*, 43, 345–355.
169. A.Stevens (2002). *Archetype Revisited: An Updated Natural History of the Self*. Routledge.

170. M.Tang, A.T.Karunanithi (2018). *Advanced Concept Maps in STEM Education*. Hershey, PA: IGI-Global.
171. H.H.Turner (1915). *A Voyage in Space*. London: Soc. for Christian Knowledge.
172. R.S.Turner (1993). Vision Studies in Germany: Helmholtz vs. Hering. *Osiris*, 8, 1,80–103.
173. G.F.J.Tyne (1977). *Saga of the Vacuum Tube*. Indianapolis, IN: H.W. Sams.
174. E.P.Wigner (1949). Invariance in Physical Theory. *Proceedings of the American Philosophical Society*, 93, 521–526.
175. E.P.Wigner (1960). The Unreasonable Effectiveness of Mathematics in the Natural Sciences.*Communications on Pure and Applied Mathematics*, 13, 1–14.
176. D.J.Wilde (2011). *Jung's Personality Theory Quantified*. Berlin: Springer.
177. B.Winston (1986). *Misunderstanding Media*. London: Routledge & Kegan Paul.
178. B.Winston (1998). *Media Technology and Society: A History From Telegraph to the Internet*. London: Routledge
179. W.Wundt (1892). *Vorlesungen über die Menschen- und Thierseele*. Hamburg: Voss.
180. W.Wundt (1896). *Grundriss der Psychologie* Leipzig: Engelmann.
181. T.Young (1802). The Bakerian Lecture: On the Theory of Light and Colours. *Philosophical Transactions of the Royal Society of London*, 92, 12–48.
182. Zhou Bi Suan Jing—Arithmetical Classic of the Gnomon and the Circular Paths of Heaven (c.100 BC). Ming dynasty copy printed in 1603. https://www.maa.org/press/periodicals/convergence/mathematical-treasures-zhoubi-suanjing

Videos and Simulations

183. C.Blasius (2014). *Tellurix: Handbetriebenes Holz-Tellurium*. HolzMechanik.de https://holzmechanik.de/holz-tellurium-tellurix.html and https://doi.org/youtu.be/GgW69jRKADo
184. Bell Laboratories (1959). *Dr. Walter Brattain on Semiconductor Physics*. https://youtu.be/TkkU77chrWg
185. Bell Laboratories (1965). *The Genesis of the Transistor*. https://youtu.be/LRJZtuqCoMw
186. W.Hammack, N.Ziech (2010). How the First Transistor Worked: Brattain and Bardeen Point-Contact Device. https://youtu.be/watch?v=RdYHljZi7ys
187. N.Heinz (2020). Homo Faciens video and open source code for simulation models in physics. Site: https://www.homofaciens.de/ ; Videos: Silicon: youtu.be/ApqFLVd0XaI, Doping: youtu.be/k0J_y0Av_f8, Diode1: youtu.be/jWh06oaG6LA, Diode2: youtu.be/mPGusBmm3XE, Transistor: youtu.be/zMNfXtwRKJs
188. S.Mathew (2017). LearnEngineering: Transistors, How do they work? BJT: youtu.be/watch?v=7ukDKVHnac4 MOSFET: youtu.be/watch?v=stM8dgcY1CA
189. SEEC - Semiconductor Electronics Education Committee (1962). *Drift and Diffusion of Minority Carriers*; Experimental demonstration for J.R.Haynes, W.B.Shockley (1949) - Investigation of Hole Injection in Transistor Action. *Physical Review*, 75, 691–691.
190. R.Soyland (2011). *Making a Spherical Audion Vacuum Tube;* part 1–4: youtu.be/oSgVGwqJ2Jk, youtu.be/iJuKY-Uo1a4, youtu.be/2kQhDIrkXNs, youtu.be/PxxLPrVbb-A

Many-Valued Logical Hexagons in a 3-Oppositional Trisimplex

Régis Angot-Pellissier

Abstract The present paper's aim is to construct a "3-oppositional trisimplex" (i.e., the *tri*-simplicial counterpart of the "logical hexagon" (1950), the most famous "oppositional *bi*-simplex") by means of "numerical sheaves" on a topological space with one non-trivial open set. First of all, we must redefine the concept of "Aristotelian 3^2-semantics" (2009) in sheaf-theoretical terms and, as a result, we explain the meaning of the "trisimplicial entities": three types of contradictions (classical negation and two non-classical ones: intuitionist and co-intuitionist), three types of simplexes (the two classical bi-simplicial ones—namely contrariety and subcontrariety—and the new one, pivotal, characteristic for trisimplexes), and three types of subalternations (classical implication and two non-classical ones). Then we shall show the relations holding between some sheaves in the terms of the above defined Aristotelian 3^2-semantics (i.e., the "oppositional qualities," or "kinds of oppositions"), in order to obtain all the hexagons of the trisimplex, whose vertices are these sheaves with these 3-oppositional relations as edges between any pair of them. In the last part, after having highlighted the paracomplete and paraconsistent nature of the two new hexagons thus emerged, we shall discuss the geometrical point of view on this trisimplex taken as a whole.

Keywords n-Opposition · Logical hexagon · Category theory · Sheaves on a topological space · Logical trisimplex · Intuitionist logic · Paraconsistent logic · Contrariety · Subcontrariety · Local sections · Global sections

Mathematics Subject Classification (2000) Primary 03B50; Secondary 03B53, 54B40

R. Angot-Pellissier (✉)
Lycée Montesquieu, Le Mans, France

J.-Y. Beziau, I. Vandoulakis (eds.), *The Exoteric Square of Opposition*, Studies in Universal Logic, https://doi.org/10.1007/978-3-030-90823-2_15

1 The Context of This Study

Following, among others, the works of Blanché ([4] in 1953 and [5] in 1966) and Béziau ([3] in 2003) on "logical hexagons," which show that Aristotle's famous "logical square" (or "square of opposition") is a mathematically incomplete form of 3-opposition (i.e., a fragment of logical hexagon), Alessio Moretti has developed a general "theory of n-opposition" (currently labeled "oppositional geometry"): his so-called modal graphs (a.k.a. "oppositional generators") allow decorations of what he has discovered as "n-oppositional figures" and which include the already known logical hexagon (i.e., 3-opposition) and the "logical cube" (i.e., 4-opposition). These extensions of Aristotle's square can be interpreted in many different logics, as fuzzy logic [9] by instance. Moretti's aim in 2004 [6] was, as in some sense Béziau had opened the way in 2003 [3], the discovery of an infinite sequence of growing geometrical n-dimensional solids (i.e., polytopes) constitutive of all the "n-oppositional figures." These solids have been characterized by Moretti as "oppositional bi-simplexes" (simplexes are so to say an n-dimensional geometrical counterpart of the positive numbers), since they fit together a (conventionally blue) "contrariety simplex" (expressing the mutual incompatibility of any two terms of a set of n terms) and a correlated (conventionally green) "subcontrariety simplex," with, between these two simplexes, "contradictions," i.e., "negation segments" (conventionally red) and "subalternations," i.e., "implication arrows" (conventionally grey). In 2008 [10] we demonstrated that a set-theoretical technique that we proposed allows producing, for any such n-opposition, i.e., for any bi-simplex seen as an "oppositional kernel," a correlated "oppositional closure," this concept being very important in so far it determines algorithmically the general shape of the whole, infinitely growing, oppositional geometry (leaving, so to say, nothing of it outside the algorithm): it is these oppositional closures which measure and model "oppositional phenomena" by determining univocally (when finite) their "oppositional complexity" (one of the most famous oppositional closures, namely the closure of the logical cube, is the 3D "oppositional tetrahexahedron," discovered in 1968 by Pierre Sauriol [11], and rediscovered independently by us [10] as "tetraicosahedron" and by Hans Smessaert [12]—this last author prefers to view it, equivalently, as a "rhombic dodecahedron" inside what he and Lorenz Demey prefer to call "logical geometry"). All this oppositional geometry, with its generators, kernels, and closures (the three being precisely put into mutual relation by our setting technique) is defined in a mathematical world logically classic (i.e., bi-valued, that is admitting only 2 truth-values: the false, "0", and the true, "1"). In order to overcome that limitation, in his 2009 PhD thesis [7], Moretti has explored what he proposed to call the "Aristotelian semantics" behind the four classical opposition relations: in game-theoretical terms, namely putting into evidence an underlying "generating algorithm" (for opposition kinds) made by a combinatorics based on the two classical Aristotelian questions: "Can two things A and B be false together?" and "Can two things A and B be true together?". If we answer those 2 questions with the help of n possible truth-values (instead of only two, as in Aristotle and his followers), this gives a so-called Aristotelian n^2-semantics (instead of the starting "Aristotelian 2^2-semantics," which generated the concept of bi-simplex), generating many forms of "weak negations" (alongside with classical negation, i.e., "contradiction"), n different simplexes (interpolating new ones in between the classical blue simplex of "contrariety" and the classical green simplex of "subcontrariety") as well as "weak implications" (alongside with classical implication, i.e., "subalternation"). The geometrical figures fitting together these elements, i.e., the n simplexes (generated by any such "Aristotelian n^2-semantics") and the negation segments and implication arrows relating two by two any pair of such n oppositional simplexes, are defined as "multisimplexes" (a.k.a. "oppositional poly-simplexes").

The aim of our paper is to explore the new structures of Moretti, in particular, the 3-oppositional trisimplex generated by the Aristotelian 3^2-semantics, with the aid of the mathematical structure of "sheaf" (i.e., based on "category theory" and "sheaf theory") and its general many-valued internal logic of topoi (i.e., based on "topos theory"). For this, we shall follow the aforementioned "setting technique" of [10], here duly adapted for sheaves. In the course of this paper, we shall show that the 3-oppositional trisimplex (a.k.a. tri-triangle) contains three hexagons whose structure we shall discuss in detail, using for that the mathematical notions of "paracompleteness" (i.e., "intuitionism") and "paraconsistency" (i.e., "co-intuitionism"), linked with the topological nature of "open" and "closed" subsets, as put into evidence and demonstrated by us in [1] and [2]. At last, we shall trace the "jewel nonagon," proposed as a way of representing the trisimplex, and discuss logically some of its oppositional-geometric features.

2 Some Numerical Sheaves on a Topological Space with One Non-trivial Open Subset

Let us choose a topological space X with a unique non-trivial open subset U. That is to say that X has only three open subsets forming a strictly increasing sequence $\emptyset \subset U \subset X$.

A sheaf on the topological space X is a presheaf with the gluing condition. But, as the gluing condition is trivial for U and X which can be only recovered by themselves as open subsets, a sheaf on X is only a presheaf with a unique section on $\emptyset$, which can be noted $*$.

So, to describe a sheaf F on X, we must give:

- the set $F(X)$ of global sections, or sections on X,
- the set $F(U)$ of local sections, or sections on U,
- a function $F(X) \to F(U)$, named restriction to U.

The category of sheaves on X is a Grothendieck topos with an internal intuitionist many-valued logic with three truth-values which are in correspondence with the three open subsets: $True = X$, $False = \emptyset$, $1/2 = U$.

In this paper we shall consider only the sheaves with value in the set $\{1, 2, 3\}$. Let i and j be two different numbers among 1, 2, or 3, in all this paper we shall note:

- i_X the sheaf where $i_X(X) = i_X(U) = \{i\}$,
- i_U the sheaf where $i_U(X) = \emptyset \subset i_U(U) = \{i\}$,
- $i_X j_U$ the sheaf where $i_X j_U(X) = \{i\} \subset i_X j_U(U) = \{i, j\}$.

3 The Aristotelian 3^2-semantics Reconsidered in This World of Sheaves

In [7], Moretti modeled the four oppositional relations (contradiction, contrariety, subcontrariety, and subalternation) of the logical square, and more generally of oppositional geometry, as an Aristotelian 2^2-semantics. The exposant "2" refers to the number of logical questions and the number "2" refers to the number of truth-values allowed for the answers. Reformulated, the two questions are the following: "Is there a non-excluded middle between two terms ?"; "Is there a non-false conjunction ?". Among the four possible answers, two "No" lead to contradiction, a "Yes" at the first with a "No" at the second lead to contrariety, a "No" at the first and a "Yes" at the second lead to subcontrariety, and the subalternation (or implication) answers two "Yes" (provided one adds an extra constraint on the order: never the first without the second—cf. [7]).

So Moretti's main idea, intended both at generating new oppositional simplexes and at introducing many-valuedness into oppositional geometry, is to allow answering these two Aristotelian questions with more possible truth-values than just the classical "yes" or "no," staying for true and false, 1 and 0 (truly speaking, as Moretti started to explore it, the number of Aristotelian *questions* can also change, but we will not discuss this in our paper, keeping only the two classical Aristotelian questions). An Aristotelian n^2-semantic is then the set (in form of ordered, 2D square-shaped network, called by Moretti "Aristotelian n^2-lattice") of all oppositional relations generated by answering these two questions in an n-valued logic. As for its "oppositional" interpretation, this network or lattice can be so to say parted into three subsets: the contradiction-type relations, weak and strong (upper triangular half of the square lattice), the simplexes (horizontal diagonal of the square lattice), and the implication-type relations (lower triangular half of the square lattice).

Here we shall study only the Aristotelian 3^2-semantic, which is the following network (called in [7, 8] "Aristotelian 3^2-lattice"):

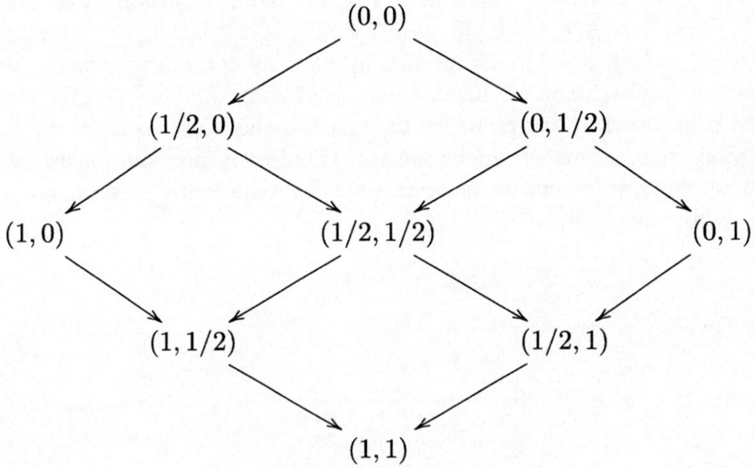

As we have said in the previous part, our numerical sheaves on X possess a three-valued logic based on the open subsets of X. Answering "1", "1/2", or "0" to a question about such sheaves corresponds to answering at which topological level it is true: at the global level X, only at the local level U, or never (i.e., at the trivial level $\emptyset$).

These considerations lead us to interpret in term of sheaves, sections, and topology the contradiction-type relations and the simplexes of the trisimplex. As for the implication-type, the straightforward meaning of inclusions of sections sets at some topological level will play.

First of all, let us consider the three negation-type relations (contradictions). The first one, "(0, 0)," is the classical contradiction, with no non-excluded middle and no true conjunction, so the global conjunction is empty and the global disjunction is the total sheaf. The relation "(1/2, 0)" allows a non-global middle term (only at level X and not at level U), but no true conjunction, so it is a global contrariety relation on X (and in fact a strict one at top level X), but with a contradiction at level U. Thus the global conjunction is empty and the local disjunction is the total sheaf but this is not the case of the global disjunction. We will note it "$C_X CDT_U$." Dually, "(0, 1/2)" permits a non-false conjunction at level U, but no middle term, so it is a local subcontrariety endowed with a contradiction at top level X. Thus the global conjunction is non-empty (even if it is empty at level X) and the global disjunction is the total sheaf. We will note it "$S_U CDT_{X-U}$."

Let us now study the three simplexes. In [7], (1, 0) and (0, 1) are proved to be respectively contrariety and subcontrariety. In our terms, they are the global contrariety "C_X" and the global subcontrariety "S_X." The more problematic (because new) simplex, i.e., "(1/2, 1/2)," can be easily explained, as for its nature, by our reasoning: it is a combined relation of contrariety, only at top level X, and of local subcontrariety, at level U; we shall note it "$C_{X-U} S_U$."

As for the three implications (subalternations), re-defining them sheaf-theoretically is even simpler than the preceding: "(1, 1/2)" is the local inclusion of section on U, noted "$\subset_U$," while "(1/2, 1)" is the inclusion of section on X noted "$\subset_{X-U}$," and "(1, 1)" is the global inclusion of sheaves, "$\subset$" (i.e., classical implication). At this point it is important to remark that the first two inclusions are not necessarily morphisms of sheaves: they would not commute with the restriction function of the sheaf. That points to the fact that the first two relations are *weak* implications whereas the *strong* implication is indeed a sheaf morphism.

With all these reflexions we can now give the Aristotelian 3^2-semantic network for the "numerical sheaves" on X:

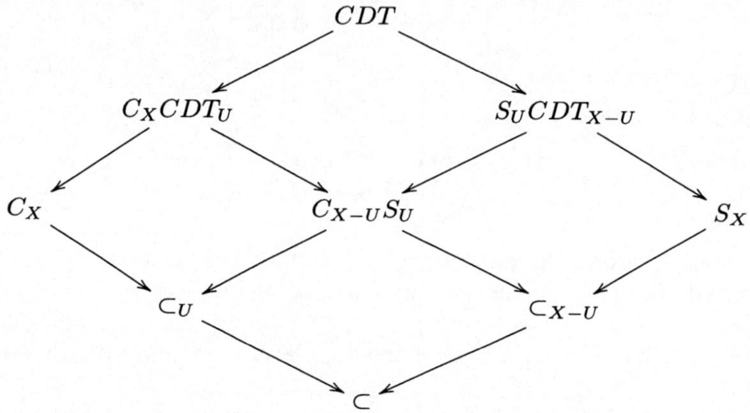

4 A Generic Trisimplex of Sheaves

As topos theory teaches, the many-valued internal logic lies on the relations between subsheaves of a determined sheaf. So let us consider now as *total* or *true sheaf* "$1_X 2_X 3_X$." We shall study the logical relations between some of its subsheaves: 1_X, 2_X, 3_X, their disjunctions $1_X 2_X$, $2_X 3_X$, and $3_X 1_X$, and some of their subsheaves, namely $1_X 2_U$, $2_X 3_U$, and $3_X 1_U$. (we could, with a symmetrical reasoning, consider instead their subsheaves $1_U 2_X$, $2_U 3_X$, and $3_U 1_X$ and obtain analogue results.)

The classical contradiction relations lie between the global sheaves, which are equivalent to the subsets of set-theory, and so can support the "setting technique" of [10]:

$1_X \cdots\cdots 2_X 3_X$ $\qquad$ $2_X \cdots\cdots 3_X 1_X$ $\qquad$ $3_X \cdots\cdots 1_X 2_X$

The $C_X CDT_U$ relation, as we saw, is a global contrary relation endowed with a local contradiction. It relies the following pairs of subsheaves:

$1_X - - - 2_X 3_U$ $\qquad$ $2_X - - - 3_X 1_U$ $\qquad$ $3_X - - - 1_X 2_U$

As we can easily observe, the global conjunction of 1_X and 2_X3_U is empty and their global disjunction is not the total sheaf. But their disjunction on level U is the total set $\{1, 2, 3\}$. The same proof runs for the two other pairs of subsheaves.

The S_UCDT_{X-U} relation is a local subcontrary relation endowed with a contradiction at the top level. Its relies the following pairs of subsheaves:

$$1_X2_U \sim\sim 2_X3_X \qquad 2_X3_U \sim\sim 3_X1_X \qquad 3_X1_U \sim\sim 1_X2_X$$

As we can easily observe, the local conjunction of 1_X2_U and 2_X3_X is non-empty because it is the local section 2, and their local disjunction is the total set $\{1, 2, 3\}$. But their conjunction at the top level X is empty and their disjunction at the top level X is also the total set $\{1, 2, 3\}$. The same proof runs for the other two pairs of subsheaves.

The contrariety simplex and the subcontrariety simplex concern the global sheaves, equivalent to sets, so they are the same as in [10]:

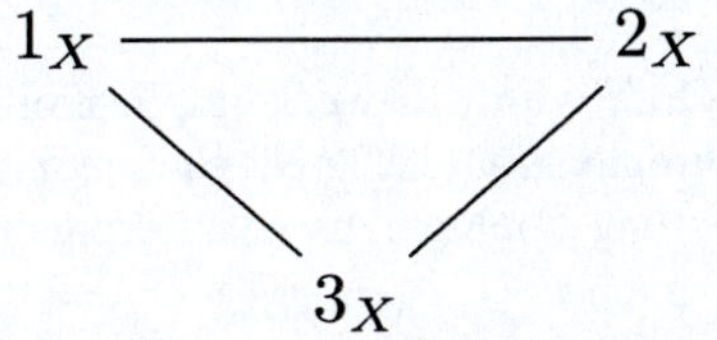

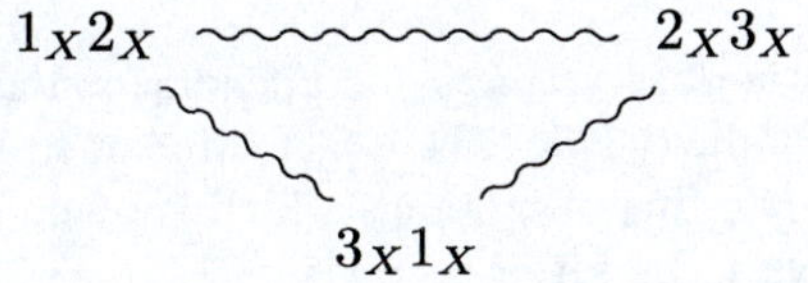

The $C_{X-U}S_U$ simplex is formed of three identical relations, each combining—relatively to each pair of subsheaves it joins—a local subcontrariety with a contrariety at top level.

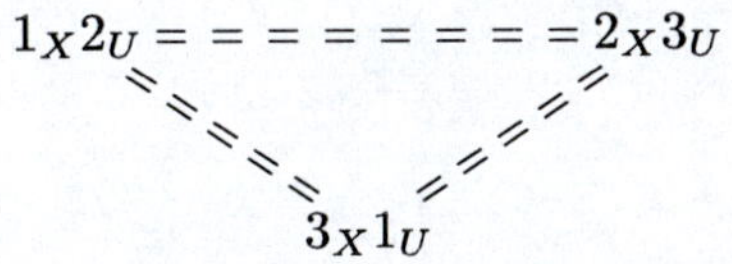

As we can easily observe, the local conjunction of 1_X2_U and 2_X3_U is non-empty because it is the local section 2 and their local disjunction is the total set $\{1, 2, 3\}$. But their conjunction at the top level X is empty and their disjunction at the top level X is not the total sheaf. The same proof runs for the two other pairs of subsheaves.

The weak implication $\subset_U$ is an inclusion of section on U, which gives the following relations:

$$1_X \underset{U}{\Longrightarrow} 1_X2_U \qquad 2_X \underset{U}{\Longrightarrow} 2_X3_U \qquad 3_X \underset{U}{\Longrightarrow} 3_X1_U$$

$$1_X \underset{U}{\Longrightarrow} 3_X1_U \qquad 2_X \underset{U}{\Longrightarrow} 1_X2_U \qquad 3_X \underset{U}{\Longrightarrow} 2_X3_U$$

We can remark that in the first of the two lines, the weak implications are in fact strong ones because the inclusions are valid also at top level X. But this is not the case in the second line: if the set of sections on U of 1_X, which is $\{1\}$, is included in the set of sections on U of 3_X1_U, which is the set $\{1, 3\}$, the set of sections on X of 1_X, which is again $\{1\}$, is not included in the set of sections on X of 3_X1_U, which is only the set $\{3\}$. So the weak implication $\subset_U$ is not always a strong implication, because it is not always a global inclusion, and therefore it is not always a morphism of sheaves.

The weak implication $\subset_{X-U}$ is an inclusion of section on X, which gives the following relations:

$$1_X2_U \underset{X-U}{\Longrightarrow} 1_X2_X \qquad 2_X3_U \underset{X-U}{\Longrightarrow} 2_X3_X \qquad 3_X1_U \underset{X-U}{\Longrightarrow} 3_X1_X$$

$$1_X2_U \underset{X-U}{\Longrightarrow} 3_X1_X \qquad 2_X3_U \underset{X-U}{\Longrightarrow} 1_X2_X \qquad 3_X1_U \underset{X-U}{\Longrightarrow} 2_X3_X$$

We can observe the same phenomenon as before with this weak implication: the first of the two lines is in fact a line of strong implications, whereas the second line is only a line of relations which are neither strong implications nor morphisms of sheaves. The set of sections on U of 1_X2_U, which is $\{1, 2\}$, is not included in the set of sections on U of 3_X1_X, which is the set $\{1, 3\}$ but the set of sections on X of 1_X2_U, which is only $\{1\}$, is indeed included in the set of sections on X of 3_X1_X, which is again the set $\{1, 3\}$.

To conclude with, the strong (i.e., classical) implications lie between the global sheaves, which are equivalent to subsets and thus are as in [10]:

$$1_X \Longrightarrow 1_X2_X \qquad 2_X \Longrightarrow 2_X3_X \qquad 3_X \Longrightarrow 3_X1_X$$

$$1_X \Longrightarrow 3_X1_X \qquad 2_X \Longrightarrow 1_X2_X \qquad 3_X \Longrightarrow 2_X3_X$$

5 Many-Valued Logical Hexagons in Our Trisimplex of Sheaves

With all these relations between our nine subsheaves, we can construct the three hexagons of the 3-oppositional trisimplex. The first hexagon is the classical "logical hexagon" (discovered independently by P. Jacoby, A. Sesmat and R. Blanché in—respectively—1950, 1951, and 1953), whose vertices are the global sheaves, which are equivalent to sets. As in [10], we have then:

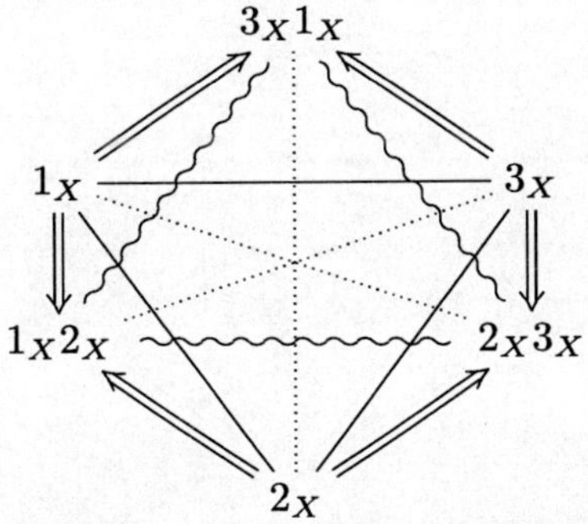

In this hexagon we meet again the traditional four kinds of opposition: contradiction (dotted line), contrariety (line), subcontrariety (waves), subalternation (arrow).

But there are two more hexagons. One is obtained by assorting the classical contrariety simplex with the new "mixed" one, i.e., with the $C_{X-U}S_U$ simplex. Here two problems do occur:

- First problem: the vertices of the mixed simplex, such as, for instance, 1_X2_U, cannot be *contradictories* (in the classical sense), pairwise, with respect to the vertices of the contrariety simplex, and then central symmetry, which is so to say the "cornerstone" of the logical hexagon (*quo* bi-simplex, i.e., bi-triangle) by fitting together two centrally symmetrical triangles, is not valid anymore as an easy and elegant visual means of expression of "contradiction." Nevertheless, as we saw, we have at our disposal two weak contradictions. And the C_XCDT_U weak contradiction is indeed the link holding between the contrariety simplex and the mixed simplex: 1_X2_U has this relation with 3_X, for example. So, in the second hexagon central symmetry will represent the C_XCDT_U *weak* contradiction.
- Second problem: the star hexagon being constructed (i.e., a logical hexagon represented without its perimeter of implication arrows), some vertices cannot be linked, as in a standard logical hexagon, by implications: 1_X and 3_X1_U, for example, because, as we saw, 1 is not a section on *X* for 3_X1_U. But, here too, as we saw, we have two weak implications, and the $\subset_U$ weak implication will rescue us here, because it can perfectly link our vertices: 1 is indeed a section on *U* of 3_X1_U, in our example.

So we have a second hexagon as follows:

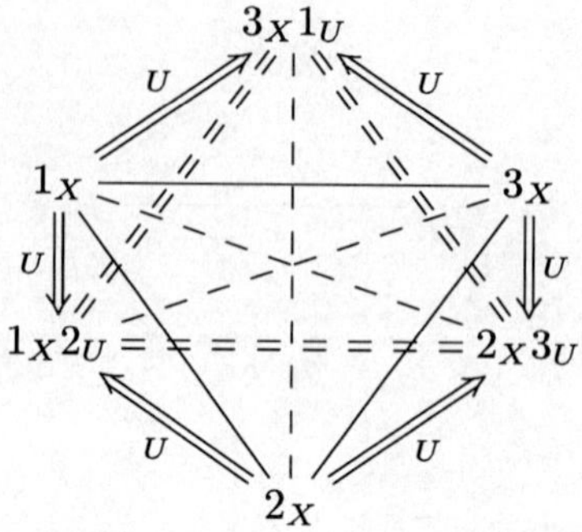

In this hexagon we have a new quartet of compatible "oppositional relations": the $C_X CDT_U$ weak contradiction, which contains a local contradiction (broken line), the classical contrariety (normal line), the mixed $C_{X-U} S_U$ relation containing a local subcontrariety (broken double line), and the $\subset_U$ local weak implication (arrow labeled U). Now, this "quartet of opposition" is fundamentally an *intuitionist* (in the sense of mathematical intuitionism, i.e., paracompleteness) quartet of opposition, for two complementary reasons. The first one is that the $C_X CDT_U$ weak contradiction is an "intuitionist negation": if $2_X 3_U$ is the "negation" of 1_X, we could see that between these two "contradictory" terms the X-section 3 is missing, and the axiom of excluded middle is therefore not valid (which is, notoriously, the main and most fundamental point in intuitionism). The second reason lies in the *local* quality of these oppositions: we could even say that these four oppositional relations are in fact the sheaf restrictions at local level U of the four traditional oppositional relations, which our previous discussion about the nature of these relations will easily show. But, as is known in topos theory, the local nature of the sheaves is the basis of the intuitionist internal logic typical of the topoi in general, the *open* subsets playing there the role of truth-values. Therefore we have here an intuitionist and sheaf-local 3-valued logical hexagon.

We can obtain the third hexagon by fitting together the new mixed simplex with the classical subcontrariety simplex. The same two problems as before will occur also now, and this time the $S_U CDT_{X-U}$ weak contradiction will be the meaning of central symmetry (in this hexagon), and the $\subset_{X-U}$ weak implication will constitute the edges of the hexagon (i.e., the perimeter made of alternated arrows). So we have, now, the following hexagon:

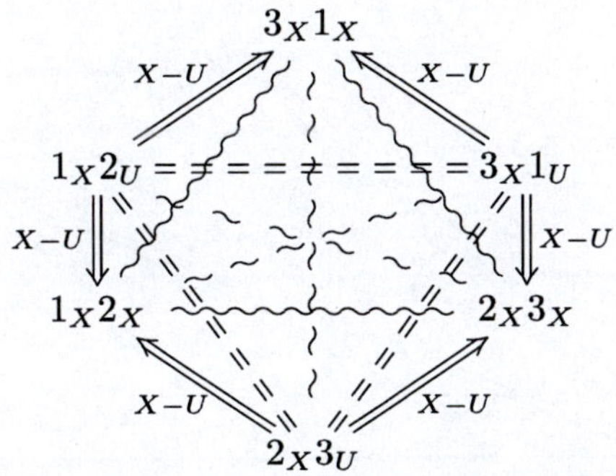

We have, with this third hexagon, a third quartet of oppositions: the S_UCDT_{X-U} weak contradiction, which contains a contradiction at top level X (tilde line), the mixed $C_{X-U}S_U$ relation containing a contrariety at top level (broken double line), the classical subcontrariety (waves), and the top level $\subset_{X-U}$ weak implication (arrow labeled $X - U$). Thus we have a top level X version of the four traditional kinds of opposition. But we must remark that our top level relations are not global relations, because the results on X do not imply the same results on U. In fact, all these relations speak about what happens in the closed subset $X \setminus U$, complementary subset in X of the open subset U. As we have shown in [1] and in [2], the closed subsets in general (i.e., topologically) have a *paraconsistent* (i.e., co-intuitionist) internal logic, and here we have the same property by this reason. The S_UCDT_{X-U} weak contradiction can be viewed as a "paraconsistent negation": 1_X2_U admits as its "negation" 2_X3_X, but their conjunction is the subsheaf 2_U, which happens to be not empty (this is a paraconsistent "glut"). Furthermore, if 1_X2_U implies straightforwardly 1_X2_X, it is more difficult to see how it can (nevertheless) weakly imply 3_X1_X, its U-section 2 not appearing in 3_X1_X unless we replace ourselves in the paraconsistent closed subset $X \setminus U$. The paraconsistent contrariety is also a $X \setminus U$ contrariety, for example, between 1_X2_U and 3_X1_U: whereas they are locally subcontraries, on the closed subset $X \setminus U$ they are as contrary as can be 1 and 3. Therefore we have here a closed paraconsistent logical hexagon.

6 The "Jewel Nonagon" as the Trisimplex

We can now end this paper with the "jewel nonagon" representation of the trisimplex, which in some sense contains in a glance all the previous discussions:

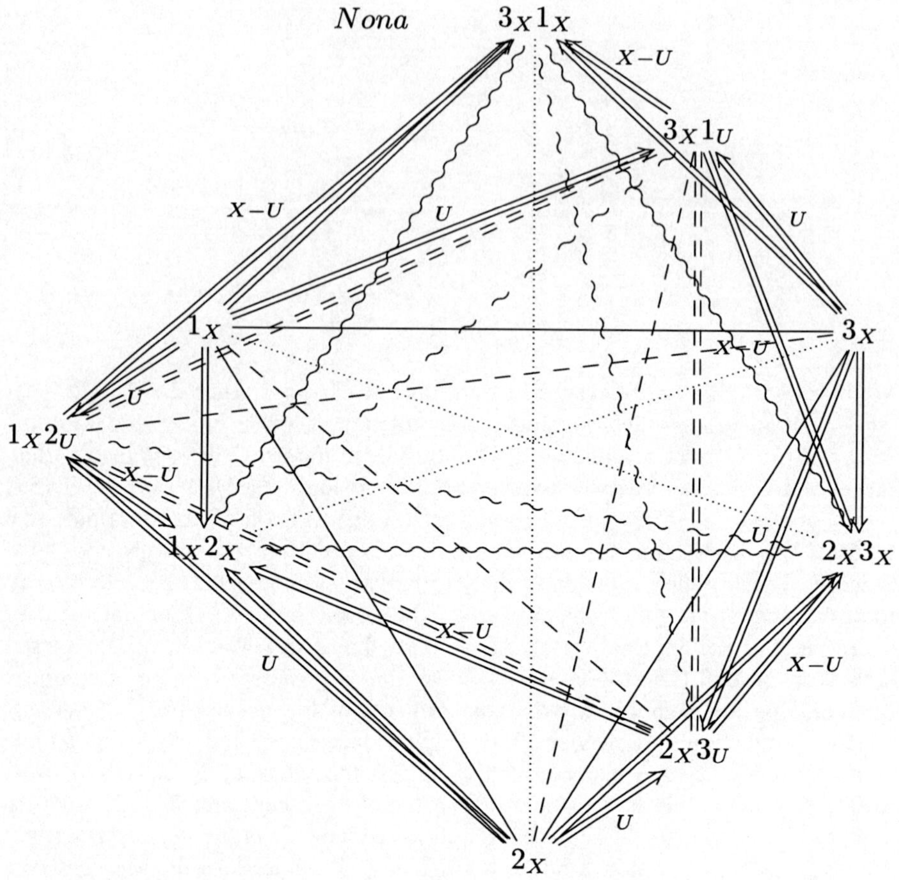

In this "Nona" representation of the jewel nonagon we have reserved central symmetry for the expression of contradiction (dotted line), and then we can see at first sight the classical logical hexagon. But we can also make the following remarks:

- The two weak contradictions (broken line and tilde line) have neither the geometrical property of meeting together (in a unique central point) nor that of passing by the center of the nonagon. In fact each forms an internal triangle, the two triangles being symmetrical along the line joining 1_X2_U and the center of the nonagon.
- The mixed simplex (broken double line) is a triangle obtained by a 90° rotation of the two other simplexes, each in one direction (i.e., clockwise and anticlockwise). Its vertices have also a mixed property relatively to the implication arrows: two arrows are coming to each of them (the U local ones) and two arrows are leaving (the $X - U$ top level ones). All this renders the fact that this mixed simplex is a local subcontrariety simplex in the intuitionist hexagon (and as

subcontrariety simplex it receives arrows) and a top level contrariety simplex for the paraconsistent hexagon (and as contrariety simplex it shoots arrows).

Remark that a symmetrical nonagon along a line joining $3_X 1_X$ to 2_X can be obtained with the other mixed simplex constituted by the subsheaves $1_U 2_X$, $2_U 3_X$, and $3_U 1_X$ (which we have left aside as redundant). And here are the only two which could be obtained by this method. Furthermore a dodecagon containing the two nonagons would have no logical meaning.

To conclude this paper, we could say that this world of sheaves illuminates the notion of 3-oppositional trisimplex. Furthermore our reasoning could apply for any 3-oppositional multisimplex (i.e., any oppositional multitriangle). It would be interesting for further research, on the one hand, to explore the case of the four-valued logical quadrisimplex, and on the other hand to develop a general sheaf technique for the decoration of the general 3-oppositional multisimplex.

References

1. Angot-Pellissier, R., "2-opposition and the topological hexagon", in J.-Y. Béziau and G. Payette (eds.), *The Square of Opposition. A General Framework for Cognition*, Peter Lang, Bern, 2012
2. Angot-Pellissier, R., "The Relation Between Logic, Set-Theory and Topos Theory as It Is Used by Alain Badiou", in A. Buchsbaum and A. Koslow (eds.), *The Road to Universal Logic*, Vol. II, Birkhäuser, Basel, 2015
3. Béziau, J.-Y., "New light on the square of opposition and its nameless corner", *Logical Investigations*, **10** (218–233), 2003
4. Blanché, R., "Sur l'opposition des concepts", *Theoria*, **19** (89–130), 1953
5. Blanché, R., *Structures intellectuelles. Essai sur l'organisation systmatique des concepts*, Vrin, Paris, 2004 (1966)
6. Moretti, A., "Geometry for Modalities ? Yes : Through *n*-Opposition Theory", in: J.-Y.Beziau, A.Costa-Leite and A.Facchini (eds.), *Aspects of Universal Logic*, Travaux de logique **N.17** (102–145), Neuchatel, 2004.
7. A. Moretti, *The Geometry of Logical Opposition*, PhD Thesis, University of Neuchatel, Switzerland, 2009
8. Moretti, A., "The Critics of Paraconsistency and of Many-Valuedness and the Geometry of Oppositions", in *Logic and Logical Philosophy*, vol. **19(1–2)** (63–94), Special Issue on Paraconsistent Logic, edited by K. Tanaka, F. Berto, E. Mares and F. Paoli, 2010.
9. Murinova, P., " Graded Structures of Opposition in Fuzzy Natural Logic", *Logica Universalis*, **14, 4** (495–522), 2020
10. Pellissier, R., " "Setting" *n*-opposition", *Logica Universalis*, **2, 2** (235–263), 2008
11. Sauriol, P., "Remarques sur la Théorie de l'hexagone logique de Blanché", *Dialogue*, **7** (374–390), 1968
12. Smessaert, H., "On the 3D visualisation of logical relations", *Logica Universalis*, **3** (2) (303–332), 2009

Tri-simplicial Contradiction: The "Pascalian 3D Simplex" for the Oppositional Tri-segment

Alessio Moretti

To Jean-Yves, Régis, and Hans
Three structuralist deep minds
Who changed forever my
Creative perception
Of mathematics

Abstract In this paper, we deal with the theory of the "oppositional poly-simplexes", producing the first complete analysis of the simplest of them: the oppositional tri-segment, the three-valued counterpart of the oppositional-geometrical red *contradiction* segment. The concept of poly-simplex has been proposed by us in 2009 (Moretti A, The geometry of logical opposition. PhD thesis, University of Neuchâtel, Switzerland, 2009), for generalizing the theory of the "oppositional *bi*-simplexes", which is the heart of our and Angot-Pellissier's "oppositional geometry" (Angot-Pellissier R, 2-opposition and the topological hexagon. In: [30], 2012; Moretti A, Geometry for modalities? Yes: through *n*-opposition theory. In: [27], 2004; Pellissier R, Logica Universalis 2:235–263, 2008). The latter is meant to be the general theory of structures like the logical hexagon (which is a bi-triangle). The *poly*-simplexes are the most straightforward way to turn any "geometry of oppositions" consequently *many*-valued, which is otherwise still a *desideratum* of the field. We start by recalling the general theoretical context: how the field was opened, around 2002, by a reflection on the foundations of paraconsistent negation (Béziau J-Y, Logical Investigations 10:218–233, 2003) and how from that has progressively emerged "oppositional geometry", the theory of the "oppositional structures", enabling to model the "oppositional complexity" of "oppositional phenomena". After recalling how emerged the idea of *poly*-simplex, we explain why time seems to have now come to explore them for real, since we have two powerful new tools: (1) Angot-Pellissier's sheaf-theoretical technique

A. Moretti (✉)
Université degli Studi eCampus, Novedrate, Italy

SSML di Varese, Varese, Italy

J.-Y. Beziau, I. Vandoulakis (eds.), *The Exoteric Square of Opposition*, Studies in Universal Logic, https://doi.org/10.1007/978-3-030-90823-2_16

(Angot-Pellissier R, Many-valued logical hexagons in a 3-oppositional trisimplex. In: this volume, 2022; Angot-Pellissier R, Many-valued logical hexagons in a 3-oppositional quadrisimplex. Draft, January 2014) for producing the many-valued oppositional *vertices* of the poly-simplexes (and evaluating their *edges*) (2) and a generalization of "Pascal's triangle", turned here into a more general "Pascalian ND simplex", whose suited "horizontal (N-1)D sections" provide a much needed and very powerful numerical-geometrical "roadmap", for constructing and exploring *any* arbitrary oppositional poly-simplex (Sect. 1). After unfolding successfully the structure of the tri-segment (Sects. 2 and 3), we make an unexpected detour (Sect. 4) by Smessaert and Demey's "logical geometry" (Smessaert H. and Demey L., Journal of Logic, Language and Information 23:527–565, 2014), composed of two "twin geometries": one for "opposition" and another for "implication". We thus develop the "implication geometry" of the tri-segment and so discover that what these authors take for a "bricolage" (called by them "Aristotelian geometry") is in fact, when considered as general "Aristotelian *combination*" in opposition*al* spaces higher than the bi-simplicial one (the one in which they remain tacitly but constantly), the mathematically optimal way, bottom-*up*, for exploring methodically the poly-simplicial space. We end (Sect. 5) by considering *applications of this tri-segment* resulted from such a tri-simplicial diffraction of the bi-simplicial contradiction segment (which adds to it paracomplete, i.e., intuitionist, and paraconsistent, i.e., co-intuitionist, features) in many-valued logics, paraconsistent logics, quantum logic, dialectics, and psychoanalysis. In particular, we show that the tri-segment, by its paracomplete substructure, models, better than did anything before it, "Lacan's square".

Keywords Pascal's triangle · Pascal's simplex · Multinomial theorem · Oppositional bi-simplex · Oppositional poly-simplex · Oppositional geometry · Logical geometry · Logical hexagon · Contradiction · Negation · Classical negation · Many-valued logics · Topos-theory · Sheaf theory · Paraconsistent logics · Paracomplete logics · Intuitionism · Co-intuitionism · Hegelian logics · Dialectics · Quantum logics · Psychoanalysis · Lacan's square · Square of sexuation

Mathematics Subject Classification (2000) Primary 18N50, Secondary 11B65, 03B50 18F20, 03B53, 03A05

1 The Context of This Study

Before studying the oppositional tri-segment (Sects. 2, 3, and 4) and its applications (Sect. 5), it will be useful to recall the context where this new strange mathematical concept has arisen in 2009. This obliges us to say something, in the following Sect. 1.1, about the genesis of "oppositional geometry". In Sect. 1.2 we will present oppositional geometry and in particular its link with the concept of "bi-simplex" (2004). Starting from Sect. 1.3, we will recall the concept of "poly-simplex" (2009). Finally, in Sect. 1.6 we will introduce the original idea of "tri-segment" (2009), meaning by that the simplest oppositional poly-simplex (poly≥3).

1.1 *The Controversy on the Foundations of Logical Negation (2003)*

Starting in 1995, some logical inquiries on the philosophical foundations of "paraconsistent logics" (the logic of "nontrivial inconsistency"; cf. [21, 35, 106, 117–118]) by Slater, Priest, Restall, Paoli, Béziau, and several other logicians and philosophers bore progressively to the front a quite ancient and almost forgotten *structure*: the "*square* of opposition" (a.k.a. "logical square", "Aristotle's square", etc.). This old structure (second century) condenses some of the main concepts of Aristotle's (384–322 BC) theory (and logic) of "opposition" (Fig. 1).

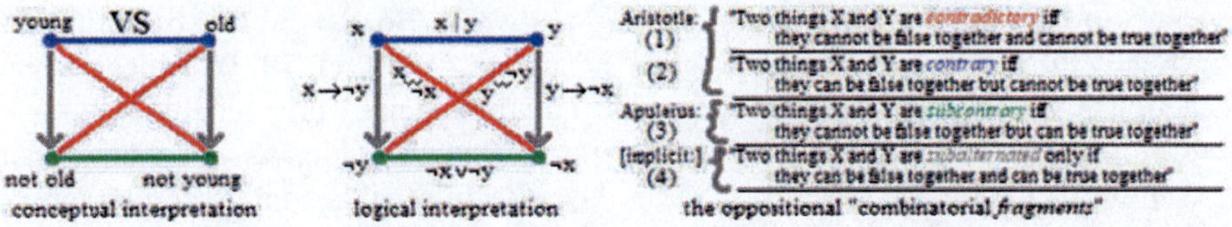

Fig. 1 The "square of opposition" (or logical square, or Aristotle's square) and some of its fundamental properties

What happened is that some scholars (and *in primis* Slater in 1995, [132]) used *abstractly* this square (i.e., without drawing it but by recalling its definitions – Fig. 1) in order to show that what paraconsistent logic (since at least 1948, with N.C.A. da Costa and his Brazilian school of paraconsistent logic, [46]) argued to be, namely, a logic provided with a new kind of "negation" operator (i.e., a "paraconsistent *negation*", a negation "~" capable of having nontrivially "A&~A"), is in fact something much less interesting than promised (by paraconsistent logicians): because "paraconsistent negation" truly speaking reveals to be something not even deserving the name "negation". So, paraconsistent logic and its paraconsistent negation would be, according to Slater (and his many followers on this point), a brutal deceit. More precisely, the claim was that, formulated in the ancient but fundamental (i.e., "transcendental") language of the old square (Aristotle's language), a conceptual language which more or less gave birth to "logic" and has been conserved (as a primitive but sound fragment) by the successive developments of logic, paraconsistent negation, seen as a binary relation, is neither a relation of "contradiction" nor a relation of "contrariety" (both being indeed two currently serious forms of negation, namely, the classical and the "intuitionist" ones). It is only (and disappointingly) a relation of "subcontrariety" (i.e., some kind, so to say, of *inverse* of contrariety, some kind of *anti*-contrariety, i.e., some form not of incompatibility but, on the contrary, of close "collaboration"!). Put more crudely, Slater's argument pretends to show that paraconsistent "negation" is in fact no other than "inclusive disjunction" (i.e., the well-known – and not negation-like! – "∨" operator of propositional logic), given that subcontrariety has precisely that meaning (cf. Fig. 1) when translated, as it can, into "propositional logic" (the logic of the "binary connective" relations). Accordingly, at least some of the *many* attempts (which I reviewed in 2010 [96]), by paraconsistent logicians, to resist Slater's devastating criticism against the very idea of their working field, paraconsistent logic, could have involved a renewed study of the logical square. But, in fact, that happened to be done almost uniquely by the paraconsistent logician and philosopher Jean-Yves Béziau (2003, [24], as a complement to [26]), who tried to answer frontally Slater's objection (i) by adopting its main argument ("paraconsistent negation *is* the square's relation of *subcontrariety*") but (ii) by reconsidering radically the very idea of logical square (and this strategy, as we will see, is *a posteriori*, without exaggeration, a bit of a stroke of genius). Béziau claimed, for short, (1) that Slater was right in his advocation of the old square concerning paraconsistency (2) *but* that "subcontrariety" is in fact not a disappointing (or marginal) but, on the contrary, *a very important relation*, indeed, such that – quite contrary to Slater's claim – it deserves *plainly* being considered *a very interesting "new" kind of nonclassical "negation"*! In a nutshell, Béziau did this by resuming the totally forgotten concept of "logical hexagon" (1950, cf. [97]), which, when speaking about the logical square, *de jure* is a mathematically unavoidable reference (but *de facto* so much and so badly underestimated by logicians, even now, that it is almost unknown by them!). This is because this hexagonal *structure* shows (cf. Blanché [33]), with mathematical certainty, nothing less than the undisputable fact that the logical square is only a problematic and misleading *fragment* of the mathematically unproblematic (but still mysterious) logical hexagon (Fig. 2).

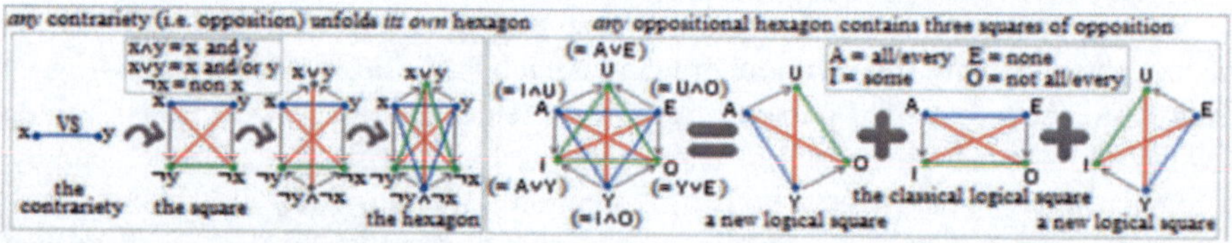

Fig. 2 From *any* "contrariety" emerges a "logical square", and hence a "logical hexagon" containing three logical squares

Then Béziau showed, by a discovery he made while reasoning, against Slater, on the possible applications of the logical hexagon to contemporary "alethic-modal logic" (Sect. 1.2), that there are not *one* but in fact *three* logical hexagons, the classical one (under its alethic-modal reading, already proposed by Blanché around 1953), which expresses classical negation, and two new (Béziauian) logical hexagons, also for alethic-modal logic, which express, respectively, relatively to alethic-modal-logical operators, not the *classical*, but the *paracomplete* (i.e., intuitionist) and the *paraconsistent* (i.e., co-intuitionist) negation (the "co-" being an important concept of "category-theory", cf. Sect. 1.4, and the "para-" being an important concept of metalogic) (Fig. 3).

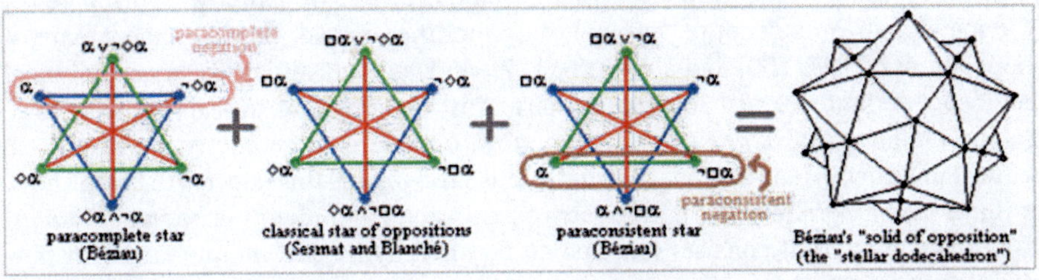

Fig. 3 Beziau's very original defense of the idea of "paraconsistent negation" (2003) against Slater's attack to it (1995)

By the proposal of that new global setting of opposition matters (i.e., notably by conjecturing, at the end of his very original – and not enough appreciated – 2003 paper, the existence of a new 3D "stellar dodecahedron" solid of opposition made by the 3D intersection of the three 2D "modal hexagons" (a 3D stellar solid, however, that Béziau was imprudent enough not to draw and check...) taking the place of Aristotle's 2D square, as the new *transcendental core* of the foundations of logic), Béziau argued he had proven the mathematical naturalness of "subcontrariety" and thus defeated in its very heart Slater's fundamental attack to the very idea of paraconsistent negation and paraconsistent logic (to this we will return on Sect. 5.3).

1.2 *The Mathematics of Opposition: There Is an Oppositional Geometry*

Béziau's 2003 paper [24] – where he introduced, as a promising conjecture, the idea of a 3D solid of opposition, larger (and more meaningful) than the old 2D square, although Béziau's very solid revealed soon mistaken in its details: not 12, but 14 vertices – unexpectedly opened, notably through Moretti (2004, [93]), Pellissier (2008) [111], and Smessaert (2009) [133], a whole new discipline devoted to the study of this kind of oppositional-geometrical solids (or polytopes): currently some (Smessaert and Demey, cf. [49, 51, 135]) call it "logical geometry", while others (ourselves and Angot-Pellissier, cf. [3, 98, 101]) call it "oppositional geometry". Whatever the name, the result, after more than 15 years of studies until now, is that there is (by now still small, but growing) a new branch (or at least a new theory) of mathematics, having "opposition" as its focus *object* of study. Epistemologically, it must be noticed that, still nowadays, this is a difficult bit to swallow for many established scholars: "opposition" was *reputed* – and so it tends to remain (mistakenly!) – to be more or less the "strict possession" of mathematical *logic* and of the "analytical philosophy" allegedly based on it. The latter, on that respect, is *reputed* to have fought victoriously against at least two tough competitors in the twentieth century: Hegelian-Marxist "dialectics" (which claimed also to be the *science* of "contradiction" and of "opposition"; Sect. 5.5) and transdisciplinary "structuralism" (which used quite often, together with instances of the mathematical concept of "group", duly modified instances of ... the square of opposition; cf. [97, 100], but also [126, 127]) and, as such, still nowadays, analytical philosophers and logicist logicians (there are!) do not want to "share their cake" (i.e., the concept of *opposition*). But by now, oppositional geometry, although still almost unknown, does indeed exist and keeps growing (as this paper will try to show, with its climax on Sect. 4.6), given its robust results and main mathematical *ideas*. As for its "body", it consists (at least so far) mainly of two infinite *series* of new mathematical structures: (1) *n*-oppositions (i.e., so to say, "*n*-oppositional *kernels*", the so-called bi-simplexes), or A*n* (Fig. 4), (2) and oppositional *closures* (one for each kernel of *n*-opposition; cf. Fig. 7 for a clear example of application of the *series* of the B*n*-structures). In fact, an important result of oppositional geometry of 2008 [111] is that *any* oppositional bi-simplex or A*n*-structure (which expresses, as such, the kernel of *n*-opposition) has its own "oppositional closure" or B*n*-structure (obtained by adding to it an "envelope"; cf. Figs. 49 and 51, which complete the kernel by adding some more vertices to it, forming its "cloud"); moreover, there also are "oppositional generators" or Γ-structures, which appeared first of all – but up to now still without a specific theorization – in modal logic (among others in Prior and Hamblin, Chellas, Hugues and Cresswell, and Popkorn; cf. Fig. 35 of [97]) but seem to have a much more general nature, still to be explored. The general character of these two series (the A*n* and B*n*) is granted mathematically by a demonstration provided by the mathematician Régis Angot-Pellissier in 2008 [111], which also gives a general mathematical handling method for oppositional geometry: this is called the "set-theoretical partition technique for *n*-oppositions" or "setting technique". Consequently, oppositional geometry is so to say ruled by this general set-theoretical method (the linguist and logician Hans Smessaert has provided in 2009 [133], independently, another one, based on "bit strings", roughly equivalent, later leading to what he and Lorenz Demey call since 2011 "logical geometry"; cf. [135]). This method allows putting into precise relation the oppositional generators (the Γ-structures, when there are), the oppositional closures (the B*n*-structures), and the oppositional kernels or *n*-oppositions (the A*n*-structures) (Fig. 5).

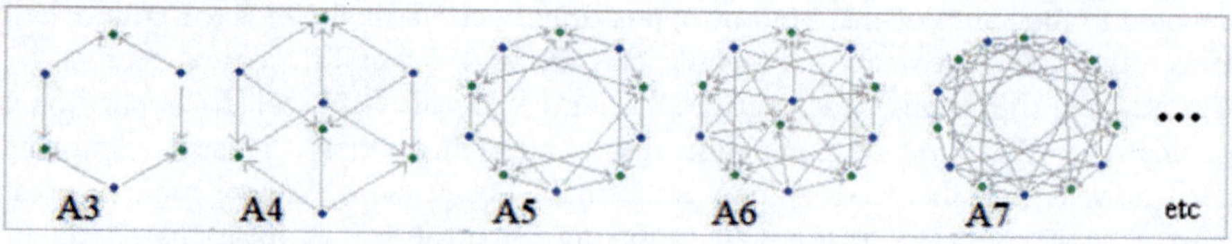

Fig. 4 The bi-simplexes of dim. $n - 1$ (hexagon, cube...) are the "kernels" (A*n*) of the *n*-opposition structures

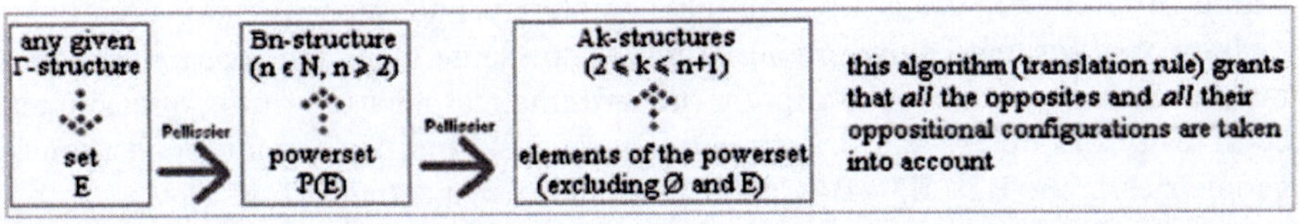

Fig. 5 Pellissier's setting method for oppositional geometry: from generators (Γ), to closures (B), to kernels (A)

Oppositional geometry, through its abstract study of "oppositions" as such, therefore offers a tool for measuring (and studying) a transversal object which can be called "oppositional *complexity*" and which in fact can appear in *every* discipline (*oppositionality* is a universal mathematical property). Let us recall two quick examples of this.

First, different phenomena (whatever their field) sharing the same degree of oppositional complexity have the same oppositional-geometrical model (or "oppositional attractor"): for instance, the "oppositional tetrahexahedron" or B4 (i.e., the complete structure of 4-opposition – i.e., its *closure* – whose *kernel* is the "oppositional cube", or bi-tetrahedron, or A4) formalizes, among many other things, propositional logic, first-order predicate logic, alethic-modal logic, and partial order-theory (Fig. 6).

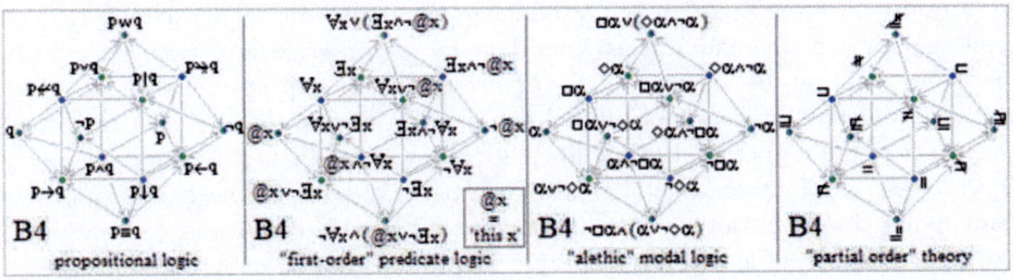

Fig. 6 The "oppositional tetrahexahedron" (B4), the attractor of 4-opposition, characterizes several different fields

But it also formalizes things much more informal, in whatever field (including humanities and art), like, for instance, some important gender issues (gender identity) and some conceptual frameworks related to sexual preference, just to give two "sexier" examples (Sect. 5.6). Remark, *en passant*, that the 3D opposition solid Béziau was *rightly* looking for (Sect. 1.1) is precisely the B4 (and more precisely the third B4-structure in Fig. 6). Béziau missed its precise structure (cf. Fig. 3) but was quite right in guessing its existence (and its three-dimensional nature).

Second, and conversely, *a same qualitative (or conceptual) phenomenon can admit (in different contexts possible for it) different degrees of oppositional complexity*. This is, for instance, the case with the mathematical object "order", studied by the fundamental branch of mathematics called "order-theory" (or "lattice theory" [48]): in different although strictly related mathematical universes (viz., discrete order, total order, partial order, set-theoretical order...), this object ("order") admits different degrees of oppositional complexity, and oppositional geometry, as a plastic tool for that, allows giving of this, at the same time, a precise *arithmetical* measure and *geometrical* description (the arithmetical measure of the oppositional complexity of "order" can be, respectively, 2, 3, 4, and 5, meaning oppositional-geometrically the B2-, B3-, B4-, and B5-structure, respectively; Fig. 7).

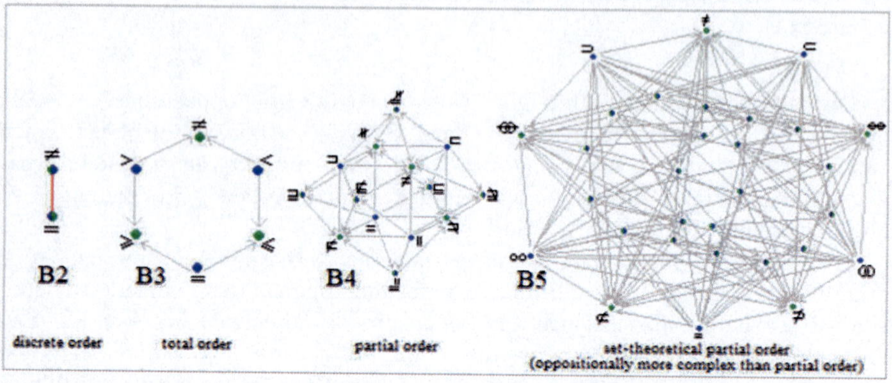

Fig. 7 Oppositional geometry, by the series of its closures (B*n*), allows *measuring* the complexity of different "orders"

It must be remarked that oppositional geometry seems to be, as any part of contemporary mathematics (this important *structuralist* feature was magistrally explained in 1968 by Piaget [112]), a converging point of several distinct mathematical "distant" areas: oppositional geometry is known to have in it important elements of graph theory [125], mathematical logic, modal logic, fractal geometry [109, 110], and knot-theory ([101]), and some facts (e.g., in the tetrahexahedron and higher similar closure structures) seem to suggest also the presence of relevant aspects of "differential topology" ([62] p. 52), to be inquired in the future (Fig. 8).

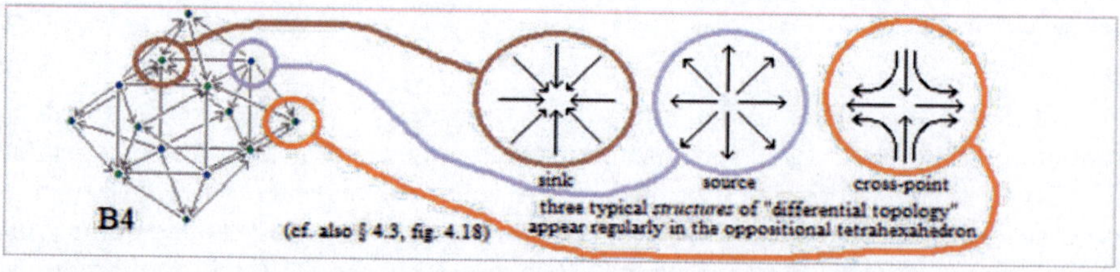

Fig. 8 There are several mathematical, "extra-logical" interesting properties in the oppositional tetrahexahedron (B4)

In our own so to say "anti-logicist" and structuralist way of understanding it (meaning by "structuralism" something like "taking seriously and therefore pushing to their still unknown limits any new combinatorial fragments"), oppositional geometry seems to be structured around the mathematical concept of "oppositional bi-simplex" (cf. [93]), and "simplexes" are *mathematical* objects usually absent from the vocabulary and the "ontological pantheon" of logic, since the latter tends to be independent from (and anterior to) "numbers", whereas simplexes are, so to say, precisely "geometrical numbers" (cf. [15]; and [45], p. 120) (Fig. 9).

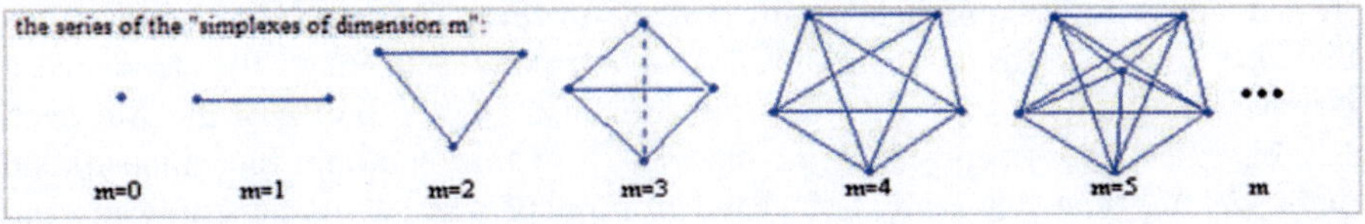

Fig. 9 "Simplexes" are so to say the geometrical counterpart of "numbers": their dimensionality grows into infinite

Historically, oppositional geometry seems to have popped out all of a sudden, unexpected, around 2004, and as such it still is "not very well accepted", as we said, for instance (and mainly), by analytical philosophy "leaders" and therefore "troops" – witness, paradigmatically, the otherwise unexplainable and unjustifiable lasting absence of *any* reference to the logical hexagon in, say, Terence Parsons' very famous, and since 1997 regularly updated (!!!), entry on the logical square in the very famous *Stanford Encyclopedia of Philosophy* [107]: remark that this valuable scholar was invited (and thereafter published) as speaker in the "First World Congress on the Square of Oppositions" (Montreux, 2007) where the logical hexagon (and the premises of oppositional geometry) was one of the main objects in explicit focus, so Parsons cannot claim to "ignore *de jure*" what he seemingly has *decided* to "ignore *de facto*". But, at the same time, despite this "analytical opposition to the geometry of oppositions" (Sect. 4.6), truly speaking, ongoing historical and epistemological studies constantly discover from time to time new elements of evidence that, although oppositional geometry has remained unveiled until very recently (i.e., 2004, with [93]), there have been, and since long, several premonitions of the existence of such a "geometry of oppositions" and even some true forerunners of it: several forgotten strange, isolated, and badly understood glimpses into oppositional-geometrical possibilities were given, among others, also by historically important thinkers like Aristotle, Apuleius, Llull, Buridanus, Lewis Carroll, De Morgan, Vasil'ev, Reichenbach, Prior, etc. (on all this cf., for instance, [97, 98, 100], but also [126, 127]).

Now, another chapter of this young and still open mathematical adventure is relative to the generalization trying to go from the aforementioned concept of bi-simplex to a new concept naturally built on top of that, namely, the concept of *oppositional poly-simplex*.

1.3 A General Extension of OG: The "Oppositional Poly-Simplexes"

In 2004, A. Costa-Leite (then a PhD student of Béziau, as soon myself) challenged oppositional geometry (then called by me "n-opposition theory", N.O.T.) to get rid, if it only could, of its founding but also constrictor notion of "bi-simplex". The (friendly but frank) reproach signified that, however (finitely) big and complex the blue (n-1)-dimensional simplex of n-contrariety and hence the geometrical n-opposition built on top of it (its "oppositional closure", the Bn of the An, Sect. 1.2, Fig. 7), it always consisted, in some sense, of four and only four oppositional main "colors" (conventionally – since the proposal of Béziau [24], generally adopted after him – blue, red, green, and black/gray). This led me to accept such radical challenge, trying to go from the concept of bi-simplex to something wider, namely, as I proposed, the more general (but at that time inexistent!) concept of oppositional "poly-simplex". In my 2009 PhD dissertation [94], in some sense I succeeded in responding this challenge (as for the key ideas for this, my intuition traces back to *at least* 2006, and I obtained the main results *no later than* in 2007). The original idea consisted of the following: (1) in trying to understand how the two oppositional "simplexes" (responsible of the four colors) so to say popped up with Aristotle's theory of opposition, and this led back to his elegant combinatorial definition (Sect. 1.1, Fig. 1 – a "combinatorial fragment"!), expressed by words, of "contrariety" and "contradiction" (completed by Apuleius with the tantamount elegant combinatorial definition – also a "combinatorial fragment"! – given by him or by the "Pseudo Apuleius", together with the square visual device, of "subcontrariety"; Fig. 1) (this was called "Aristotelian 2^2-semantics" and "Aristotelian 2^2-lattice") (Fig. 10); (2) in trying to see whether one could go beyond "bi-simpliciality" (Costa-Leite's challenge) so understood (i.e., taken as the "Aristotelian" game-theoretical "meta-level" generating the blue and the green simplexes and the red and the gray "links" between these two simplexes); and (3) that is (I proposed), in changing the number of the truth-values authorized in its generative "ask-answer" game-theoretical meta-level, observing that this kind of change seemingly generates, automatically, in an infinite variety of different possible ways, precisely extra "oppositional simplexes" (and interesting extra pairs of links between them). This was called "Aristotelian p^q-semantics" (with "q" as the number of possible different questions "Can two things ...?" and "p" as the number of possible answers "Yes/no/maybe/..." to each of these questions); it was first examined as "Aristotelian 3^2-semantics" and "3^2-lattice", which generate three instead of two simplexes (Fig. 11).

Each such Aristotelian p^q-semantics results in fact in a correlated "Aristotelian p^q-lattice", useful in so far as it gives, *a priori*, the "oppositional colors" (i.e., the possible *qualities* of opposition) of the mathematical universe under discussion. In the case where $q = 2$ remains unchanged, the variations of p generate Aristotelian 2D *square* lattices (presented as lozenges) which are bigger and bigger but remain two-dimensional. Differently, when what varies is "q", the lattice becomes increasingly many-dimensional and n-dimensionally hypercubic, a.k.a. "measure polytopic" ([45], p. 123). The q is a new, third dimension – along with p and with n

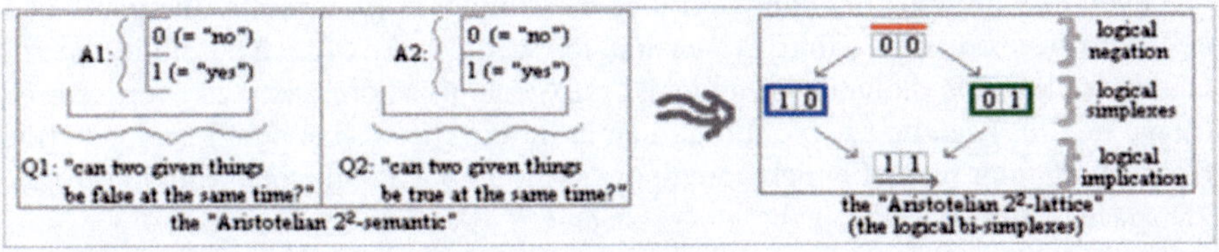

Fig. 10 The way proposed in 2009 in order to formalize Aristotle's game-theoretical generator of opposition theory

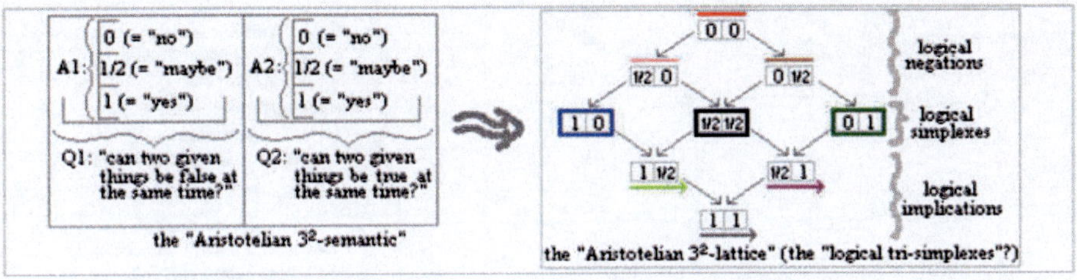

Fig. 11 The Aristotelian 3^2-semantic and its correlated 3^2-lattice giving the "kinds of opposition" of the tri-simplexes

- of *complexity growth* of the whole poly-simplicial *n*-opposition. So, in a nutshell, the changes in *p* and in *q* generate a mathematical 2D space (populated by *n*-dim measure polytopic lattices) of possible changes in the meta-level of the theory of opposition (Fig. 12).

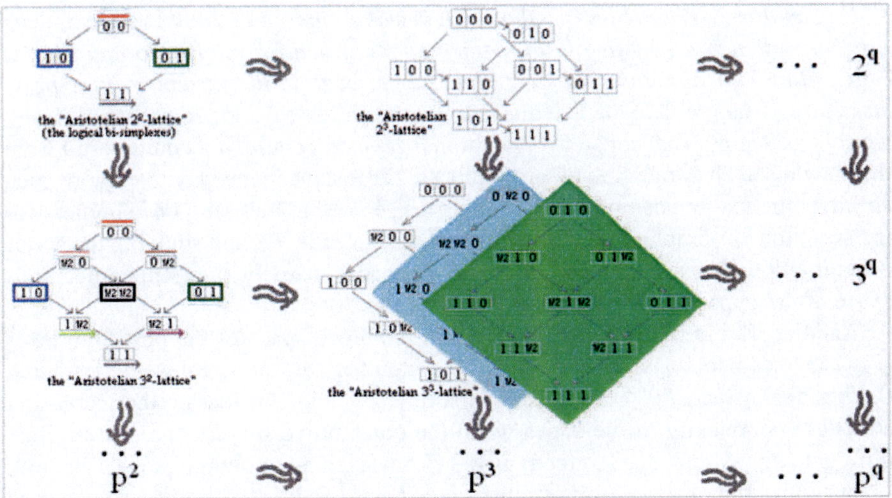

Fig. 12 An overview of the space of the possible Aristotelian p^q-lattices (for the general oppositional poly-simplexes) when "*p*" and/or "*q*" vary

Given that the growing complexity of the simplexes (indexed by the parameter "n") constitutes, as said, a third dimension, the whole space of such possible generalizations of the Aristotelian bi-simplicial case (generalizations abstractly conjectured by me in 2009) can be figured through an infinite "Aristotelian parallelepiped" (the front rectangular face of which corresponds (modulo a 3D diagonal rotation) to the 2D space of Fig. 12) (Fig. 13).

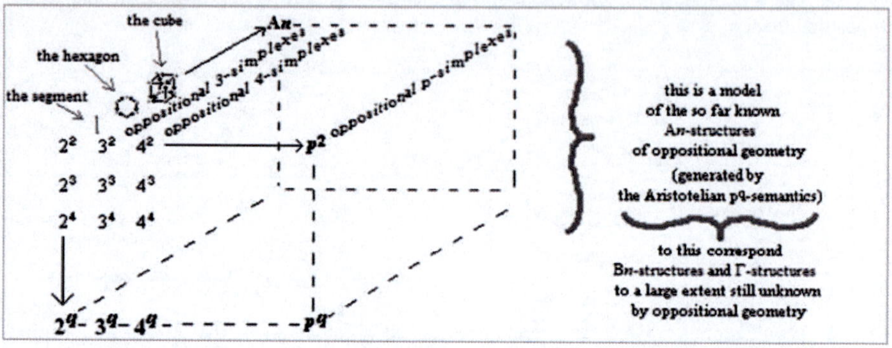

Fig. 13 The Aristotelian infinite parallelepiped for the oppositional poly-simplexes (and beyond)

Remark that in all what follows only $q = 2$ will be explored (i.e., only 2D, square-shaped Aristotelian p^2-lattices). This means that we will remain 2D at the level of the metatheory of opposition (leaving the exploration of the still intriguing "q" parameter – supposing it leads, as I believe, to something sound, meaningful, and tractable – for further studies, Sects. 4.1, 5.1 and 5.2).

It is by this "Aristotelian" method that it was proposed in 2009 to consider the existence of such a new family of mathematical structures relative to oppositions (integrating logical many-valuedness and giving birth to the structure of the poly-simplexes). The problem then, at this still very hypothetical and programmatic level, was that of seeing what concrete oppositional *geometry* could, if it could, result from this new research paradigm and program. So, for instance, one way to explore such still hypothetical oppositional poly-simplexes seemed to be fixing one simplex, for instance, the 2D simplex (i.e., the triangle, Sect. 1.2, Fig. 9), and studying the series of its infinitely growing oppositional poly-instances, namely, the (still conjectural) space of the oppositional poly-*triangles* (Fig. 14).

Another way in order to explore the still unknown space of the poly-simplexes (Fig. 13) seemed to be fixing instead the number "p" of simplexes considered (taking, for instance, $p = 3$, i.e., the *tri*-simplexes) and considering the increasing structural complexity of the series when the constitutive *simplex* (present in three different colors: blue, black, green) grows (from segment to triangle, to tetrahedron, to four-dim simplex, to five-dim simplex, etc., Sect. 1.2; Fig. 9): this was conjectured as the (hypothetical) space of the oppositional *tri*-simplexes (Fig. 15).

One of the many exciting parts of this new research line seemed to be the search for new oppositional solids. For instance, the tri-triangle seemed to be possibly

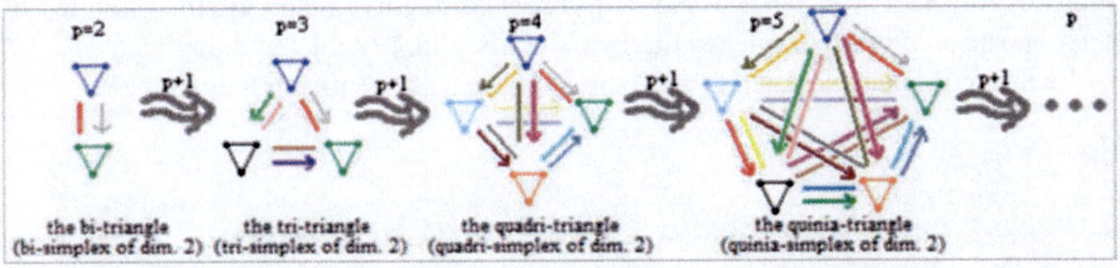

Fig. 14 The space of the oppositional poly-triangles: from the bi-triangle (i.e., the logical hexagon) to the p-triangle

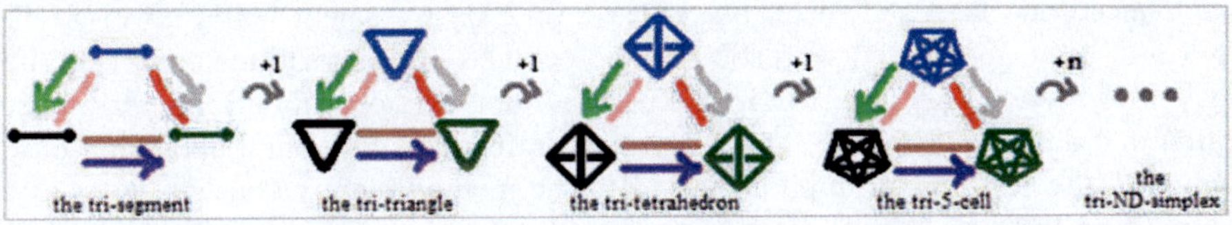

Fig. 15 The (hypothetical) space of the growing oppositional tri-simplexes (tri-segment, tri-triangle, tri-tetrahedron...)

represented by a compact 2D figure (provided one adopts curve lines, as in non-Euclidean geometry) (Fig. 16).

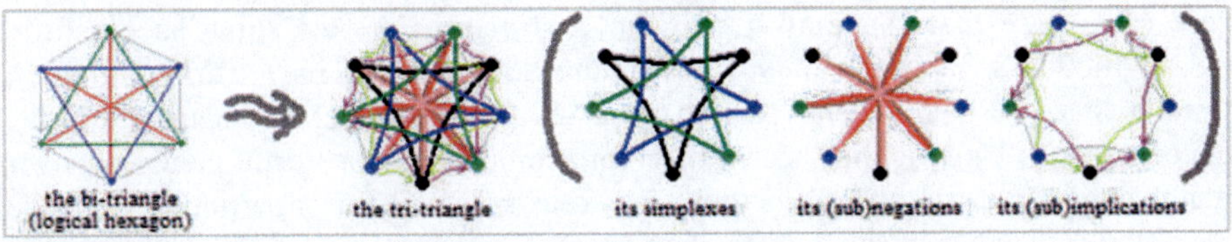

Fig. 16 From the "bi-triangle" (logical hexagon) to the "tri-triangle" (its simplexes, contradictions, and subalternations)

The idea seemed, if not worth the tiniest shadow of a postdoc (...), at least quite new, interesting, and promising, and it was proposed this could have (because of what recalled in Sect. 1.1) some non-negligible impact over the philosophy of the foundations of logic (especially with respect to nonclassical logic – paraconsistent and many-valued – as we argued in [96]).

However, the problem with this conceptual and visual conjecture of us of 2009 was that something quite important was still missing badly, namely, (1) some kind of mathematical proof (I am trained as – and I am! – a "continental philosopher"), or axiomatic construction, of the fact that these oppositional-geometrical poly-simplicial entities (of which we have seen at least the hypothetical conceptual idea, but also some possible concrete oppositional-geometrical shapes) are mathematically sound under every respect, and (2) some device (comparable to Angot-Pellissier's set-theoretical technique for generating the bi-simplicial oppositional closures, cf. Sect. 1.2, Figs. 5 and 7) in order to make, concretely, the "jungle" of the oppositional poly-simplexes mathematically real, testable, and applicable.

Tackling this problem leads us to the next Sect. 1.4 of this first chapter.

1.4 Angot-Pellissier's Sheaf-Theoretical Method for Poly-Simplexes

This needed mathematical method for the *poly*-simplexes happened to be excitedly announced and then shown on blackboard, in 2009 by Angot-Pellissier in a talk given in a four-people (!) workshop on the geometry of oppositions organized the day after my PhD defense. But it arrived in written version only 4 long years later (first in a draft in 2013 and then in a sequel draft in 2014, the first appearing only now [3], the second still unpublished [4]). The method mainly consists, so to say, in shifting from set-theory to sheaf-theory. For that recall, first of all, that sheaf-theory and topos-theory are important parts, or consequences, of "category-theory" [81, 83], which, in turn, so to say, has taken the place of "set-theory" as the main conceptual framework of general mathematics (this is called the "dynamic turn" of mathematical *structuralism*, cf. Avodey [7–9]). Recall also, secondly, that the "setting technique" for bi-simplicial oppositional geometry [111] consisted mainly in studying the possible *partitions* of a given set: the theory tells, among others, how to get to such a starting partitionable set ("Angot-Pellissier's set"), case by case (it is here that can play a role the Γ-structures, as we show successfully in a particular case study in [95]), and then how to study its *partitions*. Recall, thirdly, that a "sheaf", here, can be seen as a "topological diffraction" of this concept of "set", giving thus access to a more complex and powerful viewpoint over mathematical creativity (a set is then seen, retrospectively, as a particularly simple and "static" instance of sheaf, so to say a sheaf reduced to a point, whereas the latter, generally, has an extra topological richness [81]). *Consequently, Angot-Pellissier's new "sheafing technique"* [3] *for poly-simplicial oppositional geometry consists, mainly, in studying, instead of the partitions of a starting suited set, the possible sub-sheaves of a given starting suited sheaf (in fact, as we will see, a "numerical sheaf") and thereby in giving access to this sheaf's finer-grained "partitions" (the theory tells you how to determine this starting sheaf)*. And this starting sheaf takes into account both (1) the complexification of mathematical discourse that results (when talking about poly-simplexes) from the adoption of more than two truth-values in the Aristotelian game-theoretical algorithm (the Aristotelian p^2-semantics) generating the possible kinds of opposition relations (through the correlated Aristotelian p^2-lattice) (2) *and* the dimension of the simplex (i.e., whether it is a segment, a triangle, a tetrahedron, a five-cell, etc.).

More concretely, the first move of Angot-Pellissier's new method for oppositional poly-simplexes consists in translating in sheaf-theoretical terms the change in the number of truth-values. If "truth" (i.e., "1") is seen as a maximal starting set "X", and "false" (i.e., "0") as the (minimal) empty set "∅", the needed interpolated additional truth-values (remember the "*p*" parameter, Sect. 1.3) – like, for instance,

"$^1/_3$", "$^1/_2$", "$^2/_3$", etc. – will be constructed as (or represented by) interpolated intermediate sets "U", "V", "Y", etc., such that the first is a strict subset of "X" and that each of the following is a strict subset of the preceding ones, the empty set being by definition a strict subset of all. This constructs a suited "topos" (Fig. 17).

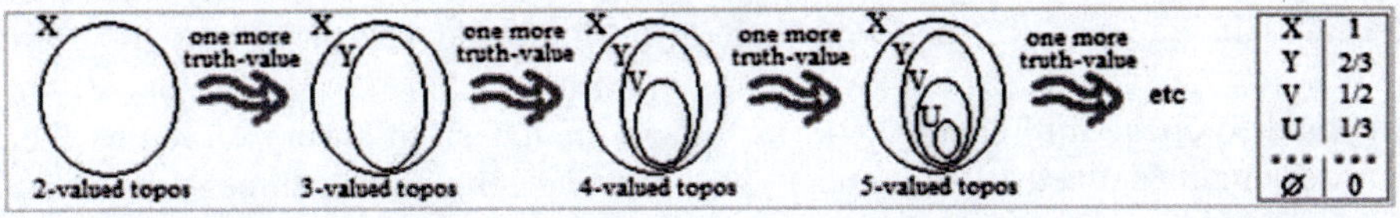

Fig. 17 The introduction of additional truth-values between "0" (false) and "1" (true) by means of topoi and sheaves

Then, another of the keys of this "sheafing method" is the redefinition in such sheaf-theoretical mathematically precise terms of the Aristotelian classical definitions of "contrariety" and subcontrariety (Sect. 1.1, Fig. 1), i.e., the two questions Q1 and Q2 of the Aristotelian p^2-semantics (cf. Sect. 1.3, Figs. 10 and 11). Provided with that, Angot-Pellissier can reconstruct, in a mathematically understandable and rigorous way, our intuitive and conjectural idea of Aristotelian p^2-semantics and p^2-lattices, by deriving any new *kind* of opposition as an articulation of two answers to the two questions relative to the sub-sheaves of the starting sheaf (we will give a step-by-step concrete example and illustration of this on Sect. 2.2). Angot-Pellissier's study, in some sense, thus confirms our general conjecture of 2009 over the possibility of theorizing with mathematical rigor the oppositional poly-simplexes. As a paradigmatic example, he finds back in [3, 4], by his new technique, the two "Aristotelian lattices" (the 3^2- and the 4^2- ones) proposed by me for the tri-simplexes and the quadri-simplexes, respectively (which he applies to the simplex "triangle") (Fig. 18).

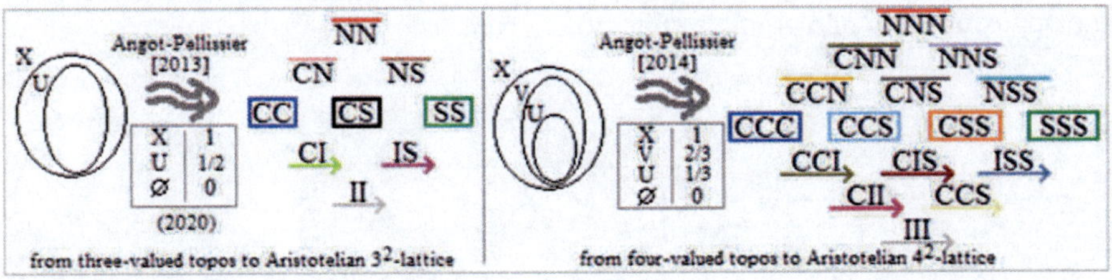

Fig. 18 Angot-Pellissier's redefinition (through sheaf-theory) of the Aristotelian 3^2- and 4^2-lattices ([2013], [2014])

An important point – to which we will come back in Sect. 5.3 – is that Angot-Pellissier, by the way, confirms by a demonstration (by *topological* arguments) that the two left-right families of "infra-negations" ("CN" on one side and "NS" on the other side in the tri-triangle's Aristotelian 3^2-lattice; "CNN", "CCN", and "CNS" on one side and "NNS", "NSS", and "CNS" on the other side in the quadri-triangle's

Aristotelian 4^2-lattice) are mathematically such that the members of the first are "paracomplete" (i.e., intuitionist) negations, while the members of the second are "paraconsistent" (i.e., co-intuitionist – cf. Sect. 1.1, Fig. 3) negations (remark that "CNS" is member of both families: it is both paracomplete and paraconsistent, behavior logically called – João Marcos *docet* – "paranormality" [sic]).

Finally, by introducing one last element, namely, the *length* of the numerical sheaves in question, by expressing them as finite indexed strings of the form "$1_j2_k3_l4_m \ldots$" (with j, k, l, m belonging to the set {∅, U, V, …, X} of the topos-theoretical truth-values), he expresses, in the sheaf's very structure (i.e., in its length, defined as numerical sheaf's string length), the dimensionality of the involved simplex of the studied poly-*simplex* (e.g., "$1_j2_k3_l$" is for triangles, "$1_j2_k3_l4_m$" is for tetrahedra, etc.). Angot-Pellissier also finds back, by his new rigorous mathematical method (i.e., in a new way), the oppositional tri-triangle I had predicted and tentatively represented in 2009 (Sect. 1.3, Figs. 14 and 16) (Fig. 19).

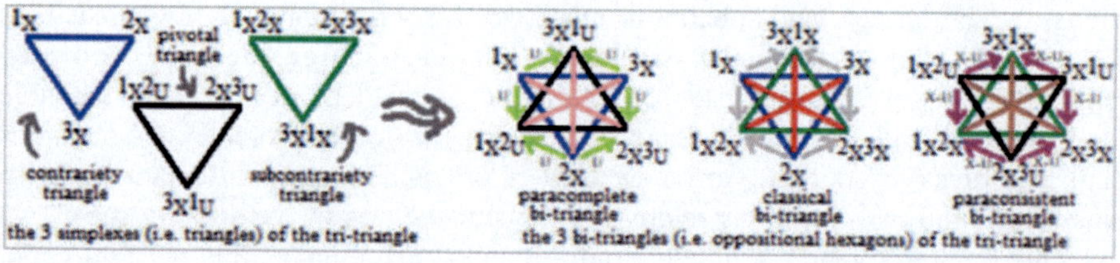

Fig. 19 Angot-Pellissier's sheaf-theoretical construction of the oppositional tri-triangle [2013] (2020)

(Angot-Pellissier also offers in [3] a new, original global 2D representation of the tri-triangle, the "nonagon", which we omit reproducing here). In the same way, in his second draft study [4], Angot-Pellissier finds back, but constructed in mathematically more rigorous (and understandable) terms than I did in 2009 [94], the oppositional quadri-triangle (Fig. 20).

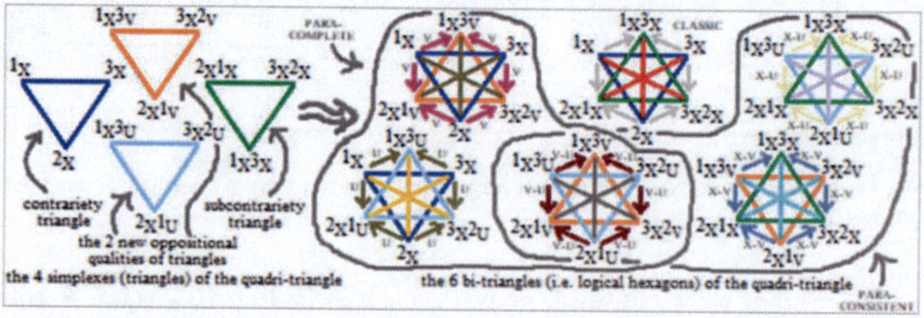

Fig. 20 Angot-Pellissier's sheaf-theoretical construction of the oppositional quadri-triangle [2014]

However, Angot-Pellissier in his two pioneering draft studies also introduces some strange and puzzling elements, which he just mentions, without discussing them much, both in his study on the tri-triangle [3] and in his study on the quadri-triangle [4]. Namely, a first element of puzzlement is that he mentions the existence, among the things generated by his new tools, of (at least) one extra triangle (lapidary judged irrelevant, because redundant) in the tri-triangle and two extra triangles (also judged irrelevant, again because redundant) in the quadri-triangle: he dismisses further discussion of this point, to us unexpected and puzzling – leaving the lucky reader (we have had the prepublication deep friendly privilege to be) rather confused – again, only mentioning that such extra triangles are oppositionally (i.e., combinatorially) equivalent to others he takes into account, and as such the redundant ones can/must (?) be neglected (Fig. 21).

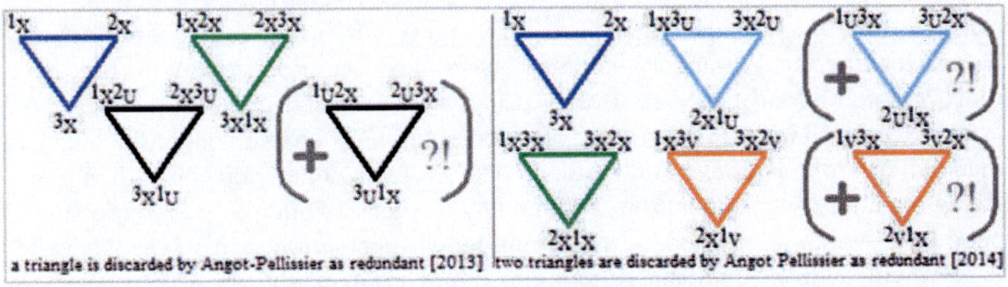

Fig. 21 Angot-Pellissier's sheaf-theoretic method reveals the existence of some "equivalent" extra triangles

Whereas with his 2008 [111] "setting" technique Angot-Pellissier gave to oppositional geometry a powerful tool such that in some sense it possibly generated all (it thus helped fixing the boundaries of the oppositionally possible/thinkable), with his 2013 [3] "sheafing" technique Angot-Pellissier did not explain what kind of "totality" this new method could lead to, speaking about oppositional poly-simplexes. As we are going to see (Sect. 1.5), the present paper proposes a first clear answer to this crucial, until now open question. A second possibly puzzling aspect in his two otherwise absolutely groundbreaking draft papers, is that Angot-Pellissier still does not provide *concrete* examples of application of the two poly-simplexes he successfully studies (successfully from a needed, *purely mathematical*, but then *not* applied point of view). So, for short, the problem, if any, with Angot-Pellissier's long waited for method, otherwise very promising and in fact quite exciting (for people interested in oppositional-geometrical research), was that (1) on one side it allowed finding more instances of poly-simplicial entities than what I predicted in 2009 (without this fact being further explained and explored by him) (2) but on the other side Angot-Pellissier himself seemingly explored (and explained) less entities than what his promising method seemed to make possible: he just selected things sufficient for confirming (as the friend he is) my analysis of 2009 (and this he did indeed successfully).

Tackling and solving these two residual problems (while benefiting decisively from the powerful and long waited for sheafing technique offered to us oppositional geometers) leads us to the next Sect. 1.5, possibly the most important, if any, of all this study.

1.5 *Our Proposal of a "Pascalian" Extra Tool for the Poly-Simplexes*

Remaining faithful to the structuralist methodology (or "ideology"! – as we explain in our theory of the "elementary structures of ideology", [102]) according to which oppositional geometry (which inquires oppositional *structures*), as any mathematical investigation, is a matter of *general mathematics* (and not of "essentially and primarily of *logic*"! – as might seem to suggest, for instance, the more sellable but misleading label "logical geometry", Sect. 4.6), we propose now to use, here, tools promising (and fit!) even if they usually are not used in "logic": repeating a gesture we dared in 2004 [93], when we successfully introduced "out of the blue", in the study of "oppositions", the mathematical n-dimensional concept of "simplex". Being a matter of "n-opposition" (with n any integer such that $n \geq 2$), oppositional geometry (which was called, at the beginning, "n-opposition theory", i.e. "N.O.T". in acronym, with n a *numerical* parameter) has inescapably to do with *numbers*, whereas "logic" (with the magnificent exception of "linear logic", which is precisely by no means – J.-Y. Girard [68–72] *docet* – a logicist tool!) normally doesn't. Recall that numbers – i.e., arithmetic – are so to say the structural "deadly threshold" of Gödel's complexity for formal systems (his famous second theorem of 1931, cf. [104]) and by that a proof of the deadly uselessness, and in fact harmfulness, of the logicist "ideology" (cf. [68]), consisting allegedly, but fruitlessly, in (keeping trying, over and over) "reducing things to logic" (logic seen – very mistakenly – as "the deepest element in mathematics"). Deepening this *fundamental relation of oppositional geometry to numbers*, we will turn now to a fundamental (and famous) *structure* of arithmetic and general "number-theory", namely, "Pascal's triangle". Remark, incidentally, that historically speaking this very important *mathematical structure* was already known in India (by the mathematician Pingala, in the second century BC) and, much later, in China (no later than in the fourteenth century). And after that, but long before its "discovery" by Blaise Pascal (1623–1662) in 1654, it had been rediscovered, independently, by the mathematicians Michael Stifel (1487–1567) in 1544 and Niccolò Tartaglia (1499–1557) in 1556. Now, as is well known even at school, "Pascal's" triangle (as it is now thus improperly called) is obtained very simply: starting from the top, with a "1", each lower integer (of the triangle) is the sum of the two integers above it (on its left and on its right, considering, by a typically structuralist move, that the absence of an integer means the presence of the number "0" . . .). Generated thus by a simple arithmetical algorithm, Pascal's triangle results in a *structure* which condensates in itself a huge number of really fundamental features of mathematical complexity, which potentially can be unfolded into infinite. Most famously, Pascal's triangle expresses, through the series of numbers in each of its horizontally left-right symmetric lines, each of the coefficients of Newton's (1642–1727) "binomial formula" for expressing the development of "(a + b)^n", whatever the natural number "n". And in fact, an important remark here (we will see why very soon) is that Pascal's triangle is so to say parallel to this Newtonian formula (a + b)^n for binomial powers (Fig. 22).

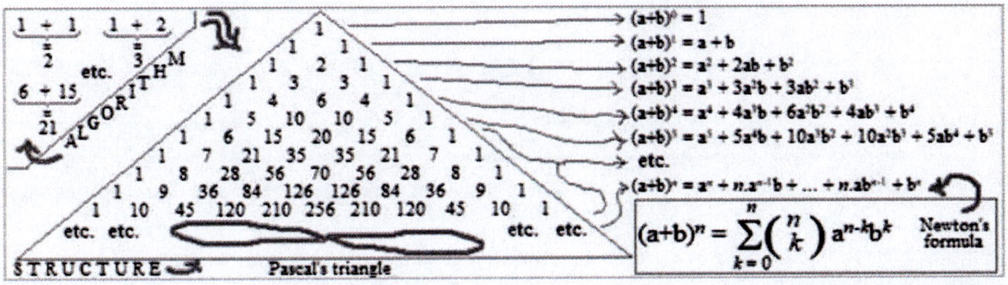

Fig. 22 An important arithmetical *structure*, "Pascal's triangle", with its algebraic counterpart, "Newton's formula"

This is perhaps the most famous application of Pascal's triangle, at least at school's level, although *by far* not the only one: for instance, reading it "diagonally" (and still top-bottom) also gives the series of the "polytopic numbers" (a.k.a. "figurate numbers", i.e., *simplicial* generalizations of the "triangular numbers", i.e., numbers – those studied by the Pythagorean – characterized by intrinsic geometrical properties). Moreover, Pascal's triangle is also known for having very strong *fractal* properties, among others, in the distribution of its numbers – be them prime numbers, or even numbers, or powers, etc. More concretely, it exhibits fractal patterns akin to "Sierpiński's gasket", cf. [109], p. 91–102, and [110], p. 85–96 (Fig. 23).

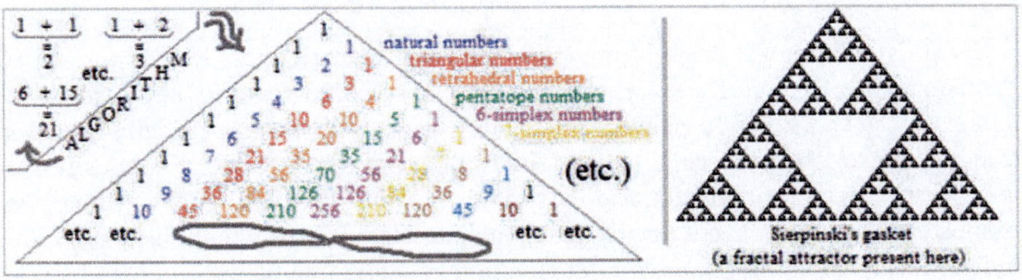

Fig. 23 Pascal's triangle also displays, "obliquely", the "polytopic numbers", and has many fractal properties

But then, what about oppositional geometry? As it happens, Pascal's triangle contains, among other *mathematical* treasures, nothing less than *all* the numerical features of the *closures* of the oppositional bi-simplexes (!!!): each of its horizontal lines, starting from the third from top, gives, line by line, the exact *full* "numericity" of the closure of one of the n-oppositions for $n \geq 2$ (with no possible exception, it is an isomorphism: no "gap" and no "glut", Sect. 3.5, Fig. 81). As such Pascal's triangle presents the two constitutive simplexes (blue and green) of any bi-simplex but also, in between them, their "cloud" (!): for short, again, *all* (bi-simplicial...) oppositional geometry is contained in Pascal's triangle (Fig. 24).

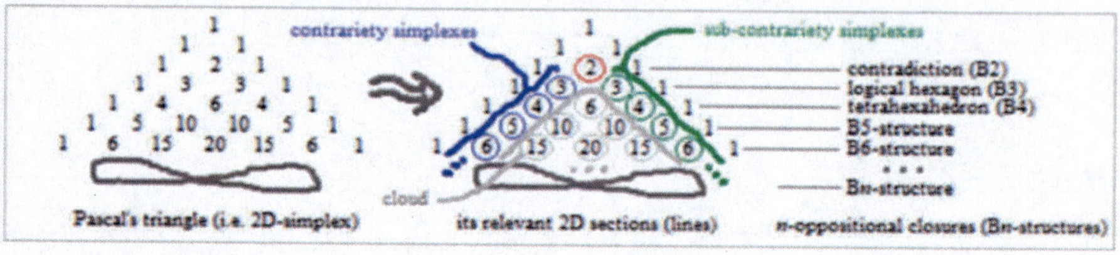

Fig. 24 Pascal's triangle gives, with its horizontal lines, *all* the numbers of (bi-simplicial) oppositional geometry!

Remark that retrospectively this is not totally surprising: since the n-oppositions are all the possible set partitions of any finite set of n elements ($n \geq 2$), as Angot-Pellissier has established in 2008 [111]. Now, it must be remarked that Newton's formula, mentioned above, also does this "oppositional job", starting from what corresponds in it to the third line from the top of Pascal's triangle (in bold the *nontrivial* oppositional elements):

$(a+b)^2 = a^2 + \mathbf{2ab} + b^2$ which are the numbers of 2-opposition (segment);

$(a+b)^3 = a^3 + \mathbf{3a^2b} + \mathbf{3ab^2} + b^3$ the numbers of 3-opposition (hexagon);

$(a+b)^4 = a^4 + \mathbf{4a^3b} + 6a^2b^2 + \mathbf{4ab^3} + b^4$ the numbers of 4-opposition (tetrahexahedron);

etc. etc.

$(a+b)^n = a^n + \boldsymbol{n.a^{n-1}b} + \ldots + \boldsymbol{n.ab^{n-1}} + b^n$ the numbers of n-opposition.

But this, although it might seem (to some good mathematical eye) *a posteriori* mathematically "natural", seems nonetheless absolutely remarkable! And as it happens, it gave us the idea (in 2018) of trying to generalize geometrically Pascal's triangle, so to obtain a general method for having an equivalent "oppositional numericity" for the oppositional *poly*-simplexes. With this very goal in mind, we propose here to introduce a new concept (new at least for us: we almost surely are rediscovering something known in contemporary – and maybe even in classical? – mathematics, as for the idea we propose here of *extending* n-*dimensionally Pascal's triangle*): something we propose to call, accordingly, the "Pascalian ND *simplex*", seen as a general geometrical-numerical *structure* such that Pascal's triangle is only a particular case of it, the case where, in "ND", $N = 2$ (*addendum*: in fact we found in Wikipedia, afterward, mention of the existence of "Pascal's simplex", which seems to be exactly what we are speaking about here – all what follows in this Sect. 1.5 has nevertheless been developed, or redeveloped, by us "out of nothing"). Let us see how to unfold this idea progressively. For a start, the very first step of this will be considering, after the well-known Pascalian "triangle" (seen as a "simplex of dimension 2"), a "tetrahedron" (seen as a "simplex of dimension 3") that we will call "Pascalian" as well, for it is made so that it has a horizontal *triangular* equilateral "basis", going downward (as the horizontally *segmental*, infinite "basis" at the infinite bottom of Pascal's triangle goes endlessly downward): by construction this horizontal triangular infinite basis of the Pascalian tetrahedron dives, step by step, into growing infinite numerical depth and complexity. Each of the remaining three non-horizontal triangular faces of this Pascalian tetrahedron will be, like Pascal's classical triangle taken as a whole, triangular and with a horizontal linear basis, step by step going down endlessly into infinitely more complex numericity (in fact, precisely the numericity of Pascal's triangle) (Fig. 25).

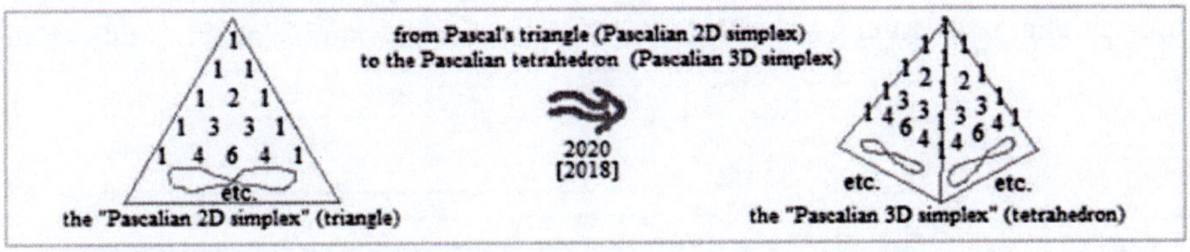

Fig. 25 Our proposal [2018]: from "Pascal's triangle" (2D simplex) to a "Pascalian tetrahedron" (3D simplex)

Now, the crucial point is that this Pascalian tetrahedron has numbers even "inside" of it, determined by a suited analog of the simple algorithm generating the numbers of Pascal's triangle. This "internal" algorithm explaining the internal numbers of the Pascalian tetrahedron can be seen, again, in (at least) two ways: (1) either as a graphical or (2) as an algebraic algorithm, the former being an extension of "Pascal's" graphic algorithm (in a nutshell, each number in a layer – i.e., horizontal triangular "section" – is the sum of the *three* numbers above it, the absence of a number being counted as number "0") and the latter being an extension of Newton's formula, such that it calculates (for each horizontal *triangle*, instead of for each horizontal *line*) the powers of the sum of *three*, instead of *two*, addenda (i.e., it calculates "$(a + b + c)^n$" instead of "$(a + b)^n$") (Fig. 26).

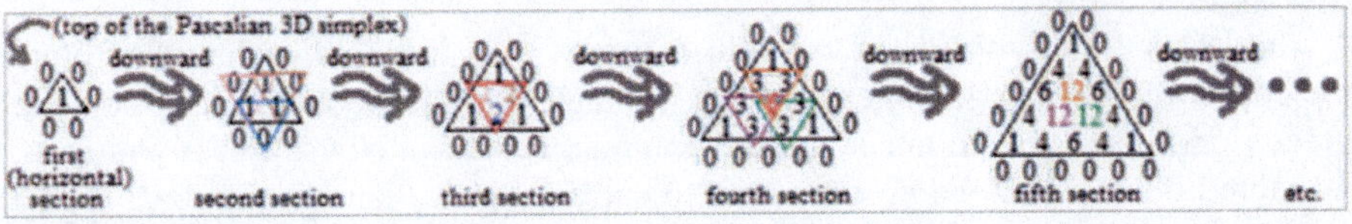

Fig. 26 The graphical 3D algorithm, extending the classical 2D one, for generating by sum the "internal numbers"

Consequently, what is immediately interesting for us, for the oppositional poly-simplexes, through this deepening move (from triangle to tetrahedron) is in fact *the generalization of the concept of "horizontal line of Pascal's triangle"*, because this happens to yield the corresponding, duly changed concept of "horizontal sections" (of the Pascalian tetrahedron) for the oppositional tri-simplexes. The Pascalian 3D simplex (a tetrahedron) happens to have, in fact, an infinite number (one simultaneously intersecting a horizontal line in every of its three lateral triangular faces) of top-bottom growing 2D horizontal "sections", which are triangles, and *these triangular sections happen to be such that they are perfectly suited for exploring ... the tri-simplexes!* The correspondence can be matched by comparing the numbers given by these triangles with the numbers given by Angot-Pellissier's sheaf-technique. In fact the more complex is the *simplex* (of the studied oppositional *tri*-simplex), the deeper you will have to go down in the 3D Pascalian simplex (the tetrahedron) for finding the adequate horizontal triangular section. We will demonstrate and explain duly our *general* claim in another paper; here it will suffice to show that it holds at the levels we are interested in and works *perfectly* with the oppositional tri-segment we will thus inquire in Sects. 2 and 3 of this study (Fig. 27).

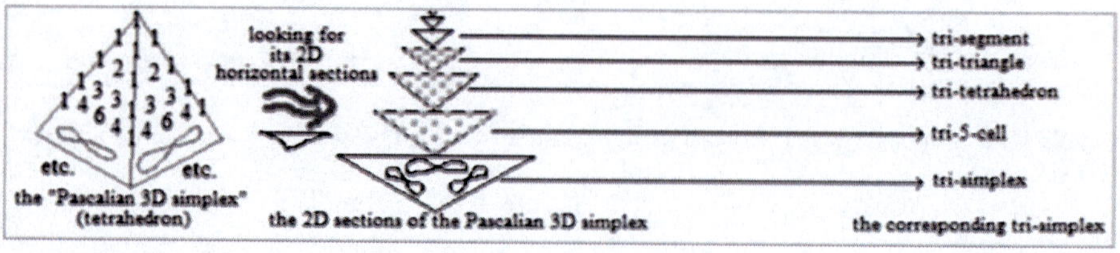

Fig. 27 The 2D simplicial sections (i.e., triangles) of the "Pascalian 3D simplex" (i.e., tetrahedron) map the tri-simplexes

Remark that, as it happens, the aforementioned Newton formula for binomial powers (i.e., the coefficients of the development of "$(a + b)^n$") still gives a parallel vision: as said, it simply becomes here $(a + b + c)^n$, i.e., what changes when going from the Pascalian 2D simplex (Pascal's triangle) to the Pascalian 3D simplex ("Pascal's tetrahedron") is, in its "Newtonian translation", the fact of having three addenda ("a", "b" and "c") instead of, classically, only two ("a" and "b"). This can be used for generating the same set of internal numbers: in fact, the formula generates all the numbers of any 2D section (triangle) of the Pascalian 3D simplex (tetrahedron).

But let us unfold this idea further. In the same way, if you now want to explore the oppositional *quadri*-simplexes (Sects. 1.3 and 1.4, Figs. 18 and 20), as we do in other working papers, you need a more complex Pascalian ND simplex: you need, no more no less, what we propose here to call the "Pascalian 4D simplex" (i.e., a 4D hyper-tetrahedron, or 4D "five-cell", i.e. a 4D "volume" delimited by five 3D tetrahedral "faces", by analogy with the tetrahedron, whose 3D volume is delimited by four 2D triangles, and by analogy with the 2D triangle whose 2D "volume", i.e. surface, is delimited by three 1D segments, its three sides, etc.). From this Pascalian 4D simplex, what you further need to consider for dealing with the oppositional *quadri*-simplexes are, again, its "horizontal sections": but these horizontal sections (i.e., sections "parallel" to the 4D simplex's infinite "horizontal" – by construction – 3D "base") now are no more 2D triangles (as they were, in the previous case, for tri-simplexes) but 3D tetrahedra; they are the "3D horizontal sections of the 4D Pascalian simplex". Incidentally, representing a 4D solid is notoriously not always straightforward [15, 45, 137]. But one way (among others) of doing it rather simply (when possible) consists, so to say, in resorting to "slicing" (i.e., in cutting this 4D solid into suited superposable slices, each of them being lower-dimensional, namely, each of them being a 3D tetrahedron, which can in turn be sliced, this time in terms of stacks of 2D triangles, as previously with the 2D sections of the Pascalian 3D simplex) (Fig. 28).

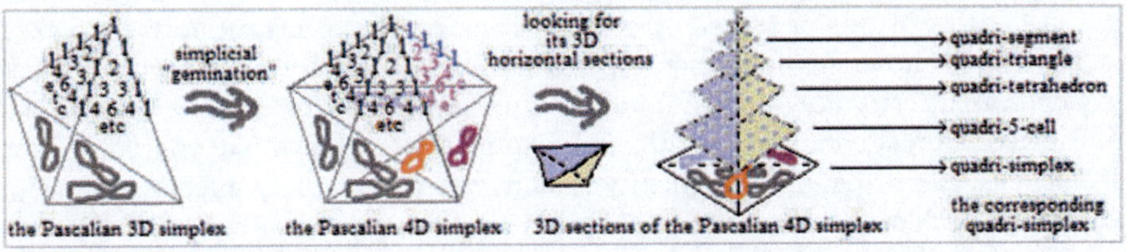

Fig. 28 The 3D simplicial sections (i.e., tetrahdra) of the "Pascalian 4D simplex" (i.e., 5-cell) map the quadri-simplexes

We will not study here its sections (we do it in other working papers on oppositional quadri-simplexes). Just remark that, as it happens, the 4D counterpart of Newton's formula still exists here, and it is now the development of "$(a + b + c + d)^n$" (the addenda become *four*, instead of *three*).

Now, as we will prove it elsewhere, this "Pascalian" geometrical-arithmetical algorithm for the oppositional *poly*-simplexes goes, always working perfectly, into infinite! And this both in its geometrical-arithmetical *stricto sensu* Pascalian aspect and in its (corresponding) algebraic Newtonian aspect: on one side the Pascalian ND solid, with its (n-1)-D horizontal sections, and on the other side the generalized Newtonian formula $(a + b + \ldots + z)^n$, where, as in all previous cases, the N addenda between parentheses stay for the degree of the oppositional *poly*-simplex (i.e., the numerical value of the "poly-"), whereas the exponent "n" stays for the quality of the *simplex* ($n = 2$ for the segment, $n = 3$ for the triangle, $n = 4$ for the tetrahedron... n for the (n-1)-dimensional oppositional simplex). *The important, general result for oppositional geometry is that, in each case, equivalently, the numbers produced by such a generalized Pascalian ND simplex (or by the correlated generalized Newtonian formula) and those generated by Angot-Pellissier's sheaf theoretic method for the oppositional poly-simplexes* [3, 4] *do match perfectly and give the "poly-simplicial oppositional closure"* (*addendum*: as it seems – and as we found, only afterward, in Wikipedia – these two correlated things we are speaking about are already known in mathematics, namely, as "Pascal's simplex" and as the "multinomial theorem"; but on the one side we rediscovered them by ourselves, and on the other we still found no exact bibliographic references for this).

Remark that, conversely, the clear understanding of the (very wide!) scope of Angot-Pellissier's until know mysterious sheafing method (i.e., understanding that, when duly followed – as we will explain on a precise case in Sect. 2.4 – i.e., with the "Pascalian ND simplex", it can lead to the oppositional closures of the poly-simplexes) shows that his own concrete analyses ([3] and [4], respectively) of the tri-triangle and quadri-triangle happen to show only a small fragment (and therefore not the oppositional *closure*) of the real structure to be brought to light in both cases (we will show this in future studies).

Remark also that given the aforementioned correlation between the oppositional poly-simplexes à la Angot-Pellissier and the generalized Pascalian ND simplex (and its correlated generalized Newtonian formula "$(a + b + c + \cdots)^n$"), the latter (Pascal and/or Newton), besides giving us an Ariadne thread (as we are going to see in the rest of this paper, starting from Sect. 2.4) for studying poly-simplexes in full rigor, allows us doing some quite useful preliminary action with respect to inquiring directly poly-oppositional structures (poly≥3): having *a synoptic view* of the geometrical complexity of the poly-simplexes (provided one concedes, as it seems reasonable, that this complexity is somehow measured by the number of the vertices of each of these structures) (Fig. 29).

etc	etc	etc	etc	etc	etc	etc	etc	etc	etc
decaseptem-	272	...	...	...	...	...	...	...	etc
...	...	...	...	...	...	...	...	...	etc
sexa-	30	210	1.290	7.770	46.650	279.930	...	...	etc
quinque-	20	120	620	3.120	15.620	78.120	390.620	...	etc
quadri-	12	● 60	252	1.020	4.092	16.380	65.532	262.140	etc
tri-	6	● 24	78	240	726	2.184	6.538	19.680	etc
bi-	■ 2	■ 6	■ 14	● 30	● 62	● 126	254	510	etc
poly- / -simplex									etc

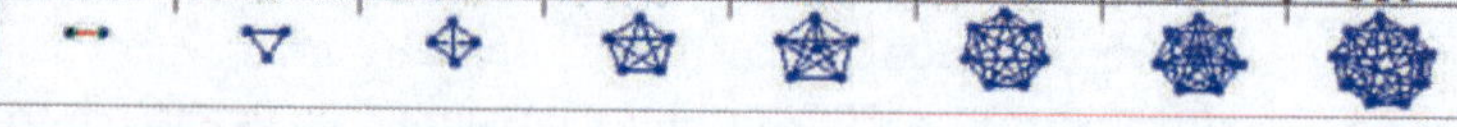

Fig. 29 Synoptic view of the complexity degree of the poly-simplicial oppositional geometry (number of vertices)

The structures with a blue square are those already studied, while the ones with a red disk are those which have been studied only partly (as, for instance, tri-triangles and quadri-triangles by Angot-Pellissier [3, 4] or the bi-simplicial closures B5-7 by myself [94, 95]). The figure highlights, by a diagonal cut, the left-bottom domain of the general poly-simplicial space having no more than around 270 vertices. So, the above synoptic view strongly suggests that *the easiest and most reasonable thing to do next*, thanks to our fresh two new tools, in order to explore the poly-simplicial space would be, at present, to study the oppositional "tri-segment" (characterized, as we are going to explain in Sect. 2.4, by the complexity degree 6). And this is precisely what we are going to do in the rest of this paper.

But before starting the inquiry on the tri-segment, let us now have one last preliminary retrospective look to the "old style" study of oppositional tri-*segments*: recalling what is known so far about them, and then (as we will inquire in Sect. 2) what we can try, from now on, to learn about them by a deeper and more accurate investigation.

1.6 Flashback: The Primitive Idea of Oppositional Tri-segment (2009)

As we saw, Angot-Pellissier's two very important draft studies of 2013 and 2014 [3, 4] concerned poly-*triangles* (viz., the *tri*-triangle and the *quadri*-triangle). But in my 2009 PhD dissertation [94], I also took into account as simplexes (for the poly-simplexes), before triangles, segments. There the "poly-segments" were imagined, roughly, as being some kind of diffraction of the logical square, since the latter *seems* to be a 2-opposition and is precisely based on two simplexes, a blue and a green segment (we come back to this in Sect. 2.5). In the case of the first higher poly-segment (poly$\geq$3), the tri-segment, it was imagined as made of the classical blue-green logical square, plus two interpolated new ones, a blue-black and a black-green squares – but of course in a way different from that in which three logical squares merge to form a logical hexagon (Sect. 1.1, Fig. 2). Therefore it was thought we could have (1) *a blue segment* of contrariety; (2) *a green segment* of subcontrariety; (3) and, interpolated (in the 3D space), *a black new segment* (interpolated) of a "pivotal" simplex. This was my guess as for having a "tri-segment", the second element of the series of the poly-*segments* and the first element of the series of the *tri*-simplexes. It was seen as made basically out of three simplexes of dimension 2 (i.e., three segments): the two classical ones (the blue and the green horizontal segments of the logical square) and a new one (the third simplex), black (Fig. 30).

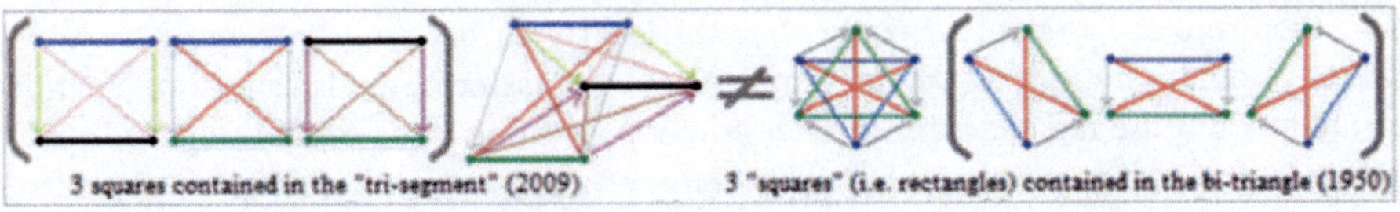

Fig. 30 How was imagined to be, in 2009, the hypothetical structure of "oppositional tri-segment"

Being committed to three truth-values (say: "0", "½" and "1", Sect. 1.3) one had to try to understand how three-valued logic can intervene here, if it does (we afford this in Sect. 3.6). Based on their definition through the Aristotelian 3^2-semantics and its correlated 3^2-lattice, the three-valued propositional connectives, allegedly embodied in this oppositional tri-segment, were defined (quite experimentally) by means of an "extensional definition" (Fig. 31).

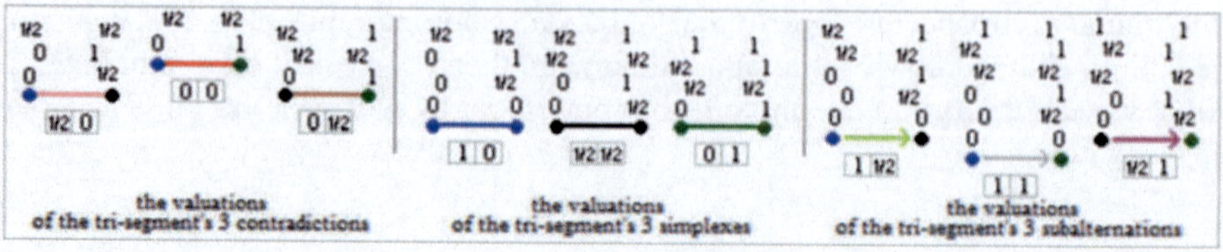

Fig. 31 2009 Conjecture over the possible valuations of the tri-segment's negations, simplexes, and implications

As for the global "valuations" we proposed something like the following three (Fig. 32).

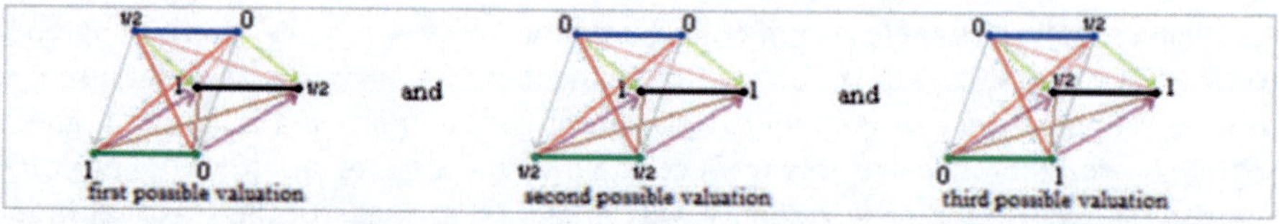

Fig. 32 The problem of the "valuation", with three truth-values, of the (still hypothetical!) tri-segment (2009)

All this, in order to think the "tri-segment", was intuitively inspired by the logical square.

But in 2012 Angot-Pellissier [1] demonstrated (or more precisely: he gave a deep mathematical explanation and clarification of a fact until then "known" too confusedly) that 2-opposition, until then not clearly elucidated in its oppositional-geometrical specificity, is in fact not a blue segment of two-contrariety (which would have *automatically* generated, by central symmetry of the contradictories, a correlated green segment of subcontrariety and therefore the logical square – and therefore the logical hexagon, Sect. 1.1, Fig. 2) but simply the red segment of (any) contradiction. "2-opposition", Angot-Pellissier demonstrated implacably, is tantamount the red segment of contradiction, with no trace of any other possible oppositional color (no blue segment!). This was strange with respect to the idea, otherwise good working, of bi-simplex, but was so. And the logical square, which exhibits a blue segment of contrariety, is therefore (as already established around 1950 by three different people!) only a fragment of the logical hexagon (i.e., 3-opposition, Sect. 1.1, Fig. 2). Angot-Pellissier clarified why contrariety, with its blue simplexes, can begin only with the triangle: *there is no contrariety smaller than 3-contrariety*, as in fact Aristotle knew already. This important and long needed clarifying result by Angot-Pellissier (on which we will come back shortly, from another viewpoint, in the introduction of ch.3) implied, among others, that *my proposal of a tri-segment was at least quite mistaken in the way I had tried it, if not totally unthinkable (as tri-segment) per se.*

So, having recalled the current context (state of the art) starting from this point, in the remaining of the present study we will try to come back to this issue of the classical (and "not complex") segment of (bi-simplicial?) contradiction (seen, as we will explain, as being nevertheless rightfully a "bi-segment", *cum grano salis*) and to the question of knowing whether it can admit (as proposed in 2009) a tri-simplicial counterpart: something like the oppositional tri-segment... As a very last preliminary remark, notice that this research results, therefore, in a voluntary provisory blindness (in this study) with respect to tri-*triangles*, whereas with bi-simplicial oppositional geometry important things clearly begin, so to say, precisely *only* with contrariety (blue) triangles. But that price once paid the expected non-negligible gain is that our present study will be simpler and therefore easier (as suggested in the synoptic view of Sect. 1.5, Fig. 29), which will be no luxury, as we will see (poly-simplexes are "wild"!). Again, the loss, of course, is that oppositional segments (and tri-segments as well) do express much less than oppositional triangles, as well as tri-triangles, and higher poly-simplexes (we study tri-triangles and higher poly-simplexes in other studies, already ongoing). Then what can be expressed by bi-oppositional segments (and by their hypothetical poly-simplicial diffractions)? (1) Neither the relation of two *independent* "atoms", neither, alternatively, the mutual opposition (i.e., contrariety) of four things: both things need a 4-oppositional A4 bi-tetrahedron (and its B4 closure); (2) nor even the simple contrariety of *two* things (antonymy) (this needs a 3-oppositional A3 bi-triangle (which is its own B3 closure, Sect. 1.1, Fig. 2)); but (3) a 2-oppositional B2 (red) segment, and *a fortiori* its hypothetical poly-simplicial diffractions, can nevertheless express something not so trivial (Sect. 1.1): the *negation* of a given element (and in fact, as we are going to see, Sect. 3.5, also this element's *affirmation*!) and, again, possibly the "diffractions" of this concept of negation/affirmation... (Fig. 33).

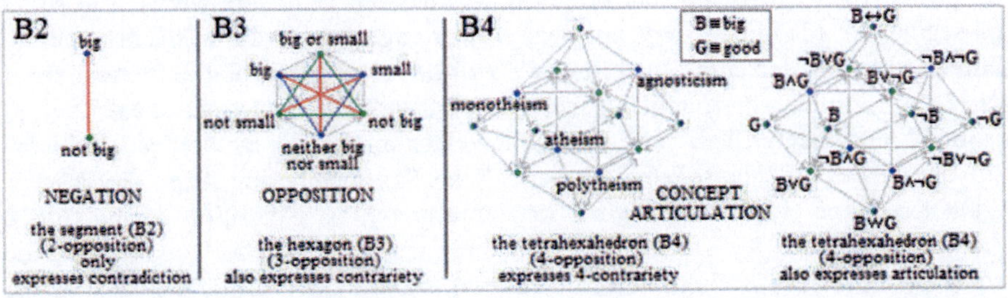

Fig. 33 Segment, triangle, tetrahedron: the increasing expressive power of the bi-*simplicial* oppositional structures

So let us now try to study, in the next two Sects. 2 and 3, this still mysterious structure anew, with the help of the two new tools which are Angot-Pellissier's sheaf-theoretical technique (Sect. 1.4) and our own new concept of "Pascalian ND simplex" (Sect. 1.5).

2 Studying with These New Tools the Oppositional Tri-segment

In the previous Sect. 1, we recalled the main ingredients, historical, conceptual, and methodological of the context of our present inquiry on poly-simplexes. In this Sect. 2, we are going to study anew the simplest poly-simplex (poly≥3), i.e., the tri-segment (Sect. 1.6), reaching non-negligible new elements of knowledge.

2.1 Oppositional Sub-sheaves of the Tri-segment: Which Are Vertices?

In order to explore the concept of tri-simplex, we start by resorting to Angot-Pellissier's sheaf-theoretical technique (Sect. 1.4), but limiting it to the study of a smaller object than what he considered in his two seminal papers on the subject [3, 4]: *not triangles* (i.e., the tri-triangle and the quadri-triangle, in Angot-Pellissier) *but segments* (here: the *tri*-segment). How to do that?

Two things must be recalled: (1) in Angot-Pellissier's sheaf-theoretical method, one parameter is the *number of truth-values*; (2) the other is the *simplex* considered; more precisely Angot-Pellissier (i) considered *three* truth-values (in [3]) and *four* truth-values (in [4]); (ii) he considered triangles (in both [3, 4]); he thus remained inside the study of oppositional poly-*triangles*.

For us, the way of applying his sheaf-theoretical method to the study of oppositional tri-*segments* (for which he gives no hint) will then consist in the following: (1) we will consider *three* truth-values (because here we want to study *tri*-segments); (2) but we will not deal with triangles, but with *segments* (we are interested in tri-*segments*): therefore, we will not resort to a total numerical sheaf "1x2x3x" (suited for triangles), but to a shorter total numerical sheaf "1x2x" (suited for segments). The "job", then, will consist in working methodically with the sub-sheaves of this shorter total numerical sheaf (remembering Angot-Pellissier's 2008 lesson of [111]: "opposition, i.e., contrariety, is a partition of the true"). Accordingly, since we are dealing with three truth-values, the "topos" (i.e., the category-theory tool ruling many-valuedness) in our study will be the same as the one in Angot-Pellissier's first study [3], namely, one with three levels (three strictly nested sets): the total set "X", one strict *open* subset of it "U", and the empty set "∅" (as such contained in any other set). By construction these three elements are therefore strictly ordered: X ⊃ U ⊃ ∅ (Fig. 34).

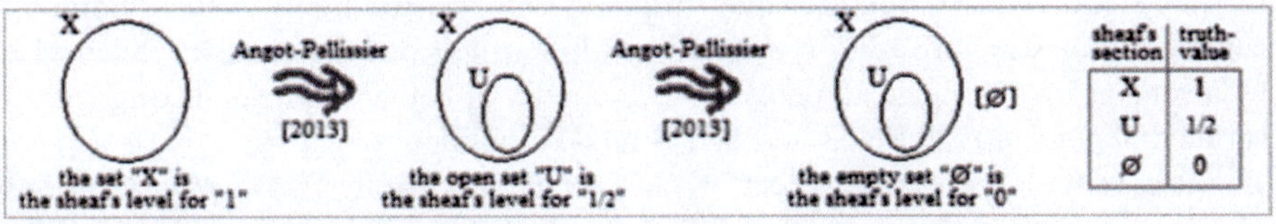

Fig. 34 Expressing, *via* the sheafing technique, the three-valuedness of the space of the oppositional *tri*-simplexes

Now, this, applied to the starting total numerical sheaf (for segments) "1_X2_X", will give that *all in all* there are, as total number of possible sub-sheaves of this total sheaf, $3^2 = 9$ possible terms (including here those particular sub-sheaves which are the total sheaf 1_X2_X itself and the null sheaf $1_\varnothing 2_\varnothing$). Put into a lattice, they result in a familiar lozenge-shaped structure (Fig. 35).

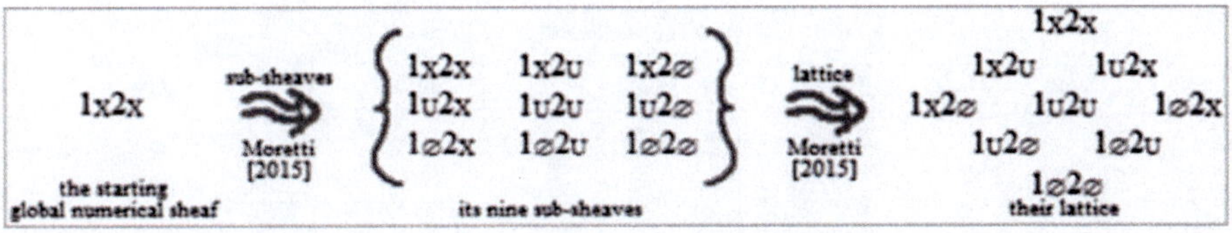

Fig. 35 Distribution of the nine sub-sheaves (i.e., the oppositional-geometrical vertices) of the tri-segment in a lattice

But one must beware: the lattice at the right hand of Fig. 35 looks like the Aristotelian 3^2-lattice of the tri-simplexes (Sect. 1.3, Fig. 11, and Sect. 1.4, Fig. 18), to which we come back in the Sect. 2.2, but truly speaking the lattice here is something totally different (and, as said in Sect. 1.4, this *totality* of the sub-sheaves is thinkable, thanks to Angot-Pellissier, but it must be remarked – in order to understand where we are going to in this study – that this is something that Angot-Pellissier did not study yet as such; notice in particular that the "extended indicial notation" here, admitting "∅" as an explicit index alongside with "X" and "U", is ours – in Angot-Pellissier's notation, our "$1_U2_\varnothing$" is "1_U", our "$1_\varnothing 2_X$" is "2_X", etc.).

The next important step, as in the case of Angot-Pellissier's set-theoretical method for the oppositional *bi*-simplexes (Sect. 1.2, Fig. 5), consists in keeping out from this set of possibilities those which are *trivial with respect to oppositional geometry*. In fact, the sub-sheaf 1_X2_X is trivial: it is an analog of "T" (the "verum" and of the "universal set"). And the sub-sheaf $1_\varnothing 2_\varnothing$ is also trivial: it is an analog of "⊥" (the "falsum", the null-element of logic, and of the "empty set", the null-element of set-theory). In oppositional geometry, which – again – in a sense is a matter of *partitioning methodically* a "cake", the "whole cake" and "no cake" are *trivial partitioning situations*, and as such, by construction, they are put outside the structural game. In fact in the bi-simplicial space, the equivalent of these tri-simplicial two points (1_X2_X and $1_\varnothing 2_\varnothing$) does exist, mathematically speaking, but, for any oppositional structure, they implode "by construction" to the symmetry center of that structure (this important result – ruling out the reproach otherwise recurrent, as in [87], against oppositional geometry of being "Boolean incomplete" – is due independently to Smessaert and Angot-Pellissier).

Let us remark that the sub-sheaf "1_U2_U" can be seen, of itself, as a bit mysterious of its own as well, so far: for, seen with our "extended indicial notation", it resembles quite much the two trivial sheaves 1_X2_X and $1_\varnothing 2_\varnothing$ (i.e., it bears on both digits of its numerical string, "1" and "2", the same index, viz., "U"), without however being trivial itself at least as long as we know (we will have to come back later, in Sect. 2.4, to this rather important and strange point). In any case, what seems to be sure so far is that in what follows the sub-sheaves 1_X2_X and $1_\varnothing 2_\varnothing$ must, by construction, be neglected (as "oppositionally trivial"). So our study will, from now on, concern no more than seven oppositional sub-sheaves (i.e., supposedly, seven oppositional-geometrical vertices) over the starting nine possible ones (Fig. 36).

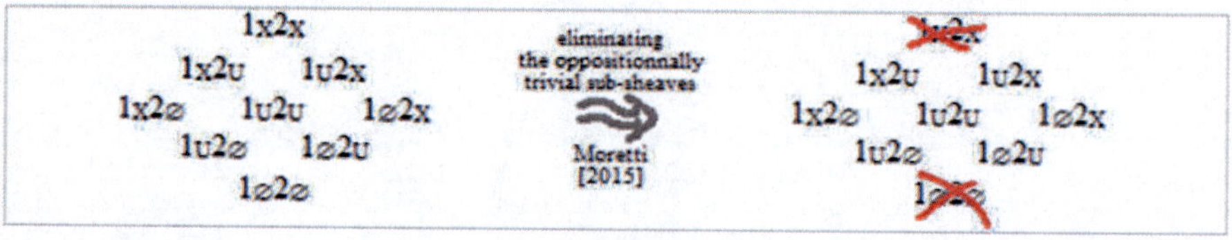

Fig. 36 From the lattice of all the nine sub-sheaves of "1_X2_X" to the set of its seven presumed nontrivial sub-sheaves

Now that we have the entities supposed to be the vertices, what must be considered next, if we want to be able to study the oppositional *geometry* of this (i.e., of the tri-segment), is the "lines" uniting each possible pair of these seven vertices (including here the pairs made of a vertex with itself! – this is also something Angot-Pellissier did not in his two draft studies [3, 4]). These are *all* the possible (oppositional) relations between the (oppositional) vertices of the tri-segment, its "oppositional colors" (and this brings us to the oppositional *closure*).

2.2 *The Oppositional Relations Between the Sub-sheaves: Edges!*

The theory, outlined in my PhD [94] and confirmed and deepened mathematically by Angot-Pellissier's method [3], tells that for tri-simplexes, there are nine possible qualities of oppositional relations, which are, visually, nine "oppositional colors" (Sect. 1.3, Fig. 11, Sect. 1.4, Fig. 18). This is also true of tri-segments, which are a particular case of the general concept of tri-*simplex*. The nine relations predicted by the Aristotelian 3^2-lattice are the following (2009 style on the left, 2013 style on the middle, and 2020 style on the right of Fig. 37) (Fig. 37).

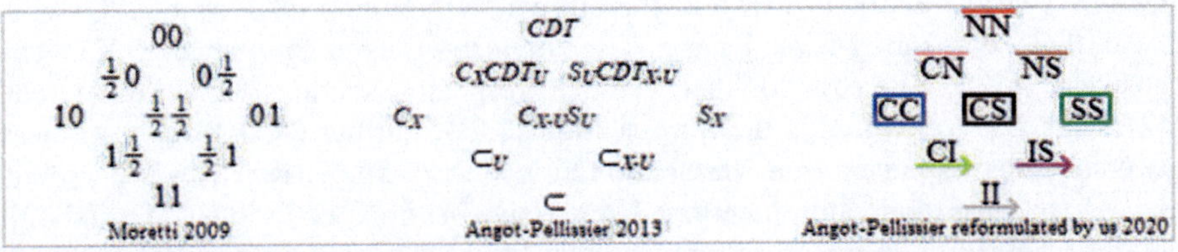

Fig. 37 The lattice of the nine "oppositional qualities" of the tri-simplexes (2009, 2013, 2020)

Now, and this is a major theoretical advance since 2009, Angot-Pellissier's method not only allows *generating* mathematically *all* the vertices of the poly-simplexes as sub-sheaves of a starting total numerical sheaf (as we will see soon, the "all" is in fact clear only now thanks to the Pascalian method, Sect. 1.5 and 2.4), but it also allows *calculating*, for *any* given pair of such sub-sheaves, the *precise* kind of oppositional relation (among the predicted nine possible ones) that holds between the two elements of the considered pair. Combinatorially speaking, regarding only the geometrical side (i.e., making momentarily abstraction of colors), there are exactly 7!, i.e. $7 + 6 + 5 + 4 + 3 + 2 + 1 = 28$ such possible pairs (including the seven pairs of identical elements, which will be twisted segments, i.e., "curls"). In the following, we will give, on the one side, at least an example of calculation for each of the three main oppositional kinds (i.e. negations, simplexes, arrows) and then, on the other side, the complete list of the colors of each of the 28 vertex-vertex relations (of which 21 are edges properly said – i.e., edges – and 7 are reflexive "curls").

Remark that I will use the notation I propose on the right side of Fig. 37 (equivalent to, but different from, Angot-Pellissier's original one, in particular more precise relatively to the two nonclassical implications CI and IS, cf. [3]).

Let us start from negations (i.e., the red, pink, and brown oppositional relations). For instance, let us compare the vertices $1_X 2_U$ and $1_\varnothing 2_X$. We must ask for this pair of nontrivial sub-sheaves the two Angot-Pellissierian meta-questions, first at the lower sheaf-theoretical level U and then at the higher level X (of the three-valued topos we use): the respective four answers to these $2 + 2 = 4$ questions will give us, combined two by two, two literals, *x* and *u* (each taken among the set {N,C,S,I}, respectively, for "negation", "contrariety", "subcontrariety", "implication"), and the noncommutative concatenation "*xu*" of these two literals will give us precisely the tri-simplicial oppositional quality of the relation (or segment) under examination (viz., one among the NN, CN, NS, CC, CS, SS, CI, IS, II). So (call this meta-question "Q1/U") "Can these two sub-sheaves have a false *(inclusive) disjunction* at level U?" The answer (call it "A1/U"), here, is "0" (i.e., "No", because their sections on U are, respectively, {1,2} and {2}, so the set-theoretical *union* ∪ of the two is {1,2}, which is the "*total* section", which as such cannot be false). Further (second meta-question, "Q2/U"), "Can these two sub-sheaves have a true *conjunction* at level U?" The answer ("A2/U"), here, is "1" (i.e., "Yes", because the set-theoretical *intersection* ∩ of their aforementioned respective sections on

U, i.e., {1,2} and {2}, is {2}, and therefore it is non-empty). Further ("Q1/X"), "Can these two sub-sheaves have a false *(inclusive) disjunction* at level X?" The answer ("A1/X"), here, is "0" (because their respective sections on X, i.e., {1} and {2}, have a *union* which is the *total* section {1,2}). Further ("Q2/X"), "Can these two sub-sheaves have a true *conjunction* at level X?" The answer ("A2/X"), here, is "0" (because their aforementioned respective sections on X, i.e., {1} and {2}, have an *intersection* which is empty). The last two answers (i.e., those at level X), "0" and "0", determine the first literal, "*x*" (the one standing, on top of the tri-simplicial oppositional quality we are determining, for the level X) as "N" (for "negation": recall that in the Aristotelian 3^2-semantics negation is precisely [0|0], cf. Fig. 10); the first two answers (i.e., those at level U), "0" and "1", determine the second literal, "*u*" (the one standing, at the bottom of the tri-simplicial oppositional quality we are now determining, for the level U) as "S" (for "subcontrariety": recall that subcontrariety is defined as [0|1], cf. Fig. 10). So, together these two ordered literals "N" and "S" give, concatenated (i.e., as the noncommutative string "*xu*"), "NS", which, as shown by the Aristotelian lattice on the right of Fig. 37, is the brown paraconsistent negation. The same kind of reasoning holds for (and only for) the further two (commutative) pairs of vertices, "1_X2_U and 1_U2_X" and "$1_X2_\varnothing$ and 1_U2_X" (thus, all in all, in the tri-segment there are three brown segments of paraconsistent negation NS). A similar reasoning establishes that the CN relation (the pink paracomplete negation) holds for (and only for) the three (commutative) pairs of vertices "$1_X2_\varnothing$ and $1_\varnothing2_U$", "$1_\varnothing2_U$ and $1_U2_\varnothing$", and "$1_U2_\varnothing$ and $1_\varnothing2_X$" (so there are three pink segments in the tri-segment). Finally, a similar reasoning establishes that the NN relation (the red classical negation) holds only between the two vertices "$1_X2_\varnothing$ and $1_\varnothing2_X$" (thus there is only one red segment in the tri-segment). So we have seen here $3 + 3 + 1 = 7$ over the 28 segments of the tri-segment.

Let us now see simplicial colors (among blue, black, and green). For instance, let us compare the vertices 1_X2_U and $1_\varnothing2_U$. So, as previously (Q1/U), "Can these two sub-sheaves have a false *(inclusive) disjunction* at level U?" The answer (A1/U) here is "0". Further (Q2/U), "Can these two sub-sheaves have a true *conjunction* at level U?" The answer (A2/U) here is "1". Further (Q1/X), "Can these two sub-sheaves have a false *(inclusive) disjunction* at level X?" The answer (A1/X) here is "1". Further (Q2/X), "Can these two sub-sheaves have a true *conjunction* at level X?" The answer (A2/X) here is "0". So, the last two answers (level X), "1" and "0", determine the first literal as "C" (contrariety); the first two answers (level U), "0" and "1", determine the second literal as "S" (for "subcontrariety"). Together these two literals give, concatenated, "CS", which is the black pivotal simplicial relation (a mixture of contrariety at level X-U and subcontrariety at level U, cf. Fig. 37). The same reasoning holds for (and only for) the further ten (commutative) pairs of vertices – "1_X2_U and $1_U2_\varnothing$", "$1_\varnothing2_U$ and 1_U2_X", and "$1_U2_\varnothing$ and 1_U2_X" – and all the remaining seven pairs containing at least one occurrence of the vertex "1_U2_U" (thus, all in all, there are $3 + 7 = 10$ black segments, one of which is in fact the black non-arrowed reflexive curl "1_U2_U and 1_U2_U"). Remark that neither the blue (i.e., "CC", contrariety) nor the green (i.e., "SS", subcontrariety) tri-simplicial oppositional relation (i.e., the two "simplicial colors" other than black) does emerge here as 1 of the 21 segments or 7 curls between the possible pairs of the 7 vertices. So we have seen here 10 over the 21 segments and 1 among the 7 curls of the tri-segment.

Finally, let us see implication arrows (we know they can be gray, light green, and violet). For instance, let us compare the commutative pair of vertices $1_U2_\varnothing$ and $1_X2_\varnothing$. So (Q1/U), "Can these two sub-sheaves have a false (inclusive) disjunction at level U?" The answer (A1/U) here is "1". Further (Q2/U), "Can these two sub-sheaves have a true conjunction at level U?" The answer (A2/U) here is "1". Further (Q1/X), "Can these two sub-sheaves have a false disjunction at level X?" The answer (A1/X) here is "1". Further (Q2/X), "Can these two sub-sheaves have a true conjunction at level X?" The answer (A2/X) here is "0". So, the last two answers (level X), "1" and "0", determine the first literal as "C"; the first two answers (level U), "1" and "1", determine the second literal as "I" (for "implication"). Together these two (orderly) literals give, concatenated, "CI", which is the light green paracomplete implication (a mixture of contrariety at level X-U and implication at level U). Notice that the direction of the arrow (which can even be two-sided) is determined, further, by *comparing the relevant sections* (for CI this is the sections at level U): the implication then goes from the shorter to the greater section: e.g., $1 \rightarrow 12$, $1 \longleftrightarrow 1$, $12 \leftarrow 2$, etc. (recall that since Angot-Pellissier's proposal in 2008 [111], "12" means "$1 \vee 2$", i.e., "either 1 or 2 or both is true"). In the case under examination the sections on U being, respectively, {1} and {1}, the light green relation CI takes the form of a biconditional. The same relation holds between the three (commutative) pairs "$1_\varnothing 2_U$ and $1_\varnothing 2_X$", "$1_\varnothing 2_U$ and $1_\varnothing 2_U$" (this is a curl), and "$1_U2_\varnothing$ and $1_U2_\varnothing$" (another curl). So, all in all in the tri-segment there are, in light green, two segments and two curls. A similar reasoning establishes that the IS relation (the violet paracomplete implication) holds, here also as biconditional, for the two pairs of vertices "$1_X2_\varnothing$ and 1_X2_U" and "$1_\varnothing 2_X$ and $1_U2_\varnothing$", as well as for the two reflexive pairs "1_X2_U and 1_X2_U" and "1_U2_X and 1_U2_X" (thus two violet arrowed curls). Finally, a similar reasoning establishes that the II relation (the gray classical implication) holds (only) for the two reflexive pairs of vertices "$1_X2_\varnothing$ and $1_X2_\varnothing$" and "$1_\varnothing 2_X$ and $1_\varnothing 2_X$" (thus two gray arrowed curls). So we have seen here, as for tri-simplicial implication arrows (of 3 colors), 4 over the 21 segments and 6 over the 7 curls. In sum, 7 (negations) + 11 (simplicial segments) + 10 (implications) = 28. *Le compte est bon.*

Incidentally, remark that to see more directly the link existing with the Aristotelian 3^2-semantics (i.e., with its terms like "[1|½]", etc.), you can read *vertically* the four numbers of the aforementioned *xu* code (obtained as in the examples just described), as in a 2×2 square matrix, where the two numbers of x are put on top of the two numbers of u: then, a left (resp. right) column of this 2×2 square matrix containing a same number "j" ($j \in \{0,1\}$) gives, as half of the Aristotelian code (relative to that column), "[j|" (resp. "|j]"), while a left (resp. right) column made of two different numbers "j" and "k" ($j,k \in \{0,1\}$, $j \neq k$) gives "[½|" (resp. "|½]").

Now, further focusing on each of the seven vertices we determined (Sect. 2.1), the one after the other, and on the just calculated list of the precise quality of each of their 28 possible mutual two-terms relations, these results can be tentatively displayed in a synoptic way (i.e., in a unique picture), vertex after vertex, in a row, in the following way (Fig. 38).

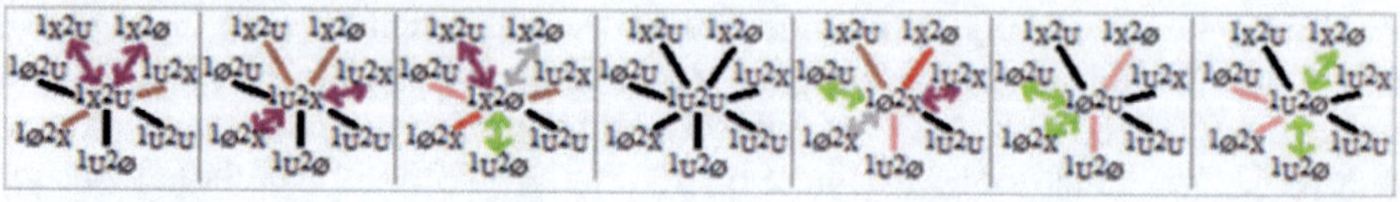

Fig. 38 Synoptical view of the kinds of tri-simplicial opposition relation each vertex has with any possible vertex

As we remarked, two over the nine possible tri-simplicial "colors" are in fact absent here: the blue and the green. This reduces the Aristotelian 3^2-lattice of the tri-*segments* (Fig. 39).

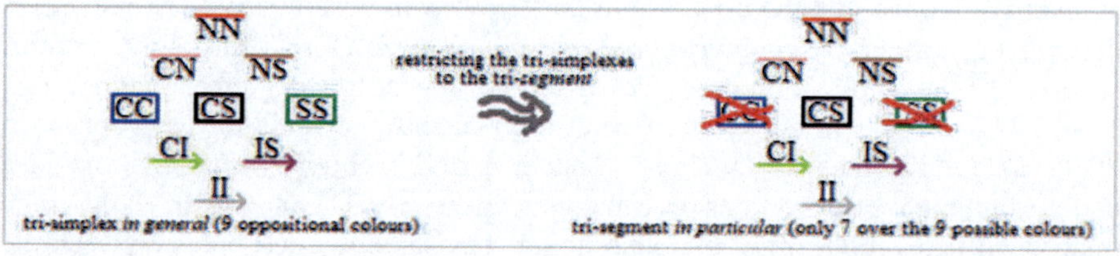

Fig. 39 The oppositional tri-*segment* has only seven of the nine possible colors of the oppositional tri-*simplexes*

But this absence (of two over the nine oppositional colors: blue and green) is normal: poly-*segments* are based on segments (and not on triangle, tetrahedron, five-cell or higher, cf. end of Sect. 1.6, Fig. 33); as already known by wordy reasonings by Aristotle and elucidated by Angot-Pellissier [1], contrariety and subcontrariety need triangles to emerge. So tri-segments are a very particular (and primitive) case of the tri-simplexes (the blue and the green colors will appear as soon as triangles do intervene, that is, in tri-triangles – as Angot-Pellissier's 2013 study [3] has in fact precisely confirmed). Remark that this establishes not only *that*, but also (in part) *why*, my model of tri-segment of 2009 (Sect. 1.6, Figs. 30, 31, and 32) was mistaken.

In our trip toward the global oppositional geometry of the tri-segment, several points come next. One is that of determining the "logical" meaning of each of the oppositional colors. Angot-Pellissier [3] has provided elements of answer to that. Recall that in the bi-simplexes the meanings of colors are clear and related to the connectives of propositional calculus (Sect. 1.1, Fig. 1), although it has taken time to understand – thanks Smessaert – two important additional things to be signaled here, namely, (1) the exact nature of the Aristotelian subalternation (to which we come in a few lines) and (2) the existence, in some sense, of one more oppositional "joker" color: orange for *the "no-relation" relation* – in some sense the structuralist null-element. Now, in the tri-simplexes what seems to be expected is – we will not discuss it here – that *three*-valued connectives (rather than two-valued) will intervene somehow in a similar way (we will try to come back to this in Sect. 5.2).

Before going on, we must recall a quite important point, Smessaert's lesson on subalternation (in some sense seemingly at the origin of his and Demey's idea of calling "*logical geometry*", cf. [49, 135], the field he and several others – including myself – are investigating since years, if not centuries). In 2009 (Sect. 1.3) I interpreted the "[1|1]" (which in some sense I created!), in the Aristotelian lattice, as meaning "logical implication" (i.e., the Aristotelian-Apuleian classical "subalternation"). I signaled that to do this a restriction of the combinatorial "yes yes" (i.e., "[1|1]") definition was necessary; otherwise, we would have had "logical equivalence" instead of "logical implication". This point remained strange and awkward (this sudden asymmetry in Aristotle and Apuleius' otherwise so elegant oppositional combinatorics). But Smessaert, later, showed, in his studies where he proposed the idea of a larger "*logical* geometry" [134, 135], that "[1|1]" is more properly to be understood as something more primitive than logical implication, namely, "noncontradiction", that is as a very general relation (of which "implication" is only a meaningful and useful *restriction*). To explain that point more deeply, he discovered that one must in fact consider a larger ask-answer game-theoretical semantics (than mine of 2009; Figs. 10 and 11), which, as it happens, generates not one but *two geometries*: "opposition geometry" (which is *more or less* "opposition*al* geometry", Sect. 1.2) and "implication geometry" (which is the new thing). And that the classical geometry of oppositions (i.e., opposition*al* geometry), called by Smessaert the new name "Aristotelian geometry", emerged, historically, as an unconscious composition of three over the four elements of "opposition geometry" (i.e., dropping precisely "noncontradiction") and one over the four elements of "implication geometry ("one-sided implication", more precisely "right-implication", taken to replace "noncontradiction") (Fig. 40).

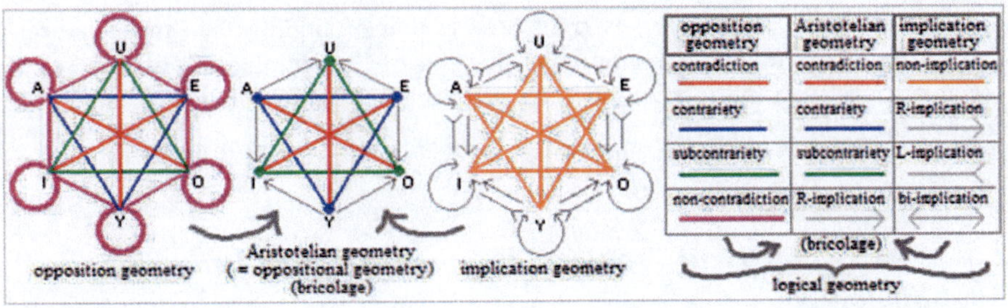

Fig. 40 Smessaert (2012): the logical hexagon (its vertices and edges) can support (at least) three different "geometries"

The discovery is important and clarifying. But, as for bi-simplexes, it does not seem to change much: one just has to be aware that in fact "I" (i.e., "[1|1]") is in fact noncontradiction and that it can be "enriched" with implication. And it justifies in no way Demey's repeated argument that high-dimensional solids (i.e., higher than 3D) (and therefore, *a fortiori*, poly-simplexes ...) are not worth being investigated (which is a *logicist* move: logicism, among others, aims at reducing all numericity to the binary "0 and 1", so that it generally takes no pleasure in exploring mathematical depth, as, for instance, poly-simplicial depth ...). In [100], where I solved the (rather difficult and unsolved since 1968) riddle of the nature of Greimas' problematic (and very famous) "semiotic square", I proposed to see Smessaert and Demey's two geometries as "*meta*-geometries", that is, as useful preconditions of the real thing: opposition*al* infinite complexity (infinite as for the dimensionality of the contrariety simplexes and as for the variability of the "poly" diffractions). *My point was, and up to now remains, that oppositional geometry is the real mathematical thing at stake and that it is by no means "a new chapter of logic" (and one that according to Demey and Smessaert should refrain from exploring higher dimensions!)*, as the label "logical geometry" strongly and recklessly suggests to a philosophically non-naïve (and non-cynical ...) eye (Sect. 4.6). Again, practically this means for us that "[1|1]" needs *interpretation* (it is more abstract than just logical implication; it is the "noncontradiction" relation, logical implication being a particular case of the more general "noncontradiction" relation). Our way of dealing with it will be "Aristotelian" (in the sense of Smessaert and Demey; Fig. 40), and this will be done, as we just saw in this Sect. 2.2, relying on Angot-Pellissier's sheaf-theoretical method for the poly-simplexes (Sect. 1.4): in front of codes containing "I" (i.e., II, CI, or IS), we will see, by examining the relevant sheaf-sections, whether the simple colored line (fluo pink, light green, or violet) can be turned into a single-sided or a double-sided arrow (in fact we will come back to this important issue in Sect. 4). In the rest of this paper, we will take into account this while refusing as inappropriate and very misleading the academically fashionable label "*logical* geometry": (1) it loses the stress put on "opposition", as a mathematical concept largely independent from logic (Sect. 1.5!!!), (2) and it falls into "logicism" (which is a deadly constant in the history of the geometry of oppositions – cf. [97] and Sect. 4.6).

The first point waiting for us right now is that of finding the "most natural" geometries (in principle "tri-simplicial" instead of "bi-simplicial") of this set of seven vertices with the relations holding between any pair of them (including the curls of the reflexive pairs).

2.3 *The Geometrical Problem: Having a Strange Pentadic Structure*

Let us try to end our inquiry still without the help of the "Pascalian ND simplexes" (Sect. 1.5). As said (Sect. 2.1, Fig. 36), from the nine numerical sub-sheaves we eliminated the oppositional-geometrical analog of T and ⊥ (i.e., 1_X2_X and $1_\varnothing 2_\varnothing$).

This leaves in our hands seven sub-sheaves, among which one (i.e., 1_U2_U), as said, seems mysterious (we saw in Sect. 2.2, Fig. 38, that the strangeness of 1_U2_U grows with its black non-arrow curl).

How can we try to find a good global geometrical expression of this reality, namely, something like *the* polygon or *the* solid (or polytope) of the tri-segment? How to display in the *n*-dimensional (2D? 3D? 4D?) oppositional-geometrical space the 7 vertices (with their 7 curls) and the 21 non-curl edges relating any pair of (nonidentical) vertices? (Fig. 41).

Fig. 41 How to display at best, in the oppositional-geometrical space, these seven non-trivial vertices of the tri-segment?

Given that in some sense the tri-segment can be seen as a transformation (tri-simplicial “oppositional *diffraction*”) of the (red) segment (of contradiction), we can try to singularize this starting classical red segment $1_X2_\varnothing$—$1_\varnothing 2_X$ (of which the tri-segment seems to be a diffraction *and* a conservative extension), by putting it so to say on a “radial position” (i.e., visual-metaphorically, as the axle of a wheel). Remark that we can try to color this segment’s vertices, but we do not know how to color the five others (there are no simplexes here): so we will use gray points. But then several different alternative dispositions of the five remaining sub-sheaves are possible, and, as it happens, it does not seem to be easy to find a particular configuration more convincing and “mathematically natural” than the other possible ones. Remark that since the self-relations (of any vertex to itself) happen, here, to resort not to only one (gray), as in the bi-simplicial space, but to three possible “arrow colors” (gray, light green, and violet and seemingly even the *non-arrow* black), we cannot go on keeping them implicit (as they usually are, *pace* Smessaert, in bi-simplicial oppositional “Aristotelian” geometry, where they usually are not drawn, since they are tautological): so to say following, at least partially, the suggestion of Smessaert inside his and Demey’s aforementioned “logical geometry” (Sect. 2.2, Fig. 40), we will do better here by choosing to *represent them as well explicitly, as “curls”* (so, on this point our research on the tri-segment goes in the direction of logical geometry). Remark, again, that over the seven curls, six are implication arrows (two gray, two light green, and two violet), while one (the black CS of the 1_U2_U vertex...) is not (CS contains no “I” in its code): as said, this adds more suspicion or puzzlement of us regarding the strangeness or, in the best case, the mathematical *singularity* (but then: why and how?), of the 1_U2_U nontrivial sub-sheaf and vertex (Sect. 2.1) (Fig. 42).

Trying to clarify this complex oppositional-geometrical (and chromatic) riddle, we can try to decompose it (*divide et impera*), hoping to lower its complexity, by

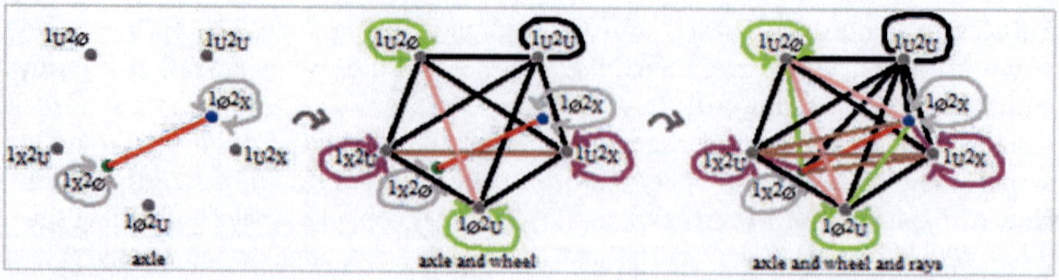

Fig. 42 Trying to represent the tri-segment like "axle and wheel", putting into light the red segment of contradiction

detaching so to say the pentadic circular "tire" from its radial "axle" (the latter can be cut in two halves) (Fig. 43).

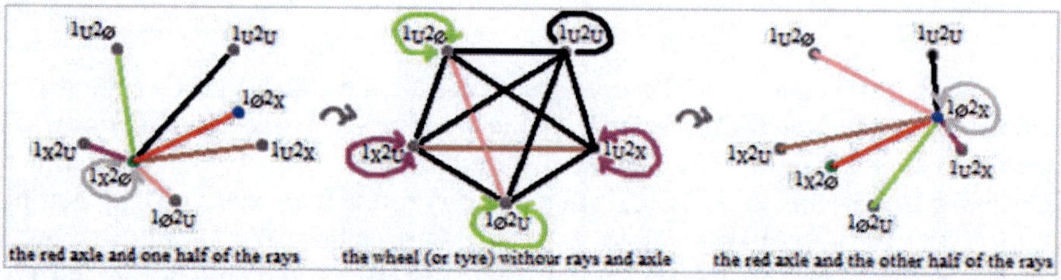

Fig. 43 Trying to decompose into parts the oppositional-*geometrical* problem of finding *the* tri-segment

Remark that this representation, and more fundamentally this structure, whatever its tentative representation, might in some sense already seem exciting *per se*, for in some sense it seems to lead us to forms or patterns rather unexpected with respect to the usual standards of oppositional geometry (mainly by having here something *pentadic*, and so to say numerically seven-based, in an *a priori* non-heptadic and non-pentadic context of mathematical analysis: the tri(3)-segment(2)). This is mathematically surprising with respect to what known so far about bi-simplicial (i.e., two-valued) oppositional geometry, which tends to be reasonably simple and symmetric. In other words, we might be tempted to *accept* as (unexpectedly) "typical" of the still mysterious and maybe durably exotic universe of the "tri-segments" this intractable pentadic flavor, if one wants to represent the classical (red) contradiction segment as the starting term of a tri-simplicial "oppositional diffraction". But truly speaking, this structure in some sense remains hard to interpret, at this stage (the "logic – *cum grano salis*! – of tri-simplicial diffraction" is not at all clear here), and this can be seen as being no specifically good sign; notably we can remain puzzled about the particular status of the still mysterious 1_U2_U term: for it clearly seems to introduce a strong geometrical (and chromatic) disequilibrium that, supposing it is justified and meaningful – but then why? –

we do not yet understand by resorting to Angot-Pellissier's sheafing technique. In other words, the so far reached structure seems to lack badly *symmetry*, and this seems to be notably influenced by the presence of a large number of black segments (which, again, converge on the still mysterious 1_U2_U sub-sheaf, Sect. 2.2, Fig. 38). Of course, we can try to invert geometrical representation priorities (in the hope of finding unseen, better oppositional-geometrical arrangements): for instance, by putting tentatively the red contradiction segment in positions other than axial (but then: why? and how?), so to let 1_U2_U take such a geometrically and chromatically (for short: oppositionally) prominent place. But no clear better alternative (to the pentadic "wheel and axle") seems, even at that price, to emerge. On that respect we might say that apart when considering "seven" to be a hexagon with an extra inner point, or as a blue seven-contrariety simplex (or 7-opposition), seven seems, fundamentally, to be (at least as long as we know) no easy number to let emerge "from itself" a geometrical regular configuration provided with the kind of symmetries oppositional geometry got us used to (Fig. 44).

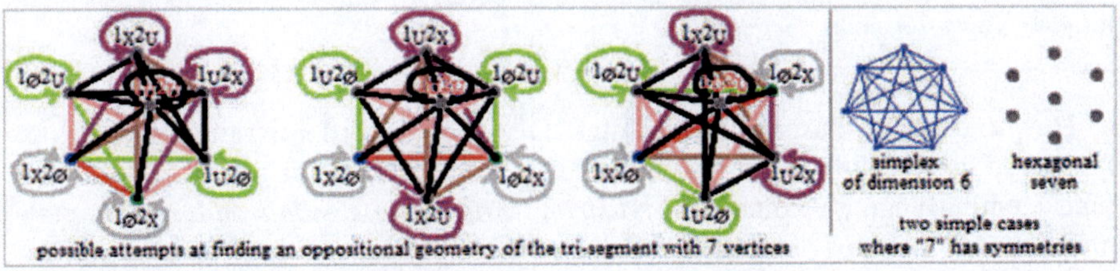

Fig. 44 The geometrical strangeness of the number seven: hard to unfold through spatial symmetries

So let us try to see, instead, in the next Sect. 2.4, whether applying to this oppositional-geometrically still mysterious putative tri-segment, so far a nut hard to crack, a totally different strategy, or better an extra piece of the puzzle, namely, the "Pascalian lens" (Sect. 1.5), can help us in finding some more fundamental (and helpful) order, regularity, and understanding of the still hypothetical oppositional tri-segment.

2.4 *The Pascalian 3D Simplex and Its "2D Section for Tri-segments"*

As we saw (Sect. 1.5, Fig. 24), Pascal's triangle matches perfectly the oppositional bi-simplexes (and most importantly, their closures, the B*n*, Sect. 1.2, Fig. 7), and we claimed to have successfully defined (with a general proof that we will give elsewhere) a generalization of Pascal's triangle to be called the "Pascalian ND *simplex*", such that it matches, case by case, the oppositional *poly*-simplexes (Sect.

1.5). So, in order to see whether we can find a way of solving the not so easy puzzle of reaching (if possible) a harmonious oppositional-geometrical structure of the tri-segment (Sect. 2.3), let us now turn back to the Pascalian *3D* simplex (i.e., a 3D tetrahedron) for the *tri*-simplexes (Sect. 1.5, Fig. 27). Among its "2D horizontal sections" (i.e., the horizontal "arithmetical triangles" this arithmetical tetrahedron is made of, when sliced horizontally) for all the tri-simplexes, the one relative to the *segments*, which gives us therefore the oppositional numericity relative to the tri-*segments*, is the third, starting from the top (Fig. 45).

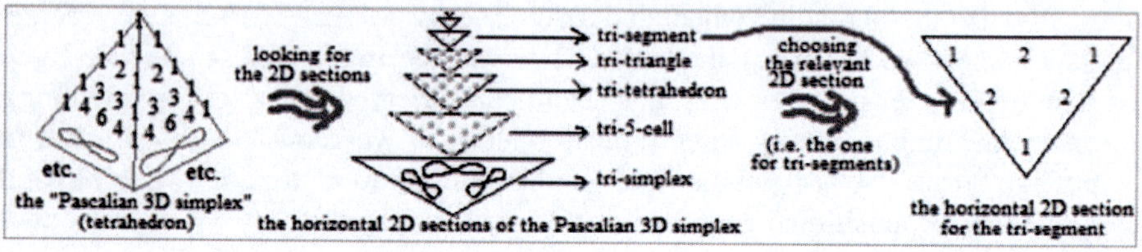

Fig. 45 The Pascalian 3D simplex (tetrahedron) for the tri-simplexes and its "2D section" (triangle) for tri-segments

How to use this "Pascalian horizontal 2D section" in order to reinterpret the useful, but so far puzzling, combinatorics of Angot-Pellissier? The idea is rather simple (and seemingly natural): one has to *distribute, as useful symbols on a roadmap*, the nine possible vertices determined through Angot-Pellissier's combinatory method, as being all the sub-sheaves of the relevant total numerical sheaf (Sect. 2.1, Fig. 35), in what we might call "Pascalian places", which happen to be qualitatively different. In this case (i.e., the tri-segment), which is rather simple (Sect. 1.5, Fig. 29), such "Pascalian" distinction essentially only runs between the three vertices (each bearing "1") on the one side and the three intermediate points (each bearing "2") on the other side. By construction, on one hand, the three Pascalian "1" correspond, respectively, to the Angot-Pellissierian sub-sheaves "$1_{\varnothing}2_{\varnothing}$", "$1_U2_U$" and "$1_X2_X$". But then, *this is big news: for, so to say, monsieur Pascal himself suggests us here nothing less than to consider the up to now problematic and mysterious numerical sub-sheaf "1_U2_U" as being an "extremum" that is something to be eliminated by (oppositional) construction together with the already known extrema "$1_{\varnothing}2_{\varnothing}$" and "$1_X2_X$"* (Sect. 2.1, Fig. 36). By construction, still, on the other hand, the three Pascalian "2" (on the Pascalian horizontal section, Fig. 45) correspond to pairs of the remaining six nontrivial sub-sheaves (the nontrivial oppositional-geometrical vertices of the tri-segment): more precisely, these three Pascalian "2" correspond, respectively, to "$1_X2_{\varnothing}$" and "$1_{\varnothing}2_X$" (on the horizontal upper side of the Pascalian section), "$1_U2_{\varnothing}$" and "$1_{\varnothing}2_U$" (on its oblique left side), and "1_X2_U" and "1_U2_X" (on its oblique right side) (Fig. 46).

Summing up, the Pascalian section helps us in solving neatly the puzzle of the term "1_U2_U": by suggesting us quite clearly to see it as being (mathematically) "on a same plane as the sub-sheaves $1_{\varnothing}2_{\varnothing}$ and 1_X2_X" and therefore to eliminate it (i.e., by

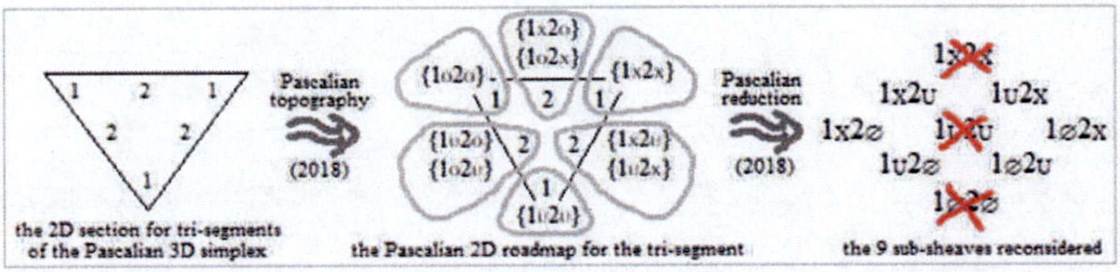

Fig. 46 The "Pascalian 2D section" as a roadmap from the nine sub-sheaves to the six vertices of the tri-segment

making "1_U2_U" implode in a way similar to the previous – bi-simplicial – implosion of the sub-sheaves "$1_\varnothing 2_\varnothing$" and "$1_X2_X$"). The result, in fact, is that not only two but in fact three over the nine sub-sheaves of the starting Angot-Pellissierian total numerical sheaf "1_X2_X" are "extrema" and must so to say be eliminated by (oppositional) construction. There remain, consequently, six nontrivial numerical sub-sheaves. Furthermore, thanks to the roadmap embodied by the Pascalian 2D horizontal section (the horizontal triangle, Figs. 45 and 46) for segments, we see that these remaining six terms distribute themselves in three precise "places", of two terms each, among which two ("$1_X2_\varnothing$" and "$1_\varnothing 2_X$") are the classic ones of bi-simplicial oppositional geometry.

But before going back, in Sect. 2.6 (and then in the next Sect. 3), to the puzzle of a global oppositional-geometrical representation of the tri-segment, let us have a quick look at the order-theoretical question possibly raised by our present "Pascalian" proposal of having not two but *three extrema* (for in some sense, order-theory deals exactly with the question of extrema, but in rigorous and systematic mathematical terms). Recall that the already mentioned order-theory (Sect. 1.2, Fig. 7) is a rich and important region of general mathematics, dealing with the most general order structures and lattice structures [48]. How can we posit ourselves safely in it relatively to our daring idea here of having not two but three extrema? Provided I am – alas – no expert on the field, if we try to think at least intuitively what can mean to have not two but three extrema, a *tentative* figuration which seems possible and hopefully helpful is maybe the following, where essentially we represent the Pascalian "2" as two "white spheres" (the three-colored spheres represent the three Pascalian extrema: "1_X2_X being green", "1_U2_U being black", and "$1_\varnothing 2_\varnothing$ being blue") (Fig. 47).

The intuition we propose to follow with this 3D order-theoretical *tentative* model seemingly can be decomposed, more classically, in terms of three 2D order-theoretical models, suggesting that in some sense this fundamental "tri-polarity" remains submitted to a *binary* transitive order ($T > Y > \bot$) (Fig. 48).

In our eyes, there seem to be at least three possibilities: (1) either there is transitivity, so that the triangular Pascalian shape (of the section) might seem illusory (in its claim of a strong ternary symmetry of what it depicts), (2) or somehow there is not transitivity, so that the triangular shape can hold on; (3) or

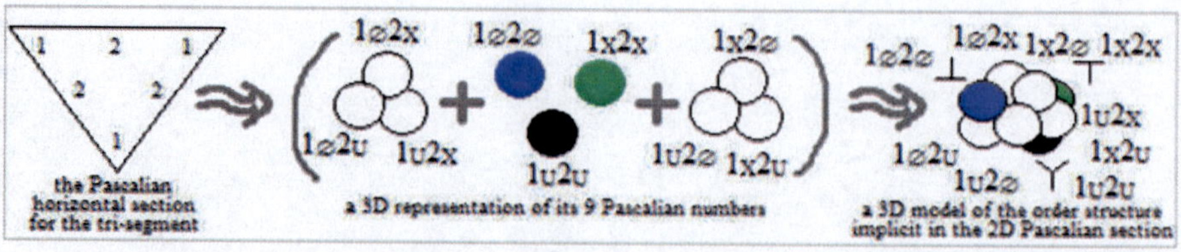

Fig. 47 A tentative 3D order-theoretical qualitative view of the Pascalian 2D section for the tri-segment

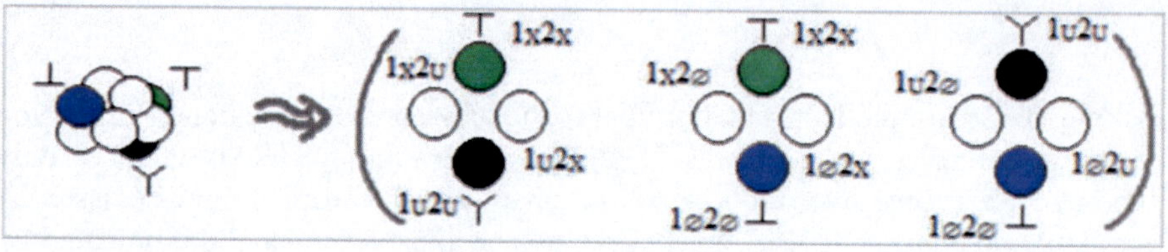

Fig. 48 Decomposing a *tentative* 3D order structure (for the tri-segment) into three more classical 2D components

(most probably), even if there is in some sense a binary order over (or embedded in) this (*per se*) ternary structure (which is the case as for truth-values as we take them here: 1 > ½ > 0), its ternarity as for its polarities "holds on". This question seems to be some order-theoretical counterpart of the question about the very idea of "oppositional tri-simplex" (in its intended innovating radicality). In our view, it could echo a famous point raised in 1975 by R. Suzsko [138] against the very idea of "many-valued logic" and the reactions (as, to mention one, [131]) this originated (we recalled and tried to discuss this in [96]). But in this paper we are not yet able to say more on it.

Let us now turn, in what follows, to our lasting problem and goal of finding *the* oppositional geometry of the tri-segment, but starting by a (last) preliminary important question on the possible structure of *oppositional colors of the six vertices* of the tri-segment.

2.5 A Point About "Points": The Oppositional Colors of the Six Vertices

Notwithstanding the undeniable material difficulty involved by this at the level of future black and white *printed* papers on the subject, and not forgetting the potential *real* pain and discrimination I thus will – alas – increase among color-blind readers (there are), I keep thinking, as the years pass, that *sua juxta principia* the *vertices* of the oppositional-geometrical structures do gain, as much as the *edges* clearly do, in getting colored: this gives to them some extra expressive power, which enhances oppositional-geometrical creative thinking. But how to proceed, thus getting more graph-theoretical (because of colors)? In a nutshell, the situation is currently the following: inside bi-simplicial oppositional geometry, coloring the vertices seems pretty fine, except for the 2-oppositional contradiction segment!

Let us recall more closely this problem. With the exception of the red segment of 2-oppositional contradiction (i.e., B2), the red segments of contradiction more generally – i.e. when they are part of an *n*-opposition strictly bigger than a 2-opposition – are by no means a problem as for coloring their two vertices. For short, (1) in the B3, each of the three red segments of contradiction it contains has one blue and one green vertex; (2) in the A4, we have the same good-functioning behavior for the four red segments of contradiction it contains; and (3) in the closure of the A4, namely, the B4, we have the A4, plus its "cloud", made of six extra vertices, two by two centrally symmetric, such that they thus let emerge three extra red segments of contradiction (so B4 has $4 + 3 = 7$): but this time each of these three extra red segments of contradiction will have at each of its two extremities a *blue-green* vertex. Put more abstractly, in the case of bi-simplicial (i.e., classical, two-valued) oppositional geometry the main use, never properly theorized, in coloring the vertices went, so far, as follows: (1) in the "oppositional kernel" (of an *n*-oppositional closure), the vertices of any simplex of contrariety are *blue* dots (related, two by two, by a blue line meaning the contrariety relation between them), and the vertices of the correlated simplex of subcontrariety are "oppositionally-dually" *green* dots (related, two by two, by a green line meaning the subcontrariety relation between them); (2) in the remaining, non-kernel part of the oppositional closure itself (and *de facto* this most of the time, so far, concerns the B4), the vertices are represented as *blue-green* dots (the mutual proportion of blue and green, in principle, can vary according to the typology of the cloud, as we study in [94], but as this enters seldom into play (so far), this point has not yet been afforded too systematically) (Fig. 49).

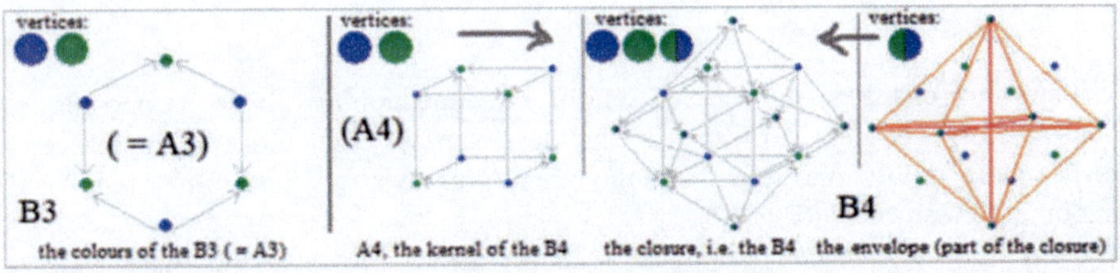

Fig. 49 By coloring the vertices one can distinguish "kernel" and "cloud" (or "envelope") of a bi-simplicial B4

This color of its two vertices, importantly, tells something about the structure where this red segment of contradiction intervenes, that is, respectively, either (i) in the kernel (i.e., an A*n*-structure) or (ii) in the cloud of a kernel (the union of these two constituting a B*n*-structure).

Now, as said, the only problem is that, when facing the question of depicting the two vertices of *a red segment of 2-opposition* (and not the two vertices of a red segment which is *a fragment of a larger* n*-oppositional solid*), two options seem possible by analogy with what precedes: representing its two vertices either as: (i) one blue and the other green (by analogy with the oppositional kernels) or (ii) depicting them as both blue-green (by analogy with the clouds). The pros and cons of this choice, so far, are not too clear. But in my own researches so far, I tended to adopt resolutely the first option. But, as we are going to show, the study of the tri-simplexes (and higher) reveals, retrospectively, that *I made the bad choice*.

The Pascalian 2D simplex (Sect. 1.5, Fig. 24) can be read as strongly suggesting that, in its third line (top-down), corresponding to bi-simplicial – *cum grano salis*! – 2-opposition, in fact *the "2" does belong to the cloud!* In other words, the 2-oppositional contradiction segment, being pre-simplicial (because in oppositional geometry *contrariety* simplexes start with triangles, cf. Sect. 2.2, Fig. 39), "lives *inside* the classical blue-green cloud" (!). More precisely, the Pascalian 2D simplex suggests, for the bi-simplexes, that their "2" belongs, in fact, not to one of the two "simplicial diagonal lines" (either the blue on the left or the green on the right or maybe even to both . . .) but to the "cloud zone". More precisely, *"2" belongs to the vertical central line of the "pivotal elements of the clouds" of the n-oppositional closures* (with *n* an even number): 2, 6, 20, 70, 256 . . . (Fig. 50).

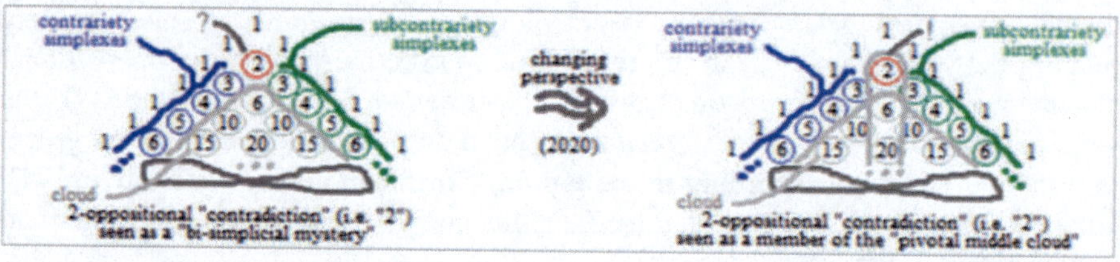

Fig. 50 Seeing, in the 2D Pascalian simplex, the 2-oppositional contradiction as an instance of bi-simplicial "cloud"

And this changes quite much as for our questioning about 2-oppositional contradictory vertices: it means, as for these vertices, that those of any B2 are, respectively, not the one "blue" and the other "green" but the one "blue-green" and the other "green-blue" (Fig. 51).

And in fact, one can and must read the same way also the more general (triangular) 2D section of the Pascalian 3D simplex (i.e., tetrahedron) for the tri-segment (Figs. 45 and 46). This means that the "Pascalian colors" of the tri-segment are in turn simple enough and must be represented in terms of (i) three pure colors (blue, black, green) for its three extrema ("$1_{\varnothing}2_{\varnothing}$", "$1_U 2_U$", "$1_X 2_X$") and (ii) the three mixed colors for the three pairs of vertices in between any pair of them: one Pascalian "2" is blue-green, another is blue-black, the third is black-green (for coloring newly the vertices of the simplexes, cf. Sect. 5.1, Fig. 125) (Fig. 52).

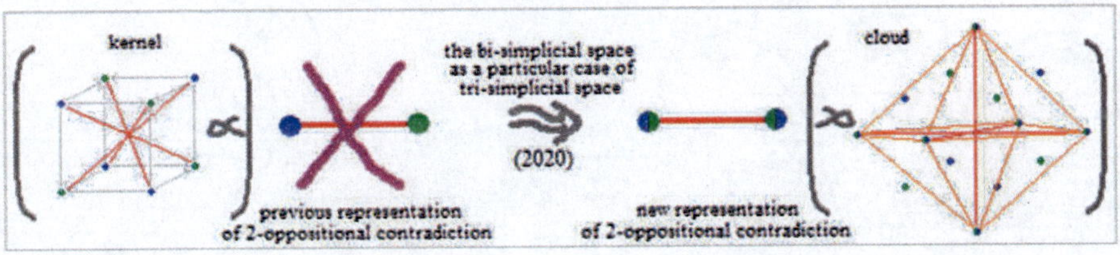

Fig. 51 Our study tells to change the way of representing the vertices of a 2-oppositional contradiction segment

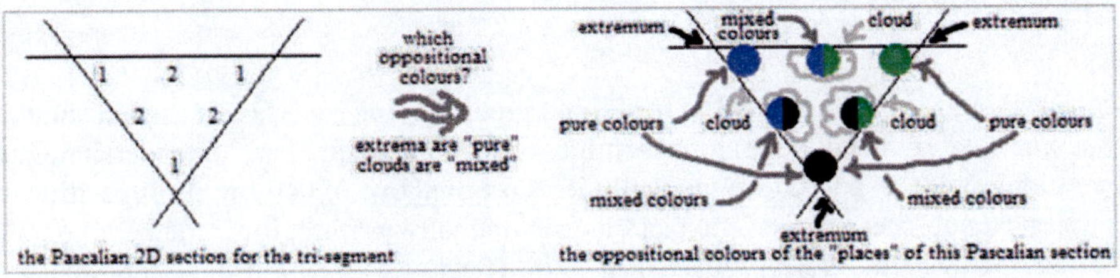

Fig. 52 The oppositional colors of the "Pascalian 2D section" for the tri-segment

So, not only the classical blue-green 2-oppositional red contradiction but each of the three "2" positions of the Pascalian 2D roadmap is in fact, in the same way, a pre-simplicial two-elements *cloud* (the simplexes, if any – and they appear for real in the tri-*triangle* – would be, for each of the three sides of the Pascalian 2D section, somewhere in between the extrema "1" and the cloud "2" positions), and *the points living in each of these three "2" cloud positions must therefore be seen as 2-oppositional negations and labeled, from the viewpoint of the oppositional colors of the extrema they live (as clouds) in between, by mixed-points* (i.e., *blue-green, green-black, and black-blue*). But if the three extrema are (as they must) identified with the three oppositional "simplicial colors" of the tri-simplexes, then it seems natural to conceive *each* of the six remaining *numerical sub-sheaves* (Sect. 2.4, Fig. 46) as composed (in its two vertex-indices) of two colors: those of its two indices! Since each pair of sub-sheaves in each Pascalian "2" represents a tri-simplicial negation (including, as we said, the classical one), this also suggests us how these diagonal negations will look like with respect, so to say, to the general spatial architecture (Sect. 2.6) of the tri-segment (Fig. 53).

To sum up, this regular and understandable combinatorial behavior suggests, very strongly, a very interesting simple but powerful relation of this *Pascalian* "oppositional chromaticity" with the *Angot-Pellissierian* numerical sub-sheaves (i.e., the six vertices, Sect. 2.4, Fig. 46) corresponding to the Pascalian view. The general rule for tri-segments (generalizable, with extensions for the chromatic expression of the simplexes, Sect. 5.1, Fig. 125, to any poly-simplex), as carried usefully by the *indices* of the Angot-Pellissierian numerical sub-sheaves, seems to be straightforward: each vertex "1_J2_K" (with J, K ∈ {∅, U, X}) will have, representing it faithfully, a dot of "type", so to say, "J-K" (as for its colors), and with

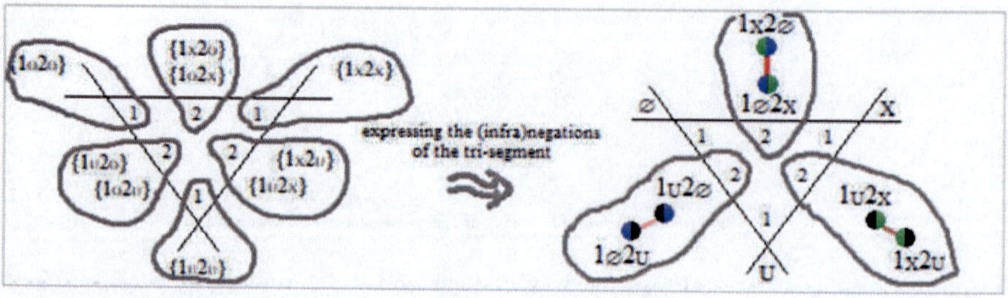

Fig. 53 The Pascalian 2D roadmap shows that the segments of the *negations* of the tri-segment "live in the clouds"

blue staying for ∅, black staying for U, and green staying for X. Here as well should become clearer the not negligible usefulness of our proposal of an "extended indicial notation" (with respect to Angot-Pellissier's original one of [3]) for his oppositional numerical sub-sheaves, i.e., the fact of systematically writing, for instance, "$1_X2_\varnothing$" instead of "1_X" or "$1_\varnothing 2_U$" instead of "2_U" (Sect. 1.5). So, the final "chromatic" reading of the Pascalian roadmap for tri-segments (but this is nicely generalizable, as we will show in other ongoing draft studies, to higher poly-simplexes) as for vertices seems to be the following (Fig. 54):

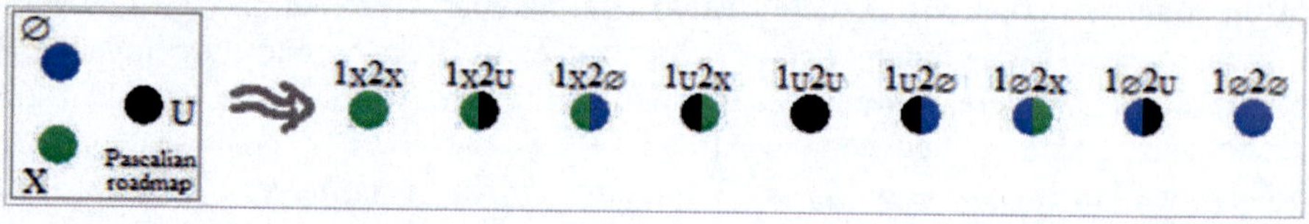

Fig. 54 Angot-Pellissier's numerical sub-sheaves (vertices) command by their full indices their "chromaticity"

With this last new tool (a theory of the oppositional chromaticity of poly-simplicial points), we can now at last come back to the main question left open at the end of Sect. 2.3, namely, the problem of finding a convincing oppositional *geometry* of the tri-segment.

2.6 Back to the Geometrical Quest of the Oppositional Structure

After having dealt with the problem of coloring the "points" (i.e., the vertices, Sect. 2.5), let us now turn to the oppositional geometry of the "lines" (the edges, i.e., the – oppositional – relations between the six nontrivial sub-sheaves of the tri-segment, Sect. 2.2). Let us come back now, with better "weapons", to the problem of expressing the global geometry of the tri-segment (Sect. 2.3). We know now that 1_U2_U is, as 1_X2_X and $1_{\varnothing}2_{\varnothing}$, an extremum and as such must be put away (Sect. 2.4, Fig. 46). This makes at once disappear as well the seven black segments which linked this vertex to any of the possible seven (including itself, with a curved reflexive "curl segment" which was strange, being the only non-arrow curl). So we have now to arrange oppositionally-geometrically not seven but six oppositional sub-sheaves (i.e., vertices), taking into account (in order to let emerge a "geometry") all the segments relying pairs of them (including, as said, reflexive pairs, which by construction yield curl arrows) which now are not $C^2{}_7 = 28$, but $C^2{}_6 = 6!/((6-2)!\,2!) = 21$. As a start, we know the relation between the two classical (i.e., bi-simplicial) vertices $1_X2_{\varnothing}$ e $1_{\varnothing}2_X$ and we know (Sect. 2.5) that each of these vertices is |green-blue|, i.e., the first is green-blue, while the second is blue-green (Fig. 55).

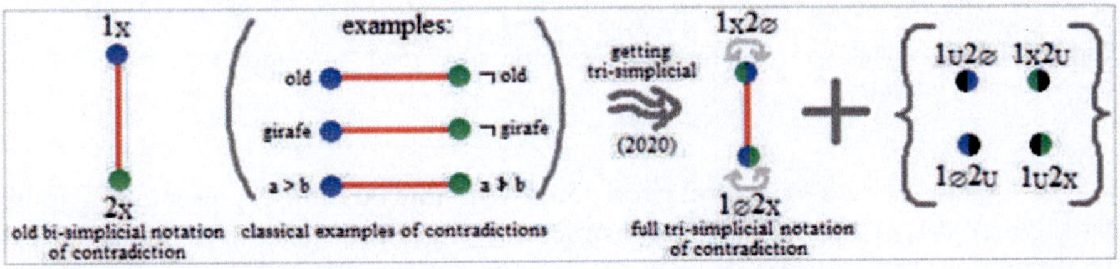

Fig. 55 The bi-simplicial starting point: the (red) segment of " contradiction" (i.e., "classical propositional negation")

How to posit the remaining four vertices (numerical sub-sheaves), of which we now know the oppositional color (Sect. 2.5), with respect to this starting pair? Let us rely, for a start, on intuitive *visual* symmetries of the symbols, namely, on those relating, for instance, 1_X2_U and 1_U2_X, and, in a similar way, $1_{\varnothing}2_U$ and $1_U2_{\varnothing}$ (i.e., symmetries relative to the *indexes* of the numerical sub-sheaves): this is interesting since all the pairs of vertices with symmetric indexes (including $1_X2_{\varnothing}$ and $1_{\varnothing}2_X$) happen to be related by kinds of negations (Sect. 2.2), so putting this into geometrical evidence (by construction), by *imposing central symmetry to these three pairs of indicially symmetric points, would keep something of the classical bi-simplicial interpretation of central symmetry as contradiction (we have here a conservative extension of bi-simplicial central symmetry)* (Fig. 56).

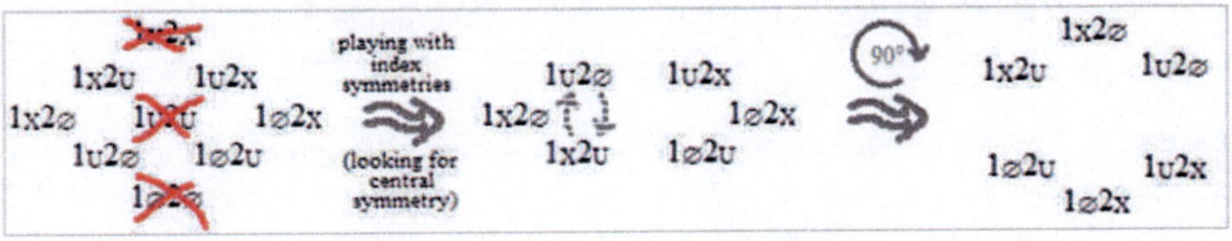

Fig. 56 Rearranging the six nontrivial sub-sheaves of the tri-segment, looking for its oppositional geometry

So let us try to put these four nonclassical vertices "around" the classical red segment (two on the left, two on the right of it); then let us draw progressively all the relevant colored segments (such as Angot-Pellissier's method of [3] has allowed us determining, cf. Sect. 2.2, Fig. 38). Thus doing, we witness at the end of the process, as one of the possible representations of the tri-segment, the emergence of a new kind of *hexagon* (Fig. 57).

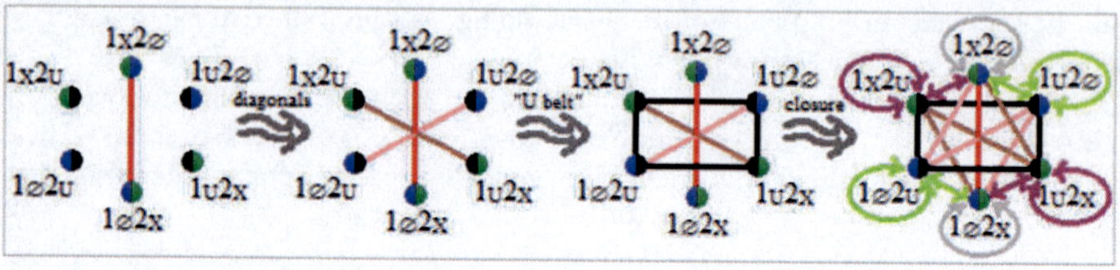

Fig. 57 Trying to let emerge, from its six vertices, a good oppositional geometry of the oppositional tri-segment

Now, there seem to be several good points with this possible representation of the tri-segment: (1) it highlights (by construction) the central position of the classical contradiction (red) segment; and (2) it imitates the bi-simplicial oppositional hexagon (i.e., the bi-triangle, Sect. 1.1, Fig. 2) (i) by putting (by construction with respect to the interpretation of central symmetry) as diagonals its "negations" (i.e., the red contradiction and the pink and the brown "infra"-contradictions) and (ii) by putting as hexagonal perimeter its "implications" (i.e., the light green and the violet "infra"-subalternations). Notice also that the final (oppositional) geometrical result toward which we provisorily tend can be seen both as a 2D and as a 3D oppositional figure (Sect. 3.3), namely, as a hexagon or as an octahedron (this was already the case with the classical bi-triangle, notably in Smessaert), and this 3D representation stresses a bit more an interesting feature of the previous hexagon: the presence of some kind of "horizontal (black) belt" (Fig. 58).

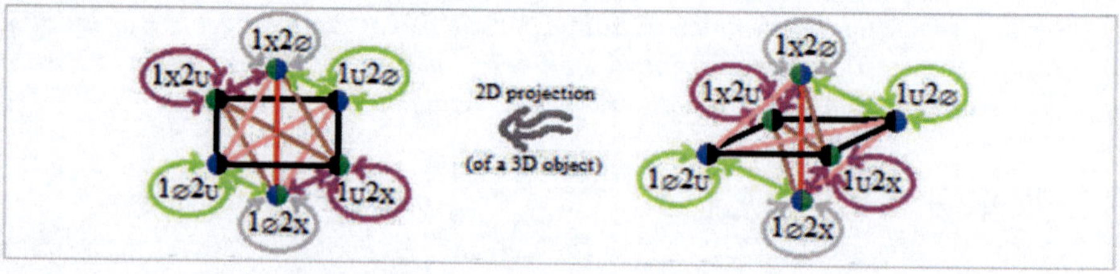

Fig. 58 The hexagon representation of the tri-segment seen as a 2D projection of an equivalent 3D octahedron

As with any 2D hexagon or 3D octahedron, one can read rather easily inside of it three (oppositionally) interesting substructures: respectively, three rectangles or (equivalently) three squares (in the hexagonal representation the components of the tri-segment are rectangles, whereas in the octahedral representation they are, equivalently, squares) (Fig. 59).

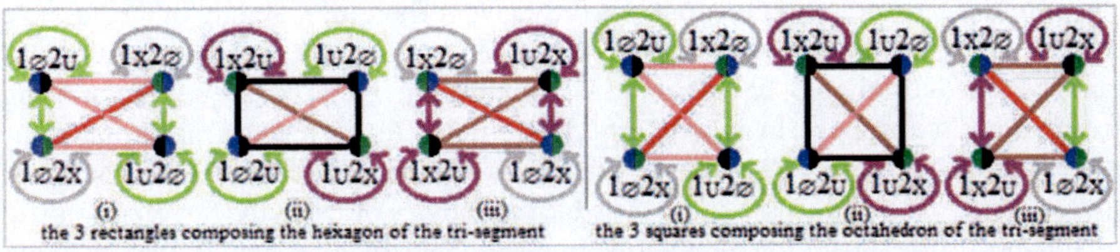

Fig. 59 Two different but equivalent views on some components of the oppositional geometry of the tri-segment

As said, our choice – among several possible other ones (Sect. 3.3) – of this particular presentation stresses the "logical square"-like expression, as diagonals, of three "negation segments" (red, pink, and brown) and the expression, as lateral vertical edges, of at least two "implication arrows" (respectively, light green and violet – in fact here biconditionals).

However, on one point at least one must beware: in this representation, differently with respect to the usual one for the bi-simplexes in oppositional geometry, central symmetry is not uniquely meaning "contradiction". And as it happens, the starting intuition of the poly-simplexes (emerged around 2006 and then presented in my PhD in 2009, [94]) consisted precisely in admitting the idea of having, in the tri-simplexes, several (in fact three) distinct symmetry centers (i.e., one for each of the three bi-simplexes composing a tri-simplex). Notice also that the comparison (which we afford in other papers) with other tri-simplexes, namely, the tri-triangle and the tri-tetrahedron, suggests that another "regular" geometrical representation of the tri-segment could be rather useful, namely, the one in terms of a 3D "trihedron" (Sect. 3.3, Fig. 71). In any case (be it hexagon, octahedron, or trihedron, Sect. 3.3), this result seems much more "regular" (and promising) than the one of Sect. 1.6, Fig. 30 (2009), on one side, and then the one of Sect. 2.3, Figs. 42 or 44, on the other side, for (1) it infirms, by correcting it strongly, my unfortunate tentative "trihedral" model of 2009 on that (Sect. 1.6); and (2) it avoids the seemingly intractable strangeness and the chromatic (and geometric) disharmony of the pentagonal-heptagonal seven-vertices model of Sect. 2.3.

Having reached, at the end of this Sect. 2, our main target (i.e., a *basic* but reliable oppositional *geometry* of the tri-segment), in the next Sect. 3, we will try to go deeper into detail with respect to the structure of the tri-segment, so to be able, notably, to start using it a little bit concretely (in Sect. 5).

3 More on the *Inner Geometry* of the Oppositional Tri-segment

At the end of the previous chapter, we reached an important point: a first, convincing approximation of the global oppositional-geometrical structure of the tri-segment, in terms of a hexagon, which is elegant and promising. But at the moment the possible functioning of this structure is not yet fully clear. Therefore, among other possible consequences, it is not yet clear which of its "parts" can be more meaningful and which ones should be seen as less interesting. As a consequence, praising complexity, we propose, in a partly *experimental* way (experimental mathematics, as defended by Mandelbrot [88]), to go in this chapter through the tri-segment's "inner jungle" in order to try to lay some possible milestones of its study, hopefully useful in the future. For the notion of "hybrid" oppositional structures and for that of "inner jungle", cf. our [101], important seminal elements of this are in Angot-Pellissier's [111], where he discovered and explained, inside the B4, what he theorized as being four equivalent instances of (previously unseen) "*weak* hexagons" in addition to the classical six instances of "strong" hexagons (until then simply known as "logical hexagons"). Our exploration will be gradual: from proximal (horizon), through fluent (circuits), and finally to global (representation optimality, inner jungle, semantic roles, and global valuation patterns).

3.1 There Are Three Possible Vertex Horizons Inside the Tri-segment

One interesting starting question is that of the possible "destinies" of any of the six terms inside the tri-segment, imagining that we "walk" from any of them toward the others: seeing things with a particular vertex's eyes (this was already in Sect. 2.2, Fig. 38). As it happens (this will be shown very soon by the emerging patterns), there seem to be exactly three pairs of such possible "proximal horizons". Let us see them. The first concerns the "classical pair" of vertices (i.e., the two extremities of the red segment): $1_X2_\varnothing$ and $1_\varnothing2_X$ (i.e., two terms centrally symmetric in the hexagonal or octahedral model of the tri-segment) (Fig. 60).

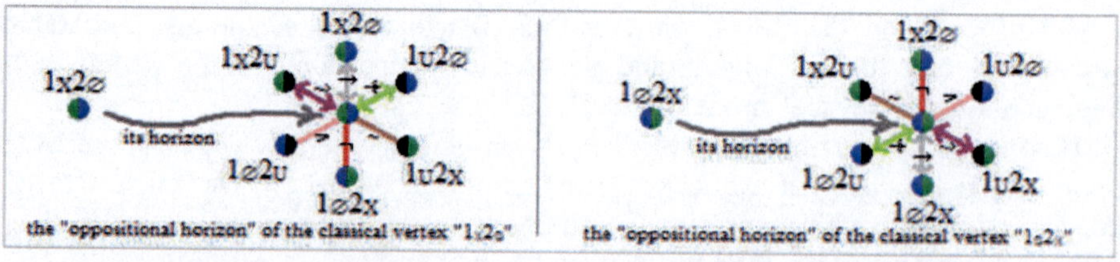

Fig. 60 Establishing the "oppositional horizon" of the classical vertices $1_X2_\varnothing$ and $1_\varnothing2_X$

The "vision" these two vertices have (i.e., their horizon) is exactly the same (i.e., the two are "chromatically congruent"), *modulo* (i.e., provided) a rotation of 180° of the two hexagonal patterns in their 2D plane (including, in this rotation of the global colored structure, even the colored vertices, but not the algebraic expression of the numerical sub-sheaves) (Fig. 61).

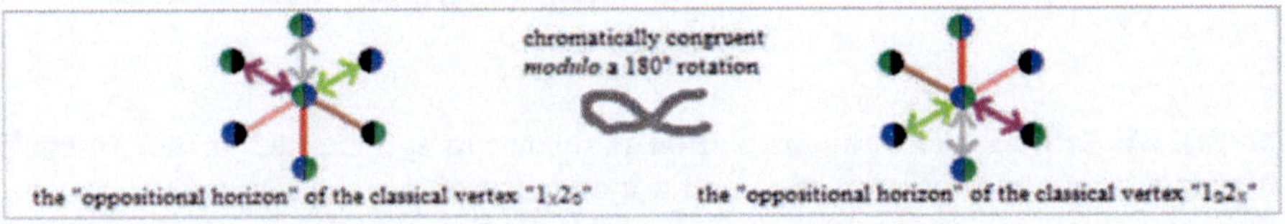

Fig. 61 These two different hexagonal patterns of "oppositional horizon" are in fact equivalent *modulo* a 180° rotation

The second possible hexagonal pattern of oppositional horizon concerns a second pair of vertices, the one expressing the Angot-Pellissierian nonclassical (and centrally symmetrical) numerical sheaves 1_U2_X and 1_X2_U. Again, their two oppositional horizons (i.e., the sets of oppositional colors they have to cross, by an oppositional segment, in order to access to any of the six vertices – i.e., including the possibility of accessing to themselves!) are in fact the same hexagonal pattern of horizon, *modulo*, again, a rotation of 180°: the two horizons are chromatically congruent (Fig. 62).

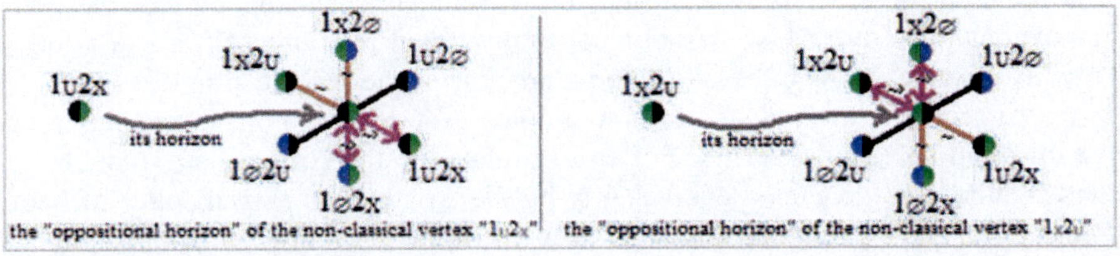

Fig. 62 The "oppositional horizon" of the vertices 1_U2_X and 1_X2_U has the same pattern, *modulo* a rotation of 180°

Finally, the third possible kind of oppositional horizon concerns the pair of the last two vertices (over the six) of the tri-segment: those expressed by the nonclassical (and centrally symmetrical) numerical sheaves $1_\varnothing2_U$ and $1_U2_\varnothing$. Again, here also the two horizons are in fact the same, *modulo* a rotation of 180° (Fig. 63).

Remark, again, that, in the three cases (i.e., the three pairs of centrally symmetrical vertices), the 180° rotation of the pattern of the hexagonal oppositional horizon comprises also a rotation of the two-color structure of the six vertices (the only thing out of rotation are, again, the algebraic expressions of the numerical sub-sheaves of these vertices, like "1_X2_U", etc., which do not move: their bicolored points do

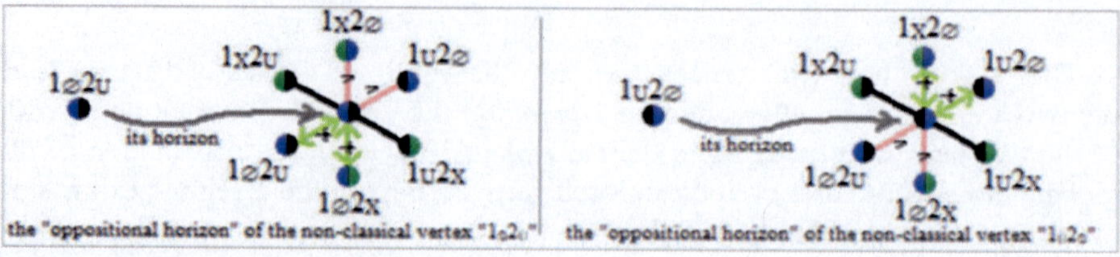

Fig. 63 The "oppositional horizon" of the vertices $1_{\emptyset}2_{U}$ and $1_{U}2_{\emptyset}$ has the same pattern, *modulo* a rotation of 180°

move). This invariance through rotation is related to symmetries in the structure (anchored, among others, in the central symmetry of the sheaves with permuted indexes, which are exactly those of the pairs, like with $1_{X}2_{U}$ and $1_{U}2_{X}$). These isometries by 180° rotation will reveal themselves quite important later (Sect. 3.6).

Having seen here what appears when a vertex is limited to "what comes next" (and that is by definition its horizon), another internal viewpoint on the complexity of this structure is, conversely, that of the possible "flows" or "inner circuits", made of arbitrarily long concatenations of oppositional segments of similar (if not identical) nature.

3.2 *Three Possible Inner Circuits of the Oppositional Tri-segment*

The principal possible "flows" inside the tri-segment, meaning by that the concatenations (also by reverse iteration) of oppositional relations of same or similar "quality", happen to be of at least three kinds: the negations, the simplexes (which, in the tri-segment, are reduced to non-simplex avatars of the black simplex), and the implications (arrows). This is of course related to the Aristotelian 3^2-lattice of the tri-simplexes (generated equivalently by the Aristotelian p^q-semantics of Sect. 1.3, Fig. 11, and by Angot-Pellissier's sheaf-theoretical method for the oppositional poly-simplexes of Sect. 1.4, Fig. 18). In fact, it respects the idea that this Aristotelian lattice is made, qualitatively speaking, of three parts: the upper triangular half (kinds of contradictions, i.e., kinds of negations), the horizontal diagonal (kinds of oppositional simplexes, taken in between classical contrariety and subcontrariety), and the lower triangular half (kinds of Smessaertian noncontradictions, classically read as kinds of subalternations, i.e., kinds of implication arrows) (Fig. 64).

So, first of all, if we now focus on "contradictions" (the classical red one as the new pink and brown ones), i.e., the three oppositional relations (among the nine) on the upper part of the Aristotelian 3^2-lattice (Sect. 1.3, Fig. 11), we find that their possible concatenations in the tri-segment constitute together a "subgraph" of the tri-segment (considered not as a solid, but as a graph, [125]), which we propose to call the "contradictions circuit" of the oppositional tri-segment (in graph theory a

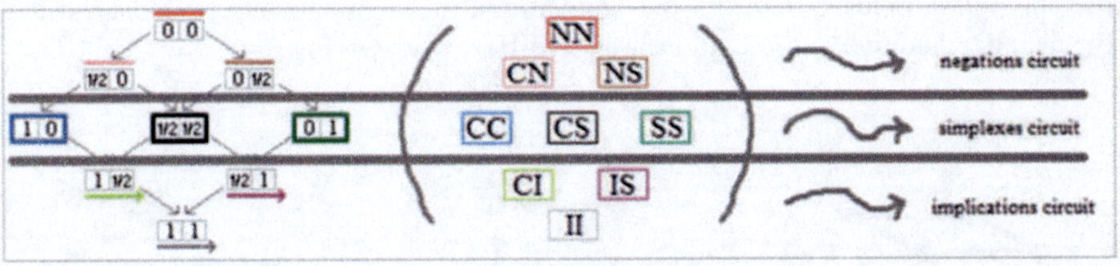

Fig. 64 The structure of the Aristotelian 3^2-lattice generates three "inner circuits" in the oppositional tri-segment

circuit is a closed line). It can be visualized equivalently in at least three different ways, if geometry is "reinjected" in this graph-theoretical structure (Fig. 65).

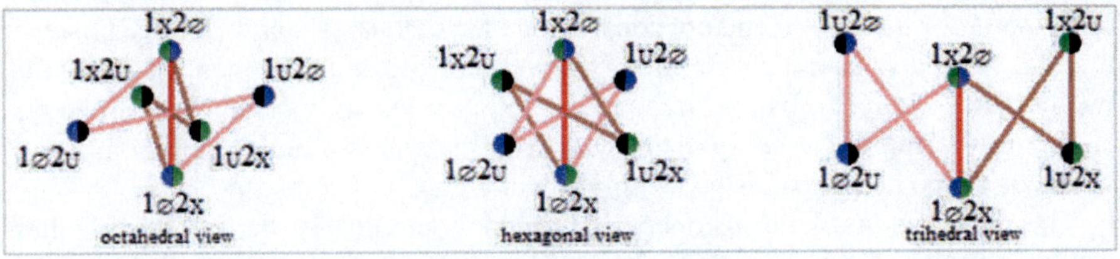

Fig. 65 Three equivalent views of the tri-chromatic "contradictions circuit" of the oppositional tri-segment

What is the use of this? Intuitively it can show, so to say, the *circulation* (hence the dynamic name "circuit" here) that "negation" (in its three different forms) can have between the six vertices of the oppositional tri-segment. As it happens, this circulation can reach all the six vertices, but not through all "passages" (i.e., segments) and not everywhere, in the circuit, by any of the three possible varieties of contradiction (classical, paracomplete, paraconsistent). The importance (if any) of this still *experimental* characterization, that will already play some role in Sect. 3.5, might (and should) become clearer in future studies of higher oppositional poly-*segments* but also, *a fortiori*, from that of any other higher *poly*-simplex. The main idea is that poly-simplexes (poly-triangles, poly-tetrahedra, etc.) grow very fast in mathematical complexity, so this kind of "index" or parameter (inner circuits) can help "navigating" conceptually (without drowning) in this otherwise *discouraging new ocean* of still mysterious shapes.

The second kind of circuit is given by *the global structure of the simplicial relations* in the oppositional tri-segment. As we saw, two of these three simplicial oppositional colors of the *general* tri-simplex (i.e., the blue and the green one) do not emerge at all in the tri-*segment* (Sect. 2.2, Fig. 39). So this "simplicial circuit" here will be not tri- but monochromatic (and more precisely black – to give an idea, in the quadri-segment it becomes bichromatic). It might therefore be called, rather, the "pivotal-simplex circuit" of the oppositional tri-segment. As previously, it can be visualized equivalently in at least three different ways (Fig. 66).

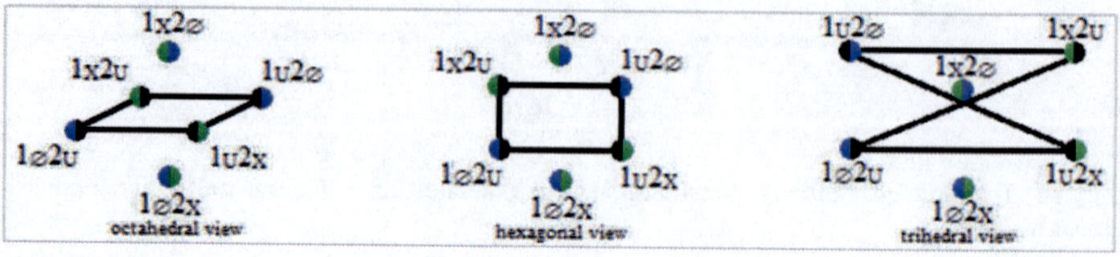

Fig. 66 Three equivalent views of the monochromatic "neo-simplicial circuit" of the oppositional tri-segment

From a graph-theoretical point of view, this subgraph, as the previous one, is a real *circuit*, for it is *closed* and not constituted of *disconnected parts*. Notice however that this time it does not reach all the six vertices of the tri-segment (it misses its two classical vertices). Its intuitive "meaning" (not yet fully clear) should become clearer in the future (in comparison with what happens in higher poly-simplexes, where its shape complexifies growingly fractalwise).

The third and last kind of inner tri-simplicial "circuit" is that of the "subalternations circuit" of the oppositional tri-segment. Remark (Sect. 2.2, Fig. 40) that Smessaert [135] has clarified that in place of subalternation (i.e., implication) one should in fact read "noncontradiction" (the latter being so to say the true "top-down *symmetrical*" *of* "contradiction" in the Aristotelian 2^2-lattice); but, as recalled, subalternation (implication) emerges on that basis as a useful and legitimate restriction of noncontradiction (by it the nondirectional "noncontradiction" relation becomes either directional or bidirectional, Sect. 2.2) and carries as such more interesting mathematical properties (the conditional or the biconditional, classical or nonclassical). So we will continue, with respect to the lower half of the 3^2-lattice, to speak of *implications* (of three different kinds) not forgetting however that these can (and must) also be seen, in some more abstract and general contexts, as three varieties of underlying Smessaertian nondirectional "noncontradiction". As previously, this third "circuit" can be visualized equivalently in at least three different ways (Fig. 67).

What emerges here is that this three-colored graph (gray, light green, violet), as such, differently from the two previous ones, is a *disconnected* one (i.e., one made of two *separated* parts, an upper and a lower one). Moreover, each of the two disconnected parts fails to be *stricto sensu* (i.e., graph-theoretically) a "circuit": each is a string (with three loops) with a noncoincident "head" and "tail". So, *stricto sensu*, it is not a *circuit* (i.e., a graph-theoretical *closed* line), unless we adopt Smessaert's discovery (related to the just aforementioned one) that the subalternations (i.e., implications) of bi-simplicial oppositional geometry (i.e., of the Aristotelian 2^2-lattice) can systematically also be read as having so to say incorporated in (or associated with) them an invisible but present "reverse

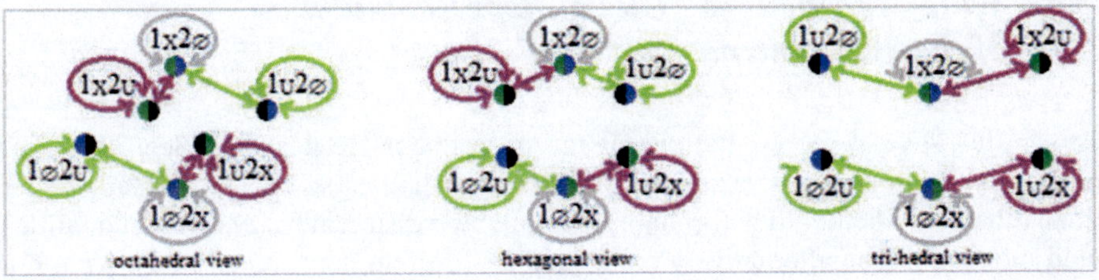

Fig. 67 Three equivalent views of the tri-chromatic "subalternations circuit" of the oppositional tri-segment

implication" (i.e., B "being implied" by A, as a reverse relation of A "implying" B, cf. Sect. 2.2, Fig. 40): for in this case (i.e., if "being implied" is also expressed graphically, near to "implying"), we can see this global graph as *two true circuits* mutually disconnected, which thing justifies (although rather trivially) keeping the term of "inner *circuit*" (Fig. 68).

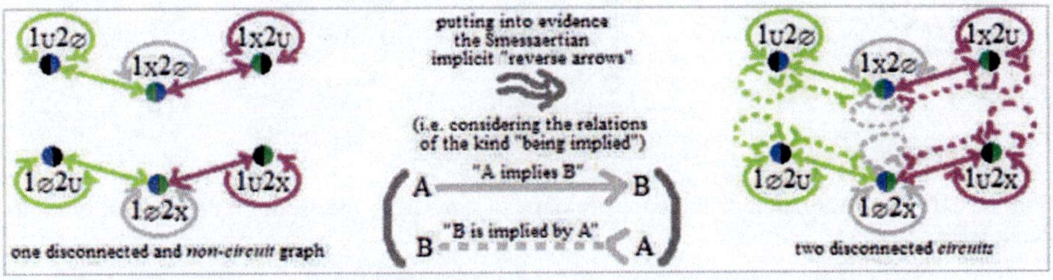

Fig. 68 The *subgraph* of subalternation (of the tri-segment) also viewed as consisting of two disconnected *circuits*

(Given that here the light green and the violet infra-implications are in fact bi-implications, i.e., they are bidirectional, the Smessaertian correlated relation will be it as well.) It will be interesting, in future studies, to examine the evolutions in pattern of this kind of circuit in other poly-simplexes (and first of all in the quadri-segment and in the tri-triangle).

Let us stress once more that these circuits, proposed by us as a new kind of hopefully meaningful oppositional-geometrical parameters of the "inner jungle" (Sect. 3.4) of the tri-segment, should become better understandable (and more clearly useful, if they will) if studied "in the long run", i.e., considering in a row this kind of features not only in the tri-segment but also in the quadri-segment, in the quinque-segment, etc. (as it happens, some of our ongoing still unpublished draft investigations on higher poly-simplexes already seem to confirm and to establish clearly the robustness of this here only conjectured point).

Having explored some inner patterns, let us now turn back to a comparison of the main different *global* representations of the tri-segment.

3.3 Which Is "the Best" Global Representation of the Tri-segment?

Given that several representations of the tri-segment are possible (Sect. 2.6, Fig. 58), the question arises relatively to knowing whether some of them are better than others or whether they (probably) simply bear different but equivalent qualities and interests, to be alternatively privileged in different contexts of study. Roughly speaking we know (at the moment) at least three ways of expressing geometrically as a whole this structure (the tri-segment) made of six vertices. Let us summarize them.

First of all, there is a 3D representation, as we saw, by means of an octahedron (Sect. 2.6, Fig. 58). The latter can be decomposed in three, two by two 3D orthogonal, inner 2D squares (more precisions on this will be given in Sect. 3.4) (Fig. 69).

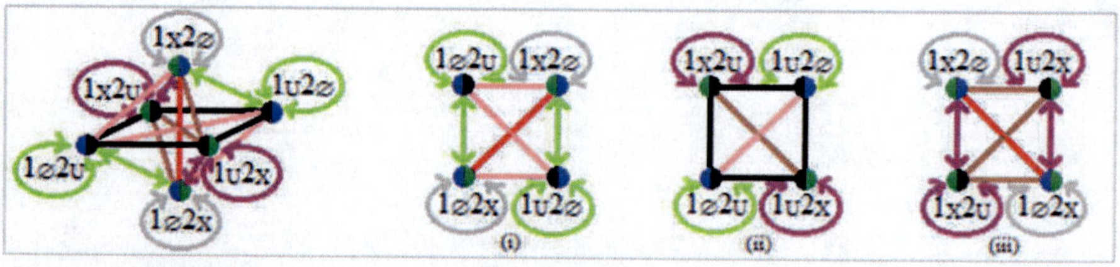

Fig. 69 The 3D expression of the oppositional tri-segment by means of an octahedron (with its three inner squares)

This octahedron-like representation helps highlighting the "oppositional diffraction" of the (here vertical) red bi-simplicial contradiction segment. This expression of the diffraction of contradiction is very symmetric: (1) with the three "negations" (red, pink, brown) in the three mutually orthogonal diagonals (which are the three 1D intersections of the three 3D orthogonal aforementioned 2D squares) (2) and with the horizontal circular "black belt" surrounding the red contradiction as a wheel with its axle. Moreover, the three 2D squares in which the octahedron can be decomposed express also by themselves some nice features, for in some sense they resemble up to a certain extent to the classical bi-simplicial logical square (Sect. 1.1, Fig. 1): in each of them their two diagonals, although different in color, are both negations, and at least four over the six vertical edges (the two light green and the two violet) are kinds of implications, but here the implications (arrows) become biconditionals (double-sided arrows, Sect. 2.2). Notice that in this octahedral representation, there is a central symmetry of the pairs of numeric subsheaves which have permutated indexes (like 1_X2_U with respect to 1_U2_X), which is a generalization that implies as a particular case the classical interpretation (as classical red contradiction) of central symmetry proper of the bi-simplexes (Sect. 2.5, Fig. 51).

Secondly, there is a 2D representation of the oppositional tri-segment, by means of a hexagon, which can also be obtained as a 2D projection of the previous (the 3D octahedron, Sect. 2.6, Fig. 58). Conversely, this hexagon can be pictured in a way (as we do here, playing knot theoretically with 3D chromatic priorities in the crossing of segments) such that it almost expresses (at least to an experimented oppositional geometer's eye) simultaneously the previous 3D octahedron. Among four possible projections of the 3D octahedron into a 2D hexagon, we choose the one which is such that the thus obtained tri-segment hexagon keeps strong enough analogies with the classical (bi-simplicial) logical hexagon (Sect. 1.1, Fig. 2) and, at the same time, it highlights the red segment of classical 2-opposition (i.e., classical contradiction) by putting it as the vertical diagonal of this hexagon. As a consequence, this "tri-segmental hexagon" can also be meaningfully decomposed into its three constitutive rectangles (as can the logical hexagon, Sect. 1.1, Fig. 2), which, as previously (with the octahedron's three squares) are partly analogous to logical squares and in fact are *fully* equivalent (i.e., have each the same four vertices and six segments) to the previous three octahedral squares (i.e., the (i), (ii), and (iii), Fig. 69) (Fig. 70).

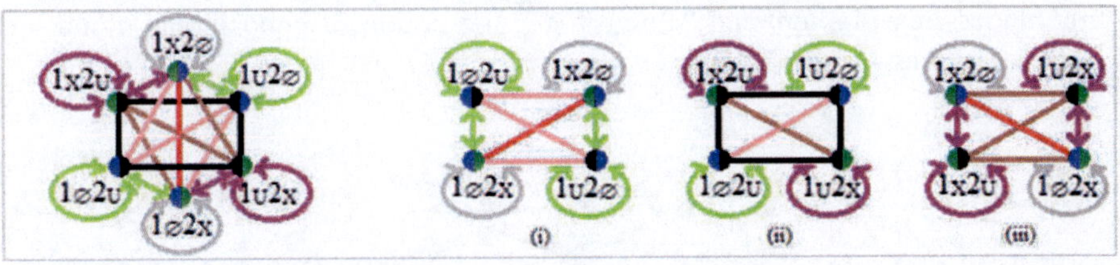

Fig. 70 The 2D expression of the oppositional tri-segment by means of a hexagon (with its three inner rectangles)

This hexagon-like representation helps (better than the 3D octahedron) in making *quick drawings* (which is very helpful to the working oppositional geometer) and as said keeps, *cum grano salis*, the good properties of the previous 3D octahedron-like representation, mainly due to the fact that it keeps the interpretation of central symmetry in terms of permutation of the indexes of the numerical sub-sheaves (e.g., in the central symmetry of 1_X2_U and 1_U2_X). In particular, we will meet gratefully enough this helpfulness when considering (in a bunch of coming other studies) higher-order poly-*segments* (and first of all with the quadri-segment and quinque-segment): when unfolding the quadri-segment, it will be very helpful to consider the tri-segments it contains as 2D hexagons.

Thirdly, there is however still another possible 3D representation of the tri-segment, by means this time of a 3D trihedron. It can be remarked that this is the same *geometrical* shape (but not the same *oppositional*-geometrical shape!) as that of our (mistaken) 2009 tentative representation of the tri-segment (Sect. 1.6, Fig. 30) (Fig. 71).

This trihedron-like representation also provides some kind of help at a fundamental level. It loses the nice property of the central symmetry of the permutations

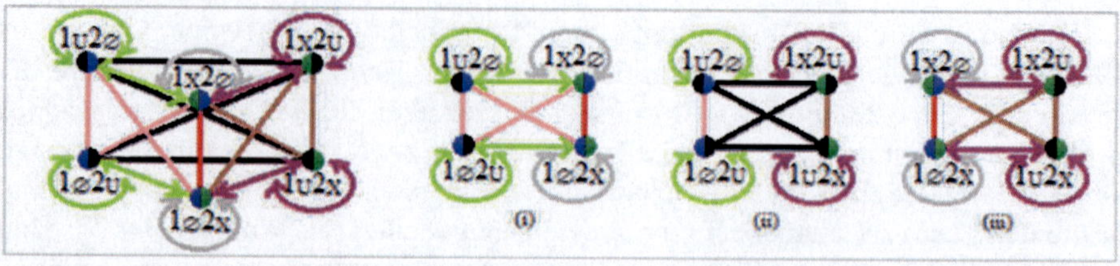

Fig. 71 A 3D expression of the oppositional tri-segment by means of a trihedron (with its three lateral rectangles)

of indexes in the numerical sub-sheaves, but, if one adopts a left-right reversal of each of the three top vertices (i.e., “1_J” and “2_K” switching in “1_J2_K” so to give the equivalent “2_K1_J” – and reversing accordingly the corresponding chromatic representation), it gains the advantage of expressing visually the deep relation existing between the tri-segment as a whole and its originating abstract horizontal 2D section of the Pascalian 3D simplex for tri-simplexes (Sect. 2.5, Fig. 53), by being now *visibly* isomorphic to it and thus giving a deeper visual intuition of the three underlying bi-simplicial “clouds” (for the notion of oppositional cloud, cf. Sects. 1.2, 2.5 and 5.1) (Fig. 72).

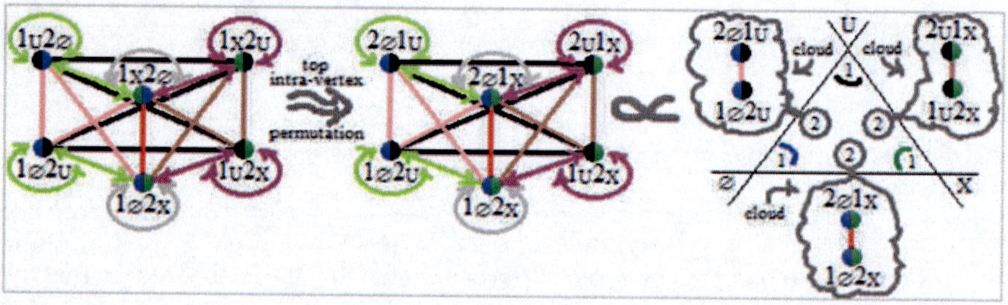

Fig. 72 The trihedral 3D representation of the oppositional tri-segment expresses something of its Pascalian 2D map

Summing up, these three possible global representations of the tri-segment seem to be fundamentally equivalent but bear different geometrical “flavors” and therefore favor different *visual intuitions*, all potentially useful. Having discussed the geometry of the possible shapes of the global structure of the tri-segment, let us now return to a deeper (but still elementary!) look on some of its possible geometrical *sub*structures.

3.4 *Inner Jungle: Possible Hybrid Substructures of the Tri-segment*

We claimed in [101] (and still claim here) that oppositional-geometrical substructures, even "hybrid" (i.e., even infra-simplicial, chromatically irregular), are interesting in the bi-simplicial oppositional geometry: we studied it first of all relatively to the complexity of what we proposed to call the "arrow-hexagons" of B4 (a new kind of generalization of the concept of logical hexagon). For instance, the useful notion of "oppositional shadow" (which appears in B4, e.g., with the counterpart, hybrid, of the non-hybrid B3 "hexagon of linear order" in the non-hybrid B4 "tetrahexahedron of partial order", Sect. 1.2, Fig. 7) can be conceptualized and studied only once one has a rich, methodical, and exhaustive typology of such hybrid substructures (the goal being of having, as tools, chromatic "markers" for describing oppositional *transformations*). The same seemingly goes for studying "oppositional-geometrical *operations*" on the bi-simplicial space: the result of several such operations (i.e., combinations of previously unrelated oppositional structures, leading then to new ones) is characterized, as by oppositional markers, by very regular hybrid structures, which are like "chromatic signatures" of various kinds of oppositional dynamic phenomena. All these things should help in the future, in building the new study of a whole "oppositional *dynamics*" (Fig. 73).

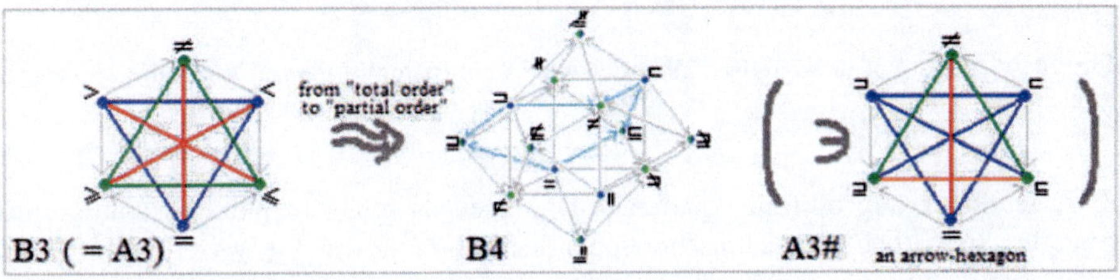

Fig. 73 Oppositional dynamics of the order relations: a B3 becomes B4, leaving a "shadow" of itself inside the B4

Now, there is no reason (other than fear of complexity) not to study this also in the oppositional-geometrical space of the poly-simplexes. And oppositional-geometrically speaking, there seem to be at least two main substructures in the "inner jungle" of the tri-segment (be it expressed as a 2D hexagon, a 3D octahedron, or a 3D trihedron, Sect. 3.3): polychrome (i.e., hybrid) *squares* and *triangles* (the study of its *segments*, even concatenated, has already been quickly evoked, Sect. 3.1 and 3.2, and will be left aside here).

Starting with the squares, we have already evoked three "regular" ones (the (i), (ii), and (iii), Sect. 2.6, Fig. 59, and Sect. 3.3, Figs. 69, 70, and 71) under two different but equivalent presentations (i.e., as contained, respectively, in the hexagon/octahedron or in the trihedron). But in the tri-segment, there are several other oppositional-geometrical squares, which are even more irregular (i.e., hybrid):

in fact, mathematically speaking any possible 4-tuple of vertices, among the six vertices of the oppositional tri-segment, can (and must) be considered an instance of such concept of hybrid square: so, combinatorially speaking, there are exactly $C^4{}_6 = 6!/(6-4)!4! = 15$ of them. So, where are (and how do they look like) the remaining 12 squares? We could view them on the trihedron: three are extensions of its "roof triangle" (adding a basement vertex to it), three are extensions of its "basement triangle" (adding a roof vertex to it), and the last six are obtained by combining two vertices of the roof with two vertices of the basement (avoiding the three cases where this gives back the three squares (i)–(iii)). But deriving our missing 12 hybrid squares from the trihedron would break any central symmetry of the numerical sub-sheaves with permuted indices in the so obtained expression of the squares. So it is preferable to derive them from the octahedron, which expresses central symmetries: this property is partly inherited by its parts (among which the squares). In it, a first group of four squares is visible on the upper half of its surface (Fig. 74).

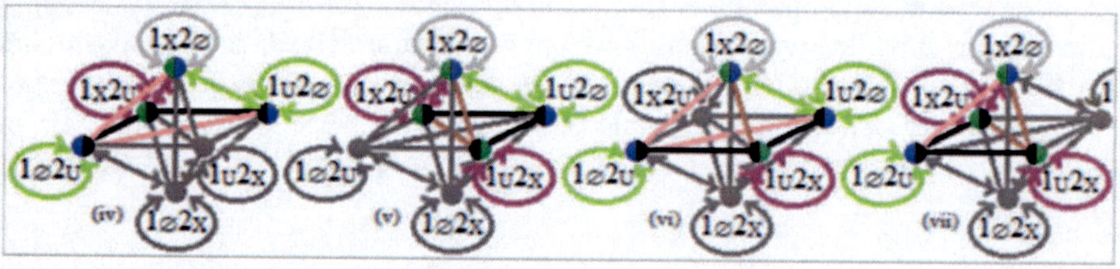

Fig. 74 Viewing 4 of the missing 12 hybrid squares on the top of the surface of the octahedral tri-segment

A second group of four squares can be seen as made of pairs of contiguous triangles such that one is on the upper half and one on the lower half of the octahedron's surface (Fig. 75).

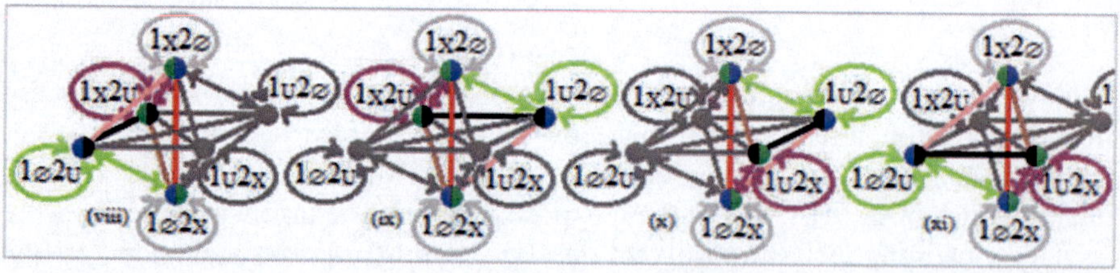

Fig. 75 Viewing 4 of the missing 12 hybrid squares on the vertical quarters of the octahedral tri-segment

And a last group of four squares can be seen on the lower half of the octahedron's surface (Fig. 76).

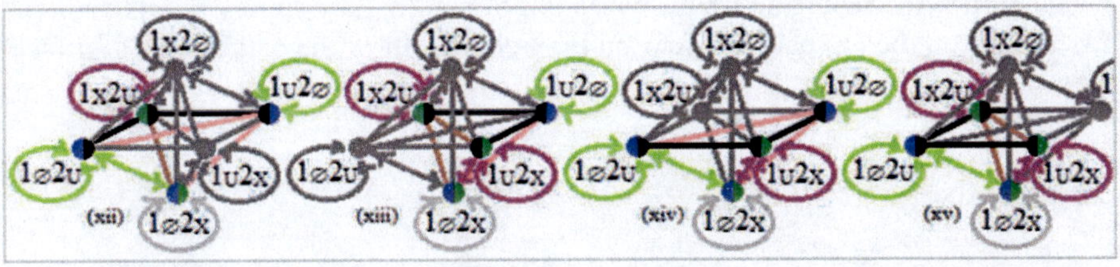

Fig. 76 Viewing 4 of the missing 12 hybrid squares on the bottom of the surface of the octahedral tri-segment

Typologically speaking, the three first squares ((i), (ii), and (iii), Sects. 2.6 and 3.3) are such that all their four vertices are two by two centrally symmetric (this regularity makes them so to say not hybrid). Differently, the 12 more hybrid squares deserve their epithet because they have only two among their four vertices which are centrally symmetric. Accordingly, these 12 can be viewed as forming 3 groups: a first group of four ((viii), (ix), (x), and (xi)) where the unique central symmetry is expressed by one red diagonal, a second group of four ((iv), (vi), (xii), and (xiv)) where the unique central symmetry is expressed by one pink diagonal, and a third group of four ((v), (vii), (xiii), and (xv)) where the unique central symmetry is expressed by one brown diagonal (it seems to be better to represent them rather as lozenges) (Fig. 77).

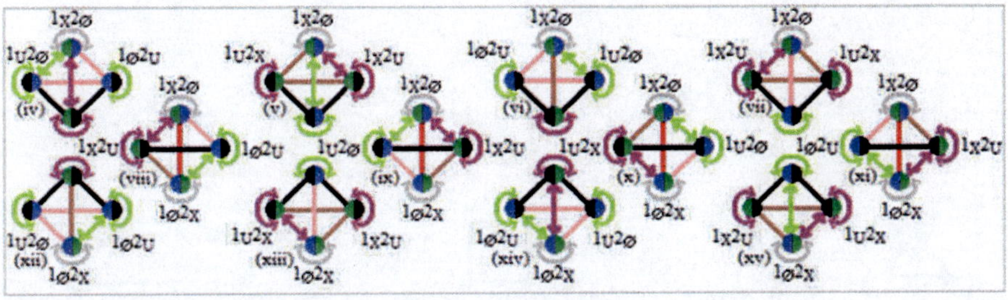

Fig. 77 The 12 broken squares (seen as lozenges) of the octahedral tri-segment

Moreover, these 12 hybrid squares are such that they divide into 6 pairs of chromatically isomorphic squares, and all such pairs of identical squares are centrally symmetric inside the octahedron (beware: their vertices undergo an *indicial* permutation). So, globally there are three different "normal" squares. And then there are six pairs of centrally symmetric hybrid squares: all in all nine different chromatic kinds of squares. These elements might be studied in the future (as "jungle") relatively to effects of oppositional shadow of the tri-segment with respect to higher tri-simplexes containing tri-segments.

If we now go to the triangles, a similar combinatorial abstract calculation as the previous tells us that there are $C^3{}_6 = 6!/(6-3)!3! = 20$ of them. Again, the octahedron helps seeing straightforwardly 8 of the 20: they simply are its eight 2D faces (four on the top half and four on the bottom half of the octahedron's surface) (Fig. 78).

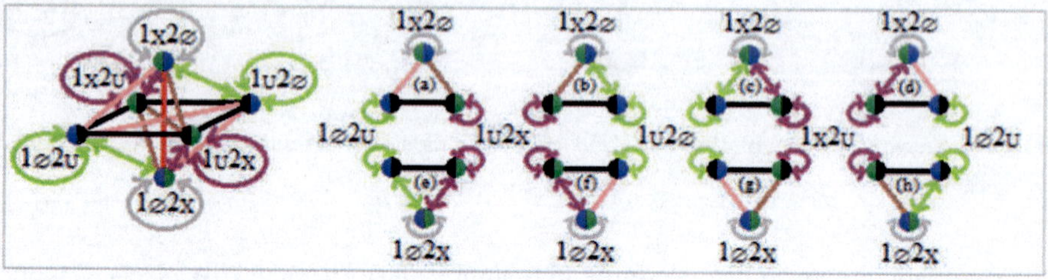

Fig. 78 Eight of the twenty chromatically hybrid (i.e., infra-simplicial) triangles of the oppositional tri-segment

Remark that these eight "surface triangles" are two by two chromatically isomorphic, and this concerns the triangles which are centrally symmetric (the central symmetry of the whole entails the mutual central symmetry of some of its parts – here as well, remember however that their respective vertices undergo an *indicial* permutation). The other 12 triangles over the 20 can be seen easily enough in the octahedron's 3 inner squares ((i), (ii), and (iii)): for there are four triangles in each of these three squares. If these three squares are seen as tetrahedra, the four triangles in each square are the four triangular faces of the equivalent tetrahedron (Fig. 79).

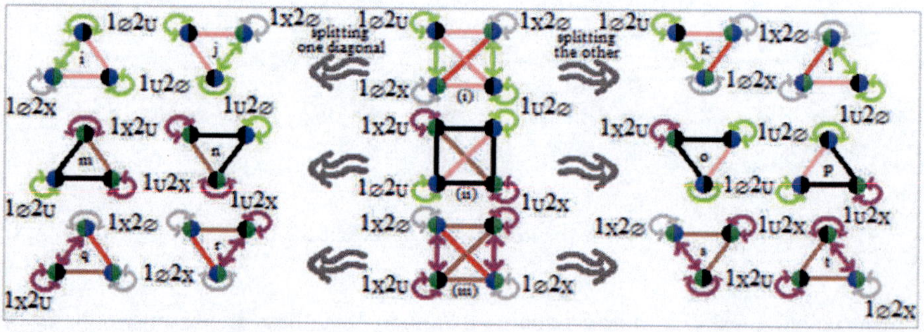

Fig. 79 Twelve chromatically hybrid (i.e., less than regular) triangles of the oppositional tri-segment

Remark that because of the full diagonal central symmetry of the global structure of each of the 3 squares (i)–(iii), these 12 triangles obtained by splitting in 2 ways each of the 3 squares (i)–(iii) let emerge only 6 different chromatic kinds of triangle (the pairs of centrally symmetric such triangles are, here as well, those chromatically isomorphic, *modulo* the reversed indices of their vertices).

As for the final typology of the 20 triangles, their chromaticity can be stressed either relatively to the edges or relatively to the curls. Eight of the 12 triangles inscribed in the 3 central squares (i)–(iii) of the octahedron (i.e., the i, j, m, n, o, p, s, t) are bichromatic with respect to edges, and 4 (i.e., the k, l, q, r) are tri-chromatic. The eight triangles of the octahedron's surface let emerge four different chromatic kinds of triangle (a and g, b and h, c and e, and d and f; Fig. 78). So, all in all in the tri-segment there are ten different pairs of isomorphic triangles such that four kinds are bichromatic and six are trichromatic with respect to edges. As for the curls, the 8 triangles (a–h) are tri-chromatic (due to the lack of central symmetric vertices), while the 12 triangles (i–t) are bichromatic (due to the presence, in each of these triangles, of two vertices centrally symmetric) (Fig. 80).

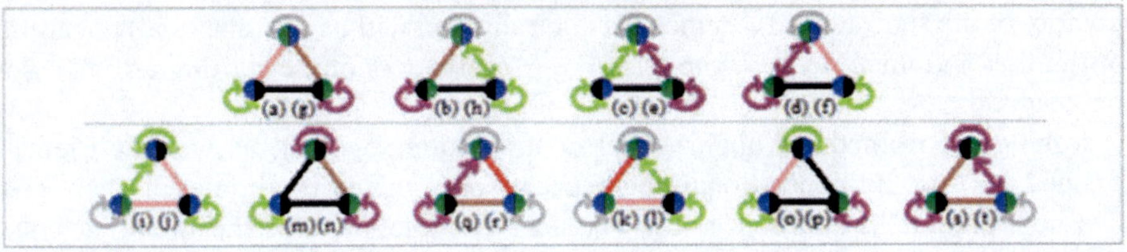

Fig. 80 General typology of the 20 inner triangles of the oppositional tri-segment: there are 10 kinds of them

We cannot say more here. But we will find some use of this already in the next Sect. 3.5: as we are going to see, at least 2 of these 30 triangles (centrally symmetric) seem to be particularly meaningful for the tri-segment taken as a whole.

3.5 *How to Put "Semantic Values" on the Vertices of the Tri-segment?*

We can face now a very important part of our global study of the concept of tri-segment: the question of *meaning*. Clarifying this is absolutely necessary in order to be able to apply this structure to something concrete whatsoever. Remark that the structural or *differential* game (in the sense launched by the structuralist linguist Saussure, cf. [112]) of the oppositional structures is in general so made that it helps *by itself* in building the meaning, by means of the very *system of the oppositions*. But here, with the tri-segment, we are handling very minimal conditions of "opposition" (we have no contrarieties! Cf. Sect. 1.6, Fig. 33), so that the meaning involved seems to have to be quite subtle (related to *varieties of contradiction*) and in that sense harder than usual to figure out.

More precisely, we are looking for something like the "semantic values" of the decorations of the six vertices of the tri-segment. Let us give a concrete example of the problem: given, as our starting point, the semantic value (or meaning) "white" (or, if you prefer, "giraffe", or "running" or anything else), what can (and in fact must) be the semantic values of the other five vertices, so that *together* they build up (without incoherence) an oppositional tri-segment? Remark that the answer seems to be composed of two qualitative halves: three over the six values of the tri-segment (i.e., three over its six vertices) are expected to be more or less "assertoric" (they "affirm"), while the other three are more or less "negative" (they "deny"). This seems clear with respect to the starting point, the classical bi-simplicial (red) segment N of 2-oppositional contradiction (of which the tri-segment is supposed to be a ternary "oppositional diffraction"): it relates two polarities, such that each is the negation of the other, but in a way such that in general (the starting) one is *concrete*, while the other (i.e., the negation of the starting one) is *vague*. So "white" or "giraffe" (or "running" or even "$2 + 2 = 5$" or whatever other possible starting meaning) will be a concrete, non-vague semantic starting point, while the other polarity of the starting (red) segment of contradiction will be the (classical) negation of the first and therefore a vague term (i.e., "all that is not the starting term"!). So far, so good.

Taking off from the 2-oppositional contradiction segment, as we saw (Sects. 1.3 and 2.2), in the oppositional-geometrical space of the tri-simplexes, there are five new colors (in addition to the classical Aristotelian 4 of the bi-simplexes). Putting aside the black one (the new simplex, which is pivotal), the four other new colors, according to the Aristotelian 3^2-lattice (generated equivalently by the Aristotelian 3^2-semantics or by Angot-Pellissier's sheaf-theoretical method), are expected to represent (1) two new forms of contradiction and (2) two new forms of (Smessaertian) noncontradiction (in fact interpretable as implications, Sect. 2.2, Fig. 40 and Sect. 3.2).

Now, as demonstrated by Angot-Pellissier [3], one of these new negations (the pink CN one) is "paracomplete" (it drops "completeness", i.e., it is "intuitionist", and it defies the principle of the excluded middle, by producing situations where you have truth-value "gaps", i.e. holes): it therefore represents a form of negation in some sense *stronger* than the classical one (it goes so far that it so to say "tears" the truth-theoretical space, cf. left side of Fig. 81), such that it is not "involutive" (i.e., with an intuitionist "NoT" negation operator, the formula or meaning "NoT NoT A" is not equivalent, in general, to "A"). The other of these two new negations (the brown NS one), as demonstrated, again by Angot-Pellissier [3], is "paraconsistent" (it drops "consistency", i.e. it is "co-intuitionist", and it defies the principle of noncontradiction, by producing situations where you have truth-value "gluts", i.e., truth-value superpositions of "1" and "0", cf. right side of Fig. 81): it represents a "negation" weaker than the classical one (and *a fortiori* weaker than the paracomplete one), so weak in fact that it might *seem* not to be a negation (cf. Slater, Sect. 1.1) (Fig. 81).

So, in the tri-segment, starting from one of the two |blue-green| vertices (Sect. 2.5, Fig. 51) of its red segment NN of classical 2-oppositional contradiction (say: the green-blue $1_X 2_\varnothing$), taken to be meaning "white" (or "giraffe", "running", etc.), one pink CN segment (of paracomplete negation) leads us therefore to a blue-black vertex ($1_\varnothing 2_U$) meaning "NoT-white" (or "NoT-giraffe", "NoT-running", etc.) and

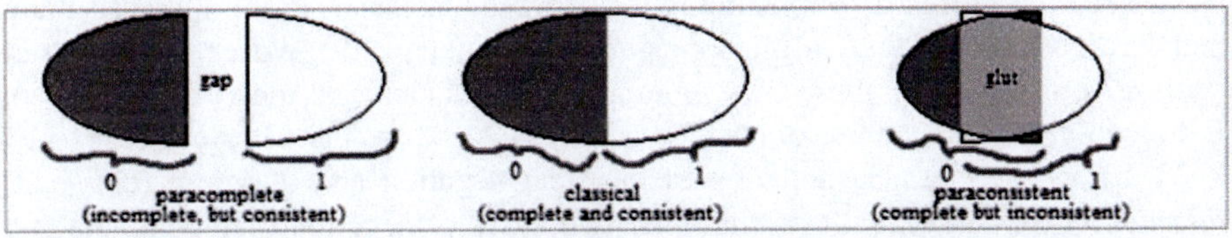

Fig. 81 Negations and truth-values: paracomplete (gaps), classical paraconsistent (gluts)

generating the possibility of a gap (at Angot-Pellissierian level X-U). And starting from the same green-blue classical vertex (i.e., $1_X2_\varnothing$), one brown NS segment (of paraconsistent negation) leads us this time to a black-green vertex (1_U2_X) meaning "nOt-white" (or "nOt-giraffe", "nOt-running", etc.), generating the possibility of a glut (at Angot-Pellissierian level U). Summing up, we thus have so far one starting vertex which expresses a positive meaning and three other vertices, related to it by a red, a pink, and a brown segments, respectively, that express, each, one of three kinds of negation (a classical, a paracomplete, or a paraconsistent one) of the starting concept (or meaning). Remark that in the octahedral tri-segment, together these three vertices, expressing negations (of our starting vertex $1_X2_\varnothing$), form a surface triangle (the "e" in the sense of Sect. 3.4, Fig. 78). What are the mutual relations of these three "negation vertices" in this triangle "e"? The two double arrows CI and IS express a weak form of equivalence between, on one side, the paracomplete and the classical negations of the starting term and, on the other side, the classical and the paraconsistent negations of the starting term. The mutual relation of the two vertices expressing nonclassical negations of the starting meaning is less evident: all we know so far is that the black CS segment means, as its name "CS" says, contrariety at level X-U and subcontrariety at level U.

What said so far covers, relatively to the expression of meaning, four over the six vertices of the tri-segment and the six edges between them: a tetrahedron, or square, the "xi" (Sect. 3.4, Fig. 75). So, what about the remaining two vertices, the numerical sub-sheaves 1_X2_U and $1_U2_\varnothing$? This can be approached in at least two ways: (i) starting from $1_X2_\varnothing$ (as previously) (ii) or starting, this time, from $1_\varnothing2_X$. If we continue starting from $1_X2_\varnothing$, the two remaining vertices are directly accessible from it by two nonclassical forms of biconditional, IS and CI, respectively. This suggests that *the two remaining vertices can be seen as partly equivalent to the starting one*. More precisely, $1_U2_\varnothing$ is "CI-equivalent" (i.e., "paracompletely equivalent") to $1_X2_\varnothing$ at level U, whereas 1_X2_U is "IS-equivalent" (i.e., "paraconsistently equivalent") to $1_X2_\varnothing$ at level X-U. As for the mutual relation of these two assertions (each represented by one of these two vertices), it is expressed by the black segment CS: again, between the two there is contrariety at level X-U and subcontrariety at level U.

Three things at least can be seen at this stage: (1) the Aristotelian 3^2-lattice suggests that they are so to speak "the (vertically) symmetrical" of negations (i.e., they are – as II, CI, and IS – in the lower triangular half of this 3^2-lattice), and, as said, under some "Aristotelian" circumstances (Sect. 2.2, Fig. 40), they even can be read as forms of implications (so is it, for instance, in the logical square and hexagon and in fact in all oppositional geometry). (2) With respect to our starting vertex ($1_X2_\varnothing$), these two remaining vertices can (and must) be read (also, if not only) as "negations of negations (of $1_X2_\varnothing$)". In the classical case (i.e., "¬") this is the classical negation of classical negation and therefore (classical) *affirmation* (or "assertion"). In the other two cases of negation (paracomplete and paraconsistent), this seems to lead, again, to *other varieties of "affirmation"* (an affirmation with a gap and one with a glut, Fig. 81). We thus have three kinds of classical and nonclassical "affirmations (of the starting $1_X2_\varnothing$)". (3) If one reads this centrally symmetric octahedral (or hexagonal) tri-segment "the other way round", i.e., starting from the vertex "$1_\varnothing 2_X$", taken (classically) as meaning the (classical) "affirmation of 'the (classical) *negation* of white'" (or of "giraffe", of "running", etc.), the two vertices up to now somehow mysterious, 1_X2_U and $1_U2_\varnothing$, must be read (because of the reasoning above and because of the central symmetry of the structure) as two nonclassical negations of it. So, they are two nonclassical negations of the classical (starting) negation ($1_\varnothing 2_X$), therefore they are two nonclassical affirmations of the starting vertex $1_X2_\varnothing$, and together with $1_X2_\varnothing$ itself they therefore seem to be *three forms of "assertion" (or noncontradiction)* of the starting meaning "white" (or of "giraffe", or of "running", etc.). Remark that together the three vertices form a "triangle of assertions" (of the starting meaning): more precisely, the triangle "c" in our notation of Sect. 3.4, Fig. 78, which is centrally symmetric to the previous "triangle of negations", i.e. "e". The 3-oppositional relations of this triangle "c" (of assertions) are, again, CI, IS, and CS. So, in some sense we have reached here two triangles, a "triangle of negations" and a correlated (and centrally symmetric to it) "triangle of assertions" (this is a first example of the apparent usefulness of hybrid substructures, Sect. 3.4). Similarly, the tetrahedron (or "square") "ix" (Fig. 75) is the structure dealing specifically with the vertex $1_\varnothing 2_X$ and its three possible negations (Fig. 82).

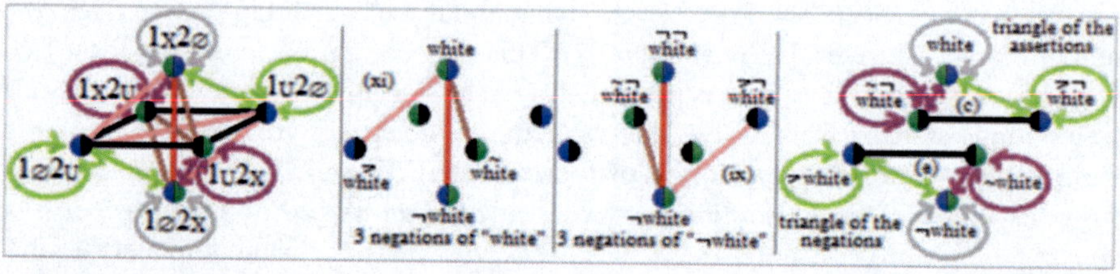

Fig. 82 The basic intuition over the fundamental "semantic values" of the oppositional tri-segment: two triangles

So, remark also that the tri-segment's inner squares reveal potentially meaningful as well in an additional way: this can be seen if one concentrates on the meaning of the three non-hybrid squares (i, ii, iii, Sect. 3.3, Fig. 69). The two vertical squares (of the octahedral tri-segment), i.e., the "i" and the "iii", seem to embody, respectively, the relations between (1) classical negation and paracomplete negation (i) and (2) classical negation and paraconsistent negation (iii). As for the horizontal square (ii), it seems to embody the relations between the four nonclassical meanings: (1) between paracomplete either affirmation or negation *and* paraconsistent affirmation or negation (the four black CS edges), (2) between paracomplete affirmation and paracomplete negation (pink diagonal CN), (3) and between paraconsistent affirmation and paraconsistent negation (brown diagonal NS).

This distribution of the six vertices into two groups of 3 (i.e., the triangles "c" and "e") seems to be *semantically stable*. The "stability" of this distribution may become clearer when one concentrates on the possible (virtual) iterations of negations in the tri-segment (and here we can see somehow at action, in its conceptual potential usefulness, the "circuit of contradictions", Sect. 3.2) (Fig. 83).

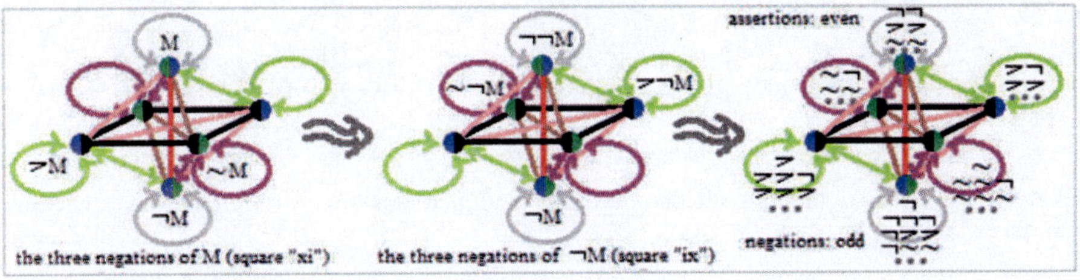

Fig. 83 The possible concatenations of iterations of the three kinds of negations reach a stable "semantic distribution"

The "triangle of assertions" (Fig. 82), with respect to any of its two possible semantic starting points (i.e., $1_X 2_\varnothing$ or $1_\varnothing 2_X$), contains three numerical sub-sheaves which have, prefixed to the semantic starting point, always an *even* number of negation signs (Fig. 83); conversely, the "triangle of denials" (Fig. 82) contains three sub-sheaves such that in each there is, prefixed to the semantic starting point, always an *odd* number of negation signs (Fig. 83). This seemingly distributes meaning over the tri-segment, in a way such that each of classical, paracomplete, and paraconsistent logic/mathematics has two *loca*, a positive (assertion) and a negative (denial) one.

Having already dealt with quite many aspects of the concept of oppositional tri-segment, one last point at least remains nevertheless to be treated before being reasonably able to start using tri-segments "for real" (in applications): understanding how they can be decorated with truth-values, that is, "valuated".

3.6 *Which Possible "Truth-Valuations" of the Global Tri-segment?*

One last crucial problem, seemingly, is that of determining how do function "valuations" (i.e., the attribution of truth-values) for the oppositional tri-segment.

Recall that for a structure of opposition, its strength comes also from coherence in that respect. And that is what works for the logical square, as well as for its many avatars or even its few components: they are useful since they admit coherent patterns of global assignments of truth-values. In order to find a solution for the problem of valuating the tri-segment, which seems to require to reduce combinatorial complexity, as we are going to recall, the inspiring image seems to be the familiar situation with the *bi*-simplexes. For here, *empirically* (cf. Mandelbrot [88]!), there seems to be something like a "law of the two hemispheres" of valuation: one-half of the valuations will be "0"; the other will be "1". This can be seen in the 1D space with the oppositional segment (B2) and in the 2D space with the oppositional hexagon (B3) (Fig. 84).

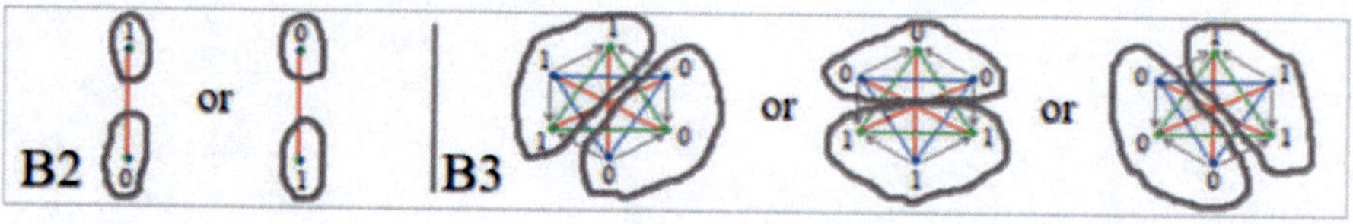

Fig. 84 "Hemisphere theory": the bi-simplexes' (and their closures') *valuation* always cuts their "surface" into two

This hypothetical "*hemi*spheral" behavior of the valuation of the *bi*-simplicial structures can also be seen in the 3D space with the oppositional tetrahexahedron (B4) (Fig. 85).

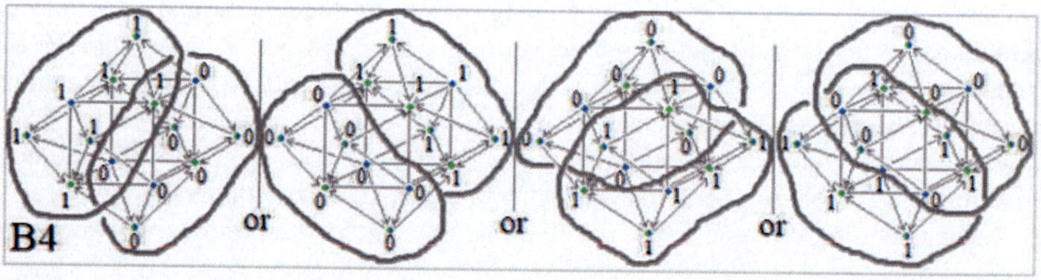

Fig. 85 The four possible valuations of the oppositional tetrahexahedron (the B4, the "closure" of the bi-tetrahedron)

Intuitively, this behavior *seems* to be general for the bi-simplexes and their closures: all the vertices of any blue simplex will have to be valued "0", except one (let us call it the "oppositional hostage"), which will have to be valued "1"; conversely, all the vertices of the correlated, centrally symmetric green simplex will have to be valued "1", except one (also an "oppositional hostage"), which will have to be valued "0" (by construction, these two hostages, the blue and the green one, are mutually centrally symmetric). Now, the structure of the bi-simplex, by construction, is such that it lets emerge (*seemingly* independently from the dimension of the starting simplex) *a partition into two of the resulting bi-simplex's n-dimensional "surface"* (the half around the green hostage, valued "0", will have to be valued "0"; the other half, around the blue hostage, valued "1", will have to be valued "1"), and this *seems* to have to be respected also by the bi-simplex's cloud (this last conjecture is not yet proven for higher Bn, but this provisory uncertainty seemingly brings not much harm here – but this point will have, of course, to be fully clarified as soon as possible). The number of vertices of the starting simplex (i.e., its dimensionality) determines how many *rotations* of this *valuation pattern* (for that precise bi-simplex and its closure) there can be (this is due to the fact that simplexes, by construction, are symmetric with respect to *all* their vertices and therefore *any* vertex can and must play the oppositional hostage). So, the valuation of bi-simplicial oppositional-geometrical figures (including their closures) admits (i) an abstract n-dimensional "hemispheral pattern" and (ii) n concrete possibilities of having this hemispheral pattern embodied on the global Bn: two for the B2 (two vertices), three for the B3 (three vertices), four for the B4 (four vertices), etc.

But our oppositional *tri*-segment is not bi-simplicial, but it is an instance of *tri*-simplex. And, by construction, dealing with tri-simplexes means dealing with *three*-valued logic. How to conceive valuation in this case, then?

Now that we seemingly know with sufficient precision the real structure of the tri-segment (but some surprises wait for us in Sect. 4), the question is: how to handle, with it, this question of its possible global valuations? Remark that from a purely combinatorial viewpoint (six vertices admitting each in abstracto three possible truth-values), we seem to have $3^6 = 729$ different possible valuations (!). But then we must remember that the B2 has *abstractly* $2^2 = 4$ valuations, but *really* it only has 2; similarly, the B3 in abstracto has $2^6 = 64$ possible valuations, but *really* it only has 3; and similarly the B4 has abstractly $2^{14} = 16.384$ valuations (!), whereas *really* it only has 4. So, *it seems natural that oppositional structures reduce drastically, by the constraints they impose by construction, the combinatorial explosiveness*. How to reduce then, comparably (reasoning the safest we can by analogy), the combinatorial complexity of the valuations of the tri-segment? Remark that the structure "tri-segment" has a strong central symmetry, meaning by that that its centrally symmetric parts are oppositionally identical, so the possible cases seem already to be reduced at least by 2. This is not much, but it suggests that symmetries play here as well as in the bi-simplexes: so there is a robust hope of finding here as well drastic reductions of combinatorial complexity. On the other hand, from the viewpoint of *analogy* (with what is known from the bi-simplexes), we have, by construction, three possible truth-values. So, a first temptation, betting on the real existence of a strong analogy with the bi-simplicial case (to be checked now), might be to have us landing here to something like a "surface *tri*partition" (instead of the surface *bi*-partition, or "*hemi*sphere theory", we seemingly have, as we have seen, with the bi-simplexes) (Fig. 86).

This solution, in fact, seems to work. It leads to see three independent "worlds": three bi-simplexes inside the tri-segment, each *bi*-valued (with three possible truth-values *only* at the global scale, *never* in the bi-simplicial substructures of the tri-simplex). And this happens to be made very clear by the "oppositional colors" of

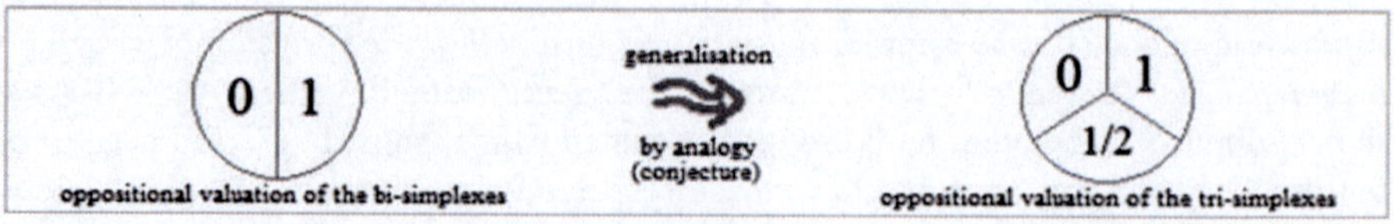

Fig. 86 An intuition on "valuation" in the tri-simplexes: from "hemisphere theory", to "sphere *n*-partition theory"

the vertices (Sect. 2.5)! The rule seems to be that each vertex, in a bi-simplicial pair (i.e., the one comprising the *cloud* where this vertex lives), can receive only either its "natural value" or its "hostage value" (i.e., the "natural value" of its co-simplex in that bi-simplex), which are precisely the two colors of the indexes of this vertex and therefore the two colors of this "point" itself! The same reasoning (with the same two colors) runs for its centrally symmetric mate, and *this gives as a major final result of our reasoning only two possible valuations of the oppositional tri-segment* (Fig. 87).

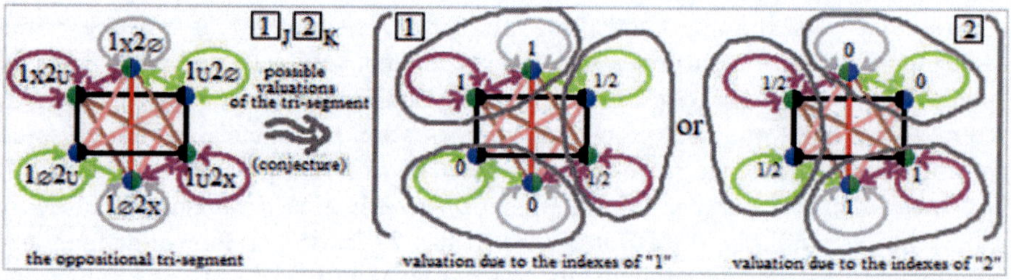

Fig. 87 Sphere *n*-partition: what seem to be the only two possible global valuations of the oppositional tri-segment

More concretely, the combinatorially natural solution to the problem of valuating the tri-segment *seems* to be that (1) |blue-green| vertices (i.e., vertices blue-green or green-blue) can only be valuated "0" or "1"; (2) |blue-black| vertices can only be valuated "1/2" or "0"; (3) |green-black| vertices can only be valuated "1" or "1/2"; (4) there is only one valuation, i.e., *only one kind of valuated pattern* (of the global tri-segment); (5) given the existence of a 2-symmetry by a 180° rotation of the 2D hexagonal representation of the tri-segment, *this valuation* (i.e., *the unique pattern) can only be "upside" or "down"*; (6) *a posteriori* this matter of fact is pretty analogous to what happens already with the opposition segment (the bi-simplicial counterpart of the tri-segment), so the solution proposed here seems to be a conservative extension; and (7) this behavior is kept in higher poly-*segments* (and in fact, *mutatis mutandis*, in higher poly-*simplexes*), so the solution proposed here seems to be a *generalizable* conservative extension. Remark that the reason why the "hostage rotation" (of the valuations) does not play here is because in the tri-segment there are not yet *simplexes* (an oppositional hostage appears as soon as, but not before, at least one simplicial *triangle* appears). Remark, then, that this seems to be a quite important new element of knowledge (if our starting conjecture holds): it seemingly rules the valuation of *any* poly-simplex! (We will handle this question as soon as we will expound, in another study, the case of the oppositional tri-triangle, of which Angot-Pellissier has opened the sheaf-theoretical exploration in [3], but without reaching its oppositional closure and without affording the question of its global valuation.)

Notice however, as well, that this valuation (Fig. 87) might – and in fact must – perplex us, in the sense that it strongly suggests to see as implication arrows some edges of the tri-segment that we have had no reason so far for seeing as implication arrows. This concerns (1) the four black CS segments, (2) the two light green CI double arrows, and (3) the two violet IS double arrows: each of these three might function – according to the two valuations – as a unidirectional implication arrow (Fig. 88).

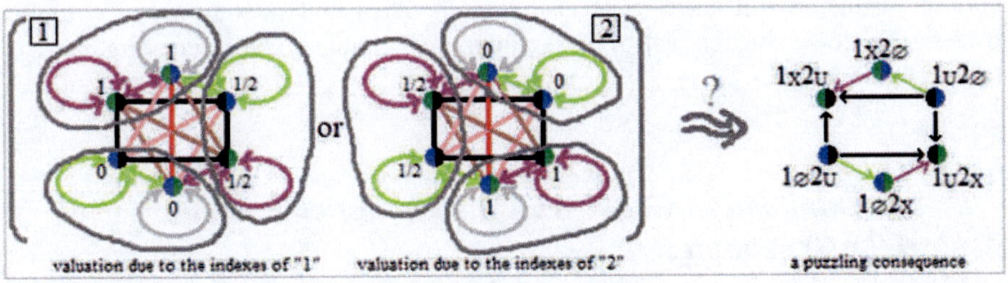

Fig. 88 Some perplexities arising from the valuation (otherwise convincing): it seems to let emerge arrows

So in some sense we have now, unexpectedly, a non-negligible new problem with "arrows". This seems to be due to the fact that our "Aristotelian strategy" (put into play in Sect. 2.2), aimed at avoiding (lazily) the "heaviness" of a potential Smessaert-like *complete analysis of implication behaviors* (at the non-Smessaertian, tri-simplicial level), possibly did not suffice: our conjecture (and our correlated lazy bet), that just looking for the implication-behavior of the edges admitting some "I" in their code might suffice for dealing, overall, with implication-like relations, was seemingly wrong (and our lazy bet is seemingly lost). So we will now face this very instructive serious problem (remember that we already faced voluntarily an instructive "laziness problem" in Sect. 2.1 and 2.4) by changing now quite radically our strategy on this point of arrows (as we changed it successfully in Sect. 2.4 with respect to trivial extrema), by having, from now on, a direct and systematic "tri-simplicial look" (not yet existing...) at the possible Smessaertian-Demeyan "*implication* geometry" of the tri-segment (Sect. 4).

4 There Is Logical Geometry *Inside* the Oppositional Tri-segment!

The result we arrived to at the end of the previous chapter is interesting, but still too puzzling with respect to *implication relations*. And it appears that the instruments used so far do not suffice yet, as we would have liked, to deal in full clarity with this massive "arrows problem". So we turn to the tools offered at the *bi*-simplicial (rather than *tri*-simplicial) level by "logical geometry" (Sect. 2.2, Fig. 40), in so far it contemplates, specifically, the existence of a whole (although small) "*implication* geometry" parallel (in Smessaert and Demey's terms) to "opposition geometry" ("opposition*al* geometry" being seen by them as a bricolage, a non-systematic and unconscious mixture of these two *scientific* characterizations, as we will reexamine in Sects. 4.5 and 4.6). So far, as long as I know, logical geometers have not yet inquired the concept of poly-simplex. Therefore we will have the honor, brave reader, you and me, to open *now* the way of this attempted "junction", right here. And this should close, at a reasonable level of understanding, our complex investigation over the concept of tri-segment, not without some interesting backfire (Sects. 4.6, 5.1 and 5.2).

4.1 *The Implication Geometry's 3^2-Semantics/Lattice of the Tri-simplex*

We want to explore the "implication geometry" (if, as we think, it exists) correlative of the opposition*al* tri-segment we investigated so far. Our "doubt" is just methodical, since this has never been done before (logical geometers have not yet taken seriously the idea of *poly*-simplex, and for a start they do not seem to favor much – euphemism – the simpler idea of *bi*-simplex). As for us, what we need for that is, first of all, generating a new kind of lattice, comparable to our game-theoretical Aristotelian 3^2-lattice (Sect. 1.3, Fig. 11), for obtaining the "*implication* kinds" of the 21 possible binary relations (edges and curls) holding between the 6 vertices of the tri-segment (Sect. 2.4, Fig. 46). Recall that at the level of the "*bi*-simplicial space" (i.e., two-valued oppositional/logical geometry), such semantics and lattice, devised by Smessaert around 2011 (Sect. 2.2), resulted of two "meta-questions", complementary of the "Aristotelian" ones (proposed by me in 2009) seen so far (Sect. 1.3, Fig. 10), but aimed at generating not "*opposition* relations" but "*implication*-like relations". Truly speaking, Smessaert investigated the strangeness of "subalternation" (i.e., the strangeness of its being, in the classical "Aristotelian quartet" (Sect. 1.1, Fig. 1), the only *asymmetric* relation), and because of his rigorous framing of this question, he discovered, probably unexpected, a whole (although small) "implication geometry" (this is a typical *structuralist* good move). Now, these two Smessaertian meta-questions are (for any pair of things A and B) (Q'1) "Is it possible to have, at the same time, A false and B true?"; (Q'2) "Is it possible to have, at the same time, A true and B false?". As one sees, they inquire not the simultaneous "truth-value *similarity*" (both false, both true) but the simultaneous "truth-value *dissimilarity*" (false the first while true the second, true the first while false the second). For short, Smessaert had the *structuralist* brilliant idea to add the study of dissimilarity to that of similarity (to a continental philosopher's eye that reminds the main methodological lesson of Plato's *Parmenides*, [113]). Now, allowing two kinds of answers, i.e., "0" or "1" (this binarity being the "*bi*-simplicial touch"!), generates four possible pairs of answers ([0|0], [0|1], [1|0], [1|1]) at these two questions (Q'1 and Q'2). These four possible "double answers" ([*x*|*y*]) distribute, similarly to what we saw with the Aristotelian 2^2-lattice (Sect. 1.3, Fig. 10), in a "Smessaertian lattice", which gives precisely the four kinds of possible "*implicative* relations" (for the bi-simplicial space) (Fig. 89).

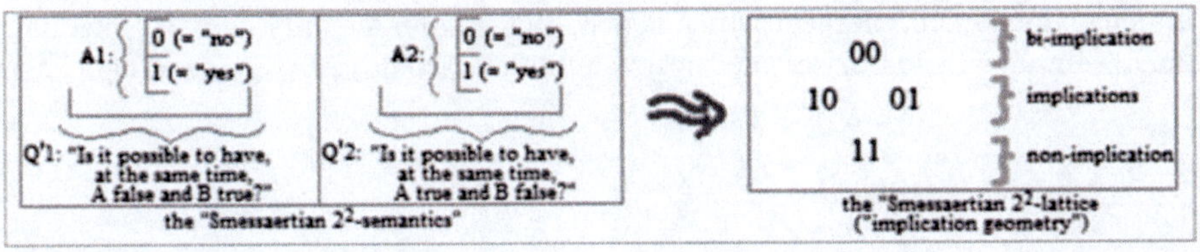

Fig. 89 The Smessaertian 2^2-semantics for the "bi-simplicial" (in our terminology) *implicative geometry*

As we recalled above (Sect. 2.2, Fig. 40), these four kinds of Smessaertian "implication relations" are (i) double-implication ([0|0]), (ii) right-implication ([1|0]), (iii) left-implication ([0|1]), and (iv) no-implication ([1|1]) (Fig. 90).

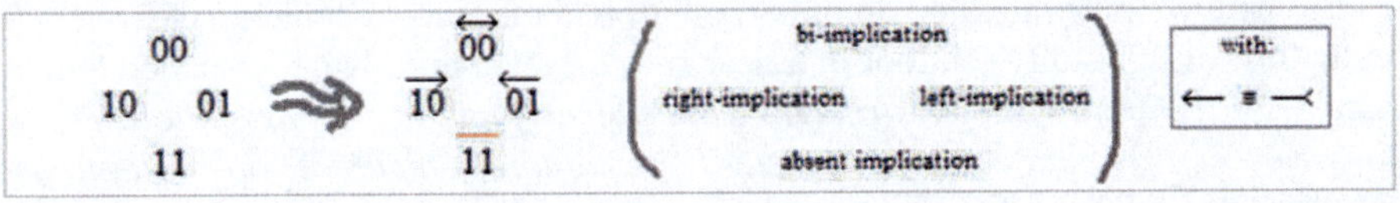

Fig. 90 The meaning of the four Smessaertian kinds of implication relations for bi-simplicial geometry

Let us introduce here, comparably with what I did with respect to Angot-Pellissier (Sect. 2.2, Fig. 37), a small terminological change (aimed at making easier the combination of logical geometry's strategy and conceptuality with our own approach based on simplexes): a useful convention (for what follows) consists in naming with a single letter each of the possible four Smessaertian kinds of implication (so to generate, once one levels up as we will be when we will move from bi-simplicial to tri-simplicial, a "code" made of *two* such literals concatenated), a swift symbolism whose utility should appear soon (Fig. 91).

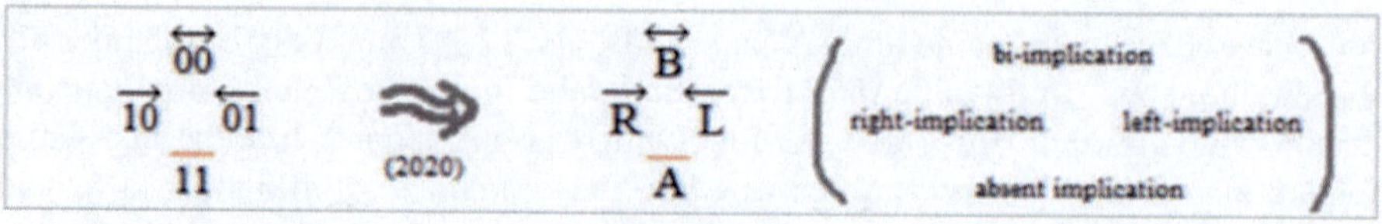

Fig. 91 A useful terminological convention (literals for building higher-level concatenations)

Starting from that, what we need now with respect to our goal (i.e., clarifying the strange "black CS emerging implications" carried by the tri-segment, notably in relation to its two valuations obtained in Sect. 3.6, Fig. 88) is something comparable to Smessaert's 2^2-semantics and its 2^2-lattice, but admitting now not two, but *three* kinds of answers, because of the adoption by us of a third truth-value, "1/2", alongside with the classical "0" and "1" (i.e., so to make this implicative-geometrical meta-lattice match, by the "implication kinds" it generates, the *tri*-simplicial and *three*-valued structure of the oppositive *tri*-segment) (Fig. 92).

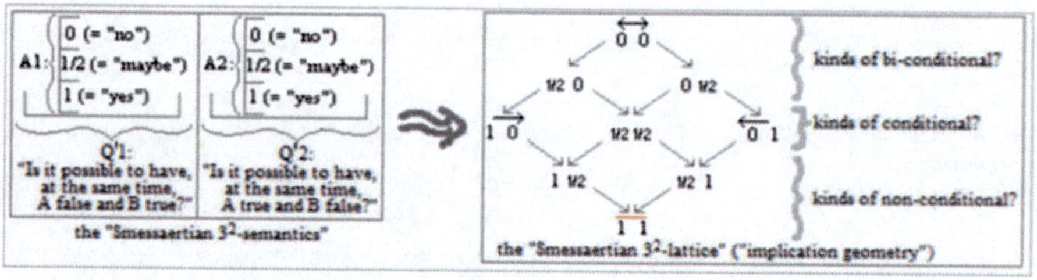

Fig. 92 The "Smessaertian" 3^2-semantics for the "tri-simplicial" implicative geometry

But here we must mention an important problem that we will have to leave open in the rest of this study (but not in a near future), relative to what we called "the q parameter" (Sect. 1.3): *some structure will be clearly missing in our approach now, for we are not (yet) able to try to ask all the remaining comparable meta-questions*, like the "Aristotelian" Q3 ("Can two things A and B be ½ together?") or like the "Smessaertian" Q'3 ("Is it possible to have, at the same time, A '0' and B '½'?"). On this we will come back later (Sects. 5.1 and 5.2).

So, back to our introduction, if not of all the possible meta-questions (q parameter), at least of a third possible kind of answer (p parameter) to them: we see that, as in the case of the Aristotelian 3^2-semantics and 3^2-lattice (Sect. 1.3, Fig. 11), there are, emerging here, five new kinds of possible answers ([1/2|0], [0|1/2], [1/2|1/2], [1|1/2], [1/2|1]), whence, by (inspiring but slippery) analogy, we would expect something like $4 + 5 = 9$ *kinds* of "implicative relations". This can be viewed, from our viewpoint at least, as a form of *tri-simplicial diffraction* of the bi-simplicial Smessaertian "*implication* geometry" (Fig. 93).

The naming, in our convention, of the five new kinds of implicative relations cannot yet be fully clear at this stage. But it should become clearer as soon as we

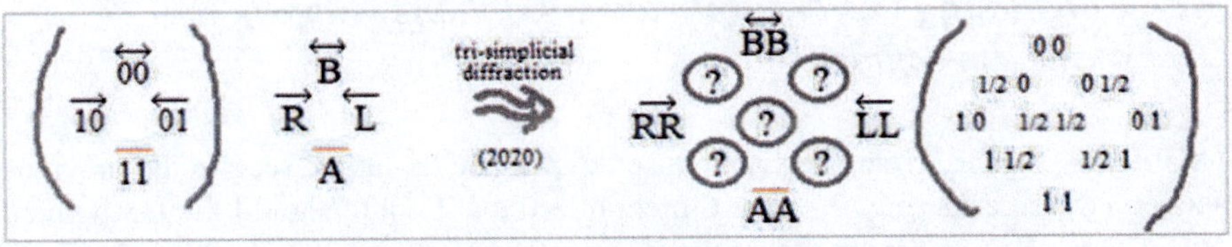

Fig. 93 A first intuition on the “tri-simplicial diffraction” of the (bi-simplicial) Smessaertian “implication geometry”

will start the calculations of the edges carrying these nine (or more?) qualities of implication and when, thus, we will see how these “implications work” (Sect. 4.2). But by analogy with the poly-simplexes, we would expect something comparable with what we have already seen (Sect. 2.2). A still more urgent problem is that of interpreting the *formal meaning* of these five new kinds of implication relations: but, again, a reasoning by analogy, always to be taken carefully (as potentially slippery, of course), can provide already now a *provisory* starting intuition on that (to be further checked by other means) (Fig. 94).

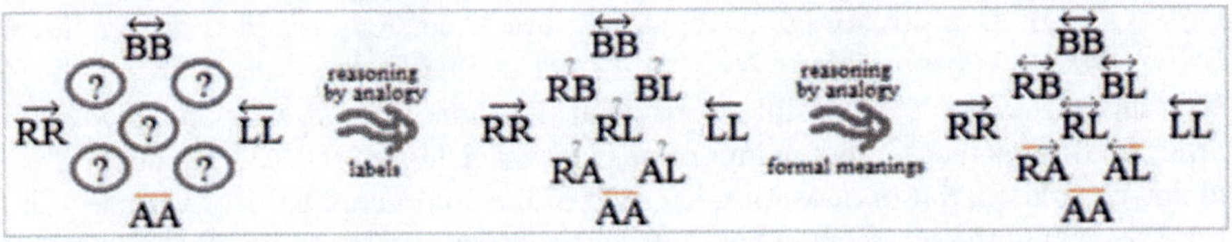

Fig. 94 Guessing, by analogy, the emerging codified labels and the formal meaning of the (at least) five new kinds

Another question arising before we get started is the following: what about the *colors* of “points” (i.e., the vertices), if any in this “implication geometry”? Here, as a provisory methodology, we will keep the “point coloring technique” of the tri-segments (Sect. 2.5). This seems not necessarily too arbitrary since the geometrical shape under examination (the tri-segment, abstraction made of its colours) has been generated by the Pascalian method of Sect. 1.5 (and the Angot-Pellissierian one, isomorphic to it), which seems to have a strong isomorphism with the “coloring of points” we adopted (Sect. 2.5, Figs. 52 and 53). But we will remain ready to revise this provisory methodology (for the coloring of points) as soon as it would appear reasonable to do that. So, given that the vertices of the tri-segment, seen as Angot-Pellissierian numerical sub-sheaves, remain the same as those seen so far (Sects. 2 and 3), we can now go to the question of the *edges*, between any pair of them, of the “implication-geometrical” version of the tri-segment we are trying to study in this Sect. 4.

4.2 *Calculating the 21 Implicative Relations (Edges) of the Tri-segment*

On the basis of the "Smessaertian" 3^2-semantics and 3^2-lattice seen in the previous section (which asks only 2 of the 6 meta-questions it could/should ask, as pointed out in Sect. 4.1), we must now calculate the 21 binary relations holding between the 6 nontrivial vertices of the tri-segment (Sect. 2.4, Fig. 46). This seemingly can be done rather easily relying anew on the Angot-Pellissierian numerical sub-sheaves introduced in the previous Sects. 1, 2, and 3, taken by pairs (hence the number 21: there are 21 possible unordered such pairs). However, an important change here is that since we are mainly dealing with *varieties of implication*, we must now also pay attention to the *order* of each such *unordered* pair of numerical sub-sheaves, i.e., distinguishing AB from BA, so to say (implication is an *order* relation). Combinatorially, taken apart the 6 reflexive pairs (as "$1x2_U$ and $1x2_U$"), the remaining 15 unordered pairs (as "$|1x2_U$ and $1x2_\varnothing|$") must be examined "both ways": so all in all $6 + (2 \times 15) = 36$ pairs must be analyzed. So, in principle to each (nonreflexive) *ordered* pair, it should be asked now the four following pairs of Smessaertian questions (Q'1 and Q'2, in both Angot-Pellissierian sheaf-levels U and X): (Q'1/U) "Is it possible to have, at the same time, A *false* and B *true* at level U?"; (Q'2/U) "Is it possible to have, at the same time, A *true* and B *false* at level U?"; (Q'1/X) "Is it possible to have, at the same time, A *false* and B *true* at level X?"; and (Q'2/X) "Is it possible to have, at the same time, A *true* and B *false* at level X?". In each case, the quartet of answers (A'1/U, A'2/U, A'1/X, and A'2/X) to this general quartet of questions, for each of the 36 ordered pairs of vertices (i.e., sub-sheaves) of the tri-segment, will give us the "implication quality" of this precise ordered pair of vertices (one among the 36), that is the "color" of the corresponding edge of the tri-segment. So let us stress again that this implies that, for each *non-ordered* pair of vertices, the quartet of questions should be applied *two* times (one for each of the *two* possible orders, A-B or B-A, of the unordered pair). But as it happens, the mutually reversed ordered pairs are in fact strictly correlated (as for the quality of implication relation generated by their answers) in a way such that the correlated reversed pairs will turn up to give *globally* (i.e., *modulo* the direction) the same answer, and this will allow in the end a very useful simplification (Sect. 4.3, Fig. 102). For short, reverse pairs just will exchange R with L and vice versa (Fig. 95).

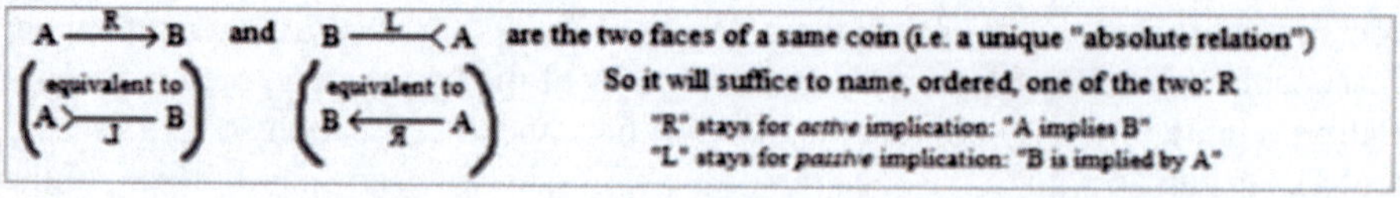

Fig. 95 Possible left-right simplifications of the "implicative relations" (eliminating hidden redundancies)

So it will suffice to calculate one of any pair of correlated pairs (of sub-sheaves), and the other will be obtained by substituting, in the result of the first, any R with L and any L with R. Therefore, we only need to test 21 pairs (instead of 36).

Let us now give just a few examples of calculations, so to nourish the intuition of the reader. Let us first consider a pair which depends on its order, the ordered pair of sub-sheaves “1_X2_U and $1_X2_\varnothing$” (taken as “A and B”). The calculation goes as follows, asking about this ordered pair the four Smessaertian questions: (Q’1/U) “Is it possible to have, at the same time, A *false* and B *true* at level U?”; the answer to this (i.e., A’1/U) is 0. (Q’2/U) “Is it possible to have, at the same time, A *true* and B *false* at level U?”; the answer to this (i.e., A’2/U) is 1. (Q’1/X) “Is it possible to have, at the same time, A *false* and B *true* at level X?”; the answer to this (i.e., A’1/X) is 0. (Q’2/X) “Is it possible to have, at the same time, A *true* and B *false* at level X?”; the answer to this (i.e., A’2/X) is 0. As a result, the string variable *x* (for concatenating orderly the two answers relative to level X, cf. Sect. 2.2), receiving here the value 00, is B (for B is defined, in the Smessaertian 2^2-semantics, as [0|0], Sect. 4.1, Fig. 91), and the variable *u* (for concatenating orderly the two answers relative to the level U, cf. Sect. 2.2), receiving here the value 01, is L (for L, in the Smessaertian 2^2-semantics, is defined as [0|1]), so here the string variable *xu* becomes BL and this is the tri-simplicial “implication kind” holding between the two ordered vertices of the tri-segment (Fig. 96).

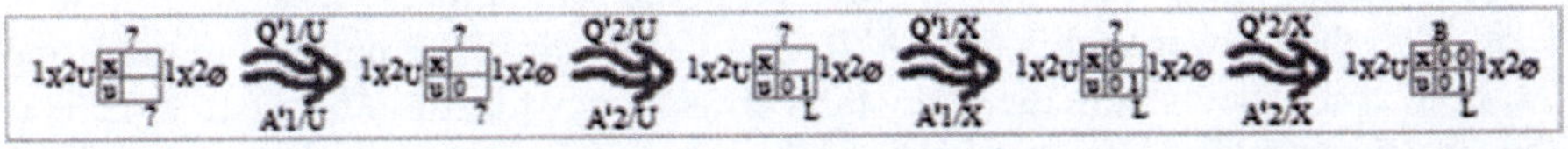

Fig. 96 How to calculate the quality of the “implication relation” of an ordered pair of numerical sub-sheaves

Let us now consider the reverse ordered pair, “$1_X2_\varnothing$ and 1_X2_U” (taken as “A and B”). The calculation for it goes as follows, asking about this ordered pair the four Smessaertian questions: (Q’1/U) “Is it possible to have, at the same time, A *false* and B *true* at level U?”; the answer to this (i.e., A’1/U) is 1. (Q’2/U) “Is it possible to have, at the same time, A *true* and B *false* at level U?”; the answer to this (i.e., A’2/U) is 0. (Q’1/X) “Is it possible to have, at the same time, A *false* and B *true* at level X?”; the answer to this (i.e., A’1/X) is 0. (Q’2/X) “Is it possible to have, at the same time, A *true* and B *false* at level X?”; the answer to this (i.e., A’2/X) is 0. As a result, *x*, being here 00, is B, and *u*, being here 10, is R, so here *xu* is BR and this is the tri-simplicial “implication kind” holding between these two ordered vertices of the tri-segment.

Remark, then, as we already said, that inverting the two vertices here just transforms “BL” in “BR”: which means that the same edge read in one direction gives BR, while read in the other it gives BL. This is because at level X-U (i.e., X minus U) the implicative relation happens here to be B (which is bidirectional), which is therefore not affected by switching the vertices, while at level U the relation happens to be L for the first two ordered vertices and R for their reversal: this means that R or L represent in fact the same "absolute relation", but it specifies (by being R or L) in which of the two possible directions it is *read*; let us call it "|R/L|", which means "L or R, just according the *directional viewpoint*" (Fig. 97).

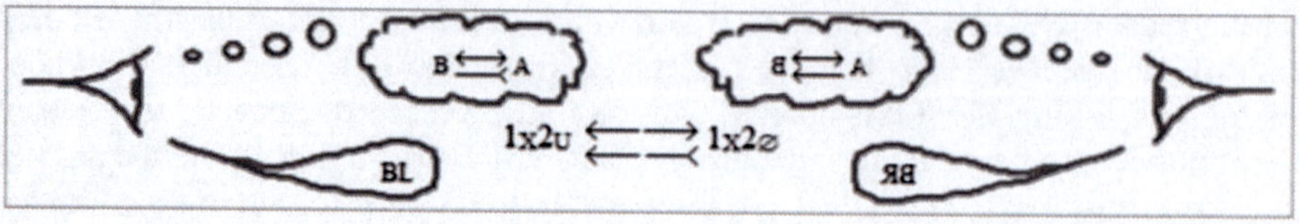

Fig. 97 The relations "R" and "L" are two directionally opposed *viewpoints* on the same unchanging reality

Let us now take a different example, that of a pair of sub-sheaves of the tri-segment which is invariant with respect to the order of its elements. Let us consider the ordered pair "1_X2_U and 1_U2_X" (taken as "A and B"). Its calculation goes like this, asking about this ordered pair the four Smessaertian questions: (Q'1/U) "Is it possible to have, at the same time, A *false* and B *true* at level U?"; the answer to this (i.e., A'1/U) is 0. (Q'2/U) "Is it possible to have, at the same time, A *true* and B *false* at level U?"; the answer to this (i.e., A'2/U) is 0. (Q'1/X) "Is it possible to have, at the same time, A *false* and B *true* at level X?"; the answer to this (i.e., A'1/X) is 1. (Q'2/X) "Is it possible to have, at the same time, A *true* and B *false* at level X?"; the answer to this (i.e., A'2/X) is 1. As a result, *x*, being here the string 11, is A, and *u*, being here the string 00, is B, so here *xu* is the string AB and this is the "implication kind" holding between the two ordered vertices of the tri-segment.

Let us now consider the same ordered pair, but reversed: "1_U2_X and 1_X2_U" (taken as "A and B"). Its calculation goes like this, asking about this ordered pair the four Smessaertian questions: (Q'1/U) "Is it possible to have, at the same time, A *false* and B *true* at level U?"; the answer to this (i.e., A'1/U) is 0. (Q'2/U) "Is it possible to have, at the same time, A *true* and B *false* at level U?"; the answer to this (i.e., A'2/U) is 0. (Q'1/X) "Is it possible to have, at the same time, A *false* and B *true* at level X?"; the answer to this (i.e., A'1/X) is 1. (Q'2/X) "Is it possible to have, at the same time, A *true* and B *false* at level X?"; the answer to this (i.e., A'2/X) is 1. As a result, *x*, being here the string 11, is A, and *u*, being here the string 00, is B, so *xu* is here the string AB. Remark that in this case the fact of reading this edge in one direction or the other makes no difference: both ways it has to be read AB.

We will not make – we are merciful – all calculations explicitly here. Instead we will give the general resulting calculation of the 21 edges (in only one direction, the result of the reverse calculation can be obtained by substituting the "R" with "L" and vice versa), in a synoptic compact format (Fig. 98).

These 21 calculations give us 11 kinds of implication relations (the BB, BL, AB, LL, AL, AR, LB, AA, LA, RA, BA). But once we exchange L with R (and not the other way round) two become redundant and we obtain thus nine kinds: BB, BR, AB, RR, AR, RB, AA, RA, and BA. If we compare these nine kinds of implication relations obtained here with the nine kinds conjectured by analogy (Sect.

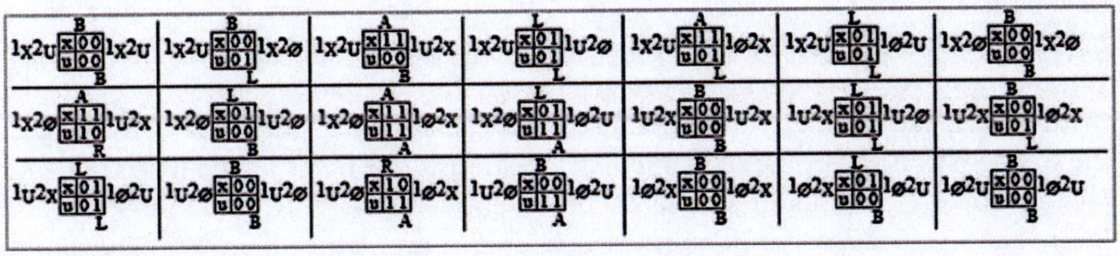

Fig. 98 Synoptic view of the calculations over the 21 ordered pairs of sub-sheaves (i.e., vertices) of the tri-segment

4.1, Fig. 94), we see that the 3^2-lattice for the tri-*segment* must be modified (we will also adopt by the way new formal symbols more fit, notably in terms of color conventions, with those of the opposition tri-segment, as we will see in Sect. 4.5). The result is well-balanced (i.e., fully symmetrical) (Fig. 99).

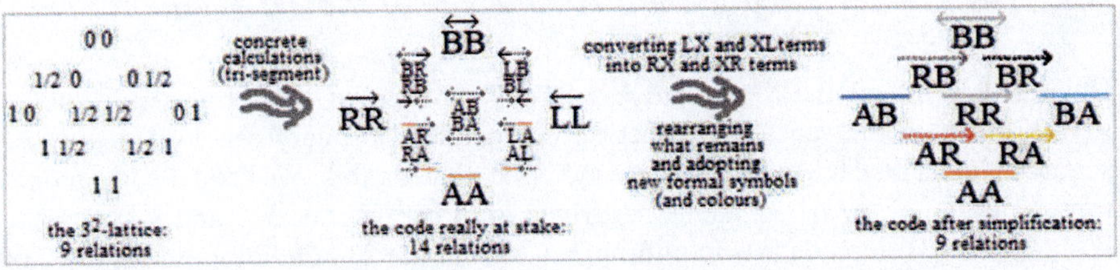

Fig. 99 A simplified version of the 3^2-lattice after the calculation of the 21 edges of the implication tri-*segment*

This change is important and will have to be understood (Sects. 5.1 and 5.2). Relying on this knowledge of the 21 edges (Fig. 98), and based on our new symbolism (cf. Figure 99), we can give now a synoptic view of the "horizons" (Sect. 3.1) of each of the six vertices of the implicative tri-segment (Fig. 100).

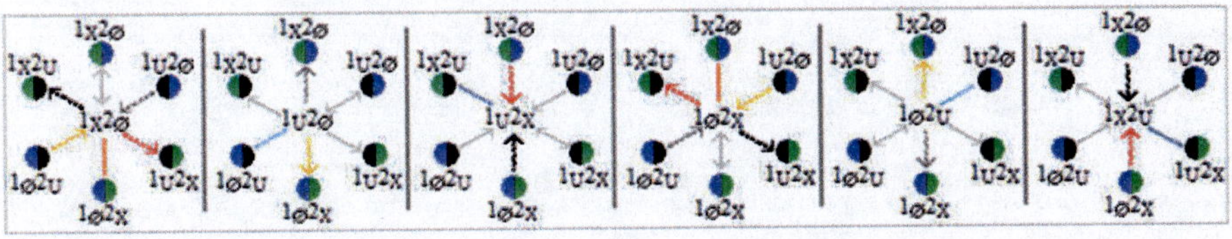

Fig. 100 Synoptical view of the kinds of tri-simplicial implicative relation each vertex has with any possible vertex

Remark that, as in Sect. 3.1, there are here three kinds of horizon, such that the vertices centrally symmetric share the same horizon *modulo* a rotation of 180° of it.

A natural question now, in comparison with what happens with the opposition qualities of the tri-*segment* (less, in number, than those of the general tri-*simplex*:

seven instead of nine, Sect. 2.2, Fig. 39), is that of knowing whether some implicative qualities of the tri-simplex are missing in the tri-segment. We are not yet able to give a systematic *a priori* answer. But this answer could seem to be that no implicative quality among those of the tri-simplex (Sect. 4.1) is missing in the tri-segment. However, comparable calculations done by us on the tri-*triangle* let emerge already two more relations there, absent here: the "LR" and the "RL" (and RL is one which appeared in our previous, conjectural lattice, Fig. 94). This point, which bears no prejudice to the rest of this study, will have to be understood and explained in the future.

Having calculated the edges of the implicative version of the tri-segment (without forgetting however that our current reasoning relies on only two over the six possible "Smessaertian" meta-questions, Sects. 4.1, 5.1, and 5.2), we can now try, in the next Sect. 4.3, to construct a global view of it.

4.3 *The Global "Implication Geometry" of the Tri-segment*

Having determined the "implicative quality" of each of the 21 possible (so to say absolute, in the sense of "direction independent") binary relations of the tri-segment, let us now try to have a more synoptic view on this. We keep the hexagonal representation of the tri-segment we arrived to in the last Sects. 2 and 3 (Sect. 2.6, Fig. 57). A first step consists in putting, on each of the 21 still colorless edges (6 of which are reflexive curls), its reading (and its color), and this can be of two kinds: (1) either bidirectional and therefore unique (when it is either BB or AA but also AB or BA); or (2) it is unidirectional (i.e., asymmetric) and then the direction in which the edge is read determines two different labels (like in RR and LL, in AR and AL, or in RB and LB, Sect. 4.2, Figs. 95 and 97). So, to begin with, we put on each edge its *two* readings (when there are 2: one with R and one with L), and we try to highlight the direction of the intended segment decoration by the direction (at times a little bit directionally strange) of the letters (Fig. 101).

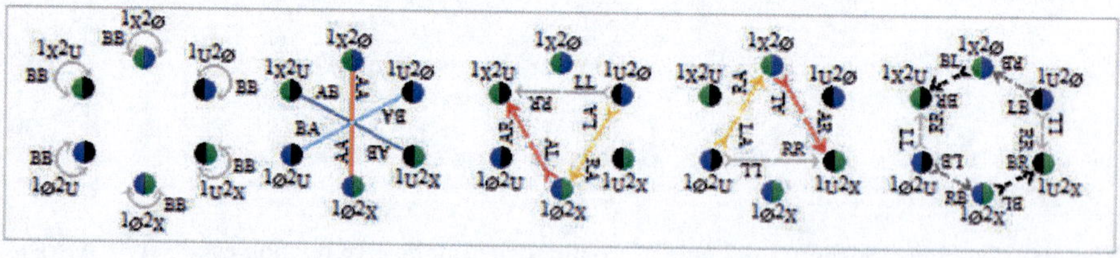

Fig. 101 A global view of the "implicative side" of the tri-segment (with *all* its kinds of "implication relations")

But then it is useful to resort to what we have understood in the previous Sect. 4.2 over the substantial (i.e., regular) coincidence "in the absolute" of all pairs of reversed labels (Figs. 95 and 97). And this will mean that we will now be able to produce another, swifter implicative (but still decomposed) decoration of the tri-segment, with only one-half of its two opposed labels on each of its 21 edges. As a rule, as said, we will privilege the so to say “direct” (or active, Fig. 95) implication (i.e., “A *implies* B”) to its “indirect” (or passive), equivalent counterpart (i.e., “B *is implied* by A”), and this means that we will always choose R instead of L, but we could have done, equivalently, the other way round. So, RR will be preferred to the correlative LL; BR and RB will be preferred to, respectively, their correlative BL and LB; RA and AR will be preferred, respectively, to their correlative LA and AL (Fig. 102).

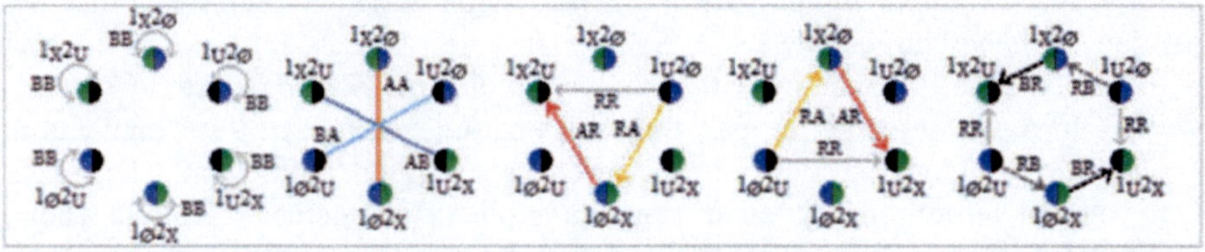

Fig. 102 Global decomposed view of the “implicative side” of the tri-segment, simplified (R and not L)

If we now gather this decomposed view into a whole, this gives the following first global representation(s) of the “implication geometry” version of the tri-segment (Fig. 103).

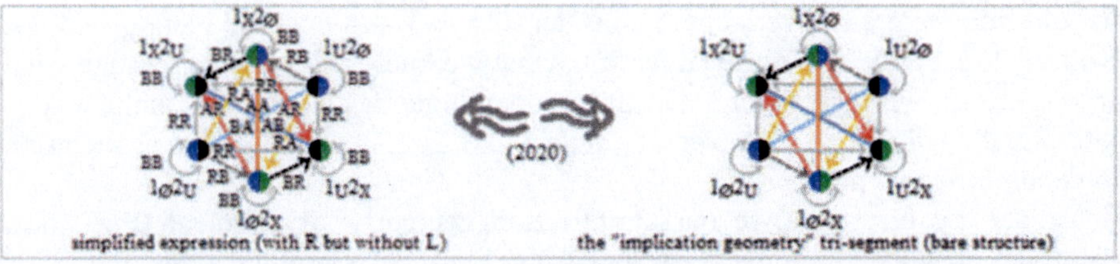

Fig. 103 First two *whole* representations of the “implication geometry” version or the tri-segment

Remark that this seems very interesting: the tri-simplicial diffraction of the contradiction segment gives us this, in some sense unexpected “arrow *complexity*”. Recall: the red segment of 2-opposition already could be seen as “implication geometry” (à la Smessaert, Fig. 104). So, at this so to say “logical-geometry stage”, we have two versions of the tri-segment, one in terms of its “opposition geometry” (on the middle), the other in terms of its “implication geometry” (on the right) (Fig. 104).

This is, so far, the – *cum grano salis* (Sect. 4.1) – complete “logical geometry” view on the tri-segment. We will however afford the question of *unifying* (if it is

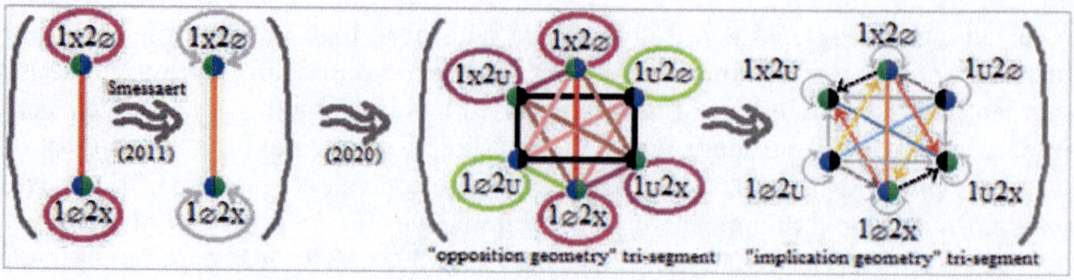

Fig. 104 Tri-simplicial diffraction of the "logical geometry" of the 2-opposition contradiction segment

possible) by a *methodical and reasoned combination* these two representations into just one (Aristotelian) later (Sect. 4.5).

But before that, by analogy with what we did in Sect. 3 (with respect to Sect. 2), we will have now to study at least the basics of the structure of this "implication geometry" tri-segment in some more depth (Sect. 4.4). And, before that, let us make here some more preliminary remarks on this implicative structure taken as a whole. Importantly (for our study), remark that this is more or less precisely what we have been looking for since the end of Sect. 3 (i.e., since the end of Sect. 3.6): finding some kind of *reasoned roadmap of the "implications" of the tri-segment* for helping us in understanding how to judge the otherwise puzzling apparent emergence (from the valuation of Sect. 3.6, Fig. 87) of unexpected *implication relations* (arrows), notably in the four black segments of (non-arrow) simplicial CS relations (Fig. 88). And in fact such a roadmap, that we now have successfully in our hands, provides us immediately at least two very important things: (1) an *exhaustive* list of the kinds of implications at stake in the tri-segment (which appears to be – although not yet the closure: we are not yet able to consider all the 3 + 6 possible meta-questions, Sects. 4.1, 5.1 and 5.2 – quite larger than what we knew and quite larger than what we expected) and (2) a rather univocal and unambiguous indication as to the way to interpret the *valuation* we arrived to (on Sect. 3.6, Figs. 87 and 88). Let us see more in detail these two points.

As for the first point, in our "implication geometry" tri-segment (Fig. 104), there are, at work, three main families – the kinds of nonimplications, the kinds of implications and the kinds of bi-implication – more precisely: (1) five kinds of "nonimplication" (i.e., all the tri-simplicial "implication geometry" relation kinds containing *at least* an occurrence of the bi-simplicial relation "A", as in AR), (2) five kinds of implication (i.e., those containing *at least* an occurrence of R, as in RA), and (3) five kinds of bi-implication (i.e., those containing *at least* one occurrence of B, as in BA). Because of the tri-valuedness of the tri-simplicial space, which implies the existence of two Angot-Pellissierian sheaf-levels U and X (Sects. 1.4 and 2.1), these three kinds are not mutually exclusive any more: their *composition* is now the rule (six over the nine cases); the "pure cases" are just particular cases of composition (three over the nine cases), namely, trivial compositions (or self-compositions), like "RR", "AA", or "BB" (Fig. 105).

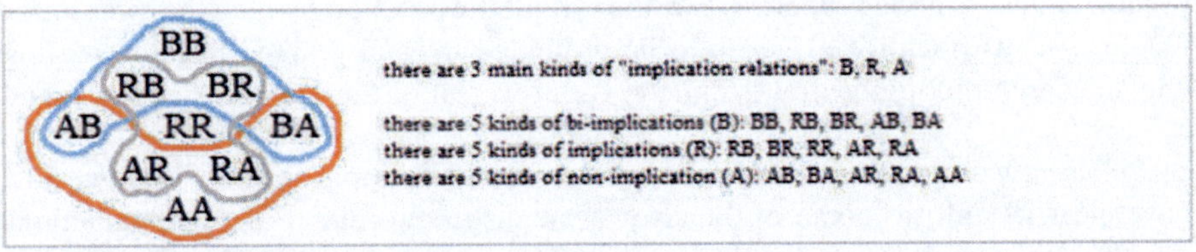

Fig. 105 There are three main "implication relations" kinds, characterized by the fact of containing B, R, and A

Remark *en passant* that the result as for implications seems to confirm that the hexagonal representation that we choose for the tri-segment is *optimal* and this can be seen here in at least three respects: (1) its three diagonals are indeed contradiction diagonals (because they are the "implication geometry" counterpart of this: the nonimplication AA and its two weakenings AB and BA); (2) as we will see in the next Sect. 4.4, Fig. 109, the three main rectangles ("squares") of this implicative hexagonal tri-segment are *very* regular with respect to its "implication geometry" relations; and (3) the balanced character of the hexagonal representation of the tri-segment, still thanks to the unveiling of its arrows, is also confirmed by the "differential topology" viewpoint (Sect. 1.2, Fig. 8, based on [62], p. 52), i.e., by the distribution of the three differential-topological kinds of vertices: two (centrally symmetric) vertices (i.e., the two |black-blue| ones) shoot each four arrows (exhibiting thus a "source" behavior), two (centrally symmetric) vertices (the two |blue-green| ones) shoot each two arrows and receive each two arrows (exhibiting thus a "saddle" behavior), two (centrally symmetric) vertices (the two |black-green| ones) receive each four arrows (exhibiting thus a "sink" behavior) (Fig. 106).

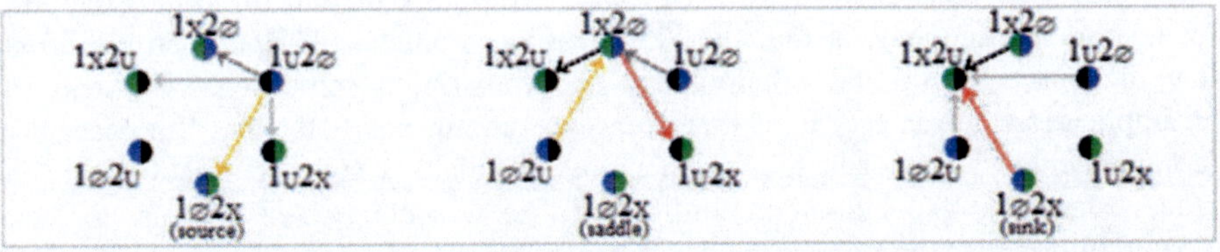

Fig. 106 Reading the "implication geometry" of the tri-segment with "differential-topology"'s eyes

Remark here that the classical position (i.e., the two |blue-green| vertices of the classical bi-simplicial 2-oppositional red segment of contradiction) seems to play a *pivotal role* in between its two symmetric "diffractions" (paracomplete and paraconsistent). In other words, the "tri-simplicial diffraction" opens so to say some kind of left-right "symmetry", orthogonal to the top-bottom affirmative-negative starting (oppositional) "symmetry" of the starting red segment of 2-oppositional contradiction. At the level of the *expression* of the tri-segment by a visual structure

(Sect. 3.3), all its good properties (i.e. the symmetries – among which the three kinds (1)–(3) we just mentioned) seem to be grounded in (and granted by) our starting choice of interpreting central symmetry as a reversal of the indices (Sects. 2.6 and 3.3).

Let us now turn to the second point (valuation and its implications). What we now have clearly confirms some of the arrows we suspected (due to the two valuations we were able to establish, for the tri-segment, at the end of Sect. 3.6, Fig. 88). So let us consider now these two valuations of the tri-segment, but this time under its implicative reading (Fig. 107).

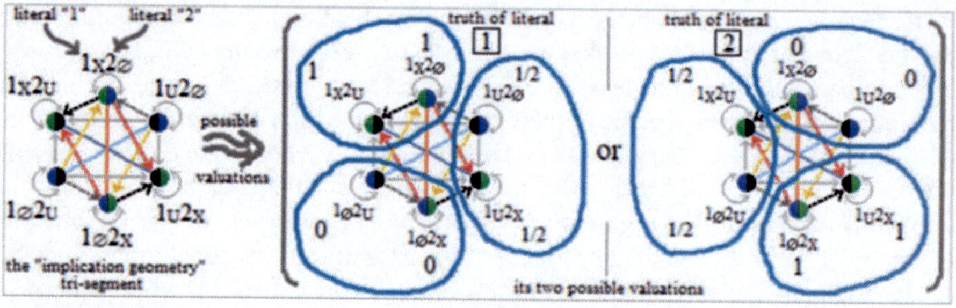

Fig. 107 The two possible valuations of the tri-segment reconsidered from the viewpoint of "implicative relations"

Here there are three remarks. The first thing to be remarked is that the naturalness of these two valuations seems to be *confirmed* here. If the "opposition(al) geometry" tri-segment so to say made clear the meaningfulness of these two valuations with respect to kinds of *negations* (particularly the diagonals), the present "implication geometry" tri-segment so to say makes clear the meaningfulness of these two same valuations with respect to the *kinds of implications*. A second important remark, then, more particularly, is that the "implication geometrical" approach confirms that the black "simplicial relation" CS can (must?) in some sense be read as an implication: no doubt about this seems to remain now. But this Smessaertian (complementary) approach tells even more: the black simplicial relation can be read, seemingly, as a *classical* implication (we will come back to this in Sect. 4.5). Additionally, the oppositional kinds of noncontradiction CI and IS (that we interpreted, à la Aristotle, as biconditionals, given the relevant Angot-Pellissierian sheaf sections, cf. Sect. 2.2, but that the two valuations pushed forward as being, rather, as the CS, possible full-fledged kinds of implications, Sect. 3.6, Fig. 88) can also be read, thanks to this Smessaertian roadmap, as *uni*directional arrows (at the implication-geometrical level): but this time not as classical RR implications (and this is coherent with their oppositional reading: the CI and the IS are, respectively, paracomplete and paraconsistent "implications", Sect. 1.4, Figs. 19 and 20). The "*opposition*-geometrical" light green CI becomes in fact, in the "implication geometry", the "*implication* geometrical" porous (in our representation here) gray implication RB, whereas the "*opposition* geometrical" violet IS becomes the "*implication* geometrical" porous black implication BR. A third and last remark with respect to our rereading of the two valuations of the tri-segment is that two more arrows, in fact, seem to emerge here: the "porous" red arrow AR and the "porous" yellow arrow RA. And this is more surprising: (1) they are parallel to (i.e., they share edges with), respectively, the *non*-arrow *negations* CN (paracomplete) and NS (paraconsistent). (2) Moreover, of all the thus possible five different kinds of implication arrows of the "implication geometry" tri-segment, these two last kinds (embodied each two times) seem rather strange, because sometimes they seem to go beyond the limits imposed by valuation; the AR, taken at face value (i.e., as an arrow), seems to lead to the strange implication (in terms of truth-values) "$1 \rightarrow 1/2$", and the RA seems to lead to the tantamount strange implication "$1/2 \rightarrow 0$". The simple, first explanation is that they are "restricted implications" (cf. [3]): moreover, they are like "water and fire", they join "R" with "A" (but in separated sheaf-levels U and X-U. (3) But for this reason, we could not clearly see them before (*which shows, again, the power and the usefulness of the Smessaertian "implication geometry"*). Remark that, as we mentioned (Sect. 4.2), in the tri-*triangle* we find relations RL and LR generating a similar "strange" issue. We will come back on this important point on Sect. 4.5.

A general final remark here, before going to the next Sect. 4.4, is that the tri-simplicial space confirms here, but also *radicalizes*, what has been seen by Smessaert in the bi-simplicial space, namely, that the relations of the "opposition geometry" and those of the "implication geometry" have partial overlaps (Sect. 4.5). This fact, as we are going to see, is – without exaggeration – hugely important (Sects. 4.5 and 4.6).

4.4 Overview of Some Inner Structures of the Implicative Tri-segment

Before summing up (with rather important consequences at stake, cf. Sects. 4.5 and 4.6), it will be useful to acquire some more understanding of the structure of the "implication geometry" version of the tri-segment (once more: we are working, however, by necessity with a restricted version of it, Sects. 4.1 and 5.1, and 5.2). In some sense, we will just try to repeat for it (quickly!) the kind of tentative categorization we proposed for the "opposition(al) geometry" version of the tri-segment in Sect. 3. We leave aside the kind of characterization in terms of "horizon" made in Sect. 3.1 (for many elements of it are already in Fig. 100 of Sect. 4.2). The question about the "inner circuits" (Sect. 3.2) of the "implication geometry" version of the tri-segment seems potentially interesting (here we will only mention it). The main idea seems to be that there are three main kinds of circuits, reflecting the three main kinds of "implication relations": bi-implication (containing *at least* a B), implication (containing *at least* an R), and nonimplication (containing *at least* an A) (Fig. 108).

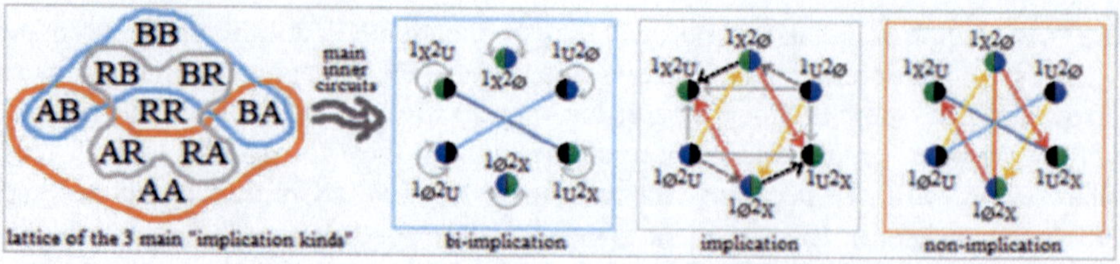

Fig. 108 The possible "inner circuits" of the "implication geometry" version of the tri-segment

A notable difference with the case of the opposition(al) inner circuits (to be reconsidered, then?) is that here, differently from there, the three inner circuits have systematic overlaps.

Another meaningful and potentially useful (although tentative) structural investigation is that consisting in looking for the hybrid "inner squares" (here: rectangles) and triangles of the implicative tri-segment. Again, our methodology here will be just to rely on what seen for the "opposition(al) geometry" of the tri-segment (Sect. 3.4). As for squares, for the same combinatorial reasons put forward in Sect. 3.4, there are here, qualitatively, $3 + 12 = 15$ of them. The three main squares (in fact: rectangles!), the (i)–(iii), are very regular and even more mutually similar than what were the three *opposition(al)* rectangles (Fig. 70) (Fig. 109).

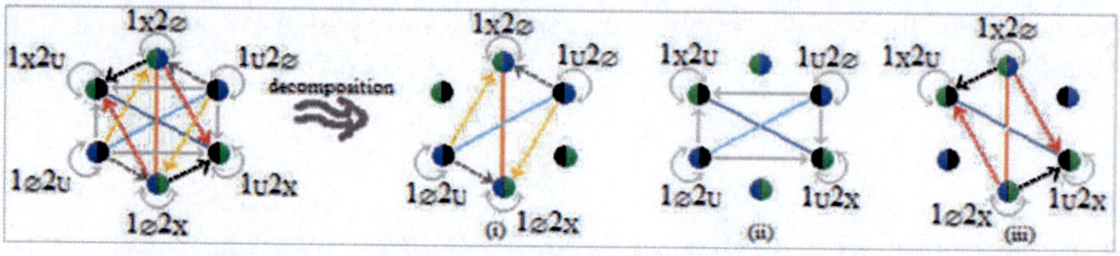

Fig. 109 The three main "inner rectangles" (or squares) of the "implication geometry" tri-segment are very regular

As for the other 12 squares, they seem to be less regular and more "hybrid" (cf. Sect. 3.4, Fig. 77), although their global combinatorial system is very regular (Fig. 110).

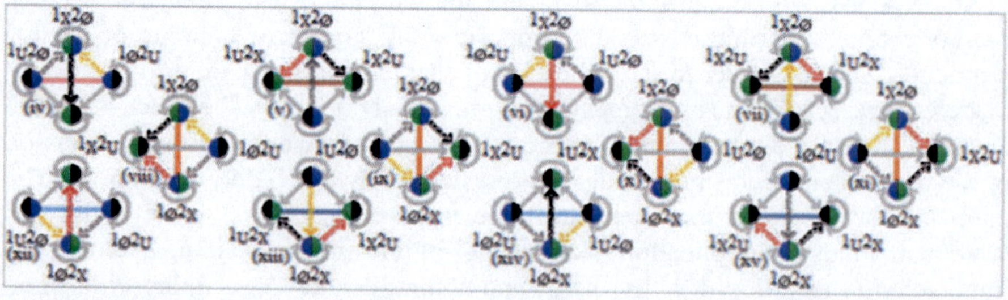

Fig. 110 The "implication geometry" version of the 12 "hybrid squares" (or tetrahedra) of the tri-segment

As for the “inner triangles”, for the same combinatorial reasons seen in the “opposition(al) geometry” tri-segment, there seem to be here 8 + 12 = 20 of them (cf. Sect. 3.4, Fig. 80). Here we only give their qualitative kinds, if needed, and the reader can easily reconstruct the rest, by referring to Sect. 3.4 (in fact the pairs of isomorphic triangles, like (a) and (g), being centrally symmetric, have inverted colors in the *readable* vertices) (Fig. 111).

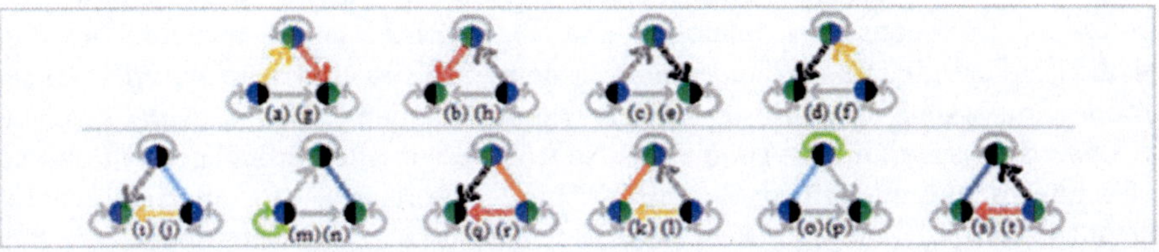

Fig. 111 The qualitative kinds of inner triangles of the “implication geometry” version of the tri-segment

En passant, a possible question is to verify how do look here the two “meaningful triangles” of Sect. 3.5, Fig. 82, recalling that these two triangles seemed meaningful to us since, as we argued, they play some kind of “semantic role” (on the “oppositional roles”, cf. Sect. 3.4): one triangle, namely, the (c), contains the three “affirmative meanings” (of the tri-segment), while the other, namely, the (e), contains the tri-segment’s three “negative meanings”. One sees that each of these two mutually isomorphic (and centrally symmetric) triangles is isomorphic to a “commutative (i.e., transitive) triangle” (*modulo*, however, the fact of having three *different* kinds of implication arrows, instead of a same kind). A general interesting feature expressed by them (and important for the tri-segment, Sect. 3.5, where in some sense we spoke *mistakenly* of “equivalences” or bidirectional light green CI and violet IS arrows) is the “strength order”, expressed by the three implication arrows, “paracomplete > classical > paraconsistent” (Fig. 112).

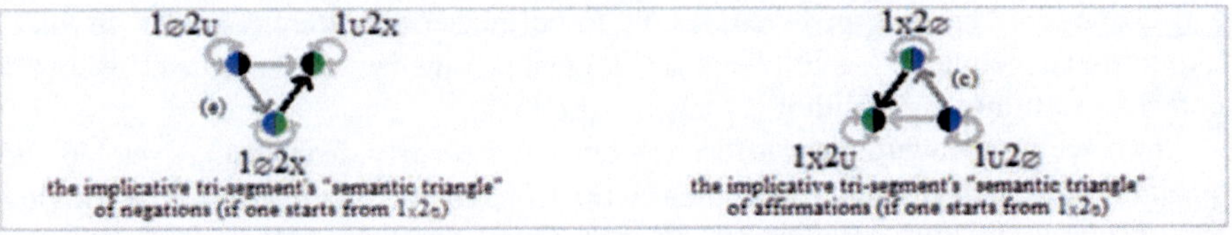

Fig. 112 “Qui peut le plus, peut le moins”: paracomplete → classical → paraconsistent in the two “semantic triangles”

So, we gained, as expected, at least the first elements of a basic understanding of this new structure (the “implicative tri-segment”). But then, if I am not mistaken (logical geometer friends will tell), we are faced now with a new main problem: some sort of “methodological schizophrenia” (or *“methodological dualism” of "logical geometry"*). In other terms, we seemingly paid quite much for obtaining our (very valuable and needed) "implication roadmap": the problem now is that of knowing what we want and/or can do with these *two* logical-geometrical "*twin* tri-segments". This means that there are at least two main possible choices (or issues) in front of us: either (i) accepting, as seem to be strongly suggesting the "logical geometers", as durable and methodologically justified the *separated, parallel twin existence* of the "opposition geometry" version and of the "implication geometry" version of the tri-segment (and this could be seen – at the level of powerful, conscious or unconscious, analogies and/or *fantasies* – as, in quantum physics, with Heisenberg's famous "uncertainty principle": a situation of *structural dualism* forever impossible to get rid of) or (ii) trying to resolutely *systematically combine* the two Smessaertian twin geometries, so to have a swifter, articulated but unique structure for the tri-segment, but also – more importantly – for any future poly-segment and poly-simplex (Sect. 4.6). But then how? Let us now try to see this point.

4.5 *Is an Aristotelian Tri-segment Possible? Yes! Meaningful? Very!*

What said at the end of the previous Sect. 4.4 is equivalent, as Smessaert and Demey in some sense have taught us these last 10 years (in their bi-simplicial-restricted logical-geometrical space), to asking the rhetorical question: "is an "*Aristotelian*" tri-segment possible?" (and their implied answer seems to me to be: "yes, but honestly . . . "). Again, the question here means: is it possible to *combine usefully* these two Smessaertian sides of the tri-segment *without losing "logical-geometrical" properties*? (i) if yes: then it would be easier, but maybe even instructive, to use this combination, instead of the two "forever parallel and substantially disjoint sides" of logical geometry; (ii) if not, then we will have to use both, in parallel, without any hope of *finding again the pre-Smessaertian "lost paradise" of a unity of the geometry of oppositions*. The position of Smessaert and Demey, if I am not mistaken, seems to be the second: they consider, *in nuce*, "Aristotelian geometry" a little bit as "logical geometry for dummies". Being a notorious dummy, my position is – alas! – the first.

So, if we now nevertheless afford (as dummies – sorry dear hostage reader) the question of understanding more radically *the* tri-segment, in some sense by looking for its possible (still hypothetical) "Aristotelian" (or maybe Pascalian) version, what we need is a methodical comparison of its two twin *sub*-geometries. In fact, this means that *all* the possible kinds of *edge overlap*, with respect to the two sub-geometries, must be studied and known: and *named* . . . . The emergence of "Aristotelian simplifications", at this level of complexity, can only be discussed after that (and it is not yet granted *a priori*). So let us now develop some tentative comparative remarks on the two Smessaertian *sub*-geometries of the (*non*-Smessaertian) supposedly unique *Aristotelian* tri-segment.

For that, let us come back, first of all, to what happens, in the tri-segment, to the pink CN and the brown NS relations. And first of all remember that they are supposed to be *oppositionally* very meaningful: they are the tri-simplicial diffractions of the red segment of 2-oppositional (i.e., bi-simplicial) contradiction; they are, so to say, the *ratio essendi* of the *tri*-segment! We just created/discovered it, in thought, with the aim of exploring "contradiction's tri-simplicial diffractions" (Sect. 1.6). So, let us then concentrate on what really happened with them: for, retrospectively, without thematizing it, we were *de facto* surprised to see these two new contradiction kinds, the paracomplete CN and the paraconsistent NS, appear: (1) in the two nonclassical diagonals (this was very satisfactory!) (2) *but also* elsewhere, namely, in the "$1_{\varnothing}2_{U}$—$1_{X}2_{\varnothing}$" and "$1_{U}2_{X}$—$1_{X}2_{\varnothing}$" edges, etc. (this was more disturbing). But then, "implication geometry", under its never seen before *tri-simplicial version*, somehow rescues us, as a roadmap, even in this respect (Sects. 4.1, 4.2, 4.3, and 4.4), by teaching us, *implicitly*, if we now just *explicitly* think of it, that there are in fact, in the tri-*simplexes* in general and in the tri-*segment* in particular, (at least) two different kinds of CN relations and similarly two different kinds of NS relations! They are (i) the CN (resp. the NS) segment *overlapping* with the BA (resp. the AB) diagonal segment (ii) *and* the two CN (resp. the two NS) segments *overlapping*, each one, with one RA (resp. one AR) segment. If one thinks of it, this just means that *"CN" (resp. "NS") means in fact not just one but two different things inside the Aristotelian tri-segment* taken not schizophrenically (Fig. 113).

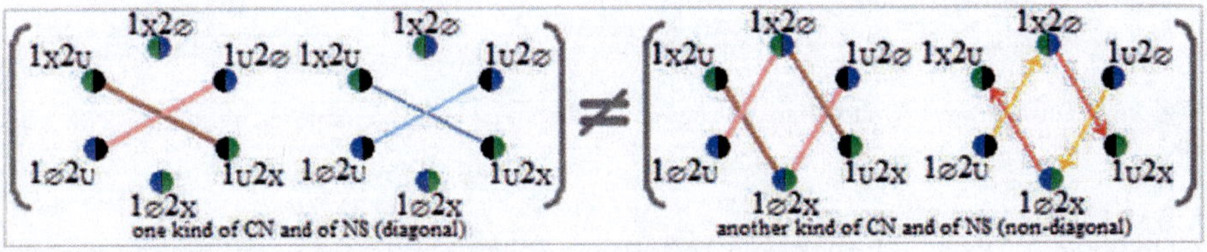

Fig. 113 In *the tri-segment globally taken* (i.e., Aristotelian!), there are, in fact, *two different kinds of CN (and NS)!*

So one can, and in fact must, (1) distinguish, *in the global (Aristotelian) tri-segment*, between (i) the "CN/BA" relation (one diagonal pink-light blue edge of the tri-segment, Fig. 113) and (ii) the "CN/RA" relation (two pink-porous yellow non-diagonal edges of the tri-segment, Fig. 113) (2) distinguish between (i') the "NS/AB" relation (one diagonal brown-ultramarine edge of the tri-segment, Fig. 113) and (ii') the "NS/AR" relation (two non-diagonal brown-porous red edges of the tri-segment, Fig. 113). Consequently, we propose to introduce here *the concept of "Aristotelian combination"* and to apply it in order to "produce" (or rather unfold) two different formal symbols for the two different kinds of CN relations, namely, the "CN/BA" and "CN/RA". For reasons to appear soon, we keep for

CN/BA just the starting representation, unchanged, of CN (i.e., a pink segment) but adopt for CN/RA a new porous *pink* arrow (porous will mean here "beware, this arrow here is *sui generis*") (Fig. 114).

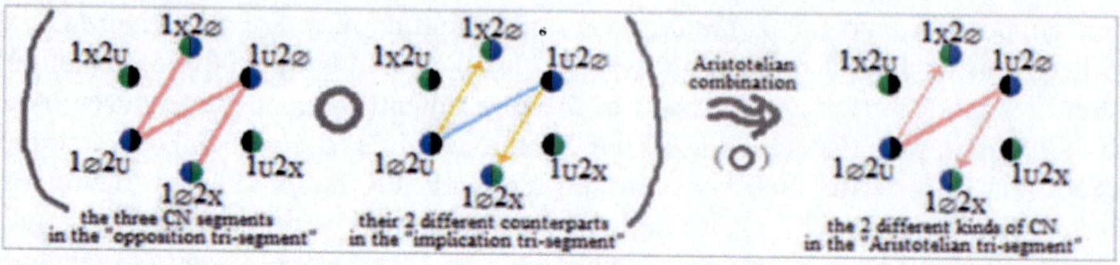

Fig. 114 Toward *the* "Aristotelian tri-segment": the *Aristotelian* composed relations "CN/BA" and "CN/RA"

Similarly, by another "Aristotelian combination", we introduce now two different forms of "NS" relations, namely, the "NS/AB" and the "NS/AR" relations. For reasons to appear soon, we just keep for NS/AB the starting representation of NS (a brown segment) but adopt for NS/AR a new porous *brown* arrow (porous will mean here as well "beware, this arrow here is *sui generis*") (Fig. 115).

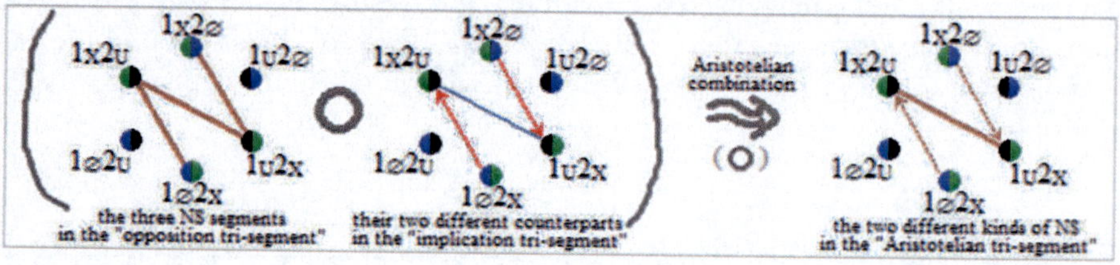

Fig. 115 Toward *the* "Aristotelian tri-segment": the *Aristotelian* composed relations "NS/AB" and "NS/AR"

Second, let us now turn, in our refreshing Aristotelian poly-simplicial dummies-journey, to the CS relation which was our main source of puzzlement about implications (Sect. 3.6, Fig. 88). If one considers the black "opposition geometry" relation CS, which is a *non-arrow* relation, as being directly "challenged" by the "implication geometry" *arrow* relation RR (which in fact seems to be none other than classical implication itself), we see that the two occupy exactly the same four edges of the tri-segment, and this is, again, the first reason of the "buzz" we did with "valuation", starting from Sect. 3.6, and which motivated the δεύτερος πλοῦς (second navigation) of this Sect. 4. In the tri-segment, things are effectively so, but *not so* in the tri-*triangle* (or higher tri-simplicial space)! There, there are two different kinds of CS and at least two different kinds of RR. Therefore, in the Aristotelian combination we operate now, we are better inspired in using, for expressing the relation CS/RR, *not* the classical gray RR arrow (which in the bi-simplicial space of Smessaert is a I/R arrow and in the tri-simplicial space of the tri-*triangle* is a II/RR arrow) but rather a suited new *black* arrow, keeping for it a combination of the arrow *shape* of the RR and of the black *color* of the CS (Fig. 116).

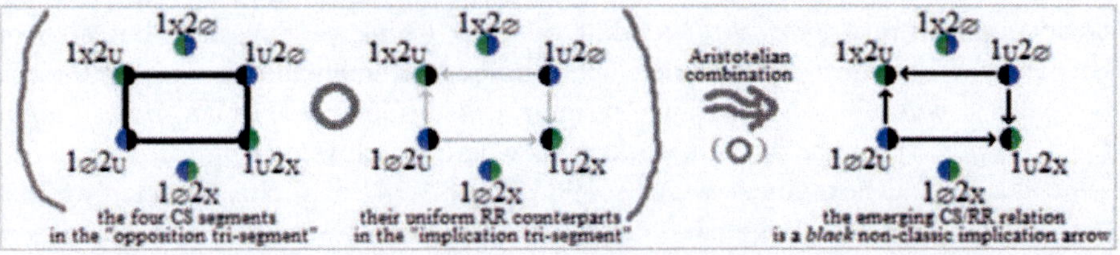

Fig. 116 Toward *the* "Aristotelian tri-segment": the *Aristotelian* composed relation "CS/RR" and its *black* arrow

Third, still dealing with implication arrows, let us now turn to the "implication geometry" BR and RB, porous black and porous gray arrows (here porous means nothing special). They seem quite interesting: they seem to be compatible with the properties stressed by the "opposition geometry" segments CI and IS (paracompleteness and paraconsistency, one sees this in their two valuations) but express, as for them (i.e., in their "implication geometry"), full-fledged implications. So we operate here a further Aristotelian combination, generating, respectively, the CI/RB and the IS/BR new arrows: each will take the *color* of its "opposition geometry" component and the *arrow* shape (and direction!) of its "implication geometry" component (Fig. 117).

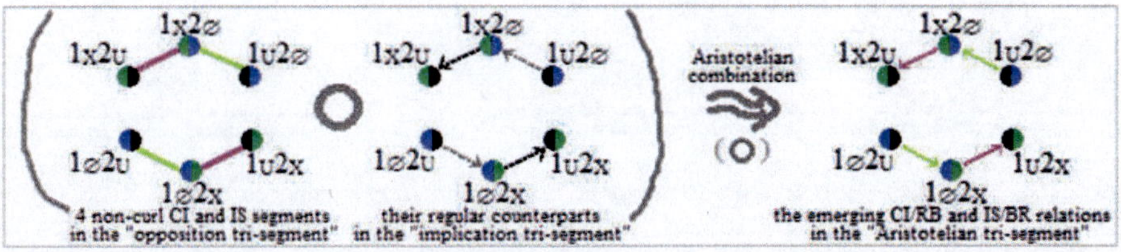

Fig. 117 Toward *the* "Aristotelian tri-segment": the *Aristotelian* composed relations CI/RB and IS/BR

But here one must remark that if any RB (respectively, any BR) arrow of the tri-*segment* is strictly coupled, as for the edge of the tri-*segment* where it takes place, with a CI (respectively, an IS) relation, *this is not true the other way round*: the CI (respectively, the IS) can also happen to overlap, in form of curls, with a light gray BB curl. This leads us to the next point.

Fourth, in fact, one must remark here that the classical biconditional BB of tri-simplicial "implication geometry" takes place in all the six curls of the tri-segment. But then this seems in some sense quite under-informative with respect to the corresponding colored curls in the Smessaertian twin "opposition geometry".

The gray classical self-implication, for sure a strong *logical* property, without any other indication, seems to be pretty tautological ("A $\leftrightarrow$ A"), and, worse, it erases the oppositive colors, not distinguishing (and not letting distinguish) between the II/BB (the tri-simplicial counterpart of the bi-simplicial case), CI/BB, and IS/BB: it transforms these three into the same (tauto-)*logical* relation. This is no good from a *structuralist* point of view (think of Saussure, but also of Blanché). What must be unfolded and studied systematically is the *differential* (i.e., the structuralist, Saussurian "système des différences"), a.k.a. *oppositional* (...), *system* (Blanché's main work [33], where he presented *philosophically* in 1966 the logical hexagon – with its 1967 sequel [34] explicitly directed against the *logicists* and the illogicists – was titled *Structures intellectuelles. Essai sur l'organisation systématique des concepts*). So, being (*oppositionally*, if not more globally *philosophically*) "Aristotelian", we want to show all *different* relations *as different* (Aristotle: "Saying the truth consists in presenting as united what is united, and as separated what is separated", i.e., "truth as adequacy", cf. [5, 144]). So we operate here one more "Aristotelian combination", to the effect of which each curl of the Aristotelian tri-segment will take from its "implication geometry" side, BB, the *shape* (double-sided arrow), but keeping also something of its "opposition geometry" side, namely, the *color* of the nonclassical CI or IS counterparts – remark that we will keep for "II/BB" the classical light gray color of the bi-simplicial I/B (Fig. 118).

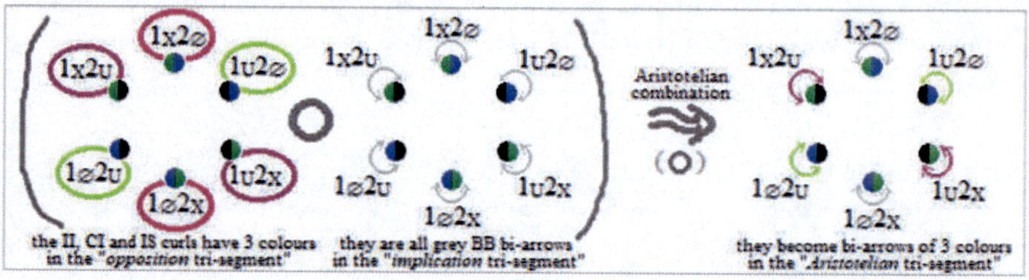

Fig. 118 Toward *the* "Aristotelian tri-segment": the *Aristotelian* composed curls II/BB, CI/BB, IS/BB

The last element to be dealt with, so to end our little journey, is the red classical diagonal of 2-oppositional contradiction (we saw already the two other diagonals, Figs. 114 and 115) which remains here, as NN/AA, exactly as it is in the bi-simplicial *Aristotelian* space, namely, N/A.

So, finally we arrive to something like a systematic "combination table", offering a global view of the 21 combined qualities of the 21 edges of the tri-segment. And this is, *pace Smessaert Demeyque*, the entry gates, if not to paradise (oy!), at least to what seems to be a full-fledged "*Aristotelian* tri-segment" (Fig. 119).

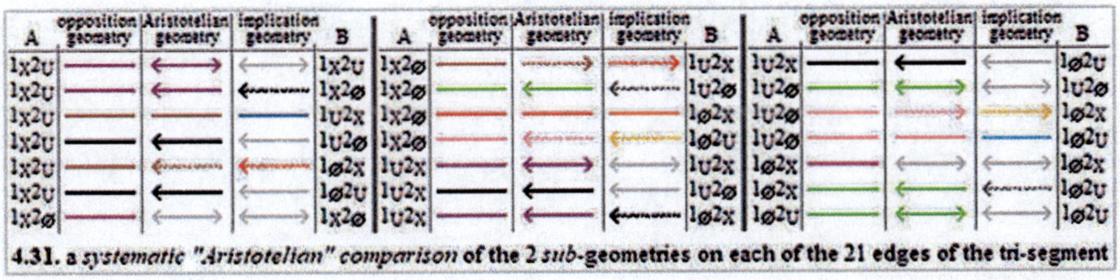

Fig. 119 A *systematic "Aristotelian" comparison* of the 2 *sub*-geometries on each of the 21 edges of the tri-segment

It seems we can therefore arrive, in full *conceptual* rigor (which is something orthogonal to the neo-Scholastic *furor axiomaticus* of the logicists), to an interesting, meaningful, and – most importantly – mathematically quite "natural" (instead of arbitrary, *ad hoc*, suboptimal, *bricolé*, etc.) "*Aristotelian* tri-segment" (Fig. 120).

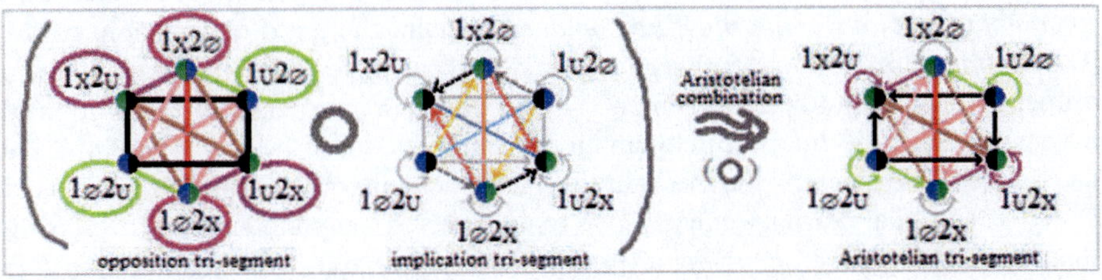

Fig. 120 Out of the two tri-segments there is, emerging, *a mathematically very natural "Aristotelian tri-segment"*!

In order to sum up (for we will spare the reader giving her/it/him a third instance (!) of inner analysis of this structure, we pretty successfully got to), we can have at least a look at the two valuations of this newborn, alive and kicking "Aristotelian tri-segment", verifying visually that its truth-value relations seem fully reasonable (i.e., conform to the different constraints laid by its two Smessaertian *sub*-geometries and the *non*logical-geometrical composition they induce by it) (Fig. 121).

The final result of this, at this level of our inquiry, is that for *any* of the 21 edges (curls included) of the tri-triangle, it seems we could find a fully reasonable and meaningful combination of its two Smessaertian *sub*-geometries (i.e., the one which was first expounded in Sect. 2.6 and the one which was first expounded in Sect. 4.3), not forgetting that we are working in a fragment: we are considering $2 + 2 = 4$, instead of the total $3 + 6 = 9$, "Aristotelian" (on truth-value identity) and "Smessaertian" (on truth-value difference) meta-questions (Sects. 4.1, 5.1, and 5.2).

If we look now for some provisory, general condensed expression of the "Aristotelian combining methodology" we are *tentatively* proposing, we can consider something like the (informal!) following: (1) one has to play with (i) shapes

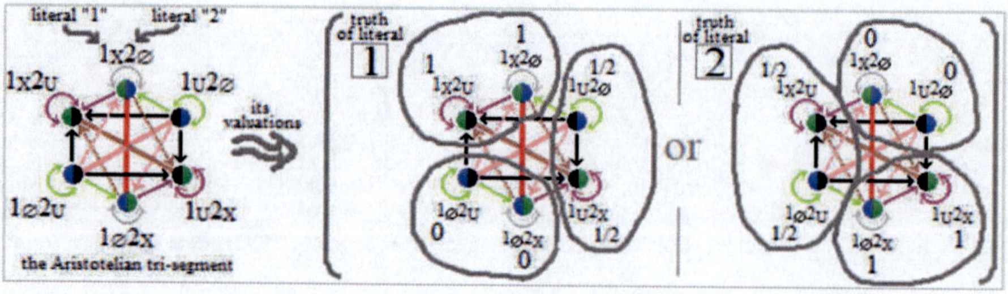

Fig. 121 The two valuations of the "Aristotelian geometry" tri-segment, articulating its two *sub*-geometries

and (ii) colors of any edge (under its expression in both Smessaertian twin and parallel *sub*-geometries); (2) one has to *combine* (rather than *substitute*!) the two versions of each edge, *creating*, for that, new formal arrangements respectful of the starting two not yet combined components; (3) the latter means, rather simply, that *one must tend to distinguish (or separate) what must be distinguished and to combine (or join) what must be combined* (again: the good old wisdom of the *Aristotelian* definition of "truth as adequacy"); (4) also, very importantly, one must prudently (and open to rearrangements) always keep in mind that the exploration of higher levels of the infinite, mathematically complex space of the poly-simplexes can induce unexpected, but meaningful, feedback effects (asking for wise and patient theoretical *re*arrangements), due to unexpected, but *mathematically natural*, theoretical *emergence phenomena* ([145], p. 19, point "d"); (5) the latter is *in line with Gödel's famous anti-formalist and anti-logicist discovery (of 1931,* [104]*) on the impossibility to rule once and forever mathematical serious things (*i.e., *numbers and higher) by a fixed Russell-Whitehead-style axiomatics* (i.e., *the impossibility of the logicist dream* – for us scary – of a world reducible *de jure* and *de facto* to compositions of "0" and "1", cf. [68]), which is *compatible with a structuralist common sense (*i.e., *remaining always open to the emergence of unexpected new forms of structure).*

But what we just saw in this Sect. 4.5 seems to be, if one now thinks of it, a small but nice enough *coup de théâtre*: Aristotelian geometry, as we just discovered and (*cum grano salis*) "proved", is not quite much a primitive (and "dummy"), imperfect version of "logical geometry", destined since 2011 to remain forever in the morn and dusty prescientific *past* (and shadows) of the now eternal light of the latter. Rather, "Aristotelian geometry" is, provided it is worked out *properly* (i.e., methodically, i.e., *in primis* with a mathematical – Platonic! – open-minded eye on the poly-simplexes!) and bottom-up (with a stress on "up"!), potentially (if *all* possible "Smessaertian" implication-questions are integrated! Cf. Sect. 4.1) the appealing *mathematical closure* of what Smessaert and Demey call "logical geometry"®. Let us now try to focus, before closing this Sect. 4, on this important question.

4.6 *"Logical Geometry" or "Poly-Simplicial Oppositional Geometry"?*

Smessaert's discovery, in 2011, of the parallel existence of two geometries, is a major discovery in our common field (be it called "oppositional geometry", "logical geometry", or something else). Our present study (Sect. 4) confirms it, if needed, and at a level of analysis that Smessaert and Demey themselves so far seemingly did not consider: that of the "poly-simplexes". Our present study, notably, seems to confirm Smessaert's and Demey's reasoning to the effect of which the classical "geometry of oppositions" (which *in some sense* contains among others Angot-Pellissier's and myself's "oppositional geometry"), which they call "Aristotelian geometry", is indeed a hybrid mixture of what they call the "opposition geometry" and the "implication geometry" (ch.4). But the *sense* in which such a "hybrid" must be understood is now to be discussed, for it appears to be possibly rather different from what Smessaert and Demey think and teach.

As I understand it, Smessaert and Demey claim that the "Aristotelian" approach to oppositions has mixed (without knowing it) two fundamental geometries (discovered 2.400 years later by Smessaert) and that *any* current researchers developing "avatars" of the traditional Aristotelian structures (such as our own "oppositional geometry", with its theory of the *bi-* and of the *poly-*simplexes) keep making the same old "Aristotelian mistake": they do not realize that, truly speaking, there are two rather different geometries at work and that the one relative to opposition (generally the one mainly investigated by naive researchers) is only one of the two. Now, "since 'logic' deals both with *opposition* (negation) and *implication*", Smessaert and Demey propose to baptize the global geometry emerging from their "opposition geometry" and their "implication geometry" with the name "*logical* geometry" (Fig. 122).

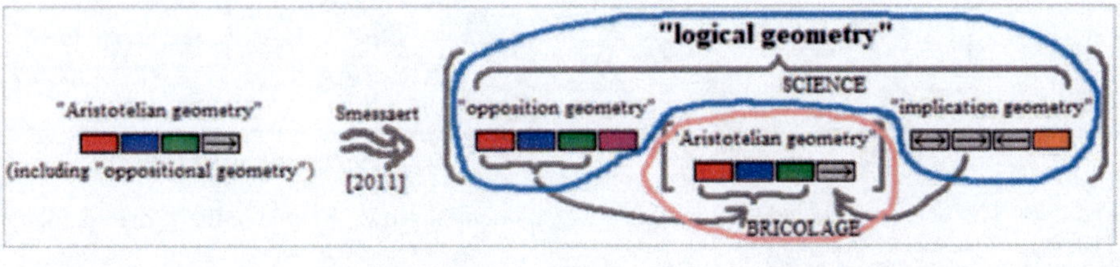

Fig. 122 "Logical geometry" explaining the *bricolage* of "Aristotelian geometry", in all its forms, past *and future*

Thus doing, however, Smessaert and Demey commit in my opinion four non-negligible mistakes: (1) they keep refusing (drastically) to take into account the importance of the presence of "bi-simplexes" (and the *geometrical* consequences this entails) in the "geometry of oppositions" (the refinement goes so far that they honor me writing papers, like [50], partly on my 2004 "*bi-tetrahedron*", "logical cube", i.e., the A4, but always calling it "Moretti's *octagon*"!); (2) they underestimate the idea that from the concept of (oppositional) *bi*-simplex emerges very naturally that of (oppositional) *poly*-simplex (and that therefore its investigations should be worth some attention, some support, or at very least some *mention*); (3) they misunderstand, in some (important!) sense, the nature of the fundamental relation between their own two geometries; (4) they try to impose *urbi et orbi – ESSLLI and JoLLI!* – as common name for the discipline of *anyone* dealing geometrically with oppositions and, as *the* real "scientific standard", the – alas – problematic label "logical geometry". These four non-negligible "mistakes", if I am not mistaken myself, are quite related. Let us try to see why.

A first important starting point is *the non-negligible "extra structure" imposed to oppositional geometry and/or logical geometry by the fact of going, as we went here, from the bi-simplicial space to the poly-simplicial space*. For, it reveals things (i.e., formal behaviors, mathematical regularities, structures) that seemingly were not so easy to perceive (and in fact seemingly were not perceived!) in the *bi*-simplicial space. But, the latter is – this point is capital and worth repetitions – *the* space where Smessaert and Demey (and therefore "logical geometry") so far remain, without however recognizing overtly that this space where "logical geometry" voluntarily remains is, in some important sense (a Pascalian sense!), a "bi-simplicial space". Why speaking of extra structure? Because our present study – although only the fragment of a future, more complete one (Sects. 4.1, 5.1, and 5.2) – dramatically unveiled that what we proposed to call the "Smessaertian" 3^2-lattice (a *structure*!) is, in fact, way more complex than its "logical-geometrical" ancestor, the official (and unique) Smessaertian 2^2-lattice. If our own simplification (with respect to R and L) is correct, the *simplified* Smessaertian 3^2-lattice at play for the tri-*segment*, as said (Sect. 4.1), is then the following not so simple extra structure (Fig. 123).

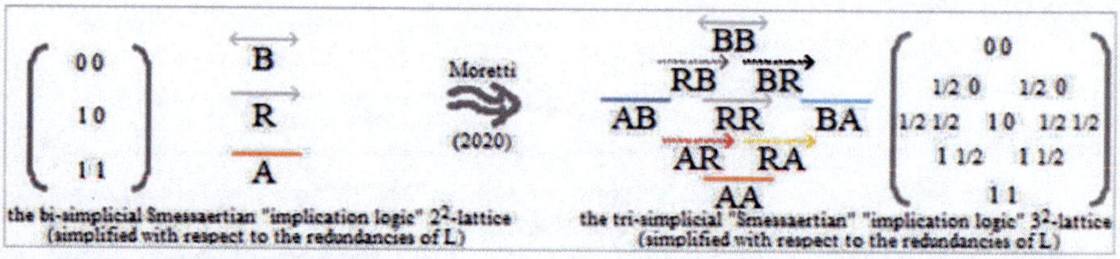

Fig. 123 The unexpected, and very meaningful, exponential growth visible in the "Smessaertian" 3^2-lattice

What one must notice here is that there is a quite big hiatus between the degree of combinatorial complexity of the two "implication geometry" structures (i.e., the 2^2-lattice and the 3^2-lattice): not only we step from three to nine qualitative kinds (once the redundant L simplified), but we pass from a situation where the three kinds (B, R, A) were strictly distinct (2^2-lattice), to a situation where six among the nine kinds are *mixed* (3^2-lattice); in the two constitutive Angot-Pellissierian sheaf-levels (U and X) of the nine kinds of the 3^2-lattice can happen to be put, side by side, very heterogeneous relations (such as "A" and "B" in AB, etc.). This extra structure, which is not even yet the whole story (it does not yet encompass all the possible "Aristotelian" and "Smessaertian" questions, Sects. 4.1, 5.1, and 5.2), introduces already a big *qualitative* jump.

Now, the bi-simplicial 2^2-case (left side of Fig. 123) seemingly gave to Smessaert and Demey the misleading idea of a stability (and simplicity) of the "strange parallelism" of the two Smessaertian *twin* geometries, a strange parallelism which in some sense, at least currently, seems to be *the essence of the methodology of "logical geometry"* (Fig. 122) (to slightly nuance this: maybe Smessaert conceives somewhere the $2 + 2 = 4$ twin questions as a *whole*, dictated by a *global* "logical" combinatorics over the possible truth-values of A and B?). Stability means the same two geometries, more or less, show up always parallel, *always in the same qualitative proportions*, for any (bi-simplicial!) Bn-structure: red segment B2, logical hexagon B3, logical tetrahexahedron (a.k.a. rhombic dodecahedron) B4, etc. Such an apparent stability (which is a stability of the two twin 2^2-lattices, the Aristotelian and the Smessaertian, both simple and unique!) seemingly suggested them, for short, that the two geometries have no deep relations (other than *mysterious* coexistence) and, most importantly, that it is potentially misleading to "unite" them, given that this happens by "surgery" (i.e., *by mutilation*, as, in fact, in the historically attested bi-simplicial "Aristotelian geometry", Sect. 2.2, Fig. 40 and, here, Fig. 122) into a unique one: for short, *bricolage* is of course useful and up to a certain extent tolerated, but suboptimal with respect to methodical (axiomatic!) science. Remark that this belief, I am ascribing them, in a (deceitful) stability is *apparently* not too harmful to them in so much their very rigorous and valuable study of many other phenomena gives them work enough (and, again, very valuable work). Still, it seems that something precious (Pascalian?) here thus dropped, at least momentarily, out of view (but not only for them: for anyone *following their logicist advice*), with potential harm for our entire discipline.

As we saw, the tri-simplicial case (of which the study is still at the very primordial beginning) reveals however (Sect. 4.5) (1) that *the "strange parallelism" of the twin geometries is not stable at all* (provided one does not stick, somewhat *geometry-blind*, to the *bi*-simplicial space) (2) and that, therefore, it is not the theoretical *terminus ad quem*, of "our common discipline", but rather the *terminus ab quo*: it is not its *all*-encompassing horizon, but just an exciting starting point!

But this, then, means quite much, speaking less abstractly: it means that the higher you go in the poly-simplexes, the more complex are, *there*, the (meaningful!) overlaps of the two Smessaertian twin *sub*-geometries, themselves more and more *geometrically* complex the way up in the higher poly-simplexes. One must stress this point: *both* geometries have, each, an exponentially growing complexity (unknown "by construction" to logical geometry), so that their *combination* (Sect. 4.5) has even more complexity (not only is it a growing series: it is a "*geometrically* growing" series!). This is quite important; it means that (1) the *combinations* of the two "parallel" Smessaertian twin *sub*-geometries seemingly are by no means "forever frozen" (as is the combinatorially morn "Aristotelian combination" proper to the bi-simplicial space, limited to putting right-implication, by "surgery", in the place of noncontradiction, Figs. 40 and 122); (2) *these combinations are, far from it, the general rule of the poly-simplicial space* (for understanding its combined *qualities*); and (3) more precisely, these combinations are the key for understanding the very nature of *each* possible edge (or curl) of *any* general opposition/implication n-dimensional polytope! For this reason, it seems to me that you cannot quite *understand* the (very important!) interplay of the "opposition geometry" with the "implication geometry" (preciously offered to us by logical geometry) by remaining in the bi-simplicial space (as until now seem to be doing voluntarily Smessaert and Demey). Because there, in the bi-simplicial space, this interplay not only does not change enough: it simply does not change at all! It remains "forever" at the level of the 2011 "Smessaertian 2^2-semantics" and of the "Smessaertian 2^2-lattice" (echoing the twin "Aristotelian" 2^2-semantics and 2^2-lattice). For this reason, very paradoxically (with respect to their otherwise quite impressive and valuable work), Smessaert and Demey's thinking about "combinations" seemingly has not changed much since its beginning (although they explore 1.000 and 1 varieties of *ad hoc*, suboptimal existing combinations and fragments, from many other – sometimes major, sometimes not – authors past or present, to which they give systematic conceptual and terminological order: but top-down!). For short, you do understand *what is really at stake with Smessaert's groundbreaking discovery* (of the twin geometries), seemingly, only in the poly-simplicial space (or in a structurally similar playground), where this interplay changes, and changes with an exponentially growing complexity!

But this, in turn, means in some sense that what Smessaert and Demey call "Aristotelian geometry" is in fact not, as they think, a *suboptimal ancestor* (because fruit of *unconscious bricolage*) of "logical geometry" (the latter being supposed to be the firm scientific standard of our general discipline) but – rather – the real thing to be studied from a mathematically serious (i.e., nonlogicist) viewpoint! Something like "Aristotelian geometry" (or any comparable equivalent name) appears, paradoxically, to be (Sect. 4.5) not the *limitation*, but the mathematical *limit* (in the powerful, positive meaning of this word), or the "mathematical closure", of what Smessaert and Demey call a little bit recklessly "*logical* geometry" and by no means the other way round! In other terms, there seems to be, here, yet another non-negligible mistake (the third), in my opinion, in Smessaert and Demey's very admirable, but also terminologically dangerously normative program, namely, a confusion between (1) the idea of "*choosing without creating*" (as it seems to be, between the two geometries, in the *bi*-simplicial space) and (2) the idea (not yet clearly assessed by Smessaert and Demey, so far they do not dare enter *poly*-simpliciality) of "*combining methodically bottom-up*", at each poly-simplicial stage (this idea which they seem to miss so far is, on the contrary, in line with Béziau's – structuralist – fertile and open idea of "universal logic", as an intended parallel with the *structuralist* idea, preoccupied of thinking about mathematical "mother structures", of "universal algebra" – "universal" being in the programs of both, universal algebra and universal logic, the fruit of *systematic infinite variations and combinations* – a powerful idea of Bernard Bolzano, cf. [36, 37, 84, 129], and Sect. 5.5). Remark, again, that Smessaert and Demey produce an impressive number of valuable studies over combinations, but so to say always top-down (i.e., *"logical geometry" clarifies with benevolence "suboptimal" materials*, mostly of the past) and not bottom-up, i.e., *not yet investigating new mathematical (oppositional!) spaces* (Fig. 124).

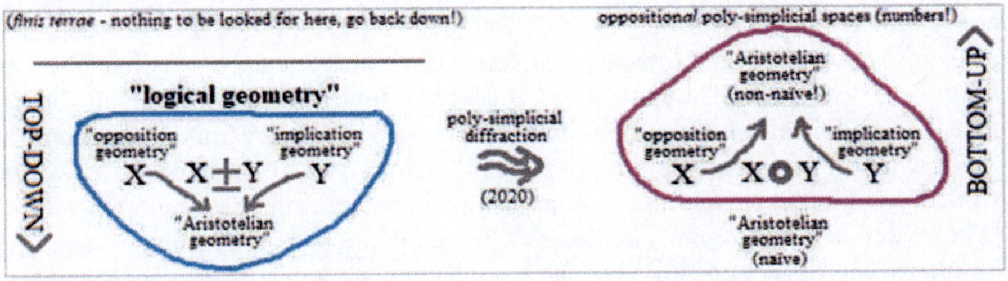

Fig. 124 "Aristotelian geometry", understood correctly, is not a "merry bricolage", but a systematic *combination!*

The mistake (the third) – if I am not mistaken myself – on the nature of "combining" the two geometries (grounded, I believe, on mistakes – the first and the second – on the bi- and poly-simplicial nature of the core of the theory of *opposition*, Sect. 1.5) introduces the fourth and last of the non-negligible mistakes we suggested to consider at the beginning of this section: that on the name to be given to our common theory. Naming it "*logical* geometry" is, for the experimented (if not employed) *general* philosopher I am, at least "surprising", if not clearly shocking. Remember that it has taken a *very* long time before something like the idea of a decent *geometry* of oppositions could be born (around 2004, [93]): and this is, without exaggeration, a major conceptual revolution, since it means that *"opposition", a very intuitive concept, common to almost all known human cultures, enters thus, unexpectedly, a mathematical legality: opposition is becoming, under our eyes, a mathematical object of its own (and a rich one)*. But one must be aware that historically at least two important thinking schools notoriously refuse(d) the simple possibility of the emergence of such a kind of revolution (for both already fought, mercilessly, against structuralism, which was very close – notably with Blanché [33, 34] – to unfolding such "elementary *structures* of opposition"): these are "dialectics" (Sect. 5.5) and the ambassador of logicism, "analytical philosophy". The latter did and does it on ground of its founding (and *never* dying) constitutional *logicism*: "things must be reduced to 'logic,' and logic is *the* tool, *the* structure, and *the* key (even of mathematics)". The sterility and harmfulness, for mathematical research, of logicist extremism (a pleonasm) should, in principle, not have to be proven, again and again, in 2020 (Zalamea's *Synthetic Philosophy of Contemporary Mathematics* of 2009 [146] is, if one looks for one, a masterpiece of intelligence in explaining, painstakingly, the seriousness of this point – but cf. also Mandelbrot [88]). But so it is not: as is known, "analytical philosophy", which (in good company with the comparable deliria of Hegelianism, positivism, Marxism, phenomenology, and many other past and future) has pretended (and still pretends!) being able to "make of philosophy a science" (!), despite being as such – beyond face lifts – a programmatic "dead horse" (we recall why in [97, 99], relying among others on [67] and [104]), still holds, by inertia and worldwide, an academic strong position of power (currently symbiotic with computer science and the growing *market* of the "smart technologies"), and as such it continues to push forward its "little gray soldiers" (a direct, nondiplomatic, but also non-flat, description of this can be found, among others, in J.-Y. Girard: [68–72]). But then is it also needed here to recall that analytical philosophy and logicism, far from developing themselves (when? how?) the *geometrical* study of "oppositions", hindered it times and times again, and by all means? Analytical philosophy produced rivers (no: oceans!) of ink about "logic", "contradiction", "negation", "implication", "tautology", "truth-values", "possible worlds", and the like. But, unless I am mistaken, one finds hardly *a single word* of Wittgenstein (& Co) about "contrariety" (which is the most characteristic concept of opposition – Sect. 1.6, Fig. 33 – and precisely *the* one requiring, for expressing *n*-contrariety and *n*-opposition, cf. Fig. 9, the concept of simplex!): and this is, paradigmatically, still the position of a Parsons, in the prestigious and "standard" *Stanford Encyclopedia of Philosophy* (Sect. 1.2); in his top-reference paper for the analytical philosophy world on the "logical square" [107], *the existence of the "logical hexagon" (1950) is not even mentioned...* (!). So, the problem with logicism is not only "ideological" in a general sense (it hinders "clumsily" the unfolding of fruitful new mathematical – and philosophical! – ideas, cf. Girard [68–72], Mandelbrot [88], and Zalamea [146]) but also in a very concrete sense: in the precise case of oppositional matters, it has proven, very specifically, times and times again, that different forms of logicism have voluntarily "killed in the egg", in a reflex of ("institutional") self-defense, several promising, embryonal developments of the *geometry* of oppositions (cf. [97, 99]). Is it necessary to recall that analytical philosophers have been dismissing (and urging others to dismiss) for more than a century, relentlessly, the square of opposition (Sect. 1.1) – and in more recent times the logical hexagon – notably because of the alleged "paradoxes of existential import" (realizing only in 2013, cf. [43], that this paradox is in fact a pseudo-paradox), which is a mathematical bad joke: *judging normatively mathematical creativity (for, here we are) from the viewpoint of logic, and not the other way round* (i.e., *judging mathematical logic for its mathematical creativity), is the world upside-down, a very bad joke, historically attested (and persisting), but still devastating.*

For these reasons, and because of its very dubious name and, if it does not change, because of what seems to be its main methodology of "schizophrenic" frozen parallelism of the two twin (micro!) geometries with respect to bottom-up pure geometrical exploration, logical geometry, *nolens volens*, and despite the impressive, increasing crop of its valuable scriptural and conceptual productions in the best journals, runs very seriously the risk of becoming one more logicist machine for killing the radicalness of the emergence of a full-fledged "*oppositional* geometry" (or, if one prefers, of a "geometry of oppositions"). Again, the paradox to be understood, and defeated, is that what Smessaert and Demey call, seemingly with soft irony, "*Aristotelian* geometry" (as meaning "inferior to *logical* geometry") is in fact not a stupid ancestor (or a bizarre fossil of the past), but it is rather the higher-order mathematical methodology (but which possibly should not be mislabeled "Aristotelian"...) to be unfolded (through an exploration of the poly-simplicial space) and followed in the future! The future of our common discipline (the geometry of oppositions), at the level of the exploration of *deep mathematical still unknown ideas* if not at that of the "academic *Zeitgeist*", will very seemingly consist, despite all logicist efforts to refrain it, in exploring systematically the overlaps of, among others, the twin *sub*-geometries of the *geometry of oppositions* and in finding techniques for expressing bottom-up (stressing the "up"!) the *autonomous* reality of the *mathematical* (and not "logical") thing. So, in my opinion the question of the name of our general and/or global common theory is very important, and it remains problematically open, "poly-simplicial opposition*al* geometry" seeming so far a much better name.

Back to the successful (although still fragmentary, Sect. 4.1) tri-simplicial diffraction of the 2-opposition*al* red segment of contradiction, if one considers (as we take now the risk of doing here – readers will have to judge) that we seem to have solved satisfactorily the last general important technical question (the tri-segment's two possible global valuations and what these imply, Sects. 3.6, 4.3 and 4.5), it seems that we are now in a position for seriously considering (only sketchily, alas) the question of the possible concrete *applications* of the oppositional tri-segment and that of the consequences that this new possible mathematical structure (to be refined and brought to its closure in the future, Sect. 4.1) has on some well-known other issues related to "contradictions".

5 *Consequences/Applications* of the Tri-segment: Some Remarks

We will, at last, be in a position of making in this concluding Sect. 5 some quick remarks on future possible applications of the tri-segment (and similar poly-segments and tri-simplexes). The concept of contradiction/negation, as distinguished from that of contrariety, has the particularity of being very important in the "exact sciences", but it is also important, in some cases, in the humanities (where contrariety seems, however, much more important). The *oppositional-geometrical* diffraction of the concept of "contradiction", a concept which can happen to generate several misleading *fantasies*, formally speaking may concern, *in primis*, three particular disciplines – many-valued logics, paraconsistent logics, and quantum logics – in so far each of these three pretends to have a very special relation to contradiction/negation. We will try to recall the issues at stake here in Sects. 5.2, 5.3, and 5.4. As for the humanities, contradiction is more or less the focus (or high spot) of at least two among the few major thinking traditions of the last one or two centuries: dialectics and psychoanalysis. We will try to recall this as well, in Sects. 5.5 and 5.6, where we will end by a surprise in the very last pages (the cherry on our oppositional "partitioning cake"). But first we will propose in Sect. 5.1 some preliminary (and necessary) remarks on the general meaning and *limitations* of our present inquiry and on its results and perspectives.

5.1 Some General Remarks on What Has Been Seen so Far in This Study

The "Renaissance" of the geometry of oppositions is taking place, mainly thanks to the intellectual and institutional efforts of Jean-Yves Béziau, since nearly 20 years (Sects. 1.1 and 1.2, [28–30]). Since more than 10 years, it has been signaled (by us, [94]) that one of the main issues at stake with the geometry of oppositions (however you prefer to name it) seems to involve oppositional "poly-simplexes" (Sect. 1.3). But for several reasons (among which – but alas not only – a natural, if not glorious, "conceptual inertia"), this message has not been received so far. The present study should have, at least, succeeded, in principle, in making loud and clear that the complexity involved in such poly-simplexes, *a posteriori*, not only exists for real (mathematically speaking) but is in fact much higher than what we perceived and therefore believed in 2009. Far from being a confused fantasy of mine, the *oppositional* poly-simplexes do exist mathematically speaking and are very promising and exciting: and we have, at last, powerful and reliable tools for dealing systematically with them, *in primis* Angot-Pellissier's sheafing technique (Sect. 1.4, [3]). But the game is much more complex, technical and rich than it was perceivable at the beginning (2007–2009). This is first of all true of the global "Pascalian structure" of the general space of the poly-simplexes (Sect. 1.5): this means that, unexpectedly, the poly-simplexes are in fact, so to say, poly-*bi*-simplexes, and this involves that there is much more "structure" than what was thought; for instance, in a tri-simplex, the three main composing bi-simplexes (whatever the *dimension* of the simplex under examination) are such that each involves its own particular instance of simplex for any of the two bi-simplicial *colors* it has, e.g., the "blue simplex" of the "blue and green" bi-simplex is not the same as the "blue simplex" of the "blue and black" bi-simplex, etc. – this is directly readable in the Pascalian roadmap, Sect. 1.5 (provided one learned how to read it), but becomes more concretely clear when one analyses, by humble and down-to-earth calculations, the tri-triangle (or higher). Remark that the natural way for coping with this (i.e., the unexpected "diffraction" of the simplexes) consists, first of all, in introducing a new element in our convention for coloring the vertices (Sect. 2.5): namely, one for coloring, with "oppositional hostages" (Sect. 3.6) inside, the vertices of the *simplexes* (the same technique can represent either the real vertices of a precise simplex or, as in Fig. 125, whole generic simplexes taken as a dimensionally undetermined whole) (Fig. 125).

But this increase in structure is also true relatively to our discovery, made in Sect. 4, that Smessaert and Demey's "logical geometry" intervenes *in* the poly-simplexes and in a way much more rich and complex than what we believed (Sect. 2.2): a

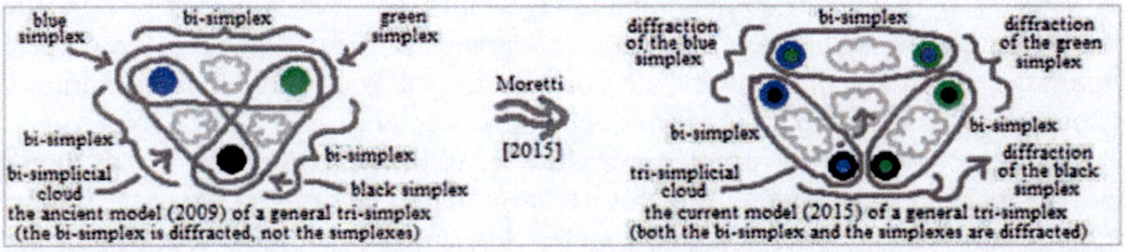

Fig. 125 The poly-simplexes behave as poly-*bi*-simplexes: each simplex becomes "diffracted" by "its" bi-simplexes

way that forced us (Sect. 4.1), from now on, to adopt logical geometry as a very important *part* of oppositional poly-simplicial geometry, namely, as generating one of its two systematic *sub*-geometries. In that respect, our choice in this paper of limiting ourselves to the most "primitive" case of poly-simplex ("poly" ≥ 3), that of the tri-segment (Sect. 1.5, Fig. 29), *a posteriori* revealed to be a rather wise but also fruitful move. For, the structure we investigated here, the tri-segment, is "simple" (it is a "three-cloud" deprived of simplexes, cf. Fig. 126), but far from trivial, and at least we obtained a rather clear, global understanding of it and, through it, a *starting* global understanding of the more general concept of *tri*-simplex and of poly-*segment* (Fig. 126).

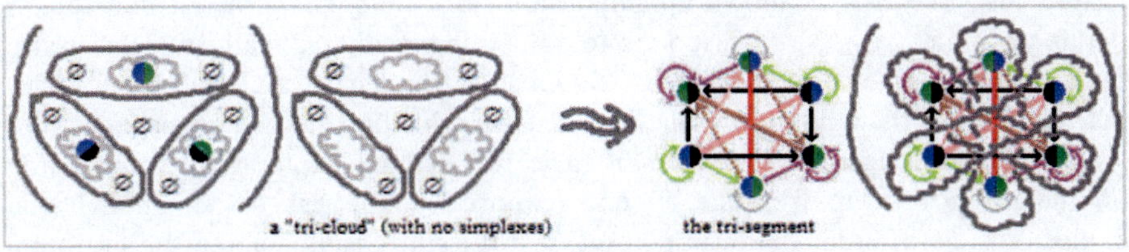

Fig. 126 The "tri-simplicial diffraction" of the red contradiction segment is a "tri-cloud" (having no simplexes)

So this choice paid in at least two respects: (1) we proved the existence (Sect. 2.6) of this until now unknown structure (conjectured by me in 2009, Sect. 1.6), the tri-segment, which is very important (for, it is *the very first full-fledged poly-simplex* and it is *the first mathematical diffraction of contradiction*); and (2) through it we made at least four rather important discoveries about: (i) what must be considered as "oppositionally extremum" (i.e., each n-simplex has not two, but n extrema!) (Sect. 2.4), (ii) how must be treated mathematically, through colors, the *vertices* of general oppositional-geometrical solids (and this is nothing less, if you think of it, than the embryo of a new chapter of graph theory!) (Sect. 2.5), (iii) how to deal successfully, for any poly-simplex, with its "valuations" (which is possibly the embrio of a new chapter of many-valued logics) (Sect. 3.6), and (iv) how must be developed systematically, from its two Smessaertian *sub*-geometries, with the concept of "Aristotelian combination" and against the trap which is laid by the misleading name (and program) of "*logical* geometry" (Sect. 4.6), something like a *non-naïve* "Aristotelian geometry" or more precisely a "poly-simplicial oppositio*nal* geometry" (Sect. 4.5, and this is possibly a new chapter of... logical geometry!).

The oppositional poly-simplexes are a successful generalization of the oppositional bi-simplexes (including, this is very important, their closures, the Bn-structures), which, as we recalled (having learned it first from Angot-Pellissier [111]), are quite important *new* mathematical tools (Sect. 1.2): the oppositional Bn-structures, generating new kinds – *oppositional* kinds! – of "equivalence classes", allow us naming, measuring, and thus thinking about "oppositional complexity", and they make of "opposition" a new mathematical *object*. The *poly*-simplexes, therefore, should enable us, from now on, to extend this conceptual and formal mathematical new power, relatively to new situations where "oppositional valuations", finer-grained than two-valued, will be needed (and this is what we will try to overview in the next Sects. 5.2, 5.3, 5.4, 5.5 and 5.6).

However, at least three further considerations must now be added to this. First, one important point to be remembered is that, so far, we nevertheless remained strictly *inside* the Aristotelian (and Smessaertian) p^2-lattices (the stress here is on the exponent "2"): this means that the "q" parameter (i.e., the number of meta-*questions*) of the general p^q-lattices (Sect. 1.3) is not yet being explored. The reason is that we still seem to lack, for this, something like an analog of the Angot-Pellissierian successful formal techniques of 2008 and 2013 [3, 111]. It might be the case that $q \geq 3$ shows up impossible. But if it will turn out that the "q parameter" (i.e., $q \geq 3$) corresponds to something mathematically real (as I still believe at the moment, given the promising results found in several draft preliminary investigations), than it seems that, at least in principle, it will be necessary to go patiently through its *a priori* non-easy, full-fledged exploration, (re-)reading *all* the p^2- poly-simplexes one by one (exposed to the risk of a combinatorial explosion...), in order to get important, still missing insights about the profound meaning of the complex concept of oppositional poly-simplicial space (Fig. 127).

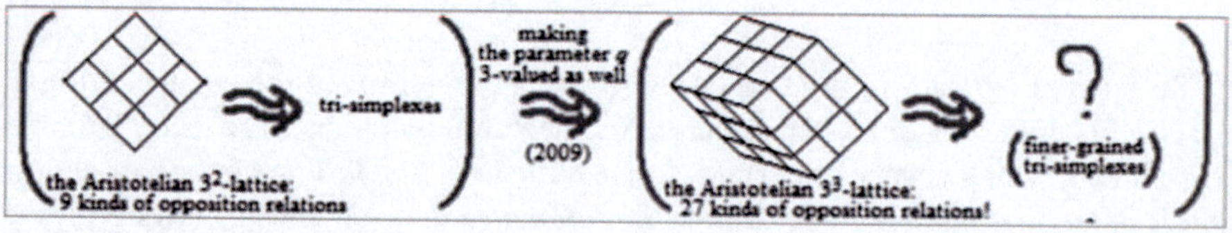

Fig. 127 From the Aristotelian (and Smessaertian) 3^2-lattice to the Aristotelian (and Smessaertian?) 3^3-lattice?

A second important point is that oppositional geometry, as we said (and as we gave further evidence for, in Sect. 1.5, but also in Sects. 2.5 and 4.3, Fig. 106), is at the crossroad of different mathematical distant "spaces". As such, its progressive, deep understanding is conditioned by future works in mathematical directions potentially quite heterogeneous and, generally, rather unpredictable. Remark, in that respect, that the theory of the poly-simplexes, although it belongs to "general mathematics" rather than to "mathematical logic" alone (Sects. 1.2 and 4.6), seems to offer interesting elements of natural conceptual radicalization to Béziau's theoretical framework (which in fact is *structuralist*, in the best sense of the word, since it looks for mathematical "mother structures", suggesting that "logic" might in fact be just *a* new one) of "universal logic" (this appears particularly in Sect. 4.5, with the notion of "Aristotelian *combination*" if it were, for instance, to be seen as a new case of "fusion").

Finally, a third important point, as we said (Sect. 1.6, Fig. 33), is that it must always be remembered here that the higher poly-simplexes will now *really* have to be explored methodically, starting from the tri-triangle (which opens the big jump into "real" poly-*simplexes*) and aiming at, as soon as possible, at least the tri-tetrahedron (so to obtain a first "poly-simplicial diffraction" of B4 (B4 being by now probably the most applied – and in that respect, *cum grano salis*, the most important – of the bi-simplicial oppositional-geometrical structures, Sect. 1.2, Fig. 6) in a way similar to the one successfully followed here for the tri-segment. Remark that we already have important and very encouraging unpublished results: (1) on higher poly-*segments* (quadri-segment and quinque-segment), the series of the poly-segments seems to offer a stable behavior and quite nice formal properties (and is the first "poly-simplicial *series*" which will be reachable soon enough), and (2) and on poly-triangles (we reached the closure of the tri-triangle and of the quadri-triangle, which are both *much* more complex than what appears in Angot-Pellissier's two precious, pioneer draft studies of 2013 [3] and 2014 [4] on the subject).

As said, in the following sections of this last Sect. 5 we are now going to try to give some quick hints and remarks on some possible applications of the tri-segment.

5.2 The Tri-segment and Many-Valued Logics: Some Remarks

As we have recalled (Sects. 1.3 and 1.4), the very idea of oppositional *poly*-simplex (of which the "geometry of oppositions" is a classical, *bi*-simplicial case) is strongly linked with the idea of *having more than two truth-values*, so the relations between the *general* theory of opposition and many-valued logics seem to be in some sense quite strong. And with respect to many-valued logics, the main results of the present study (as, for instance, the mathematical birth of the *Aristotelian* tri-segment (Sect. 4.5) but also the fundamental relevance of the Pascalian ND simplex, Sect. 1.5) seem indeed potentially important. The poly-simplexes, as we have seen, seem to open the direct and systematic study, up to now absent, of "many-valued *opposition*". However, as we have stressed several times, oppositions have shown up to be much more generally "mathematical" than specifically "logical" (Sect. 4.6). In this respect, our present inquiry seems to show that, given the now undeniable "Pascalian" side of opposition (Sect. 1.5), which puts forward not only *simplicial geometry* (the Pascalian ND *simplexes*) but also *arithmetic* (Pascal's *triangle*), well-known many-valued issues (e.g., the "MV-algebras", [20]) might have, because of the until now rather hidden or unnoticed presence in their heart of "opposition" (seemingly in terms of the, up to now, invisible *geometry* of the mutual relations of truth-values), to change something relatively to their "mathematical barycenter". For short, many-valued logic might be more "Pascalian" and/or "simplicial" than it was thought/known. In any case, it seems to me by now undisputable that the theory of the oppositional poly-simplexes offers *structure* to many-valued logics.

Currently the relations between many-valued logics and the poly-simplicial space seem, of course, to be at the very beginning of their possible exploration. On one side, precise oppositional-geometrical (poly-simplicial) studies on different known paradigmatic many-valued systems (like those of Łukasiewicz, Bochvar, Kleene..., cf. [106, 117]) should be carried out step by step. But on the other, for that, more poly-simplexes should also have been studied in *abstracto* extensively, and in particular at least some poly-*triangles*: as we recalled (Sect. 1.6, Fig. 33), the expressive (and conceptual) power of *segments* is nontrivial, but nevertheless comparatively low (with respect to triangles and higher), and higher simplexes are seemingly absolutely needed for that, starting from triangles (which open to "contrariety", a very important feature, absent in the poly-*segments*). Possibly related to these considerations, it seems to be still a little bit too early for studying easily the presence (and the action) of *many-valued connectives* inside poly-simplicial oppositional geometry (by analogy with the important presence and action of two-valued connectives in and for *bi*-simplicial oppositional geometry, cf. Sect. 1.1, Fig. 1). This important basic work still has to be done. Notice however that some non-negligible elements of knowledge in that respect seem, nevertheless, to be already emerging at the basic level of the tri-segment.

One can hope or predict that similar studies will, from now on, be carried also in the direction of what seems to be the mathematical (infinite) horizon of many-valued logics, namely, fuzzy logic (i.e., infinite-valued logic, cf. [20, 38, 75, 106, 117]). Here, several researchers (like, for instance, F. Cavaliere [41], P. Murinová [103], or D. Dubois, H. Prade and A. Rico [59], to name some recent researchers) looking for bridges between fuzziness and oppositions have already proposed many different interesting strategies: but focusing on drastic "shortcuts" (for avoiding a lethal complexity explosion), they do not seem yet to have taken in due consideration the idea (of 2009) that "many-valued oppositions" are (seemingly) to be seen, *as a systematic whole*, as poly-simplexes (Sect. 1.3). Of course, oppositional geometry, in all its variants, seems (so far) committed to *finite* numbers, whereas fuzzy logic, as remembered, is essentially an *infinite*-valued logic. But this openness of opposition theory to finite many-valuedness, through the poly-simplexes, allows, at least in principle, studying their *numerical progressions* and therefore opens, through the concept of *potential* (if not yet actual) *infinite*, the discussion about the geometrical patterns of possible infinite limits of these progressions.

Let us stress, here as well, that an important, and maybe even crucial point on that respect (fuzziness), not to be forgotten, is the potential reference of the poly-simplexes to the parameter "q" (Sects. 1.3, 4.1, and 5.1) of the Aristotelian and "Smessaertian" p^q-semantics. It seems reasonable to think, for instance, that a *real* three-valuedness (i.e., a *radically* three-valued one) would be closer to an Aristotelian (and a Smessaertian?) 3^3-lattice than to 3^2- ones. This parameter "q" ($q \geq 3$) might open to a much finer-grained approach: in a 3^3-lattice (i.e., with $q = 3$), there seem to be 27 instead of only 9 kinds of "opposition (and – *cum grano salis* – implication) qualities" (cf. Fig. 127). So, even with respect to the "spirit of fuzzy logic" (i.e., the idea of getting the more fine-grained you can and finer-grained than "false/true"), an approach to oppositional geometry based on the 3^3-lattice, rather than the 3^2-lattice, would seem, intuitively, more natural and complete. But currently the 3^3-lattice is still being investigated as a *hypothesis*, with no robust founding results already at hand on that so far. As said in Sect. 5.1, what seems to be still lacking us – although several pieces of the "q puzzle" are already there and promising – is *something like an adequate new Angot-Pellissierian mathematical tool able to cope, at the meta-level (the level of the meta-questions, precisely), with truth-values other than 0 and 1*. In fact, a further problem is that the Aristotelian and the Smessaertian possible meta-questions happen to have *different progression rates*: the former deal with "truth-value similarity" ("A and B true together", etc.), while the latter with "truth-value dissimilarity" ("A false while B true", etc.). So, to give an example, in a three-valued context (tri-simplexes), there could/should be three Aristotelian meta-questions (our Q1 and Q2, plus the new Q3: "Can A and B be ½ together?"), but six Smessaertian questions (Smessaert's Q'1 and Q'2, plus the following new four: Q'3 "Is it possible to have A 0 and B ½?"; Q'4 "Is it possible to have A ½ and B 0?"; Q'5 "Is it possible to have A ½ and B 1?"; Q'6 "Is it possible to have A 1 and B ½?"). So, the "Aristotelian" meta-questions (generating the "opposition geometries") can really be modeled, as we proposed in 2009, by the hypercubic, or measure-polytopic, p^q-lattices (Sect. 1.3, Fig. 12). And, as said, this currently seems to lack a suited Angot-Pellissierian mathematical tool. But, independently from that, the *Smessaertian* meta-questions (generating the "implication geometries") cannot be modeled as a whole (but maybe as parts?) by a p^q-lattice really parallel (i.e., with the same numerical values of p and q) to the Aristotelian one. The idea, put more clearly, is that intuitively the Aristotelian lattices should be p^p-lattices (same number of qualities of questions and of qualities of answers, given that both depend directly on the truth-values: the Aristotelian meta-questions are reflexive), whereas the Smessaertian lattices should be p^q-lattices where $q = p^2 - p$ (they embody the total number of possible nonredundant binary relations between p truth-values, i.e., p^2, minus the number of the reflexive ones, i.e., p, which are exactly the Aristotelian ones). The overall situation just described can be visualized (and in principle explained, by a simple combinatorial reasoning), again, with still one more instance of the series of the *simplexes* (here in their graph-theoretical suit of "complete graphs", or "cliques", cf. [125], p. 7), now interpreted in terms of possible binary relations between pairs of truth-values (Fig. 128).

The result is that if one aims at being "many-valuedly complete" (in the sense of being n-valued also in the meta-level – echoing, maybe, Suszko's concern about the danger of a "fake many-valuedness", [138]), this seems to cause a combinatorial explosiveness that might hang upon the researcher's head like a sword of Damocles.

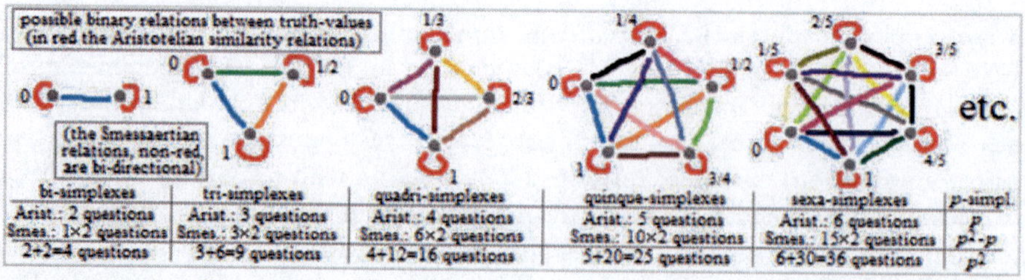

Fig. 128 The possible "meta-questions" conform to the graph-theoretical counterpart of the simplexes, the "cliques"

Again, a viable solution could maybe consist in decomposing the Smessaertian lattice into a multitude of more viable sub-lattices, supposing this is possible and meaningful. In that respect, what we have done in this study resembles a first step in between the current total absence of theory and a theory which could be complete (but at which combinatorial expensiveness' cost is still unclear).

Let us finally recall that, as we saw, tri-simplicial oppositional geometry (starting with the tri-segment) seems to show that there is an *intrinsic* deep link between three-valuedness (and higher!) and the metalogical triad "paracomplete, classical, paraconsistent" (Sects. 1.1 and 3.5, Fig. 81). And in fact, although we must stop here speaking directly about many-valued logics, in the following two sections, we are nevertheless going to have to look more specifically to particular cases of three-valuedness: in paraconsistent logics (Sect. 5.3) and in quantum logics (Sect. 5.4).

5.3 The Tri-segment and Paraconsistent Logics: Some Remarks

A second domain of the formal sciences where contradiction/negation is explicitly meant to be of the highest importance is, we have recalled, paraconsistent logics (programmatically "the mathematics of nontrivial self-*contradiction*", i.e., of nontrivial "A∧¬A", for some, but not all A, cf. [22]). And as for the latter, the relevance for it of the present study should already appear clearly in relation to our opening Sect. 1.1 (on the "Slater dispute"), as well as in relation to other comparable considerations we made all over the rest of our study (Sect. 3.5, Fig. 81). Remark that poly-simplicial oppositional geometry, notably its Angot-Pellissierian sheaf-theoretical version (Sect. 1.4, Figs. 19 and 20), seems to *deeply* confirm the rightness of Béziau's fundamental line of defense (2003, [24]) of paraconsistency against the rude charge of Slater (Sect. 1.1, and [132]). Oppositional geometry does it by rediscovering, over and over, the deep mutual relations of paracompleteness and paraconsistency (a.k.a. the relations between intuitionism and co-intuitionism, put into evidence, among others, by Béziau), seen as oppositional diffractions of "classicality": intuitionism being considered as mathematically fully natural (although not as mainstream as classicality), co-intuitionism (i.e., paraconsistency) should be as well (at the Slaterian price, however, of cleaning itself of fantasy elements). Inside (and outside) oppositional geometry, this is done by works like Angot-Pellissier's [1, 2], consisting in putting into precise link "topology" (a fundamental approach to "space" in which "distances" and "shapes rigidity" do not intervene) and (bi-simplicial as poly-simplicial) oppositional geometry, notably with his "topos construction" of tri-simplicial tri-valuedness (Sect. 1.4, Fig. 17 and Sect. 2.1, Fig. 34) and by recalling an old but profound idea (found, e.g., in V.A. Smirnov's [136] commenting N.A. Vasil'ev's [143] – Angot-Pellissier read a draft translation I made of it from Russian) according to which, fundamentally, paracompleteness is proper to any "*open* topology", while paraconsistency, its "dual", is proper to any "*closed* topology" ("open" means "*not possessing* its own frontier", while "closed" means "*comprising in itself* its own frontier"). The tri-segment we arrived at in this study, interestingly, seems to confirm and to summarize these very important ideas by means of the well-displayed interplay of its three diagonals (and of each of the three pairs of numerical sub-sheaves which are these diagonals' vertices): for short, each diagonal embodies one of the three kinds of this fundamental logical-mathematical "trio" (Fig. 129).

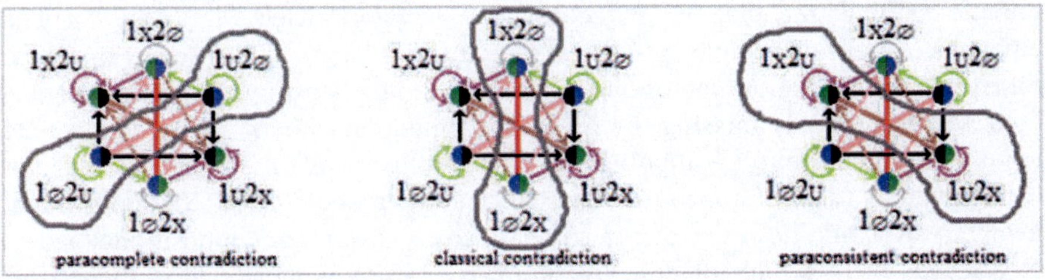

Fig. 129 The Aristotelian tri-segment as a visualization of paracompleteness, classicality and paraconsistency

This deep relation of the tri-segment to this important "metalogical" triad can be suggested more intuitively by combining graphically the two valuations of the Aristotelian tri-segment (Sect. 4.5, Fig. 121) with the symbolic (intuitive) expression we proposed of the concepts of gap and glut (Sect. 3.5, Fig. 81). One sees, then, vertex by vertex, what happens when the two valuations of the tri-segment, i.e., respectively, the supposed truth of the literal "1" and the supposed truth of the literal "2", switch the one into the other. By switching valuations, (i) classicality oscillates between true and false, (ii) whereas paracompleteness oscillates between gap and false, (iii) and paraconsistency oscillates between true and glut. *This seems to be, in some sense, one of the fundamental meanings of the tri-segment: by the tri-simplicial diffraction of the red segment of 2-oppositional contradiction it "opens" the classical concept of contradiction, seen as oscillation between true and false, adding to it two new different ways of oscillating, one paracomplete and the other paraconsistent* (Fig. 130).

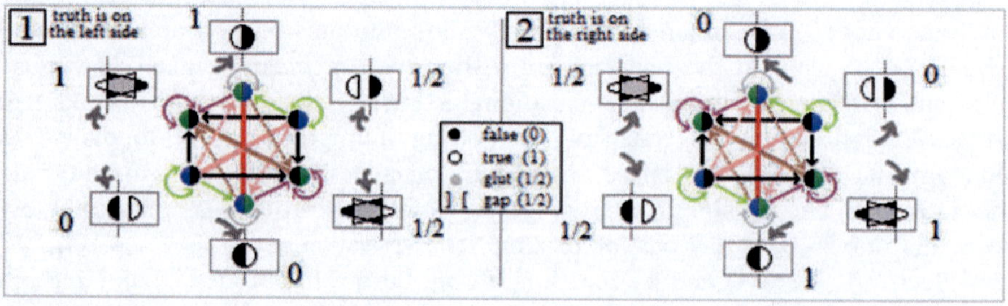

Fig. 130 The two possible valuations of the tri-segment (1 and 2) express the possible plays with "gaps" and "gluts"

This deep, natural link between intuitionism and paraconsistency, as said, had been studied among others by Béziau since years, but principally in modal logic (paradigmatically in [23–25]). He claimed in [23] (seemingly with reason!) that S5 (the "universal system", i.e., the most classical and standard system of modal logic), *very* paradoxically, can be seen as being already a full-fledged "paraconsistent logic" (!), since its apparently innocuous modal operator "$\Diamond\neg$" (traditionally read as "possibly not", equivalent to the negation of necessity, "$\neg\Box$") in fact also expresses, unseen but real, the gist of "paraconsistent *negation*" (Sect. 1.1, Fig. 3). The emergence of other "demonstrations" of the same fundamental idea seems to appear transversally (throughout logics *and* mathematics) in at least five different domains (and we are probably missing, by ignorance, important others): (1) mathematical logic, (2) modal logic, (3) topology, (4) many-valued logic, (5) and, last but not least (given its high relevance for discussing "contradiction" as such), oppositional geometry. Remark, however, that in some sense *poly-simplicial* oppositional geometry seems even to add to these five domains (which comprise it as their fifth) a sixth domain: (6) for *one, and perhaps even more fundamental, new kind of line of defense of the idea of a mathematical naturalness of paraconsistency, is, I believe, the oppositional "Pascalian ND simplex" itself* (and therefore arithmetic?); inside the "Pascalian roadmap" for the oppositional poly-simplexes (Sect. 1.5), paraconsistency's naturalness becomes even *visible*, in its being, so to say, one of the "fractal branches" – in the sense of Sierpiński's gasket, which is correlated (in several ways) with Pascal's triangle (Sect. 1.5, Fig. 23, cf. also [109, 110]) – relative to the *possible contradiction kinds*, of the global fractal structure. It must be remarked that the fractality lies not only on the very numerical structure of Pascal's 2D triangle but also on the fact that this tri-simplicial behavior (clearly readable, for instance, in the 2D section of the Pascalian 3D simplex, Sect. 2.5, Fig. 53) can be complexified, *n*-simplicially, into infinite by the very simplicial constitutive structure of the Pascalian *ND* simplex (Fig. 131).

Thus doing, poly-simplicial oppositional geometry seems to confirm, over and over, (1) that paraconsistent logics in some sense *must not be overestimated* (by effect of the *dangerous power of the fantasy relative to having "nontrivial classical contradiction"*, leading to unjustified fantasies of formal almightiness)

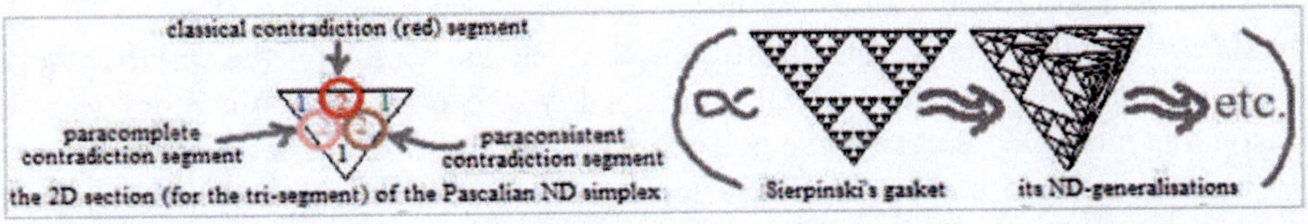

Fig. 131 The 2D section (for the tri-segment) of the Pascalian 3D simplex *shows* the "meta-logical trio" of negations

(this belief or ideal of nontrivial *classical* contradiction, Slater is right at least on this important point, would be a pure, misleading *fantasy*!) and (2) that paraconsistency is, however, indeed an important, natural, and by now even visible (!) feature of the rigorous mathematical approach, not only to negation but more generally (and deeply) to "opposition" (taken as a mathematical object, Sects. 1.2 and 4.6). In this respect, one should never forget, as (logicist) logicians tend to forget it almost "by construction", that in some important mathematical (Pascalian!) sense "contradiction" (i.e., negation) is a very meaningful but nevertheless *particular* case of "opposition" (i.e., not able to erase, or even simply dominate, the *numerical* and therefore Pascalian element of the global theoretical structure of "opposition", Sect. 1.5).

As said for many-valued logics (Sect. 5.2), it must be repeated here that future studies should also try to put at work, rereading humbly and patiently, step by step, "classical" systems and concepts of paraconsistent logic (such as those of Vasil'ev, Jaśkowski, da Costa, Belnap, Routley, Scotch, Batens, Priest, etc.), but this is not yet easy to realize, given also the already mentioned current inexistence of *full* studies of poly-triangles (i.e., published studies determining their oppositional *closure*). And this should also become partly easier than it currently can be, when something more will be understood and known about the seemingly fundamental, but still rather obscure and opaque, relations of poly-simplicial oppositional geometry and many-valued logics (Sect. 5.2).

Let us now turn to the last of the three *formal* approaches to contradiction we consider, one which seems to bear itself many-valued and paraconsistent aspects: quantum logics.

5.4 The Tri-segment and Quantum Logics: Some Remarks

A third domain of the formal sciences strongly interested by contradiction – we enter here, let it be clear, as an amateur – is "quantum logics" (QL). Since it is deeply rooted in "quantum mechanics" (QM), this also touches the question (otherwise left untouched by us in this study) of the importance of contradiction/negation (as different from contrariety) for the *experimental* natural sciences (like, for instance, interestingly, in biology, cf. Figs. 137 and 138 *infra*). As for quantum mechanics, that is microphysics, the problem is the following: this theory is said to be strange, and indeed it is, for it *seems* – as in another way, psychoanalysis (Sect. 5.6) – to challenge fundamentally, and not by choice, but by necessity, the *intuitive* very laws of logic. This is first of all, and notably, the case with "quantum leaps", that is the most elementary and small-scale known causal sequences, which appear to contain in them a strict and intractable *indeterminism* (although these quantic leaps take place in the formal framework – "Schrödinger's equation" – of the strictest statistical *determinism*). Many serious *theoretical*-physics proposals (i.e., nonexperimental, hard to test) have been done, inside physics, for coping with this rationally embarrassing mystery (i.e., the sudden irreversible loss of classical strict causality). This is the case with Everett's famous (but rigorous!) theory of the "parallel universes" (1957), also known as "many worlds (and/or many-minds) interpretation of QM" (on this, cf. [17, 18, 54, 141]), which saves strict causality, abolishes indeterminism, and makes mathematically more symmetrical and less *ad hoc* the von Neumann quantum axiomatics, but at the astonishing price of admitting that each micro-causal sequence (each quantic leap!) makes "split" the universe into two (or more) parallel universes (and this exponentially into infinite): the fractal bushy whole of these almost infinite fractal splits is called the "multiverse" (Fig. 132).

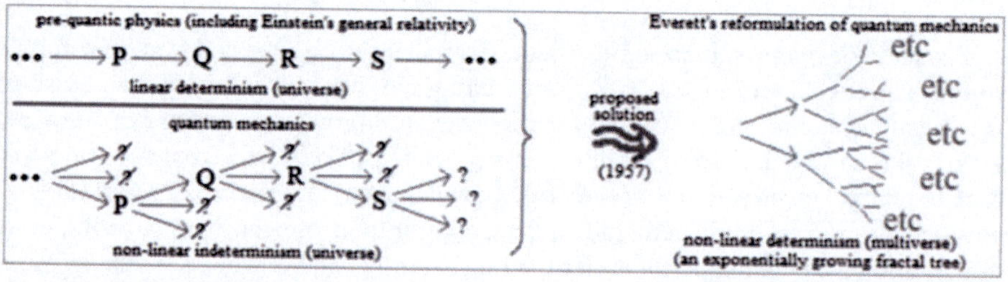

Fig. 132 Linear determinism, branching quantic indeterminism, fractal multi-linear (multi-versum) determinism

Quantic intuitive strangeness – mainly due to the scientific naturalization of some strange "intuitive *contradictions*" – remains in some sense not understood. But it is proven to be "real" (its predictions so far remain unchallenged) and useful in practice (it impacts reality technologically). But as such, by "weird" features like, for instance, "local reversals of time", or the spatial "non-locality", and the like (again, most of which, so to say, justify *intuitive* contradictions), QM becomes the support of many, less serious *fantasies*, generally aiming at justifying "magics" and paranormality (new age, esotericism, religion etc.), like in Jung [78] or Lupasco [86]. In some sense, these fantasies (justified or not) are related to the idea of a physical possibility of having "true *contradictions*" (the dangerous *fantasy* of some paraconsistent logicians: having "nontrivial *classical* contradictions", cf. Sects. 1.1 and 5.3).

With respect to the general theory of *oppositions* (i.e., considering not only contradiction but also contrariety), it must be remarked that microphysics in general is full of entities related with *contrarieties*: *anti*-particles, *anti*-matter, *anti*-energy, etc. One of the fathers of QM, Niels Bohr (1885–1962), is known for having, in that respect, made explicit *philosophical* reference to the Dao's "Yin-Yang" (notably relatively to his formal concept of "quantic *complementarity*"). Consequently, it seems that it would be interesting to look at QM with the new mathematical lens now offered, on "oppositions", by oppositional geometry.

In fact, long before the emergence of oppositional geometry, one way to cope with this lasting and resisting "illogical" strangeness of QM has been to develop, mathematically, something like "quantum logics", a.k.a. "QL" ([60]). Systems of QL have been proposed, at the beginning, by people like Birkhoff and von Neumann in 1936 [32], Destouches-Février in 1937 and 1951 [53], and Reichenbach in 1944 [120], and they are mostly three-valued logical systems (Łukasiewicz, one of the creators of many-valued logics, was among others motivated, by inventing three-valued logic, in modeling physical indeterminism, cf. [75, 106]). This invention/discovery of QL involved, notably, the theorization of new "truth-tables", suited for three-valued propositional connectives, among which are three-valued negations (Fig. 133).

negations				$\&_c$	V	F	A	$\&_i$	V	F	A	[illegible]	V	F	A	+	V	F	A	v_e	V	F	A	v_i	V	F	A	≡	V	F	A
p	V	F	A	V	V	F	A	V	A	A	A	V	A	V	V	V	f	V	V	V	V	V	V	V	A	V	V	V	V	F	F
Np	F	V	A	F	F	F	A	F	A	A	A	F	V	A	F	F	V	A	F	F	V	F	F	F	V	A	F	F	F	V	F
$\sim p$	F	V	V	A	A	A	A	A	A	A	A	A	V	F	A	A	V	F	A	A	V	F	A	A	V	F	A	A	F	F	V

Fig. 133 Some of the truth-tables of Destouches-Février's three-valued logic for quantum mechanics (1937, 1951)

As we saw, being *three-valued* logical systems, these early formal systems of QL are somehow related to tri-simplexes (Sect. 1.3). Our fresh knowledge of the basic features of the Aristotelian tri-segment (Sects. 2, 3, and 4), even without (the much needed and not yet available) knowledge of tri-triangles and tri-tetrahedra (Sect. 1.6, Fig. 33, Sect. 5.1), allows us, in principle, to try to analyze some features of quantum logic, at least those related to (three-valued) negation.

But QL, together with these rather simple, early three-valued propositional systems, has also resorted (notably with Birkhoff), more abstractly and powerfully, to the then new mathematics of "order" and "lattice theory" (Sect. 1.2, Fig. 7 – cf. [48, 147]): there have been investigations on *nonclassical order-theory* (with structures like "complemented ortholattices") aiming at coping with mathematically strange behaviors, as "non-distributivity" and the like (but there also are rivals to this, namely, mathematically more radical and powerful things, like "noncommutative geometries", cf. Girard [70] and Zalamea [146]). Remark that the lattices of QL, like "orthomodular lattices" and similar, are essentially nonstandard with respect to classical logic (it is precisely by this that they aim relentlessly – but difficultly, as heavily criticized in [70] – at capturing the strangeness of quantic "contradictions") (Fig. 134).

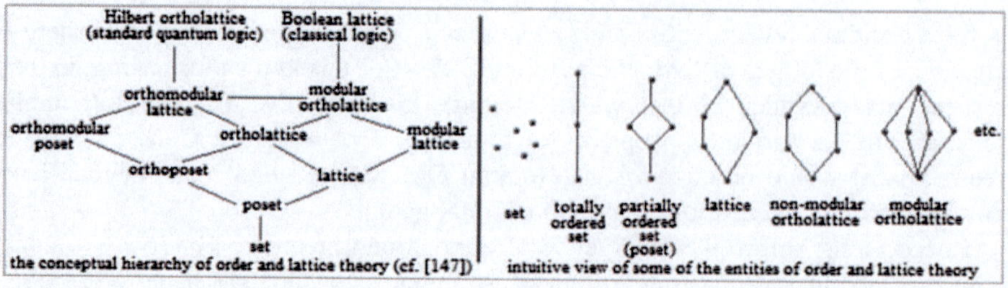

Fig. 134 The "order theory" and "lattice theory" turn of quantum logic

Currently, QL is also the basis of the hope to reach soon "quantic computers" and, through them, "quantic computations" (and, through the latter, "deep AI", cf. [76]), which also opens to some strong and astonishing *fantasies* (like the dream/nightmare of the allegedly imminent "singularity": the *almightiness* of "emergent" artificial ultra-intelligent *agents* . . .).

As said, a further idea is that of studying QM not only through three-valued logic and nonstandard lattices (QL) but also, maybe, through "oppositions". And this is also not entirely new. In the last years, some logicians and epistemologists (like Freytes, de Ronde, Bueno, and others: [19, 52, 66]) have been trying to use the "square of oppositions" (Sect. 1.1, Fig. 1) for inquiring the foundations of QM and of QL. But, *strangely enough, this has been done by them, again and again, without any reference to what is now really known about the structure "logical square"* (Sect. 1.2) – the (seemingly) logicist (Sect. 4.6) still have not understood that something mathematically serious is going on, *outside logic*, with "oppositions" (Sect. 1.5). So this line of researches seems at least suboptimal with respect (1), on one hand, to concepts like the oppositional closures of bi-simplicial n-opposition (and for a minimum, nonnegotiable start: the logical hexagon!) (2) and, on the other hand, if reference is done (as we just saw) to the use, by QL, of logical many-valuedness, to the (non) use (with respect to "*microphysical* oppositions") of *poly-simplicial* oppositional geometry (Sect. 1.3)! And, as it happens – as an intriguing example of this strangely underestimated line of thought we are arguing for here since some years – there is already notice (although still "unheard" until now) of at least one striking, possibly interesting similarity, still to be investigated and checked, between existing canonical formulations of QL (viz., those of Pavičić and Megill, cf. [91, 108]) and poly-simplicial oppositional geometry (Fig. 135).

classical implication	$a \to_0 b = a' \cup b$	(classical)
quantum implications	$a \to_1 b = a' \cup (a \cap b)$	(Sasaki)
	$a \to_2 b = b' \to_1 a'$	(Dishkant)
	$a \to_3 b = (a' \cap b) \cup (a' \cap b') \cup (a \to_1 b)$	(Kalmbach)
	$a \to_4 b = b' \to_3 a'$	(non-tollens)
	$a \to_5 b = (a \cap b) \cup (a' \cap b) \cup (a' \cap b')$	(relevance)

Fig. 135 The $1 + 5 = 6$ "implications kinds" mysteriously present in any orthomodular lattice (Pavičić and Megill)

In fact, the 1 + 5 kinds of "quantum implications", put forward by Pavičić and Megill, strongly remind us those, emerging as weakenings of the "II" relation of "opposition geometry", in the Aristotelian 4^2-lattice (i.e., in the quadri-simplexes, Sects. 1.3 and 1.4, Fig. 18) (Fig. 136).

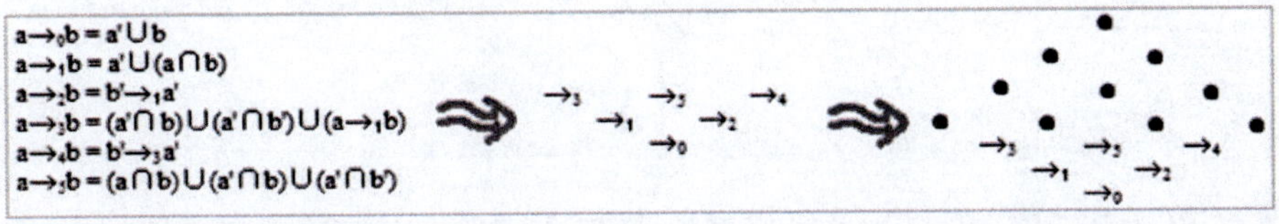

Fig. 136 The six mysterious implications of any OML seem to match quite well those of any oppositional *quadri*-simplex

Our present study, however, has clearly shown (ch. 4) that the complexity of the arrow system of an opposition*al*-geometrical poly-simplicial universe is in fact even higher than what shown by Aristotelian p^2-lattices *alone*: the twin Smessaertian p^2-lattices (for "implication geometry"), and maybe even more complex p^{p2-p} ones (Sect. 5.2, Fig. 128), are also needed to have a clear view. So it will be possible to seriously try to study this apparent correspondence between QL and poly-simplicial oppositional geometry (as we hope to do, or to see done, in future researches) only when will be inquired for themselves the *quadri*-simplexes (starting, soon enough, from the quadri-*segment*, which bottom-up unfolds the tri-segment fractally in a very elegant 3D polyhedron containing several interlaced tri-segments).

In order to try to have, nevertheless, at least a sketch of direct application of the tri-segment (and also a first application of it to biology), we can now try to rethink something of the famous thought-experiment of QM known as "Schrödinger's cat". This is a poor furry non-dog quadruped in a dangerous Austrian "black box", who is, paradoxically, provisory "dead AND alive" – because of quantic "superposition" – so long the strange quantic superposition inside the black box is not brutally abolished (quantum leap) by the intervention of an external *observer* of the black box, making a "measure" of what is inside it: a cat that, consequently, suddenly *becomes* either dead or alive, *only* immediately after this "quantic measure" has occurred – the quantic measure triggers (or not: mysteries of the quantum leaps) the opening, inside the box, of a cyanide flask. Now, in some sense a robust reflection on this seemingly requires, as a simplified standard oppositional-geometrical *starting model*: (1) either (standardly) a bi-simplicial *logical triangle* (B3) for opposing as *contraries* (and not as *contradictories*!) "dead", "alive", and "neither alive nor dead", then its tri-simplicial diffraction (the *tri*-triangle) could offer – maybe – some *starting* element of oppositional-geometrical further clarification of the quantic strangeness of the thought-experiment; (2) or (less standardly) a bi-simplicial *oppositional tetrahexahedron* (B4, the closure of the bi-tetrahedron A4) for combining, as orthogonal (and therefore freely combinable), "dead", "alive", and their respective negations, then its tri-simplicial diffraction (the tri-*tetrahedron*) could maybe offer some other *starting* element of clarification (Fig. 137).

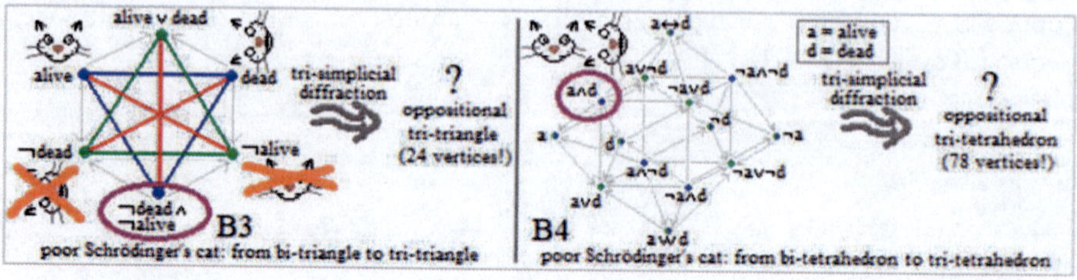

Fig. 137 A B3-structure and a B4-structure for "Schrödinger's (poor) cat"'s three or four possible existential states

Waiting for the tri-triangle (and, still later, for the tri-tetrahedron), the tri-segment can help us having at least some sort of (small and partial) "preview" of the tri-simplicial diffraction, thus investing the starting bi-simplicial oppositional description of Schrödinger's cat's conceptual experiment (but beware: the "tri-triangle", as we will show in another study, has not 6 but 24 nontrivial vertices! – cf. Sect. 1.5, Fig. 29) (Fig. 138).

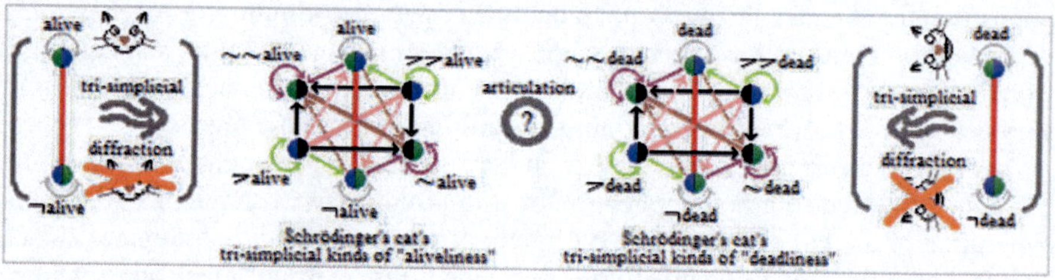

Fig. 138 "Tri-simplicial diffraction", through the tri-segment, of some "contradictory halves" (!) of Schrödinger's cat

More generally, as already said (Sects. 5.1, 5.2, and 5.3), remaining at the level of three-valued logic, it seems likely that alternative, finer-grained oppositional configurations might appear by resorting not to the Aristotelian *and* the Smessaertian twin 3^2-lattices (as here) but to the Aristotelian *and* the Smessaertian "twin" 3^3-lattices (and, in fact, heavier, Sect. 5.2, Fig. 128).

Having tried to make some remarks on possible (future) applications of the tri-simplicial diffraction of contradiction (i.e., the Aristotelian tri-segment) to some of the formal sciences, les us try to have now a quick comparable look at a possible incidence on some of the humanities: dialectics and psychoanalysis.

5.5 The Tri-segment and Hegelian-Marxian Dialectics: Some Remarks

The famous concept of "dialectics" also bears strong "*fantasy* elements" with respect to "contradiction", which is the starting object of this study. Dialectics is in fact supposed (by its partisans) to be "the *science* of contradiction". With respect to all which should be said about "dialectics" – its theory, its history, its issues, etc. (for two good overviews cf. [74, 122]) – we will try to limit ourselves to some of the main points.

First of all, there *is* some interest in considering it, for dialectics (as a doctrine, as well as a symbol) still has some strong impact on reality (notably in politics, but more generally in contemporary philosophy). However, it is notoriously difficult to define properly dialectics: dialecticians themselves (starting from Hegel) justify this as being related to the very subject matter: (1) dialectics is the conceptual and ontological "engine" of everything (according to dialecticians), and (2) it has to do (allegedly) with *the most inner structure of "being" and of "becoming"* (of anything! be it concrete or abstract), and (3) its own structure, it is said, consists mainly in defeating dynamically any *concrete* "structure". Historically, there are essentially two such dialectics (leaving aside other important theories of dialectics, generally of a very different and less "dynamic" nature, e.g., Plato's dialectics [121] or Lautman's dialectics, [16, 80, 145]. First, there is a Hegelian (1770–1831) dialectics (appeared around 1804). Second, derived from it, there is a Marxian (1818–1883) dialectics (appeared around 1841). *The important point for us is that both claim to deal, in their "kernel", with oppositions and contradictions. And both claim to be superior to (in the sense of methodologically and ontologically "more fundamental than", and "irreducible to") mathematics.* This point is crucial: among current philosophers (and activists, etc.), dialectics is still a competitor to mathematics and nourishes, in its partisans, a deep disbelief for mathematics as a reliable source of inspiration for philosophy or action in general. Later (and still nowadays) dialectics has also been put into rivalry with mathematical *logic* – which was born after it, with Boole (1815–1864), around 1847 – but the result of this second confrontation remains rather unclear ([44, 58, 89]). Logicians and analytical philosophers, like Popper (1902–1994), claim to have "demonstrated" that dialectics is unsound. But dialecticians (the remaining few) claim, not without some rigor, that *logic cannot defeat dialectics (it cannot reach it, as a target)* and that this is because logic is a very primitive, too simple thing, in which dialectics just becomes unduly frozen (dialectics is supposed to be more intrinsically lively and fundamental).

In fact, logic cannot "hit" dialectics because, truly speaking, *dialectics (i.e. Hegelian and/or Marxian) should be put into critical comparison not much with logic but with oppositional geometry, for the latter, and not the former, is indeed the "science of opposition", if any* (Sects. 1.2, 4.6). Once this point understood and adopted, the main ulterior point to be seized is that, retrospectively, *dialectics is built on some clear inaugural, deep conceptual mistakes* (inside *philosophy*) about mathematics. This was substantially proven by the great mathematician (also a philosopher, [37]) Bernard Bolzano (1781–1848) – the real discoverer, before Cantor (1845–1918), of the mathematical thinkability of the "actual infinite" [36] and one of the founders, with Cauchy (1789–1857) and Weierstraß (1815–1897), of modern mathematical analysis [129]. But it was not perceived by many, and, very dramatically, not by Marx (most of Bolzano's writings appeared posthumous around 1929, long after his death in 1848). Bolzano demonstrated, by a crystal-clear reasoning, that Kant (1724–1804) made severe mistakes in his theory of "how mathematics can/must function", a crucial theory for his project of a "criticist philosophy", which led him to difficult points ("antinomies") in his "critique of pure reason" [84]. Hegel assumed uncritically the fundamental structure of Kant's philosophy of mathematics (assuming, by that, Kant's mistakes!) but claimed that the difficulties thematized by Kant as "antinomies of pure reason" had to be understood "the other way round": here Hegel used a *binary opposition*, just reversing it (this point is crucial and paradigmatic, cf. Fig. 139), so to suggest that contradictions were not the *bad end* of thought (Kant's antinomies), but its *good beginning*! (i.e. Hegel's dialectical philosophy). On top of this, and using Fichte's (1762–1814) astonishing philosophy of (i) opposition/contradiction and (ii) of the *creation* of the object by the (absolute) subject (!), Hegel elaborated his "logic", of which "dialectics" is the second of three moments (this ternary scheme is supposed to repeat into infinite, thus unfolding *any* reality, from nothingness to "absolute Spirit").

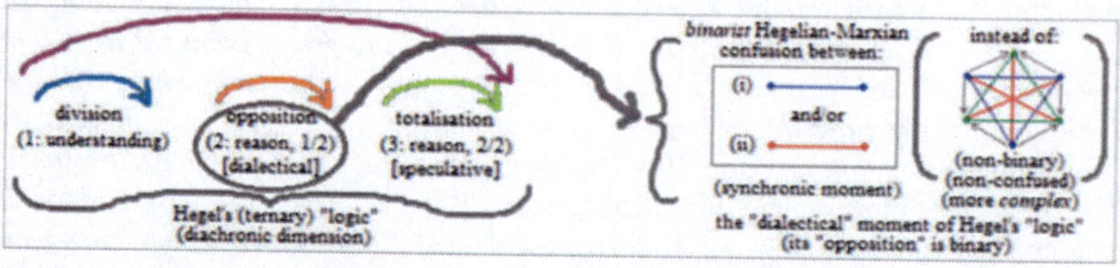

Fig. 139 "Diachronic" and "synchronic" dimensions in Hegel's "logic": the opposition in it is synchronically *binary*

Third, then, it must be understood that one of the fundamental problems with dialectics is that *current dialecticians (and already Hegel and Marx!) seem to have lost a precise intuition of the difference between "contradiction" and "contrariety"* [114]. This oblivion, quite common at the time of the birth of dialectics (and shared by schools of thought quite distant from dialectics, like analytical philosophy, phenomenology, and psychoanalysis – on the last cf. next Sect. 5.6), is conceptually deadly. Dialecticians – as, paradigmatically, Roy Bhaskar (1944–2014), cf. [31] – people obsessed by the concepts of "opposition" and "contradiction", make almost no mention of the "square of opposition", left for dead on the floor! (they discuss Priest's "dialetheism" [115, 116], but not the logical hexagon). But this confusion about contradiction and contrariety has a tremendous impact, lethal for the credibility of dialectics. On the one hand, it shows that *the dialecticians' "fundamental bet" (lost!) has been that opposition, at its root, had been theorized once and for all by Hegel (in turn based on the "firm rock" of Kant's theory of mathematics) and that no mathematical new insights on opposition might appear in the future* (Kant's similar tragicomic assertion, about logic admitting no major changes in the future, is famous). *Oppositional geometry, by its simple existence (Sect. 1.2) – and right now by the emergence of the tri-segment! – ruins this insane Hegelian view!* On the other hand, by teaching, since the discovery of the logical hexagon in 1950, that "binary *oppositions*" (i.e., *binary contrarieties*) do not exist (Sect. 1.1, Fig.2), oppositional geometry ruins *de facto* most dialectical reasonings of the past (and of the future!) which can be shown to be crucially grounded on such (alleged) *binary oppositions*: the problem being that *from a binary opposition you can deduce (as the dialecticians), by "reversal", things that you cannot from a ternary or higher contrariety.*

This becomes clearer if one realizes that, in fact, *nolens volens*, Hegelian logic bears, from a structuralist mathematically legitimate viewpoint (cf. [7–9, 112]), a "diachronic" and a "synchronic" dimension: there is, clearly, a diachronic succession of three moments, the second of which is "dialectics" properly said ([130]). But this second moment has also a fundamental "synchronic" dimension: it consists of an *opposition*, and this (Hegelian) opposition, "at some time" (think of it in terms of a mathematical "fixed point"), "exists". But in this necessarily "synchronic" dimension, dialectics commits a mistake with respect to oppositional geometry: it believes in *binary* oppositions, which it uses (by "reversing" them), whereas, as we recalled, *they do not exist* (and thus cannot be *deductively* "binarily reversed")! (Fig. 139).

This confusion appears, even nowadays, in the fact that "true dialecticians" and their admirers (Bhaskar, Ollman, etc.) speak commonly about "two things being in contradiction", whereas what happens is that their *contrariety* (a *fragment* of a larger, at least ternary one, Sect. 1.1, Fig. 2) *implies* two contradictions: each of the two contrary terms also *implies* the contradiction, i.e., the (*vague!*) negation, of the other.

What about the tri-segment, then? One of the most reputed attempts at formalizing Hegelian logic/dialectics, due to Rogowski in 1964 (cf. [124]), uses four-valued logics (the two extra truth-values stay, respectively, for "beginning to be" and "ending to be"): so, in order to see how the "oppositions" truly work in it, we would need *quadri*-simplexes (Sects. 1.3, 1.4, and 1.5). This is out of reach here (but not in the near future). What the oppositional tri-segment already shows, however, and decisively (Sects. 4.5 and 4.6), is, again, that the holy "pure contradiction" *segment* (so, even without speaking of the oppositional-geometrical bi-simplicial – and *a fortiori* poly-simplicial – "fireworks" of *n*-contrariety, starting with *triangles*) is already *mathematizable* and already leads (as draft studies on the quadri- and quinque-segments already establish) to unmistakable *mathematical, infinitely growing, highly structured complexity*, which, again, ruins the "*transcendental* flavor" of Hegel's alleged "*constant ternary* flowing of dialectics" (Fig. 140).

Similar remarks should be done, point by point, for Marx: in his own changes to dialectics (cf. [74, 105, 122]), which are real proposals, he offers new "patches"

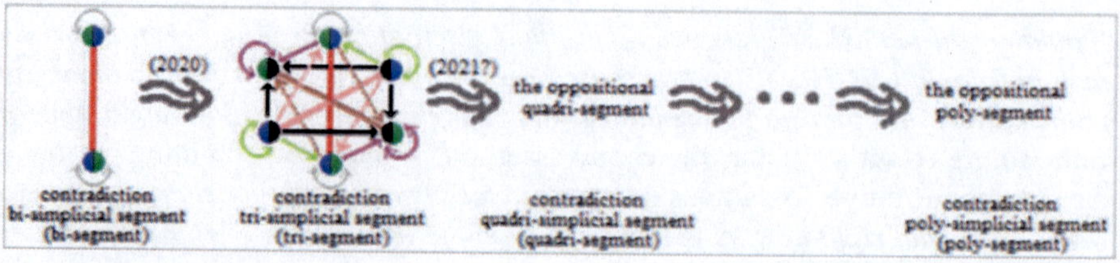

Fig. 140 "Contradiction" is a *mathematical* object: it ruins the Hegelian *anti-mathematical* fantasy on "dialectics"

which – at times – even contain bits of interesting mathematical premonitions (e.g., of catastrophe-theory, system-theory, category-theory, etc. [82, 123]). As such, they are valuable achievements of creative thought. But, they are nothing more than that, and by embedding them into *the deadly fantasy of a unique, almighty scheme, transcendental "dialectics", superior to mathematics*, Marx as well, as already Hegel (*pace* Lawvere), leaves, up to now, dialecticians and "Marxian orphans" blind and hostile to mathematical *non-unifiable* complexity (as intelligently explained, among others, by Badiou [11, 13] and by Mélès [92]).

Having made some minimal remarks on the possible impact of our successful tri-simplicial diffraction of contradiction (the tri-segment) over dialectics, let us now turn, in the last Sect. 5.6, to psychoanalysis.

5.6 The Tri-segment and Psychoanalysis: Some Remarks

Last element of our inquiry, the relation to "opposition" of "psychoanalysis" is strong, but complex. Here as well, we can only point to a few general remarks, which we will center on three main theoreticians of psychoanalysis: Freud, Lacan, and Matte Blanco.

Psychoanalysis is a study of the human mind based on the assumption of the existence of "unconscious" mental *dynamics* and "unconscious" mental *structures*, playing a *structuring* role at the level of "meanings" but also at the level of personal identity. From the viewpoint interesting us, the main contribution to this reflection made by Sigmund Freud (1856–1939), the first theoretician of psychoanalysis, can be decomposed critically in three main points. First, he discovers, and synthetically expounds in 1915 [64], that *considerable "negation/opposition problems" appear in the mysterious but real foundations of what he theorizes as the human "metapsychology"*, generated by the existence of a quite strange (but observable) "unconscious" mind, of which the metapsychology aims at giving some kind of *conceptual* axiomatics. The existence of the unconscious is shown, convincingly, to be very important, both for explaining "normal mental life" (dreams, parapraxis, acting out, ordinary psychopathology, etc.) and for "mental pathology" properly said (psychosis, severe forms of neurosis, perversion, etc.). But the unconscious is shocking, and one of the five irreducible axioms of the metapsychology says: "In the unconscious, 'negation' seems to be not working". For Freud this is shocking, but at the same time real, and therefore desperately asking for an explanation (Freud himself will never succeed in finding a satisfactory one, as he will admit in his last, posthumous, and unfinished study [65]). Secondly, in order to explore the unconscious, Freud develops a powerful theory of the mental processes and notably, in some sense, a theory of the "mental oppositions" (i.e., "complexes"), both at the individual and at the collective (and historical) level (this results, globally, in a theory of individual and collective "psychogenesis"). This is based on the idea that "mental unity" is the emerging (fragile) property of a constant *process*, rather than a "transcendental", firm starting point (like in the philosophical theories of Kant, Hegel, or Husserl). As such, Freud's theory seems to be "realist" and rather powerful, for it seems to match clinical evidence (in fact Freud's starting point) and everyday life's experience, but at the price of introducing morally shocking elements (like infantile sexuality, constitutive bisexuality, "naturalness" of murder and rape *instincts* or *fantasies*, etc.). Third, however, Freud seemingly confounds at least partly (like most people in his time and most people even now!) "contradiction" (negation) and "contrariety" (i.e., opposition properly said): what Freud talks about when he speaks about "negation not working" (in the "unconscious") seems rather to be, in fact, opposition (which also implies negation) and more precisely "contrariety" (this seems quite clear in his 1910 "remarks on the *oppositions* in primitive languages", [63]). The two (contradiction and contrariety) are of course deeply related (and showing this is one of the main tasks of the square of opposition, Sect. 1.1, Fig. 1, and, thence, of all oppositional geometry, Sect. 1.2) but must not be confused. *As it seems, contemporary psychoanalytical theory, despite its interest for mathematical developments, still has not clarified this important inaugural non-negligible confusion of Freud.* Oppositional geometry in general brings some precious light precisely on this: the articulation and the possible confusion of contrariety and contradiction, at least in the sense that if, as I think after a careful examination of it, what Freud is speaking about – when he speaks, not lightheartedly, of "negation" – is in fact (also) "contrariety", than some important obscurities and difficulties of the theorization of metapsychology by him and by his school seem to disappear, and some new, robust research lines seem to emerge, promisingly enough.

This is more or less, implicitly, the research line of Ignacio Matte Blanco (1908–1995), the author of the part of contemporary psychoanalytical theory that seems to be most deeply (and most promisingly) related to the particular mathematical discoveries of *general* poly-simplicial oppositional geometry. Since 1975 (year of the publication of his *The Unconscious as Infinite Sets. An Essay in Bi-logic*, [90, 119]), the strictly remarkable (and largely underestimated) theory of Matte Blanco proposes concrete theoretical elements for trying to go methodically in this direction. Apparently more modest than the more famous and flamboyant Lacan (cf. *infra*, [55-57, 61]), Matte Blanco has in fact made some very important theoretical proposals, where he substantially claims (convincingly!) that he has unexpectedly (i.e., not too young: aged of 67 years!) solved, through a new sort of mathematical reasoning, the main five metapsychological problems left dramatically unresolved by Freud at his death and until then reputed unsolvable by Matte Blanco himself. For short, since 1975 Matte Blanco tries to draw our attention on the fact that there seem to be important structural links, of an unexpected abstract mathematical kind, between growingly strange but meaningful psychoanalytical oppositions and "symmetrizations" (as in dreams, psychopathology, or psychosis), and the unfolding of a mental kind of hyper-geometry. I must recall that the very invention/discovery of the concept of "oppositional bi-*simplex*" was much helped, in me, by some acquaintance I had with Matte Blanco's psychoanalytical theory of the mind, formulated in terms of geometrical *n*-dimensional *simplexes*. Matte Blanco pedagogically obliges his readers to get acquainted with simplexes of different dimensions and with strange phenomena possibly resulting from the mutual projections of (mental) geometrical spaces and objects of different (perceptual) dimensionality (Fig. 141).

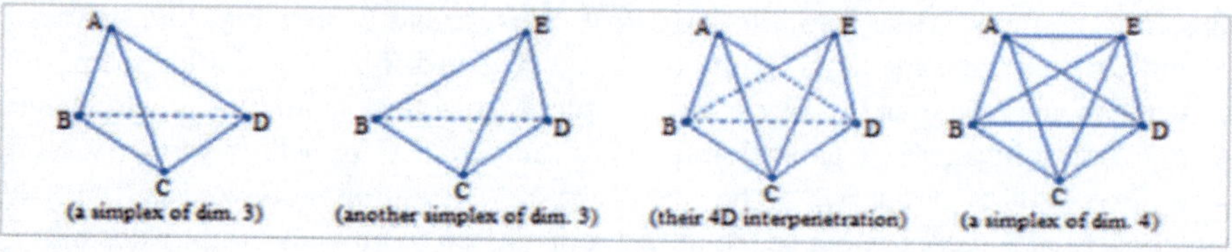

Fig. 141 Matte Blanco's study (*The Unconscious as Infinite Sets*, 1975) of the interplays of some nested simplexes

But since, as Freud, Matte Blanco seems to be much more committed with *contrariety* than with *contradiction*, we must leave an oppositional-geometrical poly-simplicial discussion of his theory (a theory of "bi-logic" and of the "bi-logical *structures*") for a context in which we will have more knowledge about poly-simplexes higher than poly-*segments* (this will start, again, with the tri-*triangle*). Since the tri-segment only deals with "contradiction", we cannot say much more here.

Fundamental metapsychological questions, similar to those explicitly left unresolved by Freud at his death in 1939, are faced with similar radicality by a third major theoretician of psychoanalysis, Jacques Lacan (1901–1981), who innovates mainly by constructing an operative link (through Saussurian differential linguistics) between Freudian investigations of the unconscious and the powerful structuralist interdisciplinary (and among others mathematical) paradigm and methodology [112]. Lacan's theoretical strength (beyond his stylistic "Gongorism", his constant provocations, and his still deranging histrionic outings), among others, seems to have been the deep understanding he progressively gained and defended (against reigning Hegelianism, for instance, but also against Marxism, Heideggerianism, and logicism – not to speak about "psychologisant" psychoanalysis that he, as well as Matte Blanco, fought relentlessly) that mathematics are so to say the most powerful, radical, and relevant key for investigating profound issues about the human mind (Lacan's concept for this is, *in fine*, the "Real", in his fundamental conceptual triad "R.S.I.", articulating Real, Symbolic, and Imaginary, which *among others* is a powerful tool for *modeling* the complex morphogenesis of *fantasies*). But the mathematics put into play by Lacan – as "mathemes": inspiring formal *images* – are such that they strongly diverge from the logicist program: they are *structure*-based (instead of *logic*-based or *deduction*-based), highly creative, and in some sense "nonstandard". In order to model meaningful (mental) "reversals" (contradictions and contrarieties) of all kinds (including meaningful *self*-contradictions), Lacan thus resorts (also) to topological structures typically "strange", like "Möbius's stripe" (an open surface, in the 3D space, with only one side!), "Klein's bottle" (a closed surface, in the 4D space, with only one side!), etc., that is mathematical *structures* such that they can help expressing the most strange and "illogical" *but natural* features of the human mind (and notably those related to the unconscious, "normal", or pathologic) (Fig. 142).

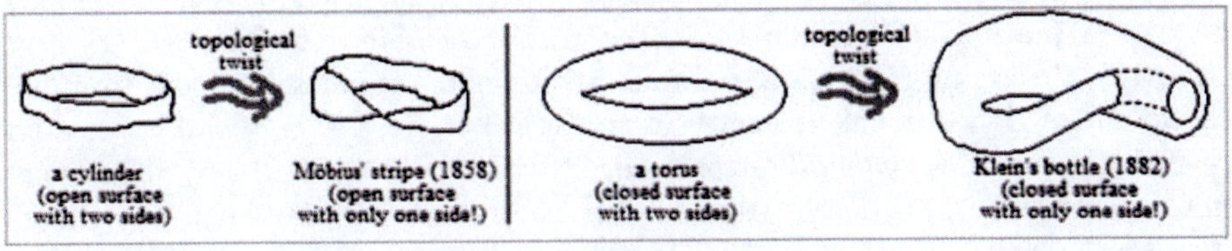

Fig. 142 "Möbius's stripe" and "Klein's bottle" provide *topological* intuitions on "opposition subversion"

In particular, one of the most famous (but also one of the most difficult to work out formally) of Lacan's innovating concepts is that of "pas-tout.e" (in French: "not-all"). Lacan arrives to it by a psychoanalytical complex (and deep) reasoning, about "sexuation", i.e., "mental gender", considered as independent from "anatomical destiny": you can have a penis but "fundamentally be" a woman, etc. (remark that nowadays Lacan's theory is, despite polemics, very appreciated, debated [14], and used, notably in gender and transgender issues). For modeling this concept of sexuation, in a nutshell, aiming at studying the unconscious relation of "man" and "woman" (as related to concepts as "universality", "exception", "enjoyment", etc.), he deforms the logical square (decorated à la Frege with quantifiers and quantified functional assertions), cutting and extracting from it one of its two red diagonals of contradiction and leaving aside what remains. More precisely, around 1972 [79] he "opens" the so obtained contradiction segment (extracted, as said, from a quantified version of the canonical logical square), by redoubling and renaming each of its two vertices, transforming it into a nonstandard new kind of formal square, his "square of sexuation" (in my opinion of oppositional geometer, one can/must think of it by analogy with the "cut and paste" construction of Möbius' stripe and Klein's bottle, Fig. 142, *supra*) (Fig. 143).

As it happens [85], this shape was inspired to Lacan by a reasoning of J. Brunschwig in 1969 [39] on the supposed psychogenetic "difficult origin" of the logical square in Aristotle, where the former argued that, historically, Aristotle

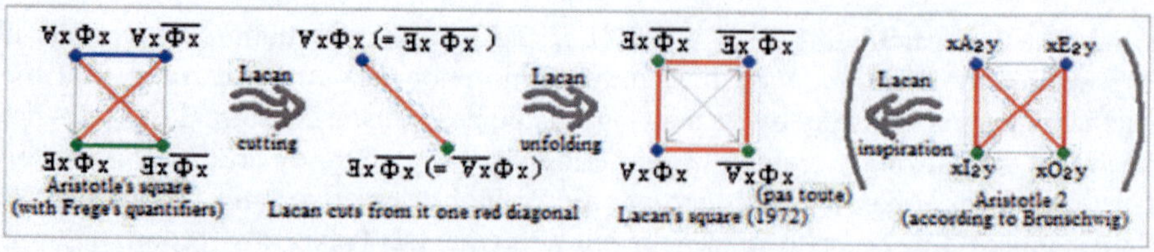

Fig. 143 From Aristotle's to Lacan's square (i.e., the "square of sexuation") through Brunschwig's square (1969)

found difficulties, allegedly "in the very heart of logic" (this is something Lacan liked!), during the invention of the square, and therefore had to hesitate between three successive kinds of (implicit!) formal squares (historically, we still do not know whether Aristotle himself developed *graphically* the full-fledged square (on Aristotle and mathematics, cf. [97], based on [77, 139]; cf. also [142]): he developed at least a preliminary version of it, the ὑπογραφή, in *Peri Hermeneias*, [6], Sect. 13). Brunschwig's square, rigorous *per se*, is nevertheless indigent from the viewpoint of oppositional geometry in so far it is a very *suboptimal* expression of formal properties *optimally* expressed, 20 years before (1950), by the logical *hexagon* (Sect. 1.1, Fig. 2). So, the fact that Lacanians, still now, keep "shielding" their Master, and themselves, with it – as did, for instance, in 2005 Grigg [73] against a benevolent but rigorous Badiou (1992) in his mathematically critical remarks on Lacan [12] – in order to claim (Grigg) that "Lacan's reasoning is *neither* a sophism *nor* a formal mistake, *because* it relies on Brunschwig's square!", is not (yet) a sign of conceptual strength (...). More interestingly, one should remark here two things: (1) Lacan aims at proving that "sexual difference" is conceptually deeper than "logical difference" (i.e., contradiction); and (2) by his square, Lacan defines sexuation through (nonstandard) contradiction. The first point is well-known, while the second seems problematic from the viewpoint of oppositional geometry: "female vs. male" is not a *contradiction*, but a *contrariety* (hence the necessity of having a logical *hexagon*, etc.).

But, here – coup de théâtre! – *our tri-segment (at least under the provisory naïve form it took in Sects. 2 and 3) seems, unexpectedly, to be quite interesting and fit, if one thinks of it, for trying to reformulate oppositional-geometrically this otherwise formally strange and perplexing current "square" expression of the Lacanian theory of sexuation* (Fig. 144).

For, not only it bears striking formal similarities with Lacan's square: more deeply, it rescues it from the aforementioned aporia of reducing dangerously "female vs. male" to a *contradiction* (instead of a *contrariety*), by its own structural richness (the tri-segment is a ... hexagon!). In fact, the Aristotelian tri-segment captures really quite much, but with mathematical rigor, of what Lacan seems "illogically" willing to capture. The tri-segment seems to have at least a quintuple advantage over the previous implicit model of this (i.e., Brunschwig-Lacan's square): (1) it is not arbitrary (while Brunschwig-Lacan's square clearly is); (2)

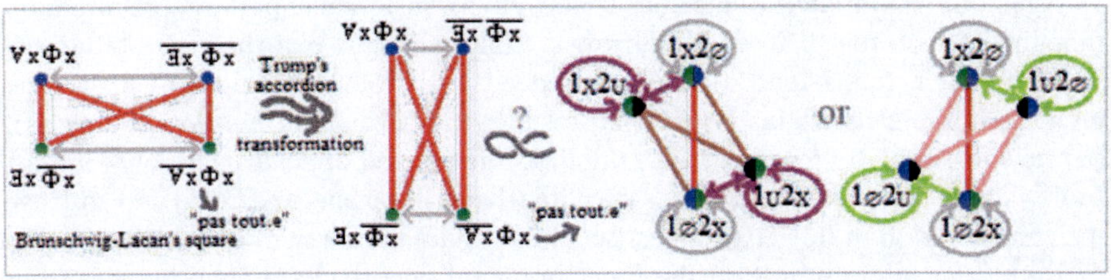

Fig. 144 The "square of sexuation", or "Lacan's square", resembles quite much a tri-segment's "inner rectangle"!

it has a large and deep *mathematical* theory behind (poly-simplicial oppositional geometry), whereas Lacan's square has not; (3) it has more structure – the (paracomplete) square, or better rectangle, is only one of its components, but even taken alone as a paracomplete rectangle, it says more (for instance, about the inner structure of the two different "contradiction diagonals" composing it) – (4) it bears *explicit* and *reasoned* reference to intuitionism (i.e., paracompleteness and its gaps), which is a major point advocated (but so far with *problems*, as pointed by Badiou in 1992 [12]) by Lacan (cf. Darmon [47]); and (5) it adds to the reference to paracompleteness a tantamount *explicit* (and intuitively important) reference to its mathematical dual (absent in Lacan!), paraconsistency (i.e., co-intuitionism).

One must remark that this last proposal of application of the tri-segment, although surely strange for some, seems interesting (and paradigmatic): quite many people looked for "Lacan's square" (including, in some sense, Lacan himself!) and still look for it (as recalled in [128]). So, finding a mathematical solution to the "riddle of the *pas tout*" would be a *fait d'armes*. If seemingly nobody found it so far, despite looking *eagerly* for it, and since years, and with all sort of formal means, this is because nobody had the idea (or, in fact, the means) of *simply looking for a mathematically rigorous, proper diffraction of the oppositional-geometrical concept of contradiction (a diffraction of the red segment)*: in terms of "poly-simplicial diffraction"! This is what we achieved in this study and in some sense seems to be quite close to what Lacan was trying, as he could, to "speak" about (Fig. 145).

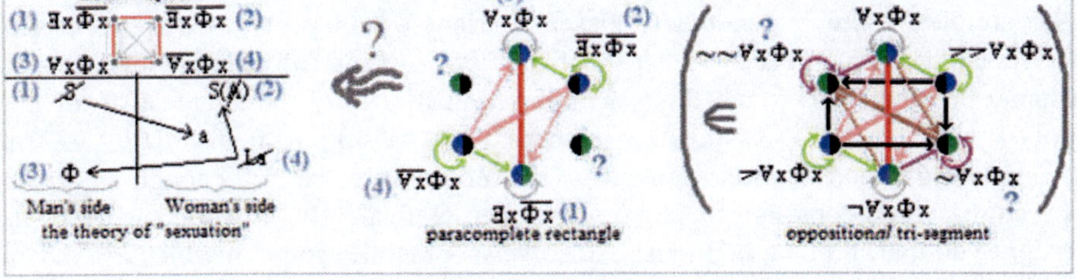

Fig. 145 The Aristotelian tri-segment as mathematical basis of Lacan's attempted "sexuation square"?

Again, the importance of the "pas-tout.e", in a considerable part of contemporary thought, is such that if it were confirmed (time will tell) that the Aristotelian tri-segment, as in fact I tend to think, is indeed a (serious) model of it, this could be a memorable rather inspiring 2020 result. And a splendid *entrée sur scène* of our newborn, small "formal artist": the first full-fledged oppositional poly-simplex (poly≥3) . . . . But here we cannot say more (that will be another story) and so leave the consideration of this last suggestion to the future curiosity of our reader.

Here we can close, at last, the long journey, brave reader, which has been our present common inquiry on the poly-simplexes in general and on the tri-segment in particular.

6 Conclusion

Our study was about the mathematical concept of "contradiction" (i.e., negation). In a nutshell, it consisted in showing that the "mathematical nature" of the concept of contradiction/negation is more geometrical (i.e., "simplicial") than "logical". Of this we gave a particularly strong new proof: (1) by establishing an astonishing (and powerful) technical reference to Pascal's triangle (which we generalized to the very useful notion of "Pascalian ND simplex") and (2) by developing a concept of "Aristotelian combination" which proves suboptimal, as for the exploration of "opposition", the logicist program and methodology of Smessaert and Demey's "logical geometry". More specifically, in our study we developed a "tri-simplicial diffraction" of contradiction (seen as, classically, bi-simplicial). This engaged us in recalling first the concept of poly-simplicial space.

We recalled the context of the emergence of the idea of poly-simplicial space that we proposed in our PhD in 2009 and that has not been much developed since. This engaged us in recalling, previously, the concept of "oppositional geometry", which is a powerful framework for explaining the structure of the "logical hexagon" (which explains, since 1950, the otherwise mysterious "logical square"), seen as a "bi-simplex" (viz., a bi-triangle). Importantly, we showed that by now the poly-simplicial space has become explorable, notably (i) thanks to Angot-Pellissier's sheaf-theoretical method for generating vertices and examining the edges between any pair of them (ii) and thanks to our concept, proposed here, of "Pascalian ND simplex", which generalizes Pascal's triangle and provides a quite useful "roadmap", complementary to the sheaf-theoretic method (of which it solves some crucial problems). Through this we explored successfully the simplest of all the poly-simplexes (poly≥3), the "tri-segment": the tri-simplicial diffraction of the classical red "contradiction segment". This turned up to be rich enough as for its structure, which consists of a 2D hexagon (or, equivalently, a 3D octahedron). In order to do that, we had, previously, to solve successfully some intermediary rather difficult steps. We thus discovered (in Sects. 2 and 3) (i) proper treatments for poly-simplicial "extrema", (ii) a general technique for coloring any poly-simplicial *vertex* (and not only the *edges* between them), (iii) and the general combinatorial laws ruling the patterns of the poly-simplicial "valuations" (i.e., the attribution of truth-values to the vertices).

In order to deal with unexpected arrows emerging in the tri-segment from its two valuations, in Sect. 4 we established an important further point: that it is possible and in fact necessary to use, additionally, Smessaert and Demey's concept of "implication geometry" (on top of which they posited the idea of a "logical geometry", encompassing the two twin geometries which are the "opposition geometry" and the "implication geometry", and therefore supposed to be the most general and best theoretical framework for dealing, *among others*, with oppositions). To do that, we developed new, suited versions of Smessaert's starting idea, thus bringing Smessaert and Demey's "logical geometry" to a level of complexity where it had never been before (the level of the poly-simplicial spaces – Smessaert and Demey's logical geometry remains, by construction, bi-simplicial, and in fact the very concept of "simplex", i.e., geometrical number, is banned from the *logicist* vocabulary of logical geometry). And this allowed us making a very interesting discovery: what these two authors take for suboptimal, namely, what they call "Aristotelian geometry" (and which is supposed to contain also our and Angot-Pellissier's "oppositional geometry"), seen as a bricolage, made unconsciously, by conceptual surgery, over the two twin halves of logical geometry (i.e., "opposition geometry" and "implication geometry"), is in fact, in the light of the emerging complexity of the poly-simplicial space, already at the "ground-zero" level of the tri-segment, a necessary and *optimal* transformation (which we baptized "Aristotelian *combination*") – not of *choice*, as they believe, but of *fusion* – allowing methodical complexity reduction. This fact allows understanding that Smessaert and Demey's logical geometry, in its programmatic negation of a mathematical autonomy (with respect to logic) of the "oppositional", is in fact suboptimal. On that respect, we also recalled the deep philosophical reasons why what we therefore take as being their *logicist* posture (inscribed in the very reckless name of their approach) is rather dangerous (and, again, suboptimal as for the exploration bottom-up of the "poly-simplicial space", which is constitutive of the key concept of "*n*-contrariety"). It is notably suboptimal with respect to a *structuralist* approach, very natural for exploring (as Blanché in [33]) the "elementary structures of opposition", but more generally "logical geometry" is suboptimal with respect to any free mathematical approach not submitted to the logicist, mathematically counterproductive agenda.

Finally, in the last part (Sect. 5), we tried to have a prospective first look at possible applications of the successful tri-simplicial diffraction of contradiction, which is the "Aristotelian" tri-segment. Before that, we discussed some current limitations of our approach, which should be, but is not yet, many-valued *also* at the meta-level (technically speaking: at the level of the *number* of the possible "meta-questions") and how it should be tried to overcome these current limitations in a near future. An important point is that, from now on, the higher poly-simplexes (and, for a start, as soon as possible, the "tri-triangle", i.e., the "tri-simplicial diffraction" of the logical hexagon) should be studied in a way comparable to the one we successfully developed in this study for the tri-segment. As for applications, we concentrated on five domains where contradiction/negation seems to play a particularly important role, both positively as a *concept* and also negatively as a *fantasy*: in the exact sciences we proposed to see this in many-valued logics, paraconsistent logics, and quantum logics; in the humanities we proposed to see this in dialectics and psychoanalysis. Among the applications we proposed, the most spectacular one seems to be the (in our view) very convincing formalization of what is traditionally known as "Lacan's square" (1972), or "square of sexuation", a structure and theory much debated and used, notably, in gender and transgender studies and which so far remained a very difficult open problem (despite the many attempts to solve it, notably by some mathematicians). We claim that this famous and strange square, generally reduced – as to its *formal* standard – to "Brunschwig's square" (1969), is in fact, when duly formulated (i.e., better than Lacan and his followers did and still do), a precise fragment (viz., its "*paracomplete* inner rectangle") of the tri-segment, and retrospectively this seems "logical": the tri-segment "diffracts" the contradiction segment, which is precisely what Lacan tried to do, but without an adequate tool. This suggests to use, in the future, the *paraconsistent* "extra structure" of the tri-segment, with respect to the *a posteriori* suboptimal Lacan-Brunschwig's square, for exploring Lacanian (and more generally: gender-theoretical) "sexuation issues" possibly not yet explored by Lacan, the Lacanians, or any of the many working on it.

Acknowledgments I wish to thank here warmly (1) my dear mother, for I owe her incredibly much, since without her love and patience, this study would not have been possible; (2) my friends Roland Bolz and Jean-François Mascari, who patiently read and commented a pre-final version of this study; (3) as well as Hans Smessaert, to which this study is co-dedicated, who tried to help me clarifying some technical points about his and Lorenz Demey's idea of "logical geometry". All mistakes remain mine.

References

1. Angot-Pellissier, R.: 2-opposition and the topological hexagon (2012). In: [30]
2. Angot-Pellissier, R.: The Relation Between Logic, Set Theory and Topos Theory as It Is Used by Alain Badiou (2015). In: [40]
3. Angot-Pellissier, R.: Many-valued logical hexagons in a 3-oppositional trisimplex (2022). In: this volume
4. Angot-Pellissier, R.: Many-valued logical hexagons in a 3-oppositional quadrisimplex. Draft (January 2014)
5. Aristotle: Metaphysics (translated by H. Lawson-Tancredi). Penguin, London (1998)
6. Aristotle: Categories and De Interpretatione (Translated with notes by J.L. Ackrill). Clarendon Aristotle Series, Oxford (1963)
7. Awodey, S.: Structure in mathematics and logic: a Categorical Perspective. Philosophia Mathematica, **4** (3), 209–237 (1996)
8. Awodey, S.: An Answer to Hellman's question: "Does category theory provide a framework for mathematical structuralism?". Philosophia Mathematica, **11**, 2 (2003)
9. Awodey, S.: Structuralism, Invariance and Univalence. Philosophia Mathematica, **22** (1), 1–11 (2014)
10. Badiou, A.: Conditions. Seuil, Paris (1992)
11. Badiou, A.: Philosophie et mathématique (1991). In: [10]

12. Badiou, A.: Sujet et infini (1992). In: [10]
13. Badiou, A.: Éloge des mathématiques. Flammarion, Paris (2015)
14. Badiou, A. and Cassin, B.: Il n'y a pas de rapport sexuel. Deux leçons sur "L'Étourdit" de Lacan. Fayard, Paris (2010)
15. Banchoff, T.F.: Beyond the Third Dimension: Geometry, Computer Graphics, and Higher Dimensions. Scientific American Library Series (1990)
16. Barot, E.: La dualité de Lautman contre la négativité de Hegel, et le paradoxe de leurs formalisations. Contribution à une enquête sur les formalisations de la dialectique. Philosophiques, **37**/1, 111–148 (Spring 2010)
17. Barrett, J.A.: The Quantum Mechanics of Minds and Worlds. Oxford University Press, Oxford (1999)
18. Barrett, J.A.: Everett's Relative-State Formulation of Quantum Mechanics. Stanford Encyclopedia of Philosophy (2018) (1998)
19. Becker Arenhart J. and Krause D.: Contradiction, Quantum Mechanics and the Square of Opposition. Logique & Analyse, Vol.59, No.235 (2016)
20. Bergmann, M.: An Introduction to Many-Valued and Fuzzy Logic. Semantics, Algebras and Derivation Systems. Cambridge UP, Cambridge (2008)
21. Berto, F.: How to Sell a Contradiction. The Logic and Metaphysics of Inconsistency. College Publications, London (2007)
22. Berto, F. and Bottai, L.: Che cos'è una contraddizione. Carocci, Roma (2015)
23. Béziau, J.-Y.: S5 is paraconsistent logic and so is first-order classical logic. Logical Investigations, **9**, 301–309 (2002)
24. Béziau, J.-Y.: New light on the square of oppositions and its nameless corner. Logical Investigations, **10**, 218–233 (2003)
25. Béziau, J.-Y.: Paraconsistent logic from a modal viewpoint. Journal of Applied Logic **3**, 7–14 (2005)
26. Béziau, J.-Y.: Paraconsistent logic! (A Reply to Slater). Sorites, **17** (2006)
27. Béziau J.-Y., Costa-Leite A. and Facchini A. (eds.): Aspects of Universal Logic, **N.17** of Travaux de logique, University of Neuchâtel (December 2004)
28. Béziau, J.-Y. and Gan-Krzywoszynska, K. (eds.): New Dimensions of the Square of Oppositions. Philosophia Verlag, Münich, (2014)
29. Béziau, J.-Y. and Jacquette, D. (eds.): Around and Beyond the Square of Opposition. Birkhäuser, Basel (2012)
30. Béziau J.-Y. and Payette G. (eds.): The Square of Opposition. A General Framework for Cognition. Peter Lang, Bern (2012)
31. Bhaskar, R.: Dialectic. The pulse of freedom. Routledge, London and New York, (2008) (1993)
32. Birkhoff, G. and von Neumann, J.: The Logic of Quantum Mechanics. Annals of Mathematics, Vol. **37**, No. 4 (October 1936)
33. Blanché, R.: Structures intellectuelles. Essai sur l'organisation systématique des concepts. Vrin, Paris (2004) (1966)
34. Blanché, R.: Raison et discours. Défense de la logique reflexive. Vrin, Paris, (2004) (1967)
35. Bobenrieth, A.: Inconsistencias ¿por qué no? Un studio filosófico sobre la lógica paraconsistente. Colcultura, Bogota (1996)
36. Bolzano, B.: Les paradoxes de l'infini (edited by H. Sinaceur). Seuil, Paris (1993)
37. Bolzano, B.: Philosophische Texte (edited by U. Neemann). Reclam, Stuttgart (1984)
38. Bouchon-Meunier, B.: La logique floue. PUF, Paris (2007) (1993)
39. Brunschwig, J.: La proposition particulière et les preuves de non-concluance chez Aristote. Cahiers pour l'Analyse, **10**, 3–26 (1969)
40. Buchsbaum A. and Koslow A. (eds.): The Road to Universal Logic, **Vol. II**, Birkhäuser, Basel (2015)
41. Cavaliere, F.: Fuzzy syllogisms, numerical square, triangle of contraries, inter-bivalence (2012). In: [29]

42. Chatti S. and Ben Aziza H. (eds.): Le carré et ses extensions: approaches théoriques, pratiques et historiques. Publications de la faculté des sciences humaines de Tunis, Université de Tunis (2015)
43. Chatti, S. and Schang, F.: The Cube, the Square and the Problem of Existential Import. History and Philosophy of Logic, **34**, 2 (2013)
44. Counet, J.-M.: La formalisation de la dialectique de Hegel. Bilan de quelques tentatives. Logique & Analyse, **218**, 205–227 (2012)
45. Coxeter, H.S.M.: Regular Polytopes (3rd edition). Dover, Mineola and New York (2020) (1963)
46. da Costa, N.C.A.: Logiques classiques et non-classiques. Essai sur les fondements de la logique (translated by J.-Y. Béziau). Masson, Paris (1997)
47. Darmon, M.: Essais sur la topologie lacanienne (nouvelle edition revue et augmentée). Éditions de l'Association freudienne, Paris (2004) (1990)
48. Davey, B.A. and Priestley, H.A.: Introduction to Lattices and Order (2nd edition). Cambridge University Press, Cambridge (2010) (1990)
49. Demey, L.: Metalogic, Metalanguage and Logical Geometry. Logique & Analyse, **62**, 248, 453–578 (2019)
50. Demey, L. and Smessaert, H.: Aristotelian and Duality Relations Beyond the Square of Oppositions. In: Chapman P., Stapleton G., Moktefi A. Perez-Kriz S. and Bellucci F. (eds.): Diagrammatic Representation and Inference. Lecture Notes in Artificial Intelligence (LNAI), 10871, 640–656 (2018)
51. Demey, L. and Smessaert, H.: Combinatorial Bitstring Semantics for Arbitrary Logical Fragments. Journal of Philosophical Logic, **47**/2, 325–363 (2018)
52. de Ronde, C., Freytes, H. and Domenech, G.: Quantum Mechanics and the Interpretation of the Orthomodular Square of Opposition (2014). In: [28]
53. Destouches-Février, P.: La structure des theories physiques. PUF, Paris (1951)
54. Deutsch, D.: The Fabric of Reality. Penguin Books, London (1997)
55. Dor, J.: Introduction à la lecture de Lacan – 1. L'inconscient structuré comme un langage. Denoël, Paris (1985)
56. Dor, J.: Introduction à la lecture de Lacan – 2. La structure du sujet. Denoël, Paris (1992)
57. Dreyfuss, J.-P., Jadin, J.-M. and Ritter, M.: Écritures de l'inconscient. De la lettre à la topologie. Arcanes, Strasbourg (2001)
58. Dubarle, D. and Doz, A.: Logique et dialectique. Larousse, Paris (1972)
59. Dubois, D., Prade, H. and Rico, A.: Structures of Opposition and Comparisons: Boolean and Gradual Cases. Logica Universalis, **14,** 1, 115–149 (2020)
60. Engesser K., Gabbay D. and Lehmann D. (eds.): Handbook of Quantum Logic and Quantum Structures: Quantum Logic. Elsevier, Amsterdam (2008)
61. Fierens, C.: Lecture du sinthome. Érès, Toulouse (2018)
62. Flegg, H.G.: From Geometry to Topology. Dover, Mineola and New York (2001) (1974)
63. Freud, S.: Über den Gegensinn der Urworter (1910). In: Freud, S.: Studienausgabe. Bd. IV. Psychologische Schriften, Fischer Verlag, Frankfurt/Main (1989) (1970)
64. Freud, S.: Die metapsychologische Schriften von 1915. In: Freud, S.: Studienausgabe. Bd.III. Psychologie des Unbewußten, Fischer Verlag, Frankfurt/Main (1989) (1975)
65. Freud, S.: Abrégé de psychanalyse. PUF, Paris (1998) (1946†)
66. Freytes, H., de Ronde, C. and Domenech, G.: The Square of Opposition in Orthomodular Logic (2012). In: [29]
67. Gärdenfors, P.: Conceptual Spaces. The Geometry of Thought. MIT Press, Cambridge MA (2004) (2000)
68. Girard, J.-Y.: Le champ du signe ou la faillite du réductionnisme (1989). In: [104]
69. Girard, J.-Y.: La machine de Turing: de la calculabilité à la complexité (1995). In: [140]
70. Girard, J.-Y.: The Blind Spot. Lectures on Logic. European Mathematical Society, Berlin (2011) (2006)
71. Girard, J.-Y.: La logique 2.0. Online draft (26 September 2018)
72. Girard, J.-Y.: Un tract anti-système. Online draft (27 November 2019)

73. Grigg, R.: Lacan and Badiou: Logic of the *Pas-Tout*. Filozofski vestnik, XXVI, 2, 53–65 (2005)
74. Gurvitch, G.: Dialectique et sociologie. Flammarion, Paris (1977) (1962)
75. Haack, S.: Deviant Logic, Fuzzy Logic. Beyond the Formalism. The University of Chicago Press, Chicago and London (1996) (1974)
76. Heudin, J.C.: Comprendre le deep learning. Une introduction aux réseaux de neurones. Science-e-book, Paris (2019) (2016)
77. Hösle, V.: I fondamenti dell'aritmetica e della geometria in Platone. Vita & Pensiero, Milano (1994)
78. Jung, C.G.: Synchronicité et Paracelsica. Albin Michel, Paris (1988)
79. Lacan, J.: Le séminaire – livre XX. Encore (edited by J.-A. Miller). Seuil, Paris (1999) (1975)
80. Lautman, A.: Les mathématiques, les idées et le reel physique. Vrin, Paris (2006†)
81. Lavendhomme, R.: Lieux du sujet. Psychanalyse et mathématique. Seuil, Paris (2001)
82. Lawvere, F.W.: Unity and Identity of Opposites in Calculus and Physics. Applied Categorical Structures, **4**, 167–174 (1996)
83. Lawvere, F.W. and Schanuel, S.H.: Conceptual Mathematics. A first introduction to categories. CUP, Cambridge (2002) (1991)
84. Laz, J.: Bolzano critique de Kant. Vrin, Paris (1993)
85. Le Gaufey, G.: Le pastout de Lacan. Consistance logique, conséquences cliniques. EPEL, Paris (2014)
86. Lupasco, S.: Le principe d'antagonisme et la logique de l'énergie. Le Rocher, Monaco (1987) (1951)
87. Luzeaux, D., Sallantin, J. and Dartnell, C.: Logical extensions of Aristotle's square. Logica Universalis, **2** (1), 167–187 (2008)
88. Mandelbrot, B.B.: Fractals and the Rebirth of Experimental Mathematics (1992). In: [109]
89. Marconi, D. (ed.): La formalizzazione della dialettica. Hegel, Marx e la logica contemporanea. Rosenberg & Sellier, Torino (1979)
90. Matte Blanco, I.: The Unconscious as Infinite Sets. An Essay in Bi-logic. Karnac, London (1998) (1975)
91. Megill, N.: Orthomodular Lattices and Beyond. Online slides (2003)
92. Mélès, B.: Pratique mathématique et lectures de Hegel, de Jean Cavaillès à William Lawvere. Philosophia Scientiae, **16** (1), 153–182 (2012)
93. Moretti, A.: Geometry for Modalities? Yes: Through *n*-Opposition Theory (2004). In: [27]
94. Moretti, A.: The Geometry of Logical Opposition. PhD Thesis, University of Neuchâtel, Switzerland (2009)
95. Moretti, A.: The Geometry of Standard Deontic Logic. Logica Universalis, **3**, 1, 19–57 (2009)
96. Moretti, A.: The Critics of Paraconsistency and of Many-Valuedness and the Geometry of Oppositions. Logic and Logical Philosophy, Special Issue on Paraconsistent Logic, Guest Editors: Koji Tanaka, Francesco Berto, Edwin Mares and Francesco Paoli, **Vol.19**, N.1–2, 63–94 (2010)
97. Moretti, A.: Why the logical hexagon?. Logica Universalis, **6** (1–2), 69–107 (2012)
98. Moretti, A.: Was Lewis Carroll an Amazing Oppositional Geometer?. History and Philosophy of Logic, **35**, IV, 383–409 (2014)
99. Moretti, A.: La science-fiction comme "désajustement onirisé" et ses enjeux philosophiques actuels. In: Albrechts-Desestré, F., Blanquet, E., Gautero, J.-L. and Picholle, E. (eds.): Philosophie, science-fiction?. Éditions du Somnium, Villefranche-sur-mer (2014)
100. Moretti, A.: Le retour du refoulé: l'hexagone logique qui est derrière le carré sémiotique (2015). In: [42]
101. Moretti, A.: Arrow-Hexagons (2015). In: [40]
102. Moretti, A.: Philosophie tragique ou anti-philosophie? La géométrie oppositionnelle et les structures élémentaires de l'idéologie. Revista Trágica: estudos de Filosofia da Imanência, **V.12**, n.3, 52–90 (2019)
103. Murinová, P.: Graded Structures of Opposition in Fuzzy Natural Logic. Logica Universalis, **14**, 4, 495–522 (2020).

104. Nagel, E., Newman, J.R., Gödel, K. and Girard, J.-Y.: Le théorème de Gödel. Seuil (translated into French by J.B. Scherrer), Paris (1997) (1931, 1958, 1989)
105. Ollman, B.: Dance of the dialectic. Steps in Marx's method. University of Illinois Press, Urbana, Chicago and Springfield (2003)
106. Palau, G.: Introducción filosófica a las lógicas no clásicas. Gedisa, Barcelona (2002)
107. Parsons, T.: The traditional square of opposition. Stanford Encyclopedia of Philosophy (2017) (1997)
108. Pavičić M. and Megill N.D.: Is Quantum Logic a Logic? (2008). In: [60]
109. Peitgen H.-O., Jürgens H. and Saupe D.: Fractals for the Classroom. Part One: Introduction to Fractals and Chaos, Springer, New York (1992)
110. Peitgen H.-O., Jürgens H. and Saupe D.: Fractals for the Classroom. Part Two: Complex Systems and Mandelbrot Set, Springer, New York (1992)
111. Pellissier, R.: "Setting" *n*-opposition. Logica Universalis, **2**, 2, 235–263 (2008)
112. Piaget, J.: Structuralism. Basic Books, New York (1970) (1968)
113. Plato: Parmenides (translated by S. Scolnicov). University of California Press, Berkeley (2003)
114. Pluder, V.: The limits of the square. Hegel's opposition to diagrams in its historical context (2020). In: this volume
115. Priest, G.: The Logic of Paradox. Journal of Philosophical Logic, 8, 219–241 (1979)
116. Priest, G.: In Contradiction. A Study of the Transconsistent. Clarendon Press, Oxford (2006) (1987).
117. Priest, G.: An Introduction to Non-Classical Logic. Cambridge University Press, Cambridge (2001)
118. Priest, G., Routley R. and Norman J. (eds.): Paraconsistent Logic. Essays on the Inconsistent, Philosophia Verlag, München Hamden Wien (1989)
119. Rayner, E.: Unconscious Logic. An introduction to Matte Blanco's bi-logic and its uses. Routledge, London and New York (1995)
120. Reichenbach, H.: Philosophical foundations of quantum mechanics. Dover, Mineola – New York (2013) (1944)
121. Richard, D.: L'enseignement oral de Platon. Une nouvelle interpretation du platonisme. CERF, Paris (1986)
122. Ritsert, J.: Kleines Lehrbuch der Dialektik. Primus, Darmstadt (1997)
123. Rodin, A.: Categorical Logic and Hegelian Dialectics. Online slides (22 February 2013)
124. Rogowski, L.S.: La logica direzionale e la tesi hegeliana della contradittorietà del mutamento (Italian translation from Polish) (1964). In: [89]
125. Roux, C.: Initiation à la théorie des graphes. Ellipses, Paris (2009)
126. Sauriol, P.: Remarques sur la Théorie de l'hexagone logique de Blanché. Dialogue, **7**, 374–390 (1968)
127. Sauriol, P.: La structure tétrahexaèdrique du système complet des propositions catégoriques. Dialogue, **15**, 479–501 (1976)
128. Schüler, H.M.: The Naturalness of Jacques Lacan's Logic (2020). In: this volume
129. Sebestik, J.: Logique et mathématique chez Bernard Bolzano. Vrin, Paris (1992)
130. Sève, L.: Structuralisme et dialectique. Editions sociales, Paris (1984)
131. Shramko, Y. and Wansing, H.: Suszko's thesis, inferential many-valuedness, and the notion of a logical system. Studia Logica, **88**, (2008)
132. Slater: Paraconsistent Logics?. Journal of Philosophical Logic, **24**, 451–454 (1995)
133. Smessaert, H.: On the 3D visualisation of logical relations. Logica Universalis, **3**, 2, 303–332 (2009)
134. Smessaert, H.: The classical Aristotelian hexagon versus the modern duality hexagon. Logica Universalis, **6**, 1–2 (2012)
135. Smessaert H. and Demey L.: Logical Geometries and Information in the Square of Oppositions. Journal of Logic, Language and Information, **23**/4, 527–565 (2014)
136. Smirnov, V.A.: Logicheskie idei N.A. Vasil'eva I sovremennaja logika (1989) (in Russian). In: [143]

137. Sommerville, D.M.Y.: An Introduction to the Geometry of *N* Dimensions, Dover, Mineola and New-York (2020) (1929)
138. Suszko, R.: Remarks on Łukasiewicz's three-valued logic. Bulletin of the Section of Logic, **4**, 97–90 (1975)
139. Toth, I.: Aristotele e I fondamenti assiomatici della geometria. Prolegomeni alla comprensione dei frammenti non-euclidei nel "Corpus Aristotelicum". Vita & Pensiero, Milano (1997)
140. Turing, A. and Girard, J.-Y.: La machine de Turing (translated into French by J. Basch and P. Blanchard). Seuil, Paris (1995) (1396, 1950, 1995)
141. Vaidman, L.: Many-Worlds Interpretation of Quantum Mechanics. Stanford Encyclopedia of Philosophy (2014) (2002)
142. Vandoulakis, I.M. and Denisova, T.Y.: On the Historical Transformations of the Square of Opposition as Semiotic Object. Logica Universalis, **14**, 1, 7–26 (2020)
143. Vasil'ev, N.A.: Voobrazhaemaja logika. Izbrannye Trudy (in Russian). Nauka, Moskva, (1989)
144. Wolff, F.: La vérité dans la *Métaphysique* d'Aristote. Cahiers philosophiques de Strasbourg, tome 7, 133–168 (1998)
145. Zalamea, F.: Albert Lautman et la dialectique créatrice des mathématiqus modernes (2006). In: [80]
146. Zalamea, F.: Synthetic Philosophy of Contemporary Mathematics (translated by Z. L. Fraser). Urbanomic and Sequence, Falmouth U.K. and New York U.S.A. (2012) (2009)
147. Ziegler, M.: Quantum Logic: Order Structures in Quantum Mechanics. Technical report, University of Paderborn, Germany (2005)